Properties and Applications of Graphene and Its Derivatives

Properties and Applications of Graphene and Its Derivatives

Editor

José Miguel González-Domínguez

MDPI • Basel • Beijing • Wuhan • Barcelona • Belgrade • Manchester • Tokyo • Cluj • Tianjin

Editor
José Miguel
González-Domínguez
CSIC - Instituto de
Carboquimica (ICB)
Spain

Editorial Office
MDPI
St. Alban-Anlage 66
4052 Basel, Switzerland

This is a reprint of articles from the Special Issue published online in the open access journal *Nanomaterials* (ISSN 2079-4991) (available at: https://www.mdpi.com/journal/nanomaterials/special_issues/graphene_derivatives).

For citation purposes, cite each article independently as indicated on the article page online and as indicated below:

LastName, A.A.; LastName, B.B.; LastName, C.C. Article Title. *Journal Name* **Year**, *Volume Number*, Page Range.

ISBN 978-3-0365-4783-1 (Hbk)
ISBN 978-3-0365-4784-8 (PDF)

© 2022 by the authors. Articles in this book are Open Access and distributed under the Creative Commons Attribution (CC BY) license, which allows users to download, copy and build upon published articles, as long as the author and publisher are properly credited, which ensures maximum dissemination and a wider impact of our publications.

The book as a whole is distributed by MDPI under the terms and conditions of the Creative Commons license CC BY-NC-ND.

Contents

Jose M. González-Domínguez
Editorial for "Properties and Applications of Graphene and Its Derivatives"
Reprinted from: *Nanomaterials* **2022**, *12*, 602, doi:10.3390/nano12040602 1

Ana María Díez-Pascual, Carlos Sainz-Urruela, Cristina Vallés, Soledad Vera-López and María Paz San Andrés
Tailorable Synthesis of Highly Oxidized Graphene Oxides via an Environmentally-Friendly Electrochemical Process
Reprinted from: *Nanomaterials* **2020**, *10*, 239, doi:10.3390/nano10020239 5

Chang-Seuk Lee, Su Jin Shim and Tae Hyun Kim
Scalable Preparation of Low-Defect Graphene by Urea-Assisted Liquid-Phase Shear Exfoliation of Graphite and Its Application in Doxorubicin Analysis
Reprinted from: *Nanomaterials* **2020**, *10*, 267, doi:10.3390/nano10020267 23

Alejandra Rendón-Patiño, Antonio Domenech-Carbó, Ana Primo and Hermenegildo García
Superior Electrocatalytic Activity of MoS_2-Graphene as Superlattice
Reprinted from: *Nanomaterials* **2020**, *10*, 839, doi:10.3390/nano10050839 35

Duong Duc La, Tuan Ngoc Truong, Thuan Q. Pham, Hoang Tung Vo, Nam The Tran, Tuan Anh Nguyen, Ashok Kumar Nadda, Thanh Tung Nguyen, S. Woong Chang, W. Jin Chung and D. Duc Nguyen
Scalable Fabrication of Modified Graphene Nanoplatelets as an Effective Additive for Engine Lubricant Oil
Reprinted from: *Nanomaterials* **2020**, *10*, 877, doi:10.3390/nano10050877 45

Daniel Torres, Sara Pérez-Rodríguez, David Sebastián, José Luis Pinilla, María Jesús Lázaro and Isabel Suelves
Capacitance Enhancement of Hydrothermally Reduced Graphene Oxide Nanofibers
Reprinted from: *Nanomaterials* **2020**, *10*, 1056, doi:10.3390/nano10061056 57

Alina I. Pruna, Arturo Barjola, Alfonso C. Cárcel, Beatriz Alonso and Enrique Giménez
Effect of Varying Amine Functionalities on CO_2 Capture of Carboxylated Graphene Oxide-Based Cryogels
Reprinted from: *Nanomaterials* **2020**, *10*, 1446, doi:10.3390/nano10081446 75

Adrian Petris, Ileana Cristina Vasiliu, Petronela Gheorghe, Ana Maria Iordache, Laura Ionel, Laurentiu Rusen, Stefan Iordache, Mihai Elisa, Roxana Trusca, Dumitru Ulieru, Samaneh Etemadi, Rune Wendelbo, Juan Yang and Knut Thorshaug
Graphene Oxide-Based Silico-Phosphate Composite Films for Optical Limiting of Ultrashort Near-Infrared Laser Pulses
Reprinted from: *Nanomaterials* **2020**, *10*, 1638, doi:10.3390/nano10091638 91

Minh Nhat Dang, Minh Dang Nguyen, Nguyen Khac Hiep, Phan Ngoc Hong, In Hyung Baek and Nguyen Tuan Hong
Improved Field Emission Properties of Carbon Nanostructures by Laser Surface Engineering
Reprinted from: *Nanomaterials* **2020**, *10*, 1931, doi:10.3390/nano10101931 111

Rabia Ikram, Badrul Mohamed Jan, Jana Vejpravova, M. Iqbal Choudhary and Zaira Zaman Chowdhury
Recent Advances of Graphene-Derived Nanocomposites in Water-Based Drilling Fluids
Reprinted from: *Nanomaterials* **2020**, *10*, 2004, doi:10.3390/nano10102004 123

Sardar Kashif Ur Rehman, Sabina Kumarova, Shazim Ali Memon, Muhammad Faisal Javed and Mohammed Jameel
A Review of Microscale, Rheological, Mechanical, Thermoelectrical and Piezoresistive Properties of Graphene Based Cement Composite
Reprinted from: *Nanomaterials* **2020**, *10*, 2076, doi:10.3390/nano10102004 149

N'ghaya Toulbe, Malvina S. Stroe, Monica Daescu, Radu Cercel, Alin Mogos, Daniela Dragoman, Marcela Socol, Ionel Mercioniu and Mihaela Baibarac
Reduced Graphene Oxide Sheets as Inhibitors of the Photochemical Reactions of α-Lipoic Acid in the Presence of Ag and Au Nanoparticles
Reprinted from: *Nanomaterials* **2020**, *10*, 2238, doi:10.3390/nano10112238 191

Xoan F. Sánchez-Romate, Alejandro Sans, Alberto Jiménez-Suárez, Mónica Campo, Alejandro Ureña and Silvia G. Prolongo
Highly Multifunctional GNP/Epoxy Nanocomposites: From Strain-Sensing to Joule Heating Applications
Reprinted from: *Nanomaterials* **2020**, *10*, 2431, doi:10.3390/nano10122431 207

Pietro Bellet, Matteo Gasparotto, Samuel Pressi, Anna Fortunato, Giorgia Scapin, Miriam Mba, Enzo Menna and Francesco Filippini
Graphene-Based Scaffolds for Regenerative Medicine
Reprinted from: *Nanomaterials* **2021**, *11*, 404, doi:10.3390/nano11020404 223

Fernando Rodríguez-Mas, Juan Carlos Ferrer, José Luis Alonso, Susana Fernández de Ávila and David Valiente
Reduced Graphene Oxide Inserted into PEDOT:PSS Layer to Enhance the Electrical Behaviour of Light-Emitting Diodes
Reprinted from: *Nanomaterials* **2021**, *11*, 645, doi:10.3390/nano11030645 265

Talia Tene, Marco Guevara, Andrea Valarezo, Orlando Salguero, Fabian Arias Arias, Melvin Arias, Andrea Scarcello, Lorenzo S. Caputi and Cristian Vacacela Gomez
Drying-Time Study in Graphene Oxide
Reprinted from: *Nanomaterials* **2021**, *11*, 1035, doi:10.3390/nano11041035 279

Jose M. González-Domínguez, Alejandro Baigorri, Miguel Á. Álvarez-Sánchez, Eduardo Colom, Belén Villacampa, Alejandro Ansón-Casaos, Enrique García-Bordejé, Ana M. Benito and Wolfgang K. Maser
Waterborne Graphene- and Nanocellulose-Based Inks for Functional Conductive Films and 3D Structures
Reprinted from: *Nanomaterials* **2021**, *11*, 1435, doi:10.3390/nano11061435 293

Gisya Abdi, Abdolhamid Alizadeh, Wojciech Grochala and Andrzej Szczurek
Developments in Synthesis and Potential Electronic and Magnetic Applications of Pristine and Doped Graphynes
Reprinted from: *Nanomaterials* **2021**, *11*, 2268, doi:10.3390/nano11092268 311

Talia Tene, Stefano Bellucci, Marco Guevara, Edwin Viteri, Malvin Arias Polanco, Orlando Salguero, Eder Vera-Guzmán, Sebastián Valladares, Andrea Scarcello, Francesca Alessandro, Lorenzo S. Caputi and Cristian Vacacela Gomez
Cationic Pollutant Removal from Aqueous Solution Using Reduced Graphene Oxide
Reprinted from: *Nanomaterials* **2022**, *12*, 309, doi:10.3390/nano12030309 351

Editorial

Editorial for "Properties and Applications of Graphene and Its Derivatives"

Jose M. González-Domínguez

Group of Carbon Nanostructures and Nanotechnology (G-CNN), Instituto de Carboquímica, ICB-CSIC, C/Miguel Luesma Castán 4, 50018 Zaragoza, Spain; jmgonzalez@icb.csic.es

Citation: González-Domínguez, J.M. Editorial for "Properties and Applications of Graphene and Its Derivatives". *Nanomaterials* **2022**, *12*, 602. https://doi.org/10.3390/nano12040602

Received: 9 December 2021
Accepted: 13 December 2021
Published: 11 February 2022

Publisher's Note: MDPI stays neutral with regard to jurisdictional claims in published maps and institutional affiliations.

Copyright: © 2022 by the author. Licensee MDPI, Basel, Switzerland. This article is an open access article distributed under the terms and conditions of the Creative Commons Attribution (CC BY) license (https:// creativecommons.org/licenses/by/ 4.0/).

Since the very first landmark report by Geim and Novoselov in 2004 on graphene [1], the interest of the scientific and technological communities for every form of this carbon-based nanomaterial has only grown and grown. In 2019, the hype for graphene turned fifteen years old and, after having experienced an intense infancy (with impressive results at laboratory scales in a myriad of different application fields), experts say that the following 15 years should be oriented towards its commercialization and day-to-day uses [2]. For this, some aspects are crucial and need to be understood, such as standardization or safety issues [2,3]. The broad family of graphene nanomaterials (including graphene nanoplatelets, graphene oxide, graphene quantum dots, and many more), go beyond and aim higher than mere single-layer ('pristine') graphene; thus, their potential has sparked the current Special Issue. In it, 18 contributions (distributed in 14 research articles and 4 reviews) have probably portrayed the most interesting lines as regards future and tangible uses of graphene derivatives.

Works on the fabrication of graphene, nanomaterials have appeared; for example, the electrochemical synthesis of highly oxidized graphene oxide (GO), by Díez-Pascual and co-workers [4], who aimed at maximizing its oxidation degree by tuning the electrochemical synthesis parameters. In another reference work, Lee et al. explored a somehow opposed concept: obtaining low-defect graphene, in this case by a tailored liquid-phase shear exfoliation of graphite [5]. Both papers demonstrate the potential up-scalability of their respective fabrication processes, and the utility of the as-made graphene-based nanomaterials, such as for instance, electrode modification for enhanced sensing purposes [5].

Some published papers in this Special Issue focus on the performance peculiarities of graphene-based nanomaterials in specific contexts. For example, Torres and co-workers have studied the effect of the reduction temperature in the electrochemical performance of reduced-GO-based nanofibers obtained by a hydrothermal method [6]. The authors found a promising 16-fold increase in capacitance as compared to the preceding GO nanofibers. Another entry on reduced GO has been reported by Toulbe et al. [7], who have found that this nanomaterial is capable of inhibiting the photodegradation of α-lipoic acid (an antioxidant) in the presence of Au and Ag nanoparticles, with important implications in pharmaceutical compounds. As a complementary insight into both of the aforementioned studies, Tene and co-workers present an elegant study on how the drying conditions of GO critically affect its structure and nature of defects [8]. The same authors report also on the environmental advantages of reduced GO by efficiently removing water pollutants [9].

In this Special Issue, there is also room for chemically modified or assembled graphene nanomaterials, such as those presented by La et al. [10], who have obtained oleic acid-modified graphene nanoplatelets showing excellent dispersion stability and tribological performance in lubricant oil. On the other hand, Pruna and co-workers present GO-based cryogels modified with amine moieties [11]. In particular, for ethylenediamine-functionalized cryogels, the authors found a promisingly high CO_2 uptake, with a view to future CO_2 capture technologies. If cryogels are a good example of an assembled graphene-based scaffold, and given the high interest for such 3D ensembles, the readers cannot miss the review paper presented by Bellet et al. [12]. It deals with this kind of structure; in this case, oriented towards biomedical applications (regenerative medicine), covering many aspects, from toxicity to applications, and structure–properties relationships.

However, the most reported field of application in graphene-based materials within this Special Issue, nearly half of the total contributions, has definitely been the nanocomposites one. Different nanocomposite studies for structural applications are reported, such as that authored by Sánchez-Romate et al. [13], dealing with graphene nanoplatelet-based epoxy composites with exhaustive electromechanical and electrothermal characterizations; or also two review papers with particular relevance on construction and building materials: one by Ikram et al., which covers the state-of-the art of graphene nanomaterials for water-based drilling fluids [14], and the other review by Rehman et al., gathering significant advances in graphene–cement composites [15]. Regarding electric or electronic applications of graphene nanocomposites, there are also excellent pieces of work in this Special Issue, such as that from Rendón-Patiño and co-workers (dealing with graphene–MoS$_2$ heterostructures with excellent catalytic activity towards H$_2$ and O$_2$ evolution reactions) [16], or the work by Rodríguez-Mas et al., who have successfully inserted a conductive polymer-coated reduced GO layer in an organic-based LED, increasing its current density [17]. The great potential of GO to stand as a versatile adjuvant in advanced applications is herein embodied by the works of Petris et al. (GO–silicophosphate compounds for the optical limiting of femtosecond lasers) [18], and that of González-Domínguez and co-workers (dealing with aqueous inks based on GO–nanocellulose hybrids with potential to be applied in electrode manufacturing) [19].

Last, but not least, this Special Issue contains a unique review paper on the properties and applications of Graphyne derivatives [20], emerging as the next stage of the current graphene state-of-the-art.

Conflicts of Interest: The author declares no conflict of interest.

References

1. Novoselov, K.S.; Geim, A.K.; Morozov, S.V.; Jiang, D.; Zhang, Y.; Dubonos, S.V.; Grigorieva, I.V.; Firsov, A.A. Electric Field Effect in Atomically Thin Carbon Films. *Science* **2004**, *306*, 666–669. [CrossRef] [PubMed]
2. Ye, R.; Tour, J.M. Graphene at Fifteen. *ACS Nano* **2019**, *13*, 10872–10878. [CrossRef] [PubMed]
3. Reina, G.; González-Domínguez, J.M.; Criado, A.; Vázquez, E.; Bianco, A.; Prato, M. Promises, facts and challenges for graphene in biomedical applications. *Chem. Soc. Rev.* **2017**, *46*, 4400–4416. [CrossRef] [PubMed]
4. Díez-Pascual, A.M.; Sainz-Urruela, C.; Vallés, C.; Vera-Lopez, S.; San Andrés, M.P. Tailorable Synthesis of Highly Oxidized Graphene Oxides via an Environmentally-Friendly Electrochemical Process. *Nanomaterials* **2020**, *10*, 239. [CrossRef] [PubMed]
5. Lee, C.-S.; Shim, S.J.; Kim, T.H. Scalable Preparation of Low-Defect Graphene by Urea-Assisted Liquid-Phase Shear Exfoliation of Graphite and Its Application in Doxorubicin Analysis. *Nanomaterials* **2020**, *10*, 267. [CrossRef] [PubMed]
6. Torres, D.; Pérez-Rodríguez, S.; Sebastián, D.; Pinilla, J.L.; Lázaro, M.J.; Suelves, I. Capacitance Enhancement of Hydrothermally Reduced Graphene Oxide Nanofibers. *Nanomaterials* **2020**, *10*, 1056. [CrossRef] [PubMed]
7. Toulbe, N.; Stroe, M.S.; Daescu, M.; Cercel, R.; Mogos, A.; Dragoman, D.; Socol, M.; Mercioniu, I.; Baibarac, M. Reduced Graphene Oxide Sheets as Inhibitors of the Photochemical Reactions of α-Lipoic Acid in the Presence of Ag and Au Nanoparticles. *Nanomaterials* **2020**, *10*, 2238. [CrossRef] [PubMed]
8. Tene, T.; Guevara, M.; Valarezo, A.; Salguero, O.; Arias-Arias, F.; Arias, M.; Scarcello, A.; Caputi, L.S.; Vacacela-Gomez, C. Drying-Time Study in Graphene Oxide. *Nanomaterials* **2021**, *11*, 1035. [CrossRef] [PubMed]
9. Tene, T.; Bellucci, S.; Guevara, M.; Viteri, E.; Arias Polanco, M.; Salguero, O.; Vera-Guzmán, E.; Valladares, S.; Scarcello, A.; Alessandro, F.; et al. Cationic Pollutant Removal from Aqueous Solution Using Reduced Graphene Oxide. *Nanomaterials* **2022**, *12*, 309. [CrossRef]
10. La, D.D.; Truong, T.N.; Pham, T.Q.; Vo, H.T.; Tran, N.T.; Nguyen, T.A.; Nadda, A.K.; Nguyen, T.T.; Chang, S.W.; Chung, W.J.; et al. Scalable Fabrication of Modified Graphene Nanoplatelets as an Effective Additive for Engine Lubricant Oil. *Nanomaterials* **2020**, *10*, 877. [CrossRef] [PubMed]
11. Pruna, A.I.; Barjola, A.; Cárcel, A.C.; Alonso, B.; Giménez, E. Effect of Varying Amine Functionalities on CO$_2$ Capture of Carboxylated Graphene Oxide-Based Cryogels. *Nanomaterials* **2020**, *10*, 1446. [CrossRef] [PubMed]
12. Bellet, P.; Gasparotto, M.; Pressi, S.; Fortunato, A.; Scapin, G.; Mba, M.; Menna, E.; Filippini, F. Graphene-Based Scaffolds for Regenerative Medicine. *Nanomaterials* **2021**, *11*, 404. [CrossRef] [PubMed]
13. Sánchez-Romate, X.F.; Sans, A.; Jiménez-Suárez, A.; Campo, M.; Ureña, A.; Prolongo, S.G. Highly Multifunctional GNP/Epoxy Nanocomposites: From Strain-Sensing to Joule Heating Applications. *Nanomaterials* **2020**, *10*, 2431. [CrossRef] [PubMed]

14. Ikram, R.; Jan, B.M.; Vejpravova, J.; Choudhary, M.I.; Chowdhury, Z.Z. Recent Advances of Graphene-Derived Nanocomposites in Water-Based Drilling Fluids. *Nanomaterials* **2020**, *10*, 2004. [CrossRef] [PubMed]
15. Rehman, S.K.U.; Kumarova, S.; Memon, S.A.; Javed, M.F.; Jameel, M. A Review of Microscale, Rheological, Mechanical, Thermoelectrical and Piezoresistive Properties of Graphene Based Cement Composite. *Nanomaterials* **2020**, *10*, 2076. [CrossRef] [PubMed]
16. Rendón-Patiño, A.; Domenech-Carbó, A.; Primo, A.; García, H. Superior Electrocatalytic Activity of MoS_2-Graphene as Superlattice. *Nanomaterials* **2020**, *10*, 839. [CrossRef] [PubMed]
17. Rodríguez-Mas, F.; Ferrer, J.C.; Alonso, J.L.; Fernández de Ávila, S.; Valiente, D. Reduced Graphene Oxide Inserted into PEDOT:PSS Layer to Enhance the Electrical Behaviour of Light-Emitting Diodes. *Nanomaterials* **2021**, *11*, 645. [CrossRef] [PubMed]
18. Petris, A.; Vasiliu, I.C.; Gheorghe, P.; Iordache, A.M.; Ionel, L.; Rusen, L.; Iordache, S.; Elisa, M.; Trusca, R.; Ulieru, D.; et al. Graphene Oxide-Based Silico-Phosphate Composite Films for Optical Limiting of Ultrashort Near-Infrared Laser Pulses. *Nanomaterials* **2020**, *10*, 1638. [CrossRef] [PubMed]
19. González-Domínguez, J.M.; Baigorri, A.; Álvarez-Sánchez, M.Á.; Colom, E.; Villacampa, B.; Ansón-Casaos, A.; García-Bordejé, E.; Benito, A.M.; Maser, W.K. Waterborne Graphene- and Nanocellulose-Based Inks for Functional Conductive Films and 3D Structures. *Nanomaterials* **2021**, *11*, 1435. [CrossRef] [PubMed]
20. Abdi, G.; Alizadeh, A.; Grochala, W.; Szczurek, A. Developments in Synthesis and Potential Electronic and Magnetic Applications of Pristine and Doped Graphynes. *Nanomaterials* **2021**, *11*, 2268. [CrossRef] [PubMed]

Article

Tailorable Synthesis of Highly Oxidized Graphene Oxides via an Environmentally-Friendly Electrochemical Process

Ana María Díez-Pascual [1,2,*], Carlos Sainz-Urruela [1], Cristina Vallés [3], Soledad Vera-López [1,2] and María Paz San Andrés [1,2]

1. Department of Analytical Chemistry, Physical Chemistry and Chemical Engineering, Faculty of Sciences, University of Alcalá, Alcalá de Henares, 28805 Madrid, Spain; soledad.vera@uah.es (S.V.-L.); mpaz.sanandres@uah.es (M.P.S.)
2. Institute of Chemistry Research, "Andrés M. del Río" (IQAR), University of Alcalá, Ctra. Madrid- Barcelona Km. 33.6, Alcalá de Henares, 28805 Madrid, Spain
3. Department of Materials and National Graphene Institute, University of Manchester, Oxford Road, Manchester M13 9PL, UK; cristina.valles@manchester.ac.uk
* Correspondence: am.diez@uah.es; Tel.: +34-918-856-430

Received: 29 November 2019; Accepted: 27 January 2020; Published: 29 January 2020

Abstract: Graphene oxide (GO) is an attractive alternative to graphene for many applications due to its captivating optical, chemical, and electrical characteristics. In this work, GO powders with a different amount of surface groups were synthesized from graphite via an electrochemical two-stage process. Many synthesis conditions were tried to maximize the oxidation level, and comprehensive characterization of the resulting samples was carried out via elemental analysis, microscopies (TEM, SEM, AFM), X-ray diffraction, FT-IR and Raman spectroscopies as well as electrical resistance measurements. SEM and TEM images corroborate that the electrochemical process used herein preserves the integrity of the graphene flakes, enabling to obtain large, uniform and well exfoliated GO sheets. The GOs display a wide range of C/O ratios, determined by the voltage and time of each stage as well as the electrolyte concentration, and an unprecedented minimum C/O value was obtained for the optimal conditions. FT-IR evidences strong intermolecular interactions between neighbouring oxygenated groups. The intensity ratio of D/G bands in the Raman spectra is high for samples prepared using concentrated H_2SO_4 as an electrolyte, indicative of many defects. Furthermore, these GOs exhibit smaller interlayer spacing than that expected according to their oxygen content, which suggests predominant oxidation on the flake edges. Results point out that the electrical resistance is conditioned mostly by the interlayer distance and not simply by the C/O ratio. The tuning of the oxidation level is useful for the design of GOs with tailorable structural, electrical, optical, mechanical, and thermal properties.

Keywords: graphene oxide; electrochemical synthesis; oxidation level; exfoliation degree; morphology; interlayer spacing; surface defects; electrical resistance

1. Introduction

Graphene oxide (GO), which is the oxidized form of graphene, is currently attracting a lot of interest due to its unique chemical, optical, and electronic properties that make it suitable for a broad range of uses including supercapacitors, solar cells, fuel cells, lithium batteries, biomedicine, polymer nanocomposites, and more [1–6]. It presents several surface oxygen-containing groups, mainly COOH moieties on the layer edges as well as C-O-C and OH on the basal planes (Scheme 1). Furthermore, the hydroxyl groups can bond to form epoxy groups, which leads to a distorted tetrahedron with four carbon atoms on the six-membered ring of the carbon plane. This provokes the flat graphene network to bend. Therefore, some characteristics of GO differ from those of graphene. The oxygenated groups in GO expand the interlayer distance, which boosts its skill to retain substances. The attached groups and lattice defects alter the electronic structure and act as scattering centers that reduce the electron mobility. Hence, it is typically an insulating material. Furthermore, it presents biocompatibility, surface functionalization capability, amphiphilicity, and aqueous processability [7]. Thus, after sonication in aqueous solutions, it straightforwardly exfoliates, which leads to stable colloidal suspensions [8]. This can be used for the preparation of transparent electrodes. The presence of oxygen functionalities also makes GO a good candidate for environmentally-related applications such as wastewater treatment and water purification to remove both inorganic and organic pollutants including hormones or antibiotics [9]. GO is also largely used for the synthesis of a number of derivatives to be used in environmental and energy-related applications [10].

The adjustment of the oxidation level would be useful for the design of GOs with tailorable structural, electrical, mechanical, thermal, and optical properties. For instance, the energy band gap of GO increases linearly while decreasing the C/O ratio [11], which provides an efficient way to tune the optical properties of graphene oxide-based materials. On the other hand, as the degree of oxidation increases, both the Young's modulus and mechanical strength will likely decrease monotonically due to the breaking of the sp^2 carbon network and lowering of the energetic stability for the ordered GO [12]. However, the increase in the oxidation level of GO could be beneficial for improving the mechanical properties of nanocomposites [13], especially in the case of matrices involving oxygenated groups like chitosan [14] or polyamide [15]. In such cases, the mechanical strength was enhanced with the increase in the oxidation degree due to the stronger H-bonding interactions between the GO sheets, the polymeric matrices, and the more extended chemical interphase. Other properties including the heat capacity, thermal conductivity, and specific capacitance could also be tailored by modifying the oxidation degree [16,17].

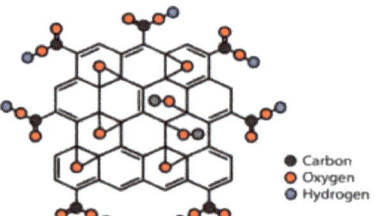

Scheme 1. Representation of the chemical structure of graphene oxide (GO).

In the 19th century, graphite oxide (also known as graphite acid) was first synthesized from graphite through oxidation using $KClO_3$ in fuming HNO_3. Since then, a lot of effort has been made in order to obtain more oxidized and exfoliated graphite oxide. Four main approaches have been described for the synthesis of GO [18]: Staudenmaier, Hofmann, Brodie, and Hummers. Both Brodie and Staudenmaier routes use $KClO_3$ and HNO_3, and present an explosion hazard as well as evolution of dangerous and carcinogenic gases such as NOx, ClO-, and ClO_2. The most widely employed is the Hummers' method, which comprises the mixing of $KMnO_4$ with a solution of $NaNO_3$, H_2SO_4, and graphite [19]. It requires a huge quantity of concentrated acid and $KMnO_4$ to guarantee enough oxidation. Furthermore, the previously mentioned approaches cause severe environmental contamination and metallic impurities on the GO layers. In particular, about a thousand-fold more water than graphite is required in the Hummers' method to get rid of the excess of H_2SO_4 and $KMnO_4$ subsequent to the oxidation reaction, which leads to an enormous amount of sewage water comprising heavy metal ions and strong acids. Furthermore, they are particularly time-consuming, since they may require more than 100 h for oxidation. Many modifications of these approaches, especially of the Hummer's one, have been published, with enhancements being constantly investigated to attain improved results at a lower cost. For example, strategies without NaNO$_3$ in order to avoid the formation of toxic gasses (NO$_2$/N$_2$O$_4$) and to make the disposal of waste water easier due to the absence of Na$^+$ and NO$_3^-$ ions have been reported [20]. GO can also be prepared from graphite oxide by means of sonication, stirring, or even by rapid freezing of an aqueous solution containing the parent oxide, which is followed by thawing of the resulting solid [21]. Sonication is a time-effective technique for completely exfoliating graphite oxide, while it can seriously impair the graphene flakes, which decreases their dimensions from μm to nm, and even leads to graphene platelets. Mechanically stirring is not such a clumsy approach, even though it involves longer periods.

Recently, electrochemical processes have been employed to synthesize carbon nanomaterials due to of their environmentally-friendliness, high efficiency, and inexpensiveness [22]. A reduced graphene with a very high C/O ratio (25.3) and high electron mobility has been synthesized within seconds by electrochemical exfoliation of graphite foil [23]. Analogously, GO has been produced by electrochemical oxidation of several graphitic materials in a variety of geometries such as pencil cores and graphite powders, foils, rods, or plates [24–27]. The reported approaches have emerged as promising and versatile alternatives to the chemical methods. By controlling the processing parameters, such as applied potentials and currents, exfoliation time, as well as electrolyte composition and temperature, GOs with different defect densities, oxygen content, graphene layer numbers, and lateral sizes have been obtained [27]. Moreover, chemical reactions with functionalizing agents can occur during the electrochemical exfoliation to produce in situ chemically functionalized (doped) GO-based materials [28]. However, the electrolysis process generally worsens the expansion and delamination of the graphitic materials, which leads to products with low oxidation and exfoliation levels, with properties that considerably differ from the GOs prepared by the conventional methods mentioned previously. Furthermore, electrochemical exfoliation methods have demonstrated poor controllability in the structural and physical properties of the resulting graphene materials, and typically lead to low-yield production [29]. Overall, despite noteworthy research efforts over recent years focusing on GO fabrication by electrochemical exfoliation from graphite, very few studies have succeeded. Therefore, further investigation in this direction is required to develop scalable, safe, rapid, and green methods. In this regard, there are two critical challenges that should be addressed. First, there is still no simple electrochemical exfoliation process that can precisely control the chemical composition and structure of the resulting graphene materials. The approaches recently reported typically lead to heterogeneous mixtures containing traces from the raw materials such as partially oxidized and fully oxidized flakes, and comprise layers with different defect content, lateral size, and number of functional groups. Second, the methods also face a compromise between the yield and the property control of the resulting materials. The development of cost-effective electrochemical exfoliation means to fabricate GOs by targeting specific applications remains an open question.

In this context, for the first time, we have developed a straightforward, green, and inexpensive electrochemical method that enables users to finely control the level of GO oxidation and exfoliation by carefully modifying the synthesis conditions, which is of large interest from an application standpoint. A wide range of experimental conditions have been tested to optimize the synthesis process in order to obtain homogeneous GOs that preserve the integrity of the graphene flakes, without remnants of the raw material and at a good yield. The aim is to maximize their oxidation and exfoliation levels, which, consequently, would improve their solubility in aqueous solutions and their aptitude to interact with hydrophilic and amphiphilic molecules such as proteins, phospholipids, DNA, and more. The idea of the method is to split the GO preparation process into two stages: a mild intercalation stage of SO$_4^{2-}$ ions within the graphite layers leading to a graphite intercalation compound (GIC) and then an oxidation/exfoliation stage of the GIC under stronger conditions. Thus, the voltage and time of both stages as well the concentration of the electrolyte in the exfoliation step varied in order to assess their effect on the degree of exfoliation, defect content, and level of oxidation of the resulting GOs. An unprecedented minimum C/O value was obtained for the optimal conditions.

The developed method provides better yields than a reference GO synthesized from graphite via a modified Hummers' method.

2. Materials and Methods

2.1. Materials and Reagents

High purity flexible graphite foil (FGF, $d_{25\ °C}$ = 1.00 g/cm^3, C: 99.5%, S <300 ppm, Cl < 50 ppm, ash < 1%, thickness 0.1 mm) was provided by Beyond Materials, Inc. (Tucson, AZ, USA) and dried in an oven at 60 °C for 48 h before use. Powdered graphite flakes (SP-1, d_{25} °C = 1.05 g/cm^3, C: 99.9%, ash < 0.5%, average size 30–150 µm) were acquired from Bay Carbon, Inc. (Michigan, MI, USA) and subjected to identical treatment. $KMnO_4$, H_2SO_4, $K_2S_2O_8$, P_2O_5, H_2O_2 (30 wt % in water), and platinum wire (ø: 0.5 mm, 99.99% trace metals basis) were obtained from Sigma-Aldrich (Madrid, Spain) and used as received. Ultrapure water was purified by a Millipore Elix 15,824 Advantage 15 UV system (Millipore, Milford, MA, USA).

2.2. Synthesis of Electrochemically Exfoliated Graphene Oxides (EGOs)

The EGOs were synthesized from FGF via an electrochemical process at room temperature (25 ± 2 °C) that comprised two stages. The first stage was carried out in an electrolysis cell, which had a slice of FGF stuck onto a tungsten wire through silver glue as an anode, a Pt wire as a cathode, and 98 wt% H_2SO_4 diluted in 100 mL of Milli-Q water as an electrolyte. A static voltage of 1 or 2 V was initially applied for 10 or 30 min, which led to formation of a turquoise-blue graphite intercalation compound (GIC). This low voltage aided to wet the carbon material and induced the mild intercalation of SO_4^{2-} ions within the interlayers of FGF [30].

Prior to the second step, the synthesized GIC was drawn out and pressed to eliminate the absorbed H_2SO_4. Then, the electrochemical oxidation was performed using the GIC as an anode, a Pt wire as a cathode, and 40, 65, or 98 wt% H_2SO_4 diluted in 100 mL of Milli-Q water as an electrolyte. A high voltage ranging from 10 to 30 V was applied for different periods between 30 and 120 s. The high bias caused the electrolytic oxidation of the GIC, which results in the formation of a yellowish oxidized graphite oxide that was readily exfoliated into graphene oxide during the voltage application. A schematic representation of the experimental setup used for the electrochemical synthesis of the EGOs and typical images of FGF, GIC, and EGO are shown in Figure 1.

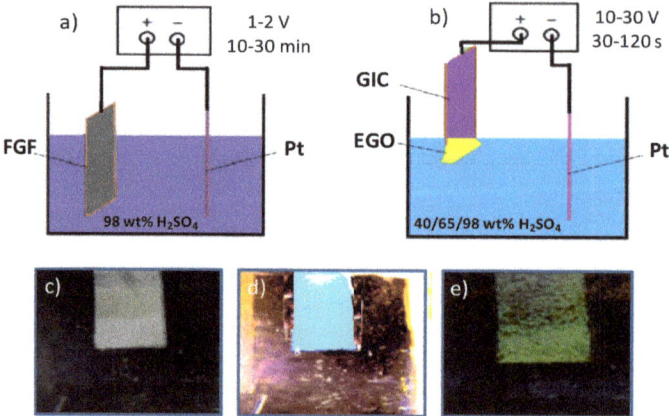

Figure 1. Top: Schematic representation of the electrolytic cells used for the synthesis of GO: (**a**) intercalation step, and (**b**) oxidation/exfoliation step. Bottom: Images of grey FGP (**c**), turquoise-blue GIC, (**d**) and yellowish synthesized EGO (**e**).

The exfoliated graphene oxide sheets were collected by filtration, cleaned with water, purified via centrifugation at 2500 rpm to remove undesirable large particles, and further exfoliated in water by ultrasonication for 30 min at a power of 140 W, which leads to well dispersed EGO. After vacuum-freezing drying, the yields of the synthesized EGOs were calculated to range from 121% to 137%. The experimental conditions, namely bias voltage and time of the intercalation and exfoliation stages, as well as concentration of the electrolyte in the exfoliation stage varied in order to determine their effect on the degree of exfoliation, defect content, and level of oxidation of the resulting EGO products. The different exfoliation conditions tested along with the nomenclature of the EGO samples obtained herein are summarized in Table 1. For comparative purposes, a GO was also synthesized from flake graphite via a modified Hummers' method, as reported elsewhere [31], in the presence of a mixture of concentrated $K_2S_2O_8$, H_2SO_4, and P_2O_5 at 80 °C for a few hours. The product was filtered, dried, and oxidized for a second time by adding H_2SO_4, $KMnO_4$, and cold water. Excess of $KMnO_4$ was decomposed through a subsequent addition of 30 wt% H_2O_2 and 5 wt% HCl aqueous solutions. Afterward, the product was filtered anew, purified, bath sonicated for half an hour at 150 W, and vacuum freeze-dried. The calculated reaction yield was 112%.

Table 1. Nomenclature, experimental conditions of the electrochemical synthesis, and characteristics of the different EGO samples.

Sample	Voltage I/II (V)	Time I (min)	Time II (s)	H_2SO_4 (wt%)	C/O Ratio	O_{EDX} (%)	I_D/I_G Ratio	d_{001} (nm)	t (nm)
GO*	–	–	–	–	2.25	31.00	1.04	0.8615	13.8
EGO 1	1.0/10	10	60	65	2.09	33.12	0.41	0.8816	12.1
EGO 2	1.0/10	10	120	65	1.98	35.41	0.47	0.8956	10.7
EGO 3	1.0/20	10	30	98	1.72	40.12	1.01	0.9034	7.14
EGO 4	1.0/20	10	60	40	1.91	36.11	0.59	0.9135	9.13
EGO 5	1.0/20	10	60	65	1.79	38.33	0.70	0.9145	7.11
EGO 6	1.0/20	10	60	98	1.54	44.80	1.37	0.9187	3.98
EGO 7	1.0/30	30	60	40	2.32	29.76	0.43	0.8594	14.4
EGO 8	1.0/30	30	120	65	2.57	26.85	0.54	0.8576	16.5
EGO 9	2.0/10	10	60	65	1.88	36.69	0.79	0.9161	8.75
EGO 10	2.0/10	10	120	65	1.81	38.40	0.69	0.9356	7.37
EGO 11	2.0/20	10	30	65	1.95	35.38	0.74	0.9021	10.2
EGO 12	2.0/20	10	30	98	1.59	43.39	1.51	0.9218	4.24
EGO 13	2.0/20	10	60	40	1.80	38.33	0.77	0.9378	7.53
EGO 14	2.0/20	10	60	65	1.67	41.31	0.87	0.9595	5.84
EGO 15	2.0/20	10	60	98	1.46	47.26	1.68	0.9230	3.79
EGO 16	2.0/20	10	120	40	1.78	38.77	0.93	0.9167	7.26
EGO 17	2.0/20	10	120	65	1.69	40.82	0.99	0.9496	5.92
EGO 18	2.0/30	10	30	65	2.49	27.71	1.22	0.8583	15.8
EGO 19	2.0/10	30	30	98	2.48	27.82	1.71	0.8472	16.0
EGO 20	2.0/20	30	120	65	2.27	30.39	1.23	0.8602	14.5
EGO 21	2.0/30	30	60	40	2.88	23.96	1.30	0.8564	18.6
EGO 22	2.0/30	30	120	65	2.96	23.31	1.39	0.8499	17.9

* Synthesized by a modified Hummers' method. I and II refer to the intercalation and exfoliation stages, respectively. The C/O ratio has been calculated from elemental analysis measurements, the percentage of oxygen (O_{EDX}) from EDX measurements, I_D/I_G is the integrated intensity ratio of the D and G peaks from the Raman spectra, t is the average flake thickness obtained from AFM measurements, and d is the interlayer spacing corresponding to the (001) reflection of GO obtained from the X-ray diffractograms.

2.3. Characterization

A LECO CHNS-932 elemental analyzer was used to perform elemental analysis measurements. Chromatograms were acquired with a 7820A Gas Chromatograph coupled with a 5975 mass spectrometry system equipped with a 30 m × 0.25 mm × 0.25 µm HP-5MS capillary column. The injector and detector temperature were 250 and 150 °C, respectively, and the oven temperature program was a ramp from 50 to 100 °C for 10 min, which is followed by another ramp until 230 °C for 15 min. The carbon/oxygen atomic ratio (C/O) was calculated according to the method described by Kaspar [32].

A SU8000 Hitachi scanning electron microscope operating at 15.0 kV and an emission current of 10 mA was employed to acquire the scanning electron microscopy (SEM) images. The microscope is fitted with an energy-dispersive X-ray (EDX) detector that allows for qualitative and quantitative analysis of the sample composition.

A Philips Tecnai 20 FEG electron microscope fitted with a LaB6 filament, working at 200 kV, was used to obtain the transmission electron microscopy (TEM) micrographs with point-to-point resolution of 0.3 nm.

Atomic force microscopy (AFM) imaging was performed using a Bruker Dimension Icon system coupled with a Nanoscope V controller, using a Peakforce QNM imaging mode and a 100 μm long monolithic silicon cantilever.

A Bruker D8 Advance diffractometer fitted with a Cu X-ray tube and a Ni K_β filter was used to carry out the X-ray diffraction (XRD) measurements, working at 40 kV and at an intensity of 40 mA.

Mid-range room temperature Fourier-transformed infrared (FT-IR) spectra were collected with a Perkin Elmer Frontier FTIR spectrophotometer fitted with an attenuated total reflectance (ATR) sampling unit and a diamond window. Thirty-two scans were recorded for each sample with 1 mW laser output and a resolution ≥ 4 cm^{-1}. Previous to the measurements, the nanomaterials were milled and mixed with KBr, and then pressed into a pellet. Atmospheric compensation was carried out.

Raman spectra were acquired at room temperature with a Renishaw Raman microscope set with a He-Ne gas laser with a wavelength of 632.8 nm and an output of 1.0 mW. To improve the signal-to-noise ratio, at least 10 scans were recorded for each sample.

The room temperature electrical resistivity of the synthesized EGOs was determined under a pressure of 600 kPa set by using an upper weight, with a KEITHLEY 2182A nanovoltmeter and a KEITHLEY 6221 current source, respectively. Prior to the measurements, each sample was positioned in a Teflon cylinder and compressed for 1 h between two stainless steel plates that acted as electrodes.

3. Results and Discussion

3.1. Oxidation Level of the Synthesized EGOs

Elemental analysis was used to determine the absolute amounts of C, H, N, and S elements of the synthesized EGOs [33], and to calculate their carbon/oxygen (C/O) atomic ratio, which is indicative of the degree of oxidation. Therefore, this shows the effectiveness of the electrochemical oxidation process of graphite. Typical C/O atomic ratios for GO lie between 2.1 to 2.9 [34]. The reference GO synthesized by the Hummer's method shows a C/O ratio of 2.25, which is in good agreement with previous studies [7]. The EGOs display a wide range of C/O ratios (1.46–2.96, Table 1), which indicates that some of them present higher and other lower degrees of oxidation than the reference, which corroborates the strong influence of the synthesis conditions on the level of oxidation of the final sample. In particular, when the voltage of the second stage is higher than 20 V, the exfoliation rate should be very fast, which results in large GO aggregates and thick flakes (i.e., ≥ 20 nm for EGO 21 and 22, as corroborated by TEM, Section 3.2). Likely, only the lateral parts of the sheets are oxidized, which results in lower oxidation degrees. A similar result is obtained when the low voltage is applied for a long time (i.e., 30 min): the FGF sheets are easily broken into small pieces during the intercalation stage and are poorly oxidized, which leads to high C/O ratios. Furthermore, too much swelling of the GIC can occur if the intercalation time is too long, which leads to a strong decrease in mechanical strength and conductivity, and, therefore, it is more difficult to be further oxidized. In contrast, syntheses performed using concentrated H_2SO_4 during the second stage, result, as expected, in a very high level of oxidation. Thus, an unprecedented minimum C/O ratio value of 1.46 has been obtained when a low bias of 2 V was applied for 10 min and the oxidation stage was performed under 20 V for 60 s using 98% H_2SO_4 as an electrolyte. This particularly highly oxidized GO should have a very high content of carboxylic acid groups, which will be discussed in a following section, preferentially located at the sheet edges, which are more easily oxidized. Nonetheless, due to its elevated oxygen content, besides the conventional epoxy/hydroxyl groups situated on the basal planes, it could also comprise other oxygen-containing functional groups such as ketone/quinone (C=O), carboxylates (O–C=O), and even peroxy groups (O–O) [35]. This sample (EGO 15) also exhibits a very high level of exfoliation according to TEM analysis. When the low voltage was decreased to 1 V, and the rest of the conditions were maintained, the C/O ratio increased to 1.54, indicating that the voltage of the first stage is also crucial to attain a good wetting of the sample and allow the intercalation of ions within the graphite sheets, thus promoting their exfoliation. Conversely, when the same conditions were applied, albeit the concentration of the electrolyte in the second stage, was reduced to 65% or 40%, the C/O ratios increased up to 1.79 and 1.91, (EGO 5 and 4, respectively), indicating a less efficient exfoliation and oxidation, as also corroborated by TEM analysis.

Very good results (C/O = 1.67) have also been attained when a low bias of 2 V was applied for 10 min and the oxidation stage was performed under 20 V for 60 s using 65% H_2SO_4 as an electrolyte (EGO 14). However, when the 20 V were only applied for 30 s, the C/O ratio noticeably increased, which indicates partial oxidation of the sheets. However, when the same voltage was applied for 120 s, the same C/O ratio was attained (1.69). This suggests that there is an optimum time for the oxidation reaction, close to 60 s, and longer oxidation periods do not lead to higher oxidation degrees because a saturation level in the number of functional groups located onto the basal planes and at the edges has been attained. This is consistent with previous theoretical studies on the distribution of oxygenated groups onto GO sheets, which reveal that epoxide and hydroxyl moieties prefer the island arrangement rather than a homogeneous distribution [36]. In the island configuration, the functional groups are very close to each other, which makes it difficult to incorporate more groups and leads to a saturation level that has been calculated to be around 9 and 8 epoxide and hydroxyl units, respectively [36]. This is also in agreement with the observations made during the electrochemical syntheses: for a given set of conditions, with increasing exfoliation time from 30 to 60 s, the yellowish areas of the EGOs increased, and, in general, the whole surface became yellow at around 60 s. However, no color changes were observed after 1 min. Only when mild conditions were applied (a low bias of 1 V for 10 min followed by 10 V for 60 s and 65% H_2SO_4 as an electrolyte), a slight decrease in the C/O ratio (~5.3%) was found when increasing the exfoliation time from 60 to 120 s.

On the other hand, it is important to highlight the very good agreement found between the C/O ratio. Hence, the percentage of C and O was detected by elemental analysis and EDX (Table 1), despite the fact that EDX is a surface technique while elemental analysis gives the value of the whole sample. Thus, the good consistency between the results obtained by the two techniques confirms that the samples are homogeneous over the entire volume. It should be noted that the C/O ratio influences structural properties like the degree of exfoliation and number of layers, amount of defects, flake size, and concentration of functional groups, among others [37], which, in turn, would tailor the band structure, hence the photoluminiscent behaviour, as well as the electrical and mechanical properties. In addition, the oxygenated moieties provide anchoring points for subsequent chemical modification [38].

3.2. Morphological Characterization of the EGOs

The surface morphology of FGF and the synthesized EGOs was investigated by SEM, and typical images of FGF and EGO 14 are compared in Figure 2. The image of FGF (Figure 2a) reveals a dense and compact surface structure, composed of large grains with different shapes, and a smooth and flat surface. This fine crystal grain structure is consistent with its crystalline nature. In contrast, the EGO (Figure 2b) shows a rough surface topography with exfoliated and well separated graphene sheets, with thicknesses ranging from 30 to 10 nm. Furthermore, the flakes are highly-wrinkled, which indicates a deformation of the graphene layers due to the linkage of the oxygenated functional groups after the electrochemical process. Furthermore, the grain pattern cannot be observed, which suggests a reduction in the level of crystallinity, while large nanosheet sizes are preserved. Overall, the images corroborate that the electrochemical process used herein maintains the integrity of the graphene flakes, and enables us to obtain large area uniform GO sheets. On the other hand, the EDX maps of EGO 14 are shown in Figure 2c,d, from which the C/O ratio was obtained (Table 1).

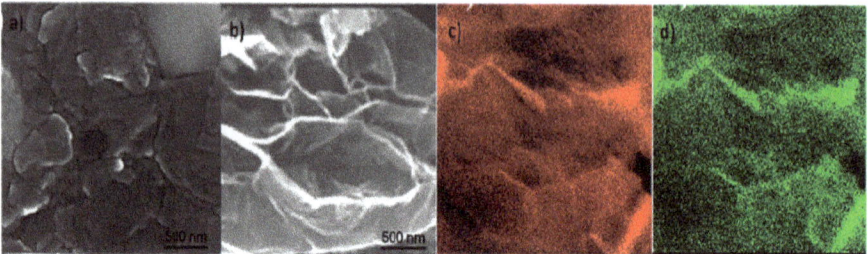

Figure 2. SEM images of pristine FGF (**a**), EGO 14, (**b**) and the EDX maps of the area shown in b for carbon (**c**) and oxygen (**d**).

Further information about the surface morphology of the samples was attained by TEM, and typical images of EGO 4, 14, 15, and 21 are compared in Figure 3. A moderate level of exfoliation can be observed for EGO 4 (Figure 3a), which was prepared by applying mild conditions: a low bias of 1V for 10 min and 60 s of exfoliation under 20 V in 40% H_2SO_4 as an electrolyte. Although no FGF is found, corroborating that it has been oxidized and exfoliated during the electrochemical process, leads to GO flakes with thicknesses <20 nm. Some small GO aggregates can be observed, which are indicative of a less efficient exfoliation process.

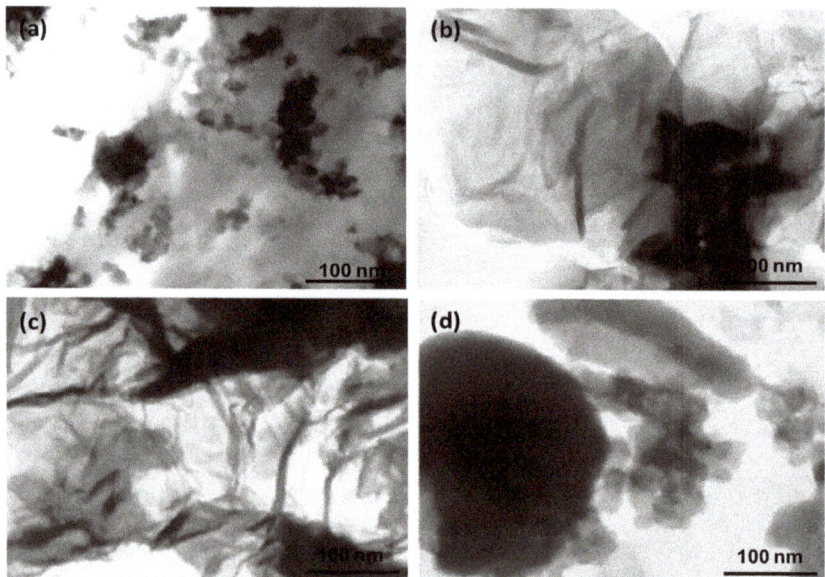

Figure 3. TEM images of EGO 4 (**a**), EGO 14 (**b**) EGO 15 (**c**), and EGO 21 (**d**).

In contrast, TEM images for EGO 14 and 15 (Figure 3b,c) clearly reveal the formation of homogeneous and good quality GO sheets, with a clear flake structure and lack of traces from the raw material. Very well exfoliated, uniform, and thin sheets can be detected, in particular for EGO 15, that exhibits highly wrinkled sheets with thicknesses <5 nm. Such a high degree of bending is consistent with the presence of a large number of oxygenated functional groups that can interact via H-bonding [35]. Furthermore, the darker areas are somehow light transparent, which also suggests a good level of exfoliation. Thus, the application of 20 V for 60 s with concentrated H_2SO_4 as an electrolyte is demonstrated to be a very effective exfoliation approach. In the case of EGO 14, prepared under the same conditions despite using 65% H_2SO_4 as an electrolyte, the sheets are slightly thicker and less bended. While they appear homogenous, their level of exfoliation is also very good and preserves the integrity of the graphene flakes. Conversely, for EGO 21 (Figure 3d), very large blocks of GO aggregates can be observed, and the darker areas reveal small, partly oxidized GO pieces or even some remaining FGF precursors. Since a high voltage of 30 V was applied during the oxidation stage, the exfoliation rate should be very fast, resulting in large GO aggregates and flakes with thicknesses ≥20 nm. Analogous morphology was found for EGO 22 prepared by applying the same high voltage during a longer time (120 s). Furthermore, during the synthesis of these samples, the low bias was applied for a long period (30 min), which likely results in a very high level of swelling. This results in poorer mechanical strength. Hence, the sheets are easily broken into small pieces difficult to be oxidized. Thus, very similar morphology, albeit with smaller aggregates and slightly thinner sheets was found for EGO 19 and 20, which were also synthesized by initially applying 2 V for 30 min.

AFM measurements were also performed to estimate the average thickness of the EGO flakes, and the results are summarized in Table 1. Theoretically, the thickness of single layer graphene is 0.345 nm. However, it has been reported that GO layers are about 1.1 nm [39]. A wide range of thickness values have been measured for the EGOs, which range between 3.8 and 18.6 nm. This defines them as few-layer GO materials according to referential nomenclature [40]. The trend observed is consistent with that obtained from both TEM and SEM observations. GOs with the highest level of exfoliation (EGO 12 and 15) display the lowest thicknesses. In general, with an increasing C/O ratio, upon decreasing the sample oxidation level, the thickness increases.

The average lateral dimensions of the EGOs were also estimated by AFM. The reference GO has an average lateral dimension of 24 μm. Samples synthesized by applying the low voltage for 30 min (EGO 7, 8, 19–22) show the smallest dimensions in the range of 4–12 μm. The FGF sheets could be broken into smaller pieces during the long intercalation stage, which results in smaller layers. Samples prepared using concentrated H_2SO_4 during the second step (EGO 3, 6, 12, 15) have lateral sizes between 10 and 22 μm, while the rest of the nanomaterials show dimensions larger than 20 μm.

3.3. FT-IR Analysis of the EGOs

IR spectroscopy was used to characterize the functional groups present onto the synthesized EGOs and get some insight about their oxidation level. GO is known to contain epoxide, hydroxyl, ketone, and even peroxy groups on the basal planes, while carboxyl moieties are located at the borders. Note that, due to the variety of oxygenated groups and the overlapping modes below 1500 cm^{-1}, the interpretation of the FT-IR spectra of GO is complex and it is not possible to unambiguously correlate specific frequencies to a particular functional group [41].

The spectrum of the reference GO (Figure 4) shows a broad peak centred at ~3735 cm^{-1} assigned to the O–H stretching vibration, another at ~1740 cm^{-1} characteristic of the C=O stretching of the COOH moieties [42], one at around 1630 cm^{-1} corresponding to the C=C stretching of aromatic rings, the other at about 1400 cm^{-1} that matches with the O–H bending vibration and bands at around 1270, and 1050 cm^{-1} associated with the C–O stretching in epoxy groups [43]. All these features are also present in the EGOs, which suggests that they have the same oxygen-containing functional groups as the reference GO. Nonetheless, noticeable changes in the peak positions and intensities can be observed among the spectra of GO and the different EGOs that point toward diverse levels of oxidation. Thus, the EGO 15 displays the wider and most intense O–H stretching band, which is in agreement with its very low C/O ratio (Table 1). This signifies a very high content of carboxylic acid groups. Furthermore, the band is displaced toward lower wave numbers, which is symptomatic of the presence of strong hydrogen bonds [44]. These are more characteristic among carboxylic acid groups. The EGO 14 also exhibits an intense band shifted toward lower frequency, while, in the reference GO, the band appears at slightly higher wave numbers, which suggests weaker/fewer H-bonds.

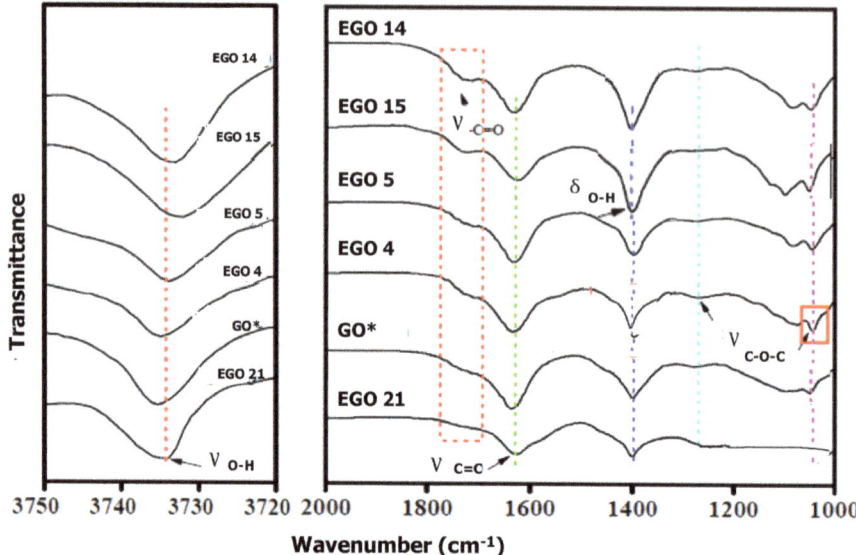

Figure 4. FT-IR spectra of the reference (GO*) and the indicated EGO samples.

A clearer trend can be observed from the O–H deformation band, which suggests the following level of oxidation: EGO 21 < GO* < EGO 4 < EGO 5 < EGO 14 < EGO 15. These results are perfectly consistent with those derived from EDX and elemental analyses. Thus, the EGO 21 prepared by applying a very high voltage during the second stage shows the lowest level of oxidation, which is in agreement with the presence of large aggregates and thick flakes as revealed by TEM. Analogous spectrum, even with a weaker band, was obtained for EGO 22, prepared under similar conditions despite applying the voltage of the second stage for a longer time and using 65% H_2SO_4 as an electrolyte. Conversely, EGO 14 and 15, prepared by applying a low bias of 2 V for 10 min, which is followed by 20 V for 60 s using 65% or 98% H_2SO_4 as an electrolyte, respectively, display the most intense band, which suggests that these are optimal conditions to attain a very high level of flake oxidation. On the other hand, EGO 4 and 5, prepared under mild conditions, display moderately intense bands, suggesting intermediate levels of oxidation. Nonetheless, the band intensity is stronger for EGO 5 compared to the other one, which corroborates that the concentration of the electrolyte strongly influences the number of oxygen-containing groups formed during the second stage.

Regarding the band at ~1740 cm^{-1} related to the C=O stretching of the COOH groups, it is clearly more intense for EGO 14 and 15, while it can hardly be detected in EGO 21 and it shows very low intensity in the reference GO. A similar trend is found for the peaks assigned to epoxy groups, which cannot be found in EGO 21, while are very intense in EGO 15. Furthermore, this sample displays a new peak at about 1135 cm^{-1} likely ascribed to ether groups that cannot be found in the other EGOs, which is consistent with its high level of oxidation [45].

3.4. Raman Analysis of the EGOs

Raman spectroscopy is a useful tool to assess the number of defects in carbon nanomaterials as well as their crystallite size and number of layers [46]. In particular, it is a widely used technique to characterize GOs with different oxidation degrees [47]. The Raman spectrum of GO shows general features at about 1360, 1600, and 2700 cm^{-1} that correlate with the D, G, and 2D peaks, respectively.

The G band corresponds to the E2g vibrational mode found in a graphite single crystal [48], which is characteristic of sp² hybridization. The D band is associated to defects, vacancies, or lattice disorders due to the binding of oxygen-functional groups and the 2D band is attributed to second order phonon processes [49]. These features are observed in the spectra of all the EGO samples with differences in the peak positions and their intensities (Figure 5). Thus, the I_D/I_G ratio (the integrated intensity ratio of the D peak and the G peak) is an indication of the quality of the GO sheets and their number of defects [50]. One of the typical defects is the double vacancy (C2), also known as 5-8-5 defect, which comprises one octagonal and two pentagonal rings [51]. In addition, the common 5-7-7-5 rings, also known as Stone-Wales (SW) defects, are a key factor in graphitic-based materials.

I_D/I_G data ranging from 1.71 to 0.41 have been calculated for the synthesized EGOs (Table 1). Clearly, EGO 19, prepared under strong oxidizing conditions (2 bias for 30 s followed by 10 V for 30 s and 98% H_2SO_4 during the second stage) shows the highest I_D/I_G value, which is followed by EGO 15, and is synthesized with a shorter swelling stage despite a longer oxidation step in 98% H_2SO_4. This sample also displays a small shift of the G band toward higher frequency (blue-shift), which is indicative of a high disorder [48]. A high I_D/I_G ratio was systematically obtained for all the EGOs prepared using 98% H_2SO_4 as an electrolyte, which is symptomatic of GO sheets with a large amount of surface defects generated in the presence of the concentrated acid. An elevated I_D/I_G value, higher than that of the reference GO, was also obtained for EGO 21 and 22, which was prepared by applying the low bias for a long period (30 min). This is likely related to their high level of swelling, which results in poorer mechanical properties. Hence, the sheets are more easily damaged and generate a larger amount of defects.

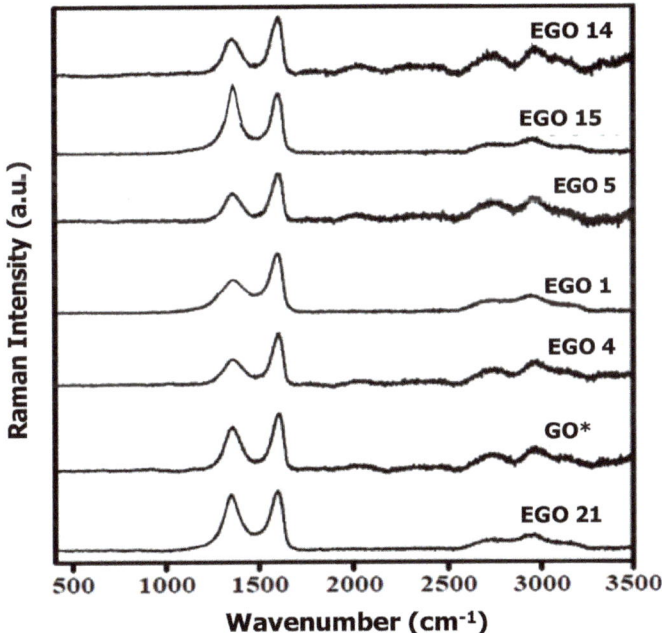

Figure 5. Raman spectra of the reference (GO*) and the indicated EGO samples.

Among the 22 EGOs synthesized, 14 show lower I_D/I_G than the reference, which indicates a superior quality of their GO flakes. For instance, an intermediate I_D/I_G value of 0.87 was attained for EGO 14. Although this sample exhibits a high level of oxidation and a good degree of exfoliation, as corroborated by TEM, its amount of surface defects is somewhat low, likely because 65% H_2SO_4 was used as an electrolyte. Lower ratios were obtained for EGO 4 and EGO 5, prepared by applying milder conditions, despite their oxidation level was also poorer. Moreover, EGO 1 prepared under the softest conditions, shows the lowest I_D/I_G value.

On the other hand, I_D/I_G has been widely used to estimate the in-plane crystallite size (L_a) in non-ordered nanostructures [48]: $L_a = 2.4 \times 10^{-10} \lambda^4 (I_D/I_G)^{-1}$. Thus, the larger the I_D/I_G ratio, the smaller L_a, that is, the crystalline regions would decrease in size as the structural disorder increases. L_a values ranging from 2.8 to 13.4 nm have been calculated for the synthesized EGO samples, which indicates a strong effect of the synthesis conditions on the crystallite size. This is in good agreement with recent works [52].

3.5. X-ray Diffraction Study of the EGOs

X-ray diffractograms were acquired to investigate possible changes in the interlayer spacing (*d*) and crystallite size depending on the synthesis conditions of the EGOs (Figure 6). As known, raw graphite shows an intense diffraction peak at $2\theta = 26.5°$, which corresponds to a *d* value of 0.336 nm taking into account the Bragg's equation. This peak was not observed in any of the synthesized EGOs, since it disappeared after oxidation, while a new peak appeared in the range of 9–12° corresponding to the (001) reflection of GO [53]. The presence of this characteristic peak corroborates that all the samples were partially or completely oxidized during the electrochemical process.

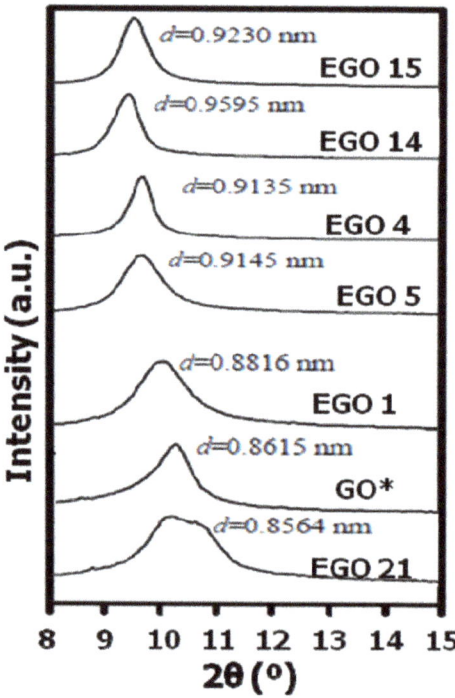

Figure 6. X-ray diffractogram of the reference (GO*) and the indicated EGO samples.

The *d*-spacing calculated for the reference GO is 0.8615 nm, about 2.6-fold bigger than the calculated for graphite. This increase in the *d*-spacing is consistent with previous works [37], which is accredited to the presence of oxygen containing groups, mainly epoxide and hydroxyl formed during the oxidation process, laying between the GO sheets, and a change in the hybridization state of the C atoms from sp^2 to sp^3 [54]. The values of *d*-spacing calculated for all the EGOs are listed in Table 1. EGO 19 shows the smallest *d* value, likely due to its low oxidation level and the fact that the oxidation stage was carried out in strong H_2SO_4. Hence, most of the oxidation likely occurred on the flake edges. Consequently, the number of oxygen-containing groups within layers is relatively small. This explanation could also account for the fact that EGO 14 presents a lager d value that EGO 15, despite the lager oxygen content of the latter sample calculated by elemental analysis and EDX (Table 1). Thus, systematically, all the samples synthesized using strong H_2SO_4 as an electrolyte display a smaller d value than expected, according to their oxygen content. The largest interlayer spacing, close to 0.96, was obtained for EGO 14, corroborating that the sample was almost completely oxidized. Longer exfoliation times resulted in slightly lower interlayer spacing (EGO 17), likely because it led to more functional groups on the flake edges. On the other hand, EGO 4 and 5, prepared under mild conditions, present intermediate d values, which are in agreement with their moderate level of oxidation, despite being higher than that of the reference GO.

The crystallite size was also estimated from the full width at half maximum (FWHM) of the diffraction peak corresponding to the (001) plane, according to the Debye-Scherrer equation [55]. The values obtained ranged from 4 to 12 nm, which were in good agreement with those determined from the Raman spectra.

3.6. Electrical Resistance of the EGOs

It is well known that there is a strong relationship between the structure of GO and its electrical conductivity. The performance of GO in electronic devices is based on its electrical properties, which are significantly influenced by the synthesis process, hence on the level of oxidation [56]. GO is reported to be electrically insulating due to the sp^3 bonds, the elevated density of oxygenated groups, which are electronegative, and the presence of defects and vacancies that lead to a gap in the electronic density of states [57]. The largest contribution to the electrical resistivity comes from the contact resistance between GO particles, which likely will have two sources: tunneling and constriction [58]. Tunneling resistance occurs in GO since it can be depicted as a conductive nucleus delimited by non-conductive regions. In contrast, constriction resistance is attributed to the narrowness of the conducting pathway due to the tiny contact region between particles. This effect would likely be less important due to the elevated aspect ratio of the GO sheets, which will tend to locate parallel to each other after the pressure is applied.

Figure 7 compares the electrical resistance of the synthesized EGOs versus their C/O ratio, which is their oxidation level, and their interlayer spacing. It can be observed that the electrical resistance decreases from about 29 to 5 kΩ while the C/O ratio increases (Figure 7a). This is indicative of a trend between the two parameters. Despite not being linear, the electrical conductivity is restored while decreasing the oxidation level. However, EGO 15, which was prepared under strong oxidizing conditions and shows the lowest C/O ratio (1.46) and the second highest I_D/I_G ratio, hence, a very large number of defects, does not exhibit the utmost resistance. Conversely, EGO 14 with a lower oxidation level and I_D/I_G ratio shows the highest resistance value. This suggests that, besides the oxidation level, other factors should govern the conduction mechanism in GO. Theoretical simulations have shown that the conductivities of few-layer graphene nanosheets are reduced when thickness is increased [59]. However, our results do not show a direct correlation between the electrical conductivity and the flake thickness measured by AFM (Table 1).

To analyze the influence of other factors, the electrical resistance was plotted against the interlayer spacing (d) calculated from X-ray diffractograms. There is a very clear relationship between both magnitudes: the electrical resistance increases steadily while increasing the interlayer distance, which indicates that the most important factor influencing the conductivity in GO is the tunneling resistance between adjacent sheets. This is consistent with earlier investigations that studied the tunneling effect in carbon nanomaterials [60]. In these materials, the tunneling resistance can be estimated according to the Simmons's formula [61], which predicts the resistance between two flat electrodes set apart by a small insulating film. Thus, the tunneling resistance increases linearly with a growth in the insulating layer thickness [62]. This is in agreement with the augment found herein with increasing interlayer spacing.

The conduction mechanisms that take place in GO sheets are not well understood yet. Frequent charge will percolate by a hopping conduction mechanism [57], which consists of the transport of charges via localized states. The percolation in random continua or the 2D Swiss-cheese model is a simple way to depict GO conductivity [63]. In this model, the localized states are arbitrarily distributed within the GO material. The model is complicated since the occupation of the localized states is governed by the confluence of the threshold voltage and the local electric field originated by the number of inserted carriers. Hopping between localized states occurs via thermal agitation.

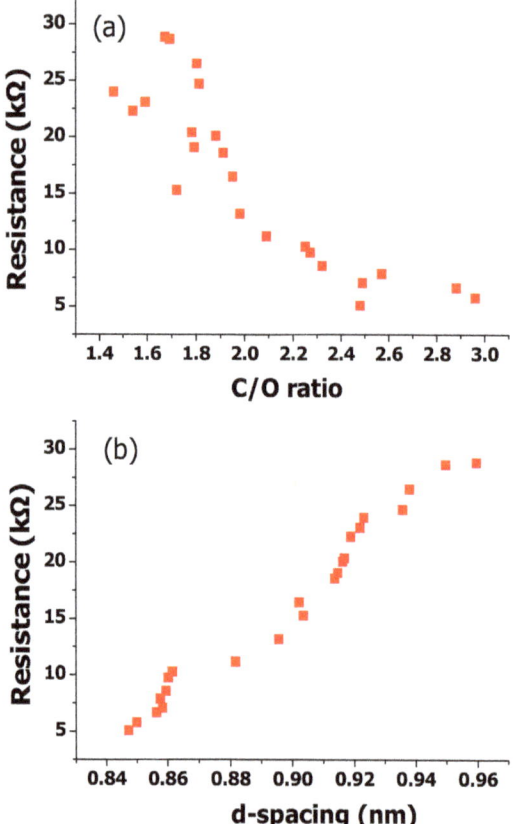

Figure 7. Electrical resistance of the synthesized EGOs versus the C/O ratio (**a**) and the d-spacing (**b**).

Experimental data reveal a certain dependence of the electrical resistance on the oxidation level. This is consistent with theoretical statistics revealing that the presence of oxygen groups influences the electronic density of states [57]. Thus, the epoxy moieties appearing in all the EGOs can displace the Fermi energy, which significantly influences the electrical resistance. Furthermore, the formation of hydrogen bonds between epoxy and hydroxyl groups could lessen the energy density close to the Fermi level, and consequently, will rise the work function of the carbon nanomaterial [57]. Previous experimental works reported that the work function of GO steadily augments when the C/O ratio increases. This resembles that of graphite for C/O higher than 3 [64]. However, our results suggest that the conductivity is likely conditioned by the type of oxygenated functional groups and their location on the GO flakes.

Therefore, despite that the majority of the models support that the electrical properties of GO are influenced by the number of oxygenated groups, hence, the oxidation level, it is clear from our experimental data that the charge transference depends mostly on the tunneling resistance between neighbouring sheets, hence on the interlayer spacing.

4. Conclusions

GO powders with a diverse amount of oxygenated surface groups have been synthesized from flexible graphite foil by an electrochemical two-step process: intercalation and oxidation/exfoliation. A wide range of synthesis conditions have been studied to optimize the oxidation level, by modifying the voltage and time of both stages as well the electrolyte concentration. For comparative purposes, a reference GO was prepared via a modified Hummers' method. The resulting GO samples have been characterized by different techniques including elemental analysis, SEM, TEM, AFM, XRD, EDX, FT-IR, and Raman spectroscopies as well as electrical resistance measurements. Elemental analysis data confirm the strong influence of the synthesis parameters on the oxidation level. An unprecedented minimum C/O ratio of 1.46 has been obtained by applying 2 V for 10 min followed by 20 V for 60 s using 98% H_2SO_4 as an electrolyte. SEM and TEM micrographs reveal the formation of good quality, large, homogeneous and well exfoliated GO sheets. FT-IR corroborates the formation of –OH, C-O-C, C=O, and –COOH surface moieties and points toward strong intermolecular interactions between adjacent groups. A high I_D/I_G ratio in the Raman spectra has been obtained for all the samples prepared using 98% H_2SO_4 as an electrolyte, which is indicative of sheets with a large number of defects. Furthermore, a high I_D/I_G is also found for samples prepared by applying the voltage of the first stage for a long period, which is related to their high level of swelling. The interlayer distance between adjacent sheets estimated from XRD was not directly proportional to the oxygen content. Samples prepared using 98% H_2SO_4 show smaller interlayer spacing than expected considering their oxidation level, which signifies that a great part of the oxidation took place on the sheet edges. Electrical resistance results do not show a clear trend with the C/O ratio nor with the flake thickness but with the interlayer spacing, which indicates that the electrical performance depends mostly on the tunneling resistance between neighbouring sheets. Overall, the optimal synthesis conditions to attain a high level of oxidation and a moderate number of defects are 2V in the intercalation stage for 10 min, which was followed by 20 V in the oxidation stage for 60 s under 65% H_2SO_4. Nonetheless, this sample displays the highest sheet resistance. Overall, the level of oxidation and exfoliation can be tailored by controlling the synthesis conditions, and should be selected depending on the specific application. The approach developed herein provides an efficient way to tune the physical properties of GO-based materials.

Author Contributions: C.S.-U. and A.M.D.-P. performed the experiments and analyzed part of the data. A.M.D.-P. designed the experiments, supervised the work, and wrote the paper. M.P.S., S.V.-L., and C.V. collaborated in the development of the experiments, in the analysis of the experimental data and in the discussion of the results. All authors have read and agreed to the published version of the manuscript.

Funding: This research was funded by the Spanish Ministry of Science, Innovation and Universities (MICIU) via Project PGC2018-093375-B-I00.

Acknowledgments: A.M.D.-P. wishes to thank the Ministry of Economy, Industry and Competitivity (MINECO) for a "Ramón y Cajal" Senior Research Fellowship (RYC-2012-11110) co-financed by the EU.

Conflicts of Interest: The authors declare no conflict of interest.

References

1. Yang, X.; Cheng, C.; Wang, Y.; Qiu, L.; Li, D. Liquid-mediated dense integration of graphene materials for compact capacitive energy storage. *Science* **2013**, *341*, 534–537. [CrossRef] [PubMed]
2. Wang, H.; Yang, Y.; Liang, Y.; Robinson, J.T.; Li, Y.; Jackson, A.; Cui, Y.; Dai, H. Graphene-wrapped sulfur particles as a rechargeable lithium-sulfur battery cathode material with high capacity and cycling stability. *Nano Lett.* **2011**, *11*, 2644–2647. [CrossRef] [PubMed]

3. Diez-Pascual, A.M.; Chen, G. Selected Papers from the 1st International Online Conference on Nanomaterials. *Nanomaterials* **2019**, *9*, 1021. [CrossRef] [PubMed]
4. Díez-Pascual, A.M.; Luceño Sánchez, J.A.; Peña Capilla, R.; García Díaz, P. Recent advances in graphene/polymer nanocomposites for applications in polymer solar cells. *Polymers* **2018**, *10*, 217. [CrossRef]
5. Salavagione, H.; Díez-Pascual, A.M.; Lázaro, E.; Vera, S.; Gomez-Fatou, M. Chemical sensors based on polymer composites with carbon nanotubes and graphene: The role of the polymer. *J. Mater. Chem.* **2014**, *2*, 14289–14328. [CrossRef]
6. Díez-Pascual, A.M.; Gómez-Fatou, M.A.; Ania, F.; Flores, A. Nanoindentation in Polymer Nanocomposites. *Prog. Mater. Sci.* **2015**, *67*, 1–94. [CrossRef]
7. Dreyer, D.R.; Park, S.; Bielawski, C.W.; Ruoff, R.S. The Chemistry of Graphene Oxide. *Chem. Soc. Rev.* **2010**, *39*, 228–240. [CrossRef]
8. Paredes, J.I.; Villar-Rodil, S.; Solís-Fernandez, P.; Fernandez-Merino, M.J.; Guardia, L.; Martínez-Alonso, A.; Tascon, J.M.D. Preparation, characterization and fundamental studies on graphenes by liquid-phase processing of graphite. *J. Alloy. Compd.* **2012**, *536*, S450–S455. [CrossRef]
9. Kyzas, G.Z.; Deliyanni, E.A.; Matis, K.A. Graphene oxide and its application as an adsorbent for wastewater treatment. *J. Chem. Technol. Biotechnol.* **2014**, *89*, 196–205. [CrossRef]
10. Chabot, V.; Higgins, D.; Yu, A.; Xiao, X.; Chen, Z.; Zhang, J. A review of graphene and graphene oxide sponge: Material synthesis and applications to energy and the environment. *Energy Environ. Sci.* **2014**, *7*, 1564–1596. [CrossRef]
11. Gupta, V.; Sharma, N.; Singh, U.; Arif, M.; Singh, A. Higher oxidation level in graphene oxide. *Optik* **2017**, *143*, 115–124. [CrossRef]
12. Liu, L.; Zhang, F.; Zhao, J.; Liu, F. Mechanical properties of graphene oxides. *Nanoscale* **2012**, *4*, 5910–5916. [CrossRef] [PubMed]
13. Valles, C.; Beckert, F.; Burk, L.; Mülhaupt, R.; Young, R.J.; Kinloch, I.A. Effect of the C/O ratio in graphene oxide materials on the reinforcement of epoxy-based nanocomposites. *J. Polym. Sci. B-Polym. Phys.* **2016**, *54*, 281–291. [CrossRef]
14. Lyn, F.H.; Peng, T.C.; Ruzniza, M.Z.; Hanani, Z.A.N. Effect of oxidation degrees of graphene oxide (GO) on the structure and physical properties of chitosan/GO composite films. *Food Packaging Shelf* **2019**, *21*, 100373. [CrossRef]
15. Scaffaro, R.; Maio, A. Influence of oxidation level of graphene oxide on the mechanical performance and photo-oxidation resistance of a polyamide 6. *Polymers* **2019**, *11*, 857. [CrossRef]
16. Bandara, N.; Esparza, Y.; Wu, J. Graphite oxide improves adhesion and water resistance of canola protein–graphite oxide hybrid adhesive. *Sci. Rep.* **2017**, *7*, 11538. [CrossRef]
17. Morimoto, N.; Kubo, T.; Nishina, Y. Tailoring the oxygen content of graphite and reduced graphene oxide for specific applications. *Sci. Rep.* **2016**, *6*, 21715. [CrossRef]
18. Park, S.; Ruoff, R.S. Chemical methods for the production of graphenes. *Nat. Nanotechnol.* **2009**, *4*, 217–224. [CrossRef]
19. Hummers, W.S.; Offeman, R.E. Preparation of graphitic oxide. *J. Am. Chem. Soc.* **1958**, *80*, 1339. [CrossRef]
20. Chen, J.; Yao, B.; Li, C.; Shi, G. An improved Hummers method for eco-friendly synthesis of graphene oxide. *Carbon* **2013**, *64*, 225–229. [CrossRef]
21. Ogino, I.; Yokoyama, Y.; Iwamura, S.; Mukai, S.R. Exfoliation of Graphite Oxide in Water without Sonication: Bridging Length Scales from Nanosheets to Macroscopic Materials. *Chem. Mater.* **2014**, *26*, 3334–3339. [CrossRef]
22. Yang, S.; Lohe, M.R.; Muellen, K.; Feng, X. New-generation graphene from electrochemical approaches: Production and applications. *Adv. Mater.* **2016**, *28*, 6213–6221. [CrossRef] [PubMed]
23. Yang, S.; Brüller, S.; Wu, Z.S.; Liu, Z.; Parvez, K.; Dong, R.; Richard, F.; Samorì, P.; Feng, X.; Müllen, K. Organic radical-assisted electrochemical exfoliation for the scalable production of high-quality graphene. *J. Am. Chem. Soc.* **2015**, *137*, 13927–13932. [CrossRef] [PubMed]
24. Liu, J.; Yang, H.; Zhen, S.G.; Poh, C.K.; Chaurasia, A.; Luo, J.; Wu, X.; Yeow, E.K.L.; Sahoo, N.G.; Lin, J.; et al. A green approach to the synthesis of high-quality graphene oxide flakes via electrochemical exfoliation of pencil core. *RSC Adv.* **2013**, *3*, 11745–11750. [CrossRef]
25. Gurzęda, B.; Florczak, P.; Kempiński, M.; Peplińska, B.; Krawczyk, P.; Jurga, S. Synthesis of graphite oxide by electrochemical oxidation in aqueous perchloric acid. *Carbon* **2016**, *100*, 540–545. [CrossRef]

26. Parvez, K.; Rincón, R.A.; Weber, N.E.; Chaa, K.C.; Venkataraman, S.S. One-step electrochemical synthesis of nitrogen and sulfur co-doped, high-quality graphene oxide. *Chem. Commun.* **2016**, *52*, 5714–5717. [CrossRef]
27. Huang, B.; Zhao, Z.; Chen, J.; Sun, Y.; Yang, X.; Wang, J.; Shen, H.; Jin, Y. Facile synthesis of an all-in-one graphene nanosheets@nickel electrode for high-power performance supercapacitor application. *RSC Adv.* **2018**, *8*, 41323–41330. [CrossRef]
28. Ustavytska, O.; Kurys, Y.; Koshechko, V.; Pokhodenko, V. One-step electrochemical preparation of multilayer graphene functionalized with nitrogen. *Nanoscale Res. Lett.* **2017**, *12*, 175. [CrossRef]
29. Ambrosi, A.; Chua, C.K.; Bonanni, A.; Pumera, M. Electrochemistry of graphene and related materials. *Chem. Rev.* **2014**, *114*, 7150–7188. [CrossRef]
30. Kang, F.; Leng, Y.; Zhang, T.Y. Influences of H2O2 on Synthesis of H2SO4-GICs. *J. Phys. Chem. Solids* **1996**, *57*, 889–892. [CrossRef]
31. Luceño-Sánchez, J.A.; Maties, G.; Gonzalez-Arellano, C.; Diez-Pascual, A.M. Synthesis and Characterization of Graphene Oxide Derivatives via Functionalization Reaction with Hexamethylene Diisocyanate. *Nanomaterials* **2018**, *8*, 870. [CrossRef] [PubMed]
32. Kaspar, T. Graphen-abgeleitete Materialien. Ph.D. Thesis, ETH Zurich, Zurich, Switzerland, 2010; pp. 50–51.
33. Salamon, A.W. The current world of nanomaterial characterization: Discussion of analytical instruments for nanomaterial characterization. *Environ. Eng. Sci.* **2013**, *30*, 101–108. [CrossRef]
34. Ahn, M.; Liu, R.; Lee, C.; Lee, W. Designing Carbon/Oxygen Ratios of Graphene Oxide Membranes for Proton Exchange Membrane Fuel Cells. *J. Nanomater.* **2019**, *2019*, 6464713. [CrossRef]
35. Hontoria-Lucas, C.; López-Peinado, A.J.; López-González, J.D.D.; Rojas-Cervantes, M.L.; Martín-Aranda, R.M. Study of oxygen-containing groups in a series of graphite oxides: Physical and chemical characterization. *Carbon* **1995**, *33*, 1585–1592. [CrossRef]
36. Shin, D.S.; Kim, H.G.; Ahn, H.S.; Jeong, H.Y.; Kim, Y.; Odkhuu, D.; Tsogbadrakh, N.; Lee, H.; Kim, B.H. Distribution of oxygen functional groups of graphene oxide obtained from low-temperature atomic layer deposition of titanium oxide. *RSC Adv.* **2017**, *7*, 13979–13984. [CrossRef]
37. McAllister, M.J.; Li, J.-L.; Adamson, D.H.; Schniepp, H.C.; Abdala, A.A.; Liu, J.; Herrera-Alonso, M.; Milius, D.L.; Car, R.; Prud'homme, R.K.; et al. Single Sheet Functionalized Graphene by Oxidation and Thermal Expansion of Graphite. *Chem. Mater.* **2007**, *19*, 4396–4404. [CrossRef]
38. Luceño-Sánchez, J.A.; Diez-Pascual, A.M. Grafting of Polypyrrole-3-carboxylic Acid to the Surface of Hexamethylene Diisocyanate-Functionalized Graphene Oxide. *Nanomaterials* **2019**, *9*, 1095. [CrossRef]
39. Schniepp, H.C.; Li, J.L.; McAllister, M.J.; Sai, H.; Herrera-Alonso, M.; Adamson, D.H.; Prud'Homme, R.K.; Car, R.; Saville, D.A.; Aksay, I.A. Functionalized single graphene sheets derived from splitting graphite oxide. *J. Phys. Chem. B* **2006**, *110*, 8535–8539. [CrossRef]
40. Bianco, A.; Cheng, H.M.; Enoki, T.; Gogotsi, Y.; Hurt, R.H.; Koratkar, N.; Kyotani, T.; Monthioux, M.; Park, C.R.; Tascon, J.M.D.; et al. All in the graphene family—A recommended nomenclature for two-dimensional carbon materials. *Carbon* **2013**, *65*, 1–6. [CrossRef]
41. Yamada, Y.; Yasuda, H.; Murota, K.; Nakamura, M.; Sodesawa, T.; Satoshi, S. Analysis of heat-treated graphite oxide by X-ray photoelectron spectroscopy. *J. Mater. Sci.* **2013**, *48*, 8171–8198. [CrossRef]
42. Luceño Sánchez, J.A.; Peña Capilla, R.; Díez-Pascual, A.M. High-Performance PEDOT:PSS/Hexamethylene Diisocyanate-Functionalized Graphene Oxide Nanocomposites: Preparation and Properties. *Polymers* **2018**, *10*, 1169. [CrossRef] [PubMed]
43. Diez-Pascual, A.M.; Diez-Vicente, A.L. Poly(propylene fumarate)/polyethylene glycol-modified graphene oxide nanocomposites for tissue engineering. *ACS Appl. Mater. Interfaces* **2016**, *8*, 7902–17914. [CrossRef]
44. Diez-Pascual, A.M.; Diez-Vicente, A.L. Multifunctional poly(glycolic acid-co-propylene fumarate) electrospun fibers reinforced with graphene oxide and hydroxyapatite nanorods. *J. Mater. Chem. B* **2017**, *5*, 4084–4096. [CrossRef]
45. Fuente, E.; Menendez, J.A.; Díez, A.M.; Suarez, D.; Montes-Moran, M.A. Infrared Spectroscopy of Carbon Materials: A Quantum Chemical Study of Model Compounds. *J. Phys. Chem. B* **2003**, *107*, 6350–6359. [CrossRef]
46. Diez-Pascual, A.M.; Naffakh, M.; González-Domínguez, J.M.; Ansón, A.; Martinez Rubi, Y.; Martínez, M.T.; Simard, B.; Gómez, M.A. High performance PEEK/carbon nanotube composites compatibilized with polysulfones-I. Structure and thermal properties. *Carbon* **2010**, *48*, 3485–3499. [CrossRef]

47. López-Díaz, D.; López Holgado, M.; García-Fierro, J.L.; Velázquez, M.M. Evolution of the raman spectrum with the chemical composition of graphene oxide. *J. Phys. Chem. C* **2017**, *121*, 20489–20497. [CrossRef]
48. Tuinstra, F.; Koenig, J.L. Raman Spectrum of Graphite. *J. Chem. Phys.* **1970**, *53*, 1126. [CrossRef]
49. Determination of riboflavin based on fluorescence quenching by graphene dispersions in polyethylene glycol. *RSC Adv.* **2016**, *6*, 19686. [CrossRef]
50. Diez-Pascual, A.M.; Valles, C.; Mateos, R.; Vera-López, S.; Kinloch, I.A.; San Andrés, M.P. Influence of surfactants of different nature and chain length on the morphology, thermal stability and sheet resistance of graphene. *Soft Matter* **2018**, *14*, 6013–6023. [CrossRef]
51. Nudin, K.N.; Ozbas, B.; Schniepp, H.C.; Prud'homme, R.K.; Aksay, I.A.; Car, R. Raman Spectra of Graphite Oxide and Functionalized Graphene Sheets. *Nano Lett.* **2008**, *8*, 36–41. [CrossRef]
52. Lavin-Lopez, M.P.; Patón-Carrero, A.; Muñoz-Garcia, N.; Enguilo, V.; Valverde, J.L.; Romero, A. The influence of graphite particle size on the synthesis of graphene-based materials and their adsorption capacity. *Coll. Surf. A* **2019**, *582*, 123935. [CrossRef]
53. Luceño Sanchez, J.A.; Diez-Pascual, A.M.; Peña Capilla, R.; García Diaz, P. The Effect of Hexamethylene Diisocyanate-Modified Graphene Oxide as a Nanofiller Material on the Properties of Conductive Polyaniline. *Polymers* **2019**, *11*, 1032. [CrossRef] [PubMed]
54. Diez-Pascual, A.M.; Hermosa Ferreira, C.; San Andrés, M.P.; Valiente, M.; Vera, S. Effect of Graphene and Graphene Oxide Dispersions in Poloxamer-407 on the Fluorescence of Riboflavin: A Comparative Study. *J. Phys. Chem. C* **2017**, *121*, 830–843. [CrossRef]
55. Díez-Pascual, A.M.; Díez-Vicente, A.L. High-Performance Aminated Poly(phenylene sulfide)/ZnO Nanocomposites for Medical Applications. *ACS Appl. Mater. Interfaces* **2014**, *6*, 10132–10145. [CrossRef] [PubMed]
56. Peng, Z.; Zhi, L.; Shijie, Z.; Guosheng, S. Recent Advances in Effective Reduction of Graphene Oxide for Highly Improved Performance Toward Electrochemical Energy Storage. *Ener. Environ. Mater.* **2018**, *1*, 5–12. [CrossRef]
57. Jiang, X.; Nisar, J.; Pathak, B.; Zhao, J.; Ahuja, R. Graphene oxide as a chemically tunable 2-D material for visible-light photocatalyst applications. *J. Catal.* **2013**, *299*, 204–209. [CrossRef]
58. Marinho, B.; Ghislandi, M.; Tkalya, E.; Koning, C.E. Electrical conductivity of compacts of graphene, multi-wall carbon nanotubes, carbon black, and graphite powder. *Powder Technol.* **2012**, *221*, 351–358. [CrossRef]
59. Fang, X.-Y.; Yu, X.-X.; Zheng, H.-M.; Jin, H.-B.; Wang, L.; Cao, M.-S. Temperature- and thickness-dependent electrical conductivity of few-layer graphene and graphene nanosheets. *Phys. Lett. A* **2015**, *379*, 2245–2251. [CrossRef]
60. Foygel, M.; Morris, R.D.; Anez, D.; French, S.; Sobolev, V.L. Theoretical and computational studies of carbon nanotube composites and suspensions: Electrical and thermal conductivity. *Phys. Rev. B* **2005**, *71*, 104201. [CrossRef]
61. Simmons, J.G. Generalized Formula for the Electric Tunnel Effect between Similar Electrodes Separated by a Thin Insulating Film. *J. Appl. Phys.* **1963**, *34*, 1793. [CrossRef]
62. Li, C.; Thostenson, E.T.; Chou, T. Dominant role of tunneling resistance in the electrical conductivity of carbon nanotube–based composites. *Appl. Phys. Lett.* **2007**, *91*, 223114. [CrossRef]
63. Jung, I.; Dikin, D.A.; Piner, R.D.; Ruoff, R.S. Tunable Electrical Conductivity of Individual Graphene Oxide Sheets Reduced at "Low" Temperatures. *Nano Lett.* **2008**, *8*, 4283–4287. [CrossRef] [PubMed]
64. Shin, H.J.; Kim, K.K.; Benayad, A.; Yoon, S.M.; Park, H.K.; Jung, I.S.; Jin, M.H.; Jeong, H.K.; Kim, J.M.; Choi, J.H.; et al. Efficient Reduction of Graphite Oxide by Sodium Borohydride and Its Effect on Electrical Conductance. *Adv. Funct. Mat.* **2009**, *19*, 1987–1992. [CrossRef]

© 2020 by the authors. Licensee MDPI, Basel, Switzerland. This article is an open access article distributed under the terms and conditions of the Creative Commons Attribution (CC BY) license (http://creativecommons.org/licenses/by/4.0/).

Article

Scalable Preparation of Low-Defect Graphene by Urea-Assisted Liquid-Phase Shear Exfoliation of Graphite and Its Application in Doxorubicin Analysis

Chang-Seuk Lee, Su Jin Shim and Tae Hyun Kim *

Department of Chemistry, Soonchunhyang University, Asan 31538, Korea; eriklee0329@sch.ac.kr (C.-S.L.); sujean0395@naver.com (S.J.S.)
* Correspondence: thkim@sch.ac.kr

Received: 1 January 2020; Accepted: 4 February 2020; Published: 5 February 2020

Abstract: The mass production of graphene is of great interest for commercialization and industrial applications. Here, we demonstrate that high-quality graphene nanosheets can be produced in large quantities by liquid-phase shear exfoliation under ambient conditions in organic solvents, such as 1-methyl-2-pyrrolidinone (NMP), with the assistance of urea as a stabilizer. We can achieve low-defect graphene (LDG) using this approach, which is relatively simple and easily available, thereby rendering it to be an efficient route for the mass production of graphene. We also demonstrate the electrochemical sensing of an LDG-modified electrode for the determination of doxorubicin (DOX). The sensor shows an enhanced electrocatalytic property towards DOX, leading to a high sensitivity (7.23×10^{-1} μM/μA) with a detection limit of 39.3 nM ($S/N = 3$).

Keywords: graphene; mass production; shear exfoliation; physical exfoliation

1. Introduction

Graphene can be applied to everything from electronics and pharmaceuticals to sensing, biomedical and energy devices [1,2]. With exceptional properties derived from the atom-thick layer of graphite, it is becoming more prevalent and more critical in industry and scientific research. Although there is a myriad of uses for this wonder material, production can often be expensive and time-consuming. Graphene can be prepared using epitaxial growth and chemical vapor deposition (CVD) [3,4]. Even with the production of high-quality graphene, however, these techniques suffer from the drawbacks of complicated procedures, the need for toxic chemicals, expensive production or limited quantity, making it difficult to scale up to mass production for commercialization and industrial-scale application. With respect to mass production, chemical exfoliation is the facile and cost-effective technique in which the graphene preparation begins with the oxidation of graphite to graphene oxide (GO) and its subsequent reduction to get reduced graphene oxide (rGO) [5,6]. However, the resulting rGO is low-quality graphene with many of the defects of oxygen-containing functional groups.

To circumvent these issues, one of the most promising techniques is the physical exfoliation of layered graphite to graphene, which can lead to the mass production of low-defect graphene (LDG) [7]. The mechanical forces in physical exfoliation, such as sonication [8,9], ball milling [10,11] and shear exfoliation [12,13] can overcome the van der Waals forces that hold the graphite layers together, enabling graphene production on a large scale at a low cost. Graphene prepared from sonication and the ball-milling method, however, has more fragmentation and defects than expected, which is attributed to the sonication-induced cavitation, or collisions among the grinding media during the milling process. On the other hand, liquid-phase shear exfoliation with a high-shear mixer is easily available and makes it relatively simple to mass-produce defectless graphene. The size and amount of graphene produced through shear exfoliation can be controlled by the shearing speed, time, pore size, diameter of the rotor, stabilizers, and species of solvent [13]. In particular, the stabilizer facilitates the exfoliation of graphite to create a single layer of graphene by penetrating into the edges of the graphite layers, acting as "molecular wedge" and preventing the self-aggregation of the graphene in water through the intersheet steric and/or electrostatic repulsion force [14–17]. However, the surface residuals of stabilizers on graphene surfaces can seriously decrease the extraordinary electrical performance. Thus, the removal of stabilizers from graphene sheets is necessary to utilize graphene in practical applications. Recently, various chemicals have been introduced as a stabilizer for liquid-phase exfoliation, due to their easy-to-remove and easy-to-control qualities, low cost, and safety [16,17]. However, there still remains a challenge to improve efficiency for the scalable preparation of graphene. Meanwhile, species of solvent are also one of the key factors for enhancing the production efficency and dispersion homogeniety of graphene. A variety of organic solvents have been utilized for the exfoliation studies. The most common solvents are isopropyl alcohol (IPA) [18], N,N-dimethylformamide (DMF) [19], dimethyl sulfoxide (DMSO) [20], 1-methyl-2-pyrrolidinone (NMP) [21], etc., because they can minimize the interfacial tension with graphene due to their surface tension being similar to that of graphene.

In this study, we demonstrated that graphite can be exfoliated to give large quantities of graphene nanosheets by the liquid-phase shear exfoliation method, using a homogenizer in NMP as an organic solvent, with the assistance of urea as a stabilizer, under ambient conditions. Remarkably, we could obtain non-oxidized LDG sheets on a large scale under optimal conditions. Furthermore, the LDG was used to modify an electrode for fabricating the electrochemical sensor of doxorubicin (DOX). This sensor showed a good sensitivity and selelctivity for the electrochemical detection of DOX. This approach has a great potential in the industrial-scale production of graphene with good quality for practical applications.

2. Experimental

2.1. Chemicals and Reagents

The chemicals and materials were purchased from Sigma-Aldrich, Inc. (St. Louis, MO, USA), as follows: natural graphite powder, NMP, phosphate buffer saline (10 mM, pH 7.4), glucose, KCl, NaCl, urea, doxorubicin HCl, tryptophan and commercial graphene. The deionized (DI) water was prepared using water purification systems (specific resistivity > 18 MΩ cm, MilliQ, Millipore Korea, Co., Ltd., aqua Max Ultra 370 (Younglin Instrument Co., Anyang, Kyounggi-do, Korea).

2.2. Instrumentation and Measurements

The shear exfoliation was performed with an L5M-A mixer homogenizer (Silverson, Co., East Longmeadow, MA, USA) with a Square Hole High Shear Screen™ workhead. The absorbance of the exfoliated LDG was measured by UV-vis spectroscopy (SINCO Co., Ltd., Seoul, Korea). The concentrations of the LDG were calculated from the standard curve for the absorbance at 660 nm of the commercial graphene solution. The Raman spectroscopy was performed using an EnSpectr R532 Raman spectrometer (Enhanced Spectrometry, Inc., San Jose, CA, USA). Transmission electron microscopy (TEM) was carried out on a JEOL 2010 device (JEOL, Inc., Peabody, MA, USA) with an accelerating voltage 200 keV. The X-Ray diffraction (XRD) patterns were recorded using a Rigaku SmartLab X-ray diffractometer (Rigaku Co., Tokyo, Japan) with Cu Kα source. Fourier transform infrared spectroscopy (FTIR) was performed with a NICOLET iS10 (Thermo Scientific Korea Ltd., Seoul, Korea). The X-ray photoelectron spectroscopy (XPS) signals were recorded using a Thermo Scientific K-Alpha XPS system (Thermo Fisher Scientific, Paisley, UK) equipped with a micro-focused monochromatic Al Kα X-ray source (1486.6 eV). The electrochemical measurements were performed with a CHI 660D electrochemical workstation (CH Instruments, Inc., Austin, TX, USA). A conventional three-electrode cell was used with a Pt wire as a counter electrode, an Ag/AgCl electrode was used as a reference electrode and a bare glassy carbon electrode (GCE, diameter = 3 mm), or modified electrode, was used as a as working electrode at room temperature.

2.3. Shear Exfoliation of Graphite by Using a Homogenizer

The LDG samples were prepared by the liquid-phase shear exfoliation method using a high-speed mixer homogenizer, as shown in Scheme 1. In a typical procedure, 1 g graphite was added into the 100 mL of aqueous urea solution (100 g/100 mL). The mixture solution was sonicated in a sonication bath (4 W) for 20 min at room temperature to dissolve the urea. After sonication, the mixture solution was exfoliated by the mixer homogenizer at a speed of 3000 rpm for 4 h. Then, to remove the urea, the exfoliated few-layered graphene was collected by centrifugation at 3000 rpm for 20 min, twice. The pellet of few-layered graphene was re-dispersed in 100 mL NMP. The resultant dispersion was mixed by the mixer homogenizer at a speed of 3000 rpm for 5 h. Finally, to obtain the exfoliated LDG, the mixture solution was centrifuged at 3000 rpm for 20 min and the supernatant was collected for the LDG solution. The LDG solution was evaporated under a vacuum to discard the solvent at 50 °C.

2.4. Preparation of Graphene Electrode for Electrochemical Sensor

Before the electrode fabrication, the GCE was polished with 1.0, 0.3, and 0.05 μm Al_2O_3 powder and washed in an ultrasonic bath for 10 min, then rinsed with DI water thoroughly. After that, 5 μL of 1 mg/mL LDG in ethanol was drop-casted onto the GCE and dried with an infrared spectroscopy (IR) lamp for a few minutes, then rinsed with DI water several times. Before the electrochemical sensing experiments, the LDG-modified GCE was activated electrochemically by cyclic voltammetry (CV) in 10 mM PBS buffer solution (pH 7.4); the potential was scanned from 0 to 1.0 V (vs. Ag/AgCl) three times with a 5 mV/s scan rate.

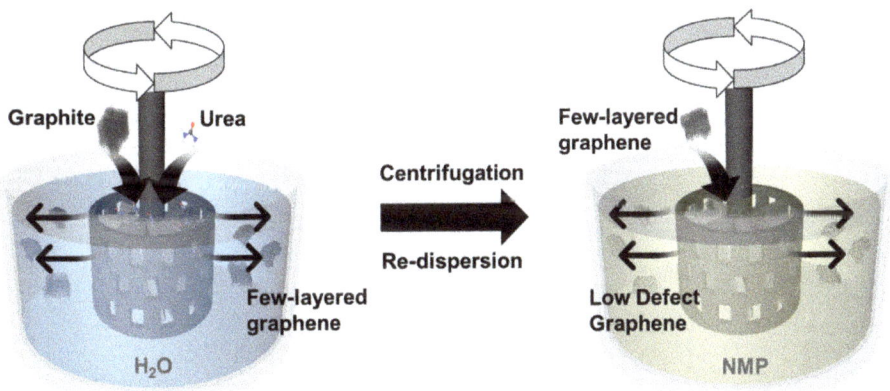

Scheme 1. Illustration of mass production of low-defect graphene (LDG) by the urea-assisted liquid-phase shear exfoliation of graphite.

3. Results and Discussion

3.1. Effect of Processing Parameters on Graphene Concentration

To confirm the production of the LDG sheets, as described in Scheme 1, Raman spectroscopy was first employed to monitor the exfoliation of graphite into graphene sheets. As shown in Figure 1A, three peaks were observed at 1339 (D band), 1573 (G band) and 2688 cm^{-1} (2D band) for both graphite and LDG. Using the ratio of peak intensities I_D/I_G in Raman spectra, the level of disorder in graphene can be characterized. As the disorder in the graphene increases, I_D/I_G displays two different behaviors, as follows: (i) at a low-defect concentration, like pure graphite materials, $I_D/I_G \simeq 1/L_d^2$ (Stage 1); (ii) at a relatively high defect concentration, like produced graphene by chemical or physical exfoliation, $I_D/I_G \simeq L_d^2$ (Stage 2), where L_d represents the mean distance between two defects [22,23]. The transition between Stage 1 and 2 is usually observed at $I_D/I_G \simeq 3$ [22]. So, in the present case, the I_D/I_G value increased as the level of defects decreased. The exfoliation caused an increase in the I_D/I_G value to 0.39 in the LDG, which is higher than that of graphite (0.16), but considerably lower than that of GO reported previously [24]. This should be attributed to the edge defects of the exfoliated samples rather than the lattice destruction, meaning that very few defects existed in the as-prepared LDG [25]. In addition, the 2D band of LDG turned out to be a relatively symmetrical shape and appeared at lower position (~10 cm^{-1}) when compared to the 2D band of graphite. A shoulder peak on G band, along with the D, G and 2D band, appeared at 1616 cm^{-1} in the LDG, which is namely D band and is evidence of few-layer graphene. These results successfully demonstrate the reduction in the number of graphene layers and the subsequent formation of few-layer graphene with a low density of defects [26].

To optimize the preparation procedure, we investigated the effects of the processing parameters, such as the shearing speed, urea concentration, shearing time in urea and shearing time in NMP on the concentration of LDG solutions and the intensity ratio of the D band and G band (I_D/I_G) using Raman spectroscopy (Figure 1B–E). The processing parameters were optimized with the maximal concentration and the highest I_D/I_G value of LDG, indicating the lowest level of defects. To test the influence of the processing parameters, only one parameter was varied, while the other parameters were kept constant. Figure 1B shows the effect of the shearing speed in the urea solution on the concentration and the I_D/I_G value of LDG dispersions. As the shearing speed increased in the range of 1000 to 5000 rpm, the concentration of LDG also increased. However, the I_D/I_G of LDG dispersions reached its maximum at 3000 rpm and decreased after that. This indicates that the highest LDG concentration with the lowest level of defects was obtained at a shearing speed of 3000 rpm. Therefore, 3000 rpm was selected as the optimal shearing speed for the preparation of LDG dispersion. A similar trend was observed in the test for the effect of the shearing speed in NMP, showing the highest I_D/I_G value of LDG at 3000 rpm, although the concentration increased with increasing shearing speed (Figure 1C). This is because the graphene size was reduced with the increase of edge defects and agglomeration, as the shear force increased with the shearing speed. Thus, 3000 rpm in NMP was enough to achieve a high yield of good quality. The effect of the urea concentration on the exfoliation efficiency of graphite was also examined (Figure 1D). The concentration and the I_D/I_G value of LDG dispersions increased with increasing concentrations of urea. This result suggests that the urea molecules could expand the interlayer space of graphite to overcome the van der Waals force between graphite layers and, thus, LDG concentration significantly increased due to the presence and adsorption of more concentrated urea molecules. A suitable concentration of urea was chosen to be 100 g/100 mL, due to the solubility of urea (107.9 g/100 mL in 20 °C). The effect of the shearing times in aqueous urea solution and NMP, respectively, were further evaluated in the range of 1 to 5 h (Figure 1E,F). The highest concentrations and I_D/I_G values of LDG dispersions were achieved at 4 h in urea and 5 h in NMP and, thus, they were chosen as the optimal shearing times. Interestingly, the graphene prepared without urea assistance showed a lower concentration and I_D/I_G value than those of LDG (Figure S1, Supplementary Materials), proving the role of urea as a stabilizer and a molecular wedge in shear exfoliation. Taken together, these results show that the highest concentration of LDG with high quality can be prepared under the following processing parameters: (i) the shearing speed in urea solution is 3000 rpm; (ii) the shearing speed in NMP is 3000 rpm; (iii) the urea concentration is 100 g/100 mL; (iv) the shearing time in urea is 4 h; (v) the shearing time in NMP is 5 h.

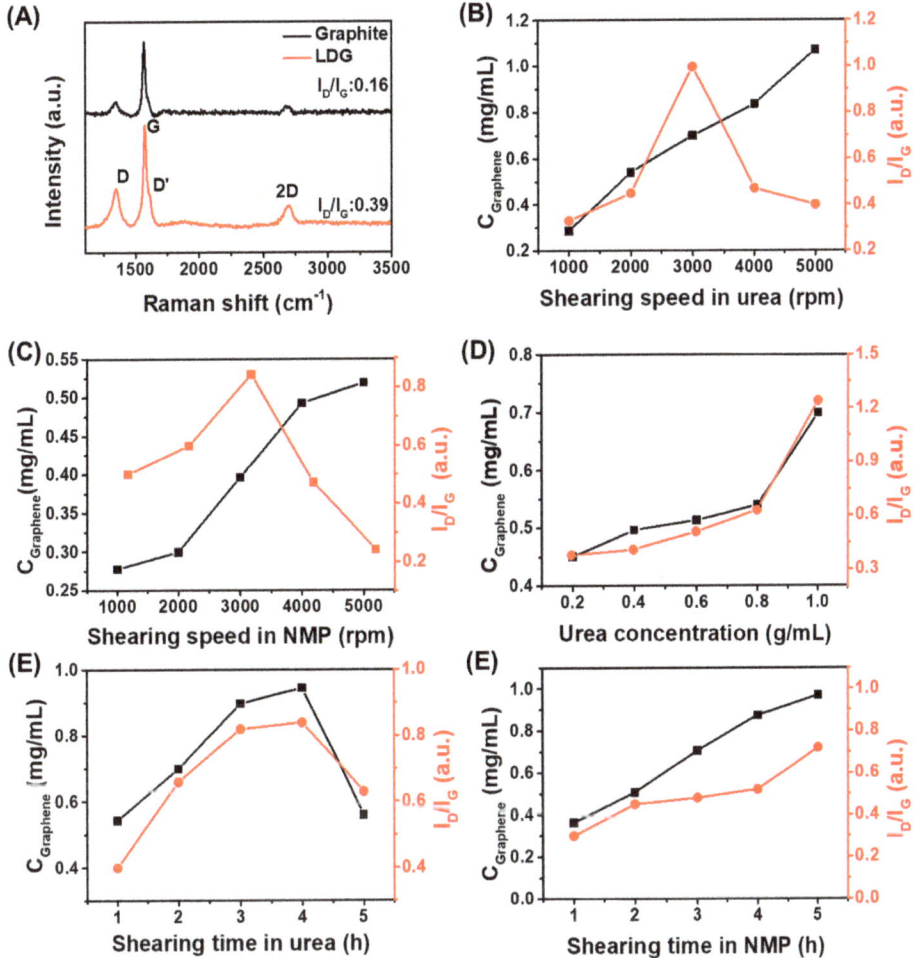

Figure 1. (**A**) Characterization of LDG by Raman spectroscopy. The influences of the processing parameters on the concentration (Left y-axis) and the I_D/I_G ratio in Raman spectroscopy of the LDG solutions (Right y-axis). (**B**) The shearing speed in urea, (**C**) the shearing speed in NMP, (**D**) the urea concentration, (**E**) the shearing time in urea, and (**F**) the shearing time in NMP.

3.2. Characterization of Low-Defect Graphene

The XRD patterns of the LDG and graphite powder were utilized to confirm the exfoliation of the graphite to graphene sheets (Figure 2A). While the XRD pattern of the graphite showed a sharp diffraction peak at $2\theta = 26.5°$, corresponding to a d-spacing of 0.34 nm, the LDG showed a broad peak around 24° and very weak diffraction peak around 26.5°, which indicates the exfoliation of graphite. The LDG also exhibited a broad peak around 10°, demonstrating that the interlayer spacing was expanded during the exfoliation [27,28]. As shown in Figure 2B,C, both TEM and SEM images also revealed the microstructures of the exfoliated graphene sheets in the LDG dispersions. The reduction in the thickness of the graphene sheets (mono-, bi- and tri-layer, Figure 2B) and their lateral size (~1000 nm) is evident compared with the graphite flakes, confirming the high degree of exfoliation of the process. The SEM image of the LDG clearly shows a thin planar structure with curled and irregular edges (Figure 2C). According to statistical analysis based on the SEM and TEM images of the LDG sheets, lateral size and layer distribution were measured in Figure S2. By measuring ~90 graphene sheets, a lateral size histogram (Figure S2A) displays that more than 84% of the graphene sheets are larger than 3 μm, which is much higher than that of liquid phase exfoliated graphene sheets by sonication [14,29]. The layer distribution (Figure S2B) shows that monolayer graphene accounts for 48.8% of the graphene sheets in the dispersion. We believe that the large sheet size and high exfoliation efficiency are related to the intercalation of urea molecules into the graphite and the liquid phase shear exfoliation process. These results suggest the successful preparation of well-exfoliated and non-oxidized graphene sheets with high quality, verifying that this approach is a nondestructive and efficient method to produce LDG sheets on a large scale.

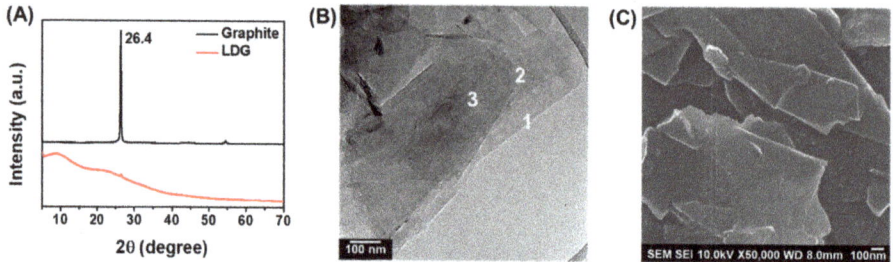

Figure 2. Characterization of LDG by (**A**) X-ray diffraction spectroscopy, (**B**) TEM, and (**C**) SEM.

XPS was performed to probe the chemical composition and binding states of the exfoliated LDG (Figure 3, Figure S3 and Table S1). The XPS survey curves of LDG, commercially available graphene and graphite show a pronounced peak at 285 eV and a weak peak at 533 eV, except for graphene oxide (GO), which displayed considerable peak heights at both 292 and 538 eV, due to the oxygen functionalities derived from chemical exfoliation (Figure 3A). Based on the results from XPS, the oxygen content of the LDG powder was estimated to be 6.4 atomic %, thus being higher than that of the graphite precursor (3.0%), but considerably lower than that of GO (32.5%) and even that of commercially available graphene (11.1%), as summarized in Table S1. The overall oxygen of LDG is associated with the edge defects of LDG in the form of oxide groups, which is in agreement with the results from Raman and XRD spectra, and consistent with the aforementioned higher crystalline size of graphene sheets [30]. The calculated C/O ratio of the LDG (14.6) is significantly higher than that of commercially available graphene (8.0) and that of GO (2.1). Figure 3B–D show the C 1s core-level spectra of graphite, LDG, and commercially available graphene, respectively. The pattern of LDG (Figure 3C) exhibits a similar structure to that of graphite (Figure 3B), revealing that the intrinsic in-plane crystal structure of graphite remains intact during the shear exfoliation treatment. The XPS data of the C 1s profiles also prove that the LDG (Figure 3C has fewer oxygen-containing functional groups (such as C-O, C=O, and COO-) than GO (Figure S1) and even commercially available graphene (Figure 3D).

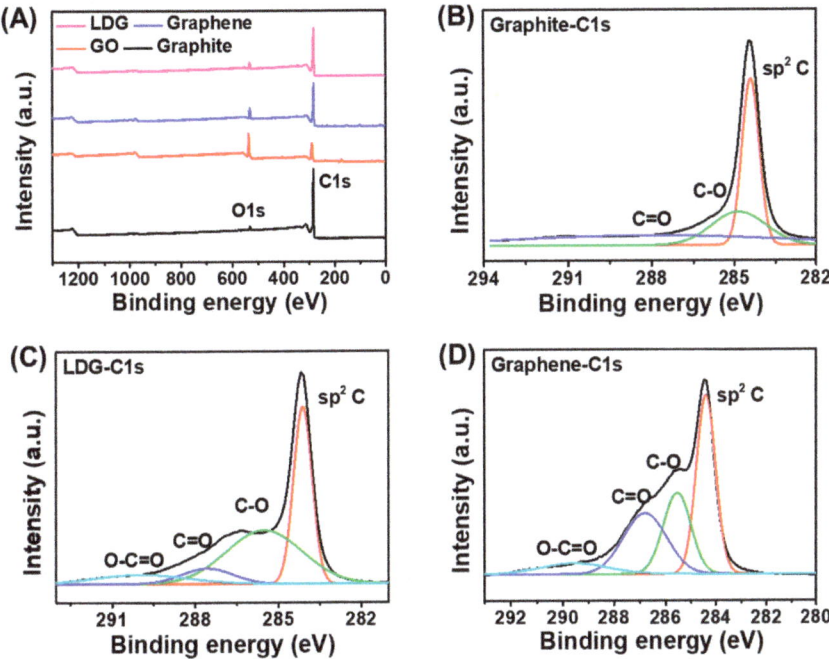

Figure 3. (**A**) XPS survey of LDG powder, commercially available graphene, graphene oxide and graphite with high-resolution C 1s spectra of (**B**) graphite, (**C**) LDG, and (**D**) commercially available graphene, respectively.

3.3. Electrochemical Detection of Doxorubicin with LDG Electrode

To assess the performance of LDG as a sensing material for practical applications, we fabricated the LDG-modified GCE (LDG-GCE) for the detection of DOX. DOX, as one of the clinically important anti-cancer drugs, has been extensively used in the treatment of various cancers. However, long-term clinical use of DOX is limited due to its side effects, such as systemic toxicity, drug resistance, and life-threatening heart damage. Therefore, the development of highly sensitive analytical techniques for DOX detection is of great significance in clinics and pharmaceutics.

Before the electrochemical sensing experiments, CV and electrochemical impedance spectroscopy (EIS) were performed to investigate the charge transfer ability of LDG-GCE in 0.1 M KCl solution containing 3 mM of $K_3Fe(CN)_6$. As shown in Figure 4A, both LDG-GCE and GCE show a pair of well-defined redox peaks. However, the peak potential difference (ΔE_p = 70 mV) of LDG-GCE is smaller than that of bare GCE (119 mV), while the peak currents for LDG-GCE are significantly higher, indicating that the LDG-GCE electrode exhibited a better electrochemical performance. The EIS experiments also showed better electron-transfer characteristics for LDG-GCE. The charge transfer resistance (R_{ct}) of LDG-GCE was estimated to be 25.9 Ω from the diameter of a semicircle in the high frequency region of a Nyquist plot (Figure 4B), which was much smaller than that of the GCE (R_{ct} = 168.8 Ω). This suggests that rapid electron transfer occurred on the LDG-GCE surface.

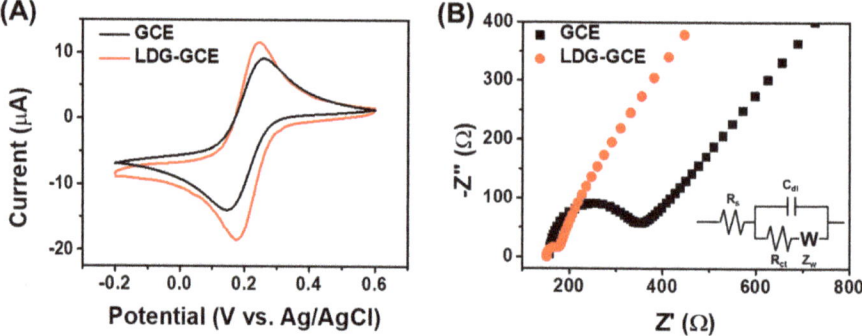

Figure 4. (**A**) CV curves of GCE (black) and LDG-GCE (red) in 0.1 M KCl containing 3 mM of $K_3Fe(CN)_6$. Scan rate: 50 mV/s. (**B**) The Nyquist plots of the GCE (black) and LDW-GCE (red) in 0.1 M KCl containing 3 mM of $K_3Fe(CN)_6$, using a frequency range of 0.1 to 10^5 Hz, a bias potential of +0.214 V and an AC amplitude of 10 mV. (Inset: Randles equivalent circuit model).

The electrochemical detection of DOX was performed with differential pulse voltammetry (DPV) and chronoamperometry in 10 mM PBS solution (pH = 7.4) containing various concentrations of DOX, using LDG-GCE. Figure 5A shows the typical DPV curves in which only one broad peak was observed at 0.366 V. Upon the addition of DOX in the range of 0.3 to 5.0 µM, the LDG-GCE exhibited an increase in the oxidation peak currents. The corresponding calibration plots for DOX showed a good linear relationship between the DPV peak currents and the concentration of DOX in the range of 0.3–3.0 µM. As is to be expected from the electrochemical performance of LDG-GCE, the sensor shows better sensitivity when compared with bare GCE (Figure 5B). The calculated sensitivity for the electrochemical detection of DOX was 7.23×10^{-1} µM/µA with LDG-GCE, which is higher than bare GCE (3.54×10^{-1} µM/µA). Moreover, the limit of detection (LOD) for LDG- GCE (39.3 nM) at an S/N ratio of 3 was also much better than that of bare GCE with LDG-GCE (254 nM) and bare GCE. These results suggest that the LDG-GCE exhibited an excellent performance with respect to the detection of DOX. To evaluate the practical applications in DOX analysis, chronoamperometric measurements were carried out in 0.1 M PBS (pH 7.4). Figure 5C shows the amperometric current–time (i–t) plots of the LDG-GCE at 0.366 V in response to the successive addition of DOX the PBS solution. The corresponding calibration curves are displayed as i (nA) = 1.06×10^{-8} + 0.00828 C (µM) in Figure 5C inset. The relationship between DOX concentration and the corresponding current is a linear one in the range of 0.3–2.7 µM with a LOD of 653 nM (S/N = 3). Table S2 summarizes and compares the analytical parameters of the LDG-GCE sensor with other electrodes, revealing that the proposed sensor shows a high sensitivity with a low LOD and suitable linear range. The selectivity experiments were performed by comparing the amperometric responses to DOX with potential interfering compounds found in biological fluids, such as glucose, urea, tryptophan, KCl, and NaCl. As shown in Figure 5D, no significant interference was observed during the amperometric detection of 3 µM DOX in the presence of 30 µM glucose, urea, KCl, NaCl, and tryptophan. This result underlines the suitable selectivity of LDG-GCE.

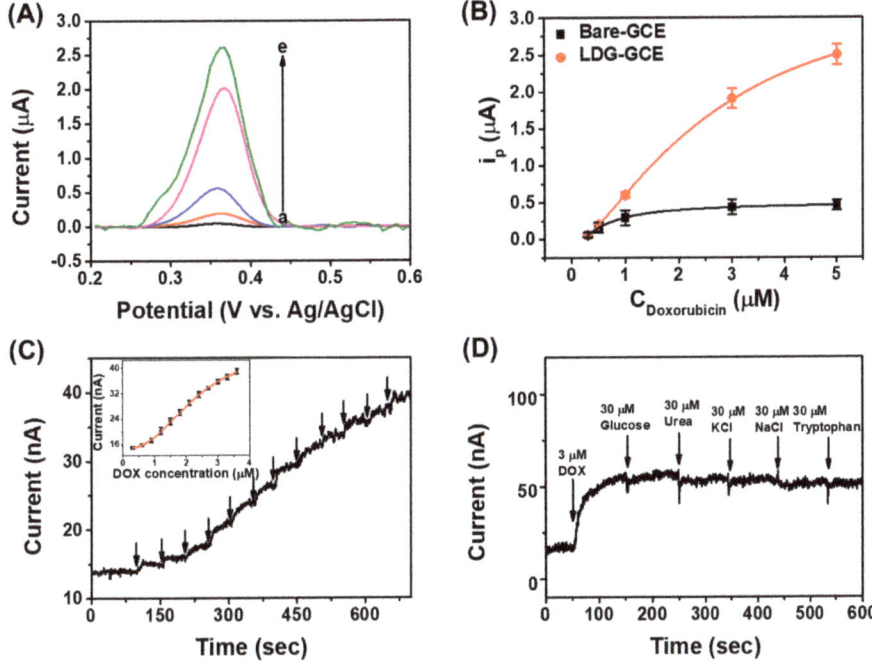

Figure 5. (A) Differential pulse voltammetry (DPV) curves of LDG-GCE in 10 mM PBS solution (pH = 7.4) containing various concentration of DOX (0.3, 0.5, 1.0, 3.0 and 5.0 μM. (B) The calibration plots of i_p vs. concentration of DOX from DPV at LDG-GCE and GCE, respectively. The average values and standard deviations were obtained based on three results. (C) The amperometric responses of LDG-GCE to the successive addition of DOX from 0.3 μM to 3.6 μM at 0.366 V vs. Ag/AgCl. (Inset: The calibration plot from the amperometric measurements) (D) The amperometric responses of LDG-GCE to the successive addition of 3 μM DOX, 30 μM glucose, 30 μM urea, 30 μM KCl, 30 μM NaCl, and 30 μM tryptophan. at 0.366 V vs. Ag/AgCl.

4. Conclusions

We have demonstrated the scalable production of graphene sheets with high quality by a simple and facile method, via NMP-mediated shear exfoliation of graphite with the assistance of urea. The successful production of LDG on a large scale was possible using a high speed mixer homogenizer by optimizing the processing parameters for the shear exfoliation, such as shearing speed, urea concentration, and shearing time by comparing the produced graphene concentration and I_D/I_G value from the Raman spectra. The characterization experiments confirmed that the graphene sheets, produced by shear exfoliation with optimized parameters, are single layered, or few-layered at least, with negligible defects. In addition, using the LDG as a modification material for an electrode, we have fabricated an LDG-GCE electrochemical sensor for DOX, demonstrating sensitive and selective DOX analysis. Therefore, the proposed approach is promising for the industrial production of graphene for numerous applications.

Supplementary Materials: The following are available online at http://www.mdpi.com/2079-4991/10/2/267/s1, Figure S1: Comparison of (A) the concentration and (B) the I_D/I_G ratio in Raman spectroscopy in graphene solution prepared with (black square dot) and without (red square dot) assistance of urea., Figure S2: (A) lateral size and (B) layer distribution of LDG calculated from 95 isolated sheets in SEM and TEM images., Figure S3: High resolution C1s spectra of GO., Table S1: XPS data of graphene oxide (GO), graphite, commercially available graphene and LDG., Table S2: Comparison of the analytical performance of LDG-GCE with other electrodes for DOX detection.

Author Contributions: Conceptualization, T.H.K.; methodology, T.H.K.; validation, C.-S.L. and T.H.K.; analysis, C.-S.L. and S.J.S.; investigation, S.J.S.; writing—original draft preparation, C.-S.L.; writing—review and editing, T.H.K.; funding acquisition, T.H.K. All authors have read and agreed to the published version of the manuscript.

Funding: This work was conducted with the support of the Korea Environment Industry & Technology Institute (KEITI) through its Ecological Imitation-based Environmental Pollution Management Technology Development Project, and funded by the Korea Ministry of Environment (MOE) (2019002800001). This work was also supported by the Soonchunhyang University Research fund.

Conflicts of Interest: The authors declare no conflict of interest.

References

1. Choi, W.; Lahiri, I.; Seelaboyina, R.; Kang, Y.S. Synthesis of Graphene and Its Applications: A Review. *Crit. Rev. Solid State Mater. Sci.* **2010**, *35*, 52–71. [CrossRef]
2. Wang, Y.; Li, Z.; Wang, J.; Li, J.; Lin, Y. Graphene and graphene oxide: Biofunctionalization and applications in biotechnology. *Trends Biotechnol.* **2011**, *29*, 205–212. [CrossRef] [PubMed]
3. Zhang, Y.; Zhang, L.; Zhou, C. Review of Chemical Vapor Deposition of Graphene and Related Applications. *Acc. Chem. Res.* **2013**, *46*, 2329–2339. [CrossRef] [PubMed]
4. Mattevi, C.; Kim, H.; Chhowalla, M. A review of chemical vapour deposition of graphene on copper. *J. Mater. Chem.* **2011**, *21*, 3324–3334. [CrossRef]
5. Zhang, L.; Liang, J.; Huang, Y.; Ma, Y.; Wang, Y.; Chen, Y. Size-controlled synthesis of graphene oxide sheets on a large scale using chemical exfoliation. *Carbon* **2009**, *47*, 3365–3368. [CrossRef]
6. Park, S.; An, J.; Potts, J.R.; Velamakanni, A.; Murali, S.; Ruoff, R.S. Hydrazine-reduction of graphite- and graphene oxide. *Carbon* **2011**, *49*, 3019–3023. [CrossRef]
7. Yi, M.; Shen, Z. A review on mechanical exfoliation for the scalable production of graphene. *J. Mater. Chem. A* **2015**, *3*, 11700–11715. [CrossRef]
8. Xia, Z.Y.; Pezzini, S.; Treossi, E.; Giambastiani, G.; Corticelli, F.; Morandi, V.; Zanelli, A.; Bellani, V.; Palermo, V. The Exfoliation of Graphene in Liquids by Electrochemical, Chemical, and Sonication-Assisted Techniques: A Nanoscale Study. *Adv. Funct. Mater.* **2013**, *23*, 4756. [CrossRef]
9. Ciesielski, A.; Samorì, P. Graphene via sonication assisted liquid-phase exfoliation. *Chem. Soc. Rev.* **2014**, *43*, 381–398. [CrossRef]
10. Zhao, W.; Fang, M.; Wu, F.; Wu, H.; Wang, L.; Chen, G. Preparation of graphene by exfoliation of graphite using wet ball milling. *J. Mater. Chem.* **2010**, *20*, 5817. [CrossRef]
11. Teng, C.; Xie, D.; Wang, J.; Yang, Z.; Ren, G.; Zhu, Y. Ultrahigh Conductive Graphene Paper Based on Ball-Milling Exfoliated Graphene. *Adv. Funct. Mater.* **2017**, *27*, 1700240. [CrossRef]
12. Varrla, E.; Paton, K.R.; Backes, C.; Harvey, A.; Smith, R.J.; McCauley, J.; Coleman, J.N. Turbulence-assisted shear exfoliation of graphene using household detergent and a kitchen blender. *Nanoscale* **2014**, *6*, 11810–11819. [CrossRef] [PubMed]
13. Paton, K.R.; Varrla, E.; Backes, C.; Smith, R.J.; Khan, U.; O'Neill, A.; Boland, C.; Lotya, M.; Istrate, O.M.; King, P.; et al. Scalable production of large quantities of defect-free few-layer graphene by shear exfoliation in liquids. *Nat. Mater.* **2014**, *13*, 624–630. [CrossRef]
14. Lotya, M.; Hernandez, Y.; King, P.J.; Smith, R.J.; Nicolosi, V.; Karlsson, L.S.; Blighe, F.M.; De, S.; Wang, Z.; McGovern, I.T.; et al. Liquid Phase Production of Graphene by Exfoliation of Graphite in Surfactant/Water Solutions. *J. Am. Chem. Soc.* **2009**, *131*, 3611–3620. [CrossRef] [PubMed]
15. Bourlinos, A.B.; Georgakilas, V.; Zboril, R.; Steriotis, T.A.; Stubos, A.K. Liquid-Phase Exfoliation of Graphite towards Solubilized Graphenes. *Small* **2009**, *5*, 1841–1845. [CrossRef] [PubMed]
16. He, P.; Zhou, C.; Tian, S.; Sun, J.; Yang, S.; Ding, G.; Xie, X.; Jiang, M. Urea-assisted aqueous exfoliation of graphite for obtaining high-quality graphene. *Chem. Commun.* **2015**, *51*, 4651–4654. [CrossRef]
17. Hou, D.; Liu, Q.; Wang, X.; Qiao, Z.; Wu, Y.; Xu, B.; Ding, S. Urea-assisted liquid-phase exfoliation of natural graphite into few-layer graphene. *Chem. Phys. Lett.* **2018**, *700*, 108–113. [CrossRef]
18. Wei, Y.; Sun, Z. Liquid-phase exfoliation of graphite for mass production of pristine few-layer graphene. *Curr. Opin. Colloid Interface Sci.* **2015**, *20*, 311–321. [CrossRef]
19. O'Neill, A.; Khan, U.; Nirmalraj, P.N.; Boland, J.; Coleman, J.N. Graphene Dispersion and Exfoliation in Low Boiling Point Solvents. *J. Phys. Chem. C* **2011**, *115*, 5422–5428. [CrossRef]

20. Du, W.; Lu, J.; Sun, P.; Zhu, Y.; Jiang, X. Organic salt-assisted liquid-phase exfoliation of graphite to produce high-quality graphene. *Chem. Phys. Lett.* **2013**, *568–569*, 198–201. [CrossRef]
21. Wei, P.; Gan, T.; Wu, K. N-methyl-2-pyrrolidone exfoliated graphene as highly sensitive analytical platform for carbendazim. *Sens. Actuators B Chem.* **2018**, *274*, 551–559. [CrossRef]
22. Ferrari, A.C.; Robertson, J. Interpretation of Raman spectra of disordered and amorphous carbon. *Phys. Rev. B* **2000**, *61*, 14095–14107. [CrossRef]
23. Brovelli, S.; Galland, C.; Viswanatha, R.; Klimov, V.I. Tuning Radiative Recombination in Cu-Doped Nanocrystals via Electrochemical Control of Surface Trapping. *Nano Lett.* **2012**, *12*, 4372–4379. [CrossRef] [PubMed]
24. Perumbilavil, S.; Sankar, P.; Priya Rose, T.; Philip, R. White light Z-scan measurements of ultrafast optical nonlinearity in reduced graphene oxide nanosheets in the 400–700 nm region. *Appl. Phys. Lett.* **2015**, *107*, 051104. [CrossRef]
25. Claramunt, S.; Varea, A.; López-Díaz, D.; Velázquez, M.M.; Cornet, A.; Cirera, A. The Importance of Interbands on the Interpretation of the Raman Spectrum of Graphene Oxide. *J. Phys. Chem. C* **2015**, *119*, 10123–10129. [CrossRef]
26. Cançado, L.G.; Jorio, A.; Ferreira, E.H.M.; Stavale, F.; Achete, C.A.; Capaz, R.B.; Moutinho, M.V.O.; Lombardo, A.; Kulmala, T.S.; Ferrari, A.C. Quantifying Defects in Graphene via Raman Spectroscopy at Different Excitation Energies. *Nano Lett.* **2011**, *11*, 3190–3196. [CrossRef]
27. Stobinski, L.; Lesiak, B.; Malolepszy, A.; Mazurkiewicz, M.; Mierzwa, B.; Zemek, J.; Jiricek, P.; Bieloshapka, I. Graphene oxide and reduced graphene oxide studied by the XRD, TEM and electron spectroscopy methods. *J. Electron Spectrosc. Relat. Phenom.* **2014**, *195*, 145–154. [CrossRef]
28. Wang, G.; Shen, X.; Wang, B.; Yao, J.; Park, J. Synthesis and characterisation of hydrophilic and organophilic graphene nanosheets. *Carbon* **2009**, *47*, 1359–1364. [CrossRef]
29. Hernandez, Y.; Nicolosi, V.; Lotya, M.; Blighe, F.M.; Sun, Z.; De, S.; McGovern, I.T.; Holland, B.; Byrne, M.; Gun'Ko, Y.K.; et al. High-yield production of graphene by liquid-phase exfoliation of graphite. *Nat. Nanotech* **2008**, *3*, 563–568. [CrossRef]
30. Li, X.; Zhang, G.; Bai, X.; Sun, X.; Wang, X.; Wang, E.; Dai, H. Highly conducting graphene sheets and Langmuir–Blodgett films. *Nat. Nanotech* **2008**, *3*, 538–542. [CrossRef]

 © 2020 by the authors. Licensee MDPI, Basel, Switzerland. This article is an open access article distributed under the terms and conditions of the Creative Commons Attribution (CC BY) license (http://creativecommons.org/licenses/by/4.0/).

Article

Superior Electrocatalytic Activity of MoS$_2$-Graphene as Superlattice

Alejandra Rendón-Patiño [1], Antonio Domenech-Carbó [2], Ana Primo [1,*] and Hermenegildo García [1,*]

[1] Instituto de Tecnología Química (CSIC-UPV) and Department of Chemistry, Consejo Superior de Investigaciones Científicas-Universitat Politècnica de Valencia, Avenida de los Naranjos s/n, 46022 Valencia, Spain; alrenpa@itq.upv.es

[2] Departamento de Química Analítica, Universitat de Valencia, Av. Del Dr. Moliner s/n, 46100 Burjassot, Spain; antonio.domenech@uv.es

* Correspondence: aprimoar@itq.upv.es (A.P.); hgarcia@qim.upv.es (H.G.)

Received: 6 April 2020; Accepted: 21 April 2020; Published: 27 April 2020

Abstract: Evidence by selected area diffraction patterns shows the successful preparation of large area (cm × cm) MoS$_2$/graphene heterojunctions in coincidence of the MoS$_2$ and graphene hexagons (superlattice). The electrodes of MoS$_2$/graphene in superlattice configuration show improved catalytic activity for H$_2$ and O$_2$ evolution with smaller overpotential of +0.34 V for the overall water splitting when compared with analogous MoS$_2$/graphene heterojunction with random stacking.

Keywords: superlattice; 2d materials; electrocatalytic

1. Introduction

There is considerable interest in developing noble metal-free electrodes that could efficiently perform water splitting due to the current change from fossil fuels to renewable electricity [1–4]. In this context, it has been reported that MoS$_2$ could be an efficient electrocatalyst for the oxygen evolution reaction (OER) replacing IrO$_2$ and Pt [5,6]. MoS$_2$ also shows electrocatalytic activity for hydrogen evolution reaction (HER) [7–9] and, therefore, it could be an ideal material to perform both redox processes of overall water splitting.

Regarding the use of MoS$_2$ as electrocatalyst, it has been shown that the assembly of this chalcogenide with graphene improves its performance as electrodes [10–12]. Graphene introduces electrical conductivity favouring electron transfer from the catalytic site on MoS$_2$ to the external circuit. For this reason, there has been considerable interest in developing different procedures for the preparation of MoS$_2$/graphene heterojunctions to be used in electrolysis [13–15]

In this context, a procedure consisting in the one-step formation of MoS$_2$ strongly interacting with graphene has been recently reported. The process is based on the pyrolysis of (NH$_4$)$_2$MoS$_4$ and chitosan as precursors of MoS$_2$ and N-doped defective graphene, respectively [16]. One of the advantages of this procedure was that chitosan allows for the formation of nanometric films on arbitrary, non-conductive substrates, such as quartz or ceramics, which, after transformation on (N)G, becomes electrically conductive and can be used as electrode [17,18]. Also related to the present study is a recent report from us on the formation of large area boron nitride/graphene heterojunctions with superlattice configuration [19].

Since both boron nitride and graphene are two-dimensional (2D) materials with very similar lattice parameters, the term superlattice refers to the coincidence of the hexagons of boron nitride overlapping those of graphene. Fundamental studies have shown that superlattice configuration of 2D materials heterojunctions can give an assembly with tunable electrical conductivity in contrast to the random heterojunctions of the two materials [20,21]. This modulation of the electron mobility through the graphene layer by the underlying boron nitride could be reflected in some unique properties of the superlattice heterojunction, particularly those that are related to electrochemistry and electrocatalysis [22–24]. In view of these precedents, it would be of interest to determine the electrocatalytic behaviour of MoS$_2$/graphene heterojunction in superlattice configuration, particularly when considering the well-known activity of MoS$_2$ for important reactions, such as HER, OER, and oxygen reduction reaction (ORR) processes [5].

This manuscript reports a novel procedure for the preparation of large area (cm × cm) films of MoS$_2$/graphene heterojunction in superlattice configuration. It will be shown that this fl-MoS$_2$/graphene material (fl meaning few layers) exhibits much lower onset potential for HER and OER than an analogous electrode prepared with random MoS$_2$/G heterojunction, illustrating the advantages of lattice matching to improve the electrochemical performance.

2. Materials and Methods

2.1. Methods-Exfoliation MoS$_2$ and Preparation of MoS$_2$/fl-G

Molybdenum sulfide exfoliation was carried out while using polystyrene as an exfoliating agent (ALDRICH). The polystyrene was dissolved in dichloromethane at a concentration of 3 mg/mL and powdered Molybdenum sulfide was mixed at a concentration of 1.5 mg/mL. The sample was sonicated while using a Sonic tip (FisherbrandTM Model 705 at 50% 700W for 5 ho per sample with a on/off sequence consisting in 1 s off and 1 s pulse and an ice bath was used to prevent solvent evaporation). After sonication, the dispersion was centrifuged at 1500 rpm for 45 min (Hettich Zentrifugen EBA 21(Hettich, Westphalia, Germany)).

The supernatant was preconcentrated and deposited on quartz films by rotation coating on 2 × 2 cm^2 quartz substrate to prepare the films (APT-POLOS spin-coater(SPS-Europe B.V., Putten, The Netherlands): 4000 rpm, 30 s). The pyrolysis of polystyrene treatment was performed using an electric oven and using the following heating program: heating at 5 °C/min at 900 °C for 2 h.

For the preparation of films with random configuration, commercial MoS$_2$ (Aldrich, St. Louis, Missouri, USA) at a concentration of 1.5 mg/mL and commercial graphene with an area of 700 m^2 (STREM CHEMICALS, Newburyport, MA, USA) at a concentration of 7.5 mg/mL were added to a 30 mg/mL polystyrene solution (Aldrich, St. Louis, MO, USA) in dichloromethane and the suspension submitted to ultrasounds using a Sonic tip at 700 W for 5 h. After sonication, the dispersion was centrifuged at 1500 rpm for 46 min. and the supernatant used to prepare films of randomly configured MoS$_2$/G heterojunction following the same procedure indicated above.

2.2. Methods-Catalytic Measurements

Electrochemical measurements were made in a conventional three electrode electrochemical cell while using a Pt disc pseudo-reference electrode and a Pt wire auxiliary electrode that accompanies the graphene modified glassy carbon (GCE) working electrode (BAS, MF 2012, area geometric 0.071 cm^2). Voltammetric experiments were carried out with a potentiostatic device CH 920c (Cambria Scientific, Llwynhendy, Llanelli, Wales, UK), using H$_2$SO$_4$ 1.0 M saturated with air as an electrolyte. The working electrodes were the MoS$_2$/fl-G films that were deposited on graphite. The graphite was deposited on the quartz films using the automatic carbon coater (sputter coater BALTEC_SCD005 (BAL-TEC AG, Schalksmühle, Germany) with carbon evaporation supplied BALTEC-CEA035).

3. Results

Upper and bottom layers of S atoms sandwiching and internal Mo IV layer constitute the crystal structure of MoS$_2$. The geometrical arrangement of the three layers is such that top views define hexagons with alternating edges S and Mo atoms that are located at different planes. These hexagons are similar in size to those of G and, therefore, a superlattice configuration can be possible for MoS$_2$-G heterojunctions when there is a coincidence of the two hexagonal arrangements. Figure 1 illustrates an ideal model of MoS$_2$ structure and graphene layer assembled in the superlattice configuration.

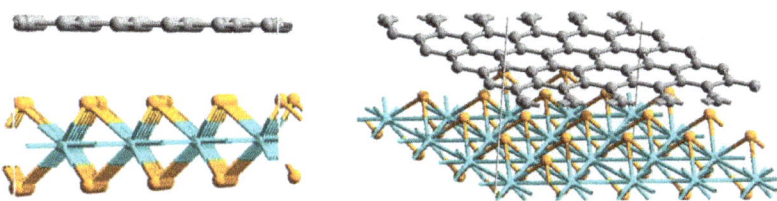

Figure 1. Models showing the structures of graphene and MoS$_2$ and their heterojunction in superlattice configuration.

There are precedents in the literature showing the possibility to obtain MoS$_2$/Graphene heterojunction with lattice matching geometry [25,26]. It was the leading hypothesis of the present study that the adaptation of this method could also serve to obtain MoS$_2$/G superlattice since it has been previously reported the preparation of large area films of BN/G heterojunctions as superlattice [19].

Scheme 1 illustrates the procedure followed in the present study. As it can be seen there, the process starts with commercial MoS$_2$ crystals that are subsequently exfoliated in viscous halogenated solvent while using polystyrene as promoter. After exfoliation, the residual bulk MoS$_2$ crystals can be removed from the single and few layers MoS$_2$ sheets by decanting the supernatant. Figure 2a presents AFM (Atomic Force Microscope) images of films of polystyrene containing MoS$_2$. As expected in view of the plastic characteristic of polystyrene, these films were smooth and have a thickness about 200 nm (see profile in panel C of Figure 2). The presence of MoS$_2$ in these films was not apparent from AFM at this moment due to the thickness. In a subsequent step, these films of polystyrene embedding MoS$_2$ were submitted to pyrolysis in the absence of oxygen at 900 °C. These conditions have been previously reported to transform polystyrene into thin films of few layers defective of graphene [27]. One of the main advantages of the present procedure is its reproducibility and the possibility to prepare large surface areas.

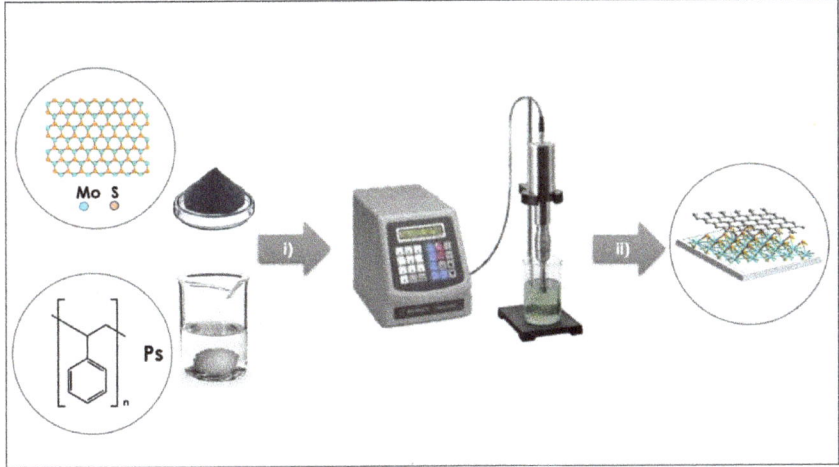

Scheme 1. The preparation procedure of MoS$_2$/G in superlattice configuration. Starting from bulk MoS$_2$ particles and a solution of polystyrene (Ps) in CH$_2$Cl$_2$. Sonication of the dispersion followed by sedimentation of bulk MoS$_2$ particles results in a dispersion of exfoliated MoS$_2$ and Ps in CH$_2$Cl$_2$ (i). Subsequent pyrolysis results in fl-MoS$_2$/G (ii).

A considerable shrinkage in the film thickness accompanies this transformation, as it can be seen in Figure 2b, where films of about 4 nm thickness corresponding to graphene can be seen (Figure 2d). The frontal image of MoS$_2$/G film shows also the presence of MoS$_2$ particles on this continuous graphene film. Measurement by AFM of a statistically relevant number of MoS$_2$ particles on graphene indicate that the lateral size distribution ranges from 20 to 140 nm with an average about 80 nm and the thickness of the MoS$_2$ is from 2 to 8 nm, with an average about 2. Figure S1 in the supporting information illustrates the corresponding histograms of the lateral dimension and height. When considering that the interlayer distance of MoS$_2$ is 0.615 nm, this average thickness corresponds to less than four MoS$_2$ layers.

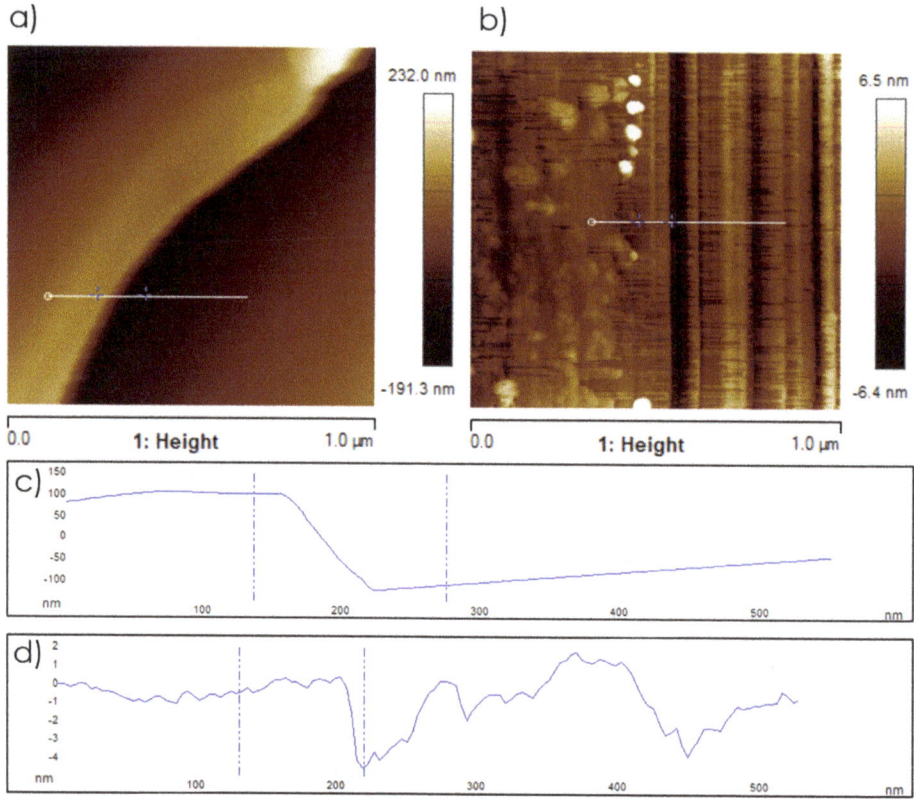

Figure 2. AFM image of Ps film on quartz containing MoS$_2$ before (**a**) and after (**b**) pyrolysis. The profiles c and d show the variation in height along the white lines drawn in a (**c**) and b (**d**).

The SEM images of these films show a smooth surface with low apparent roughness and without cracks or pinholes. Figure 3a shows one of these representative SEM images. The TEM of these films were obtained by scratching a bit of these films. As an example, Figure 3b shows a representative TEM image of MoS$_2$/G film after detachment from the quartz support.

The composition of MoS$_2$ was confirmed by EDS elemental mapping, determining an overlap of Mo and S in 1:2 atomic ratio. Figure S2 in the supporting information provides a summary of the Mo, S, O (from graphene defects), and C elemental distribution for a representative TEM image. These TEM images show a graphene layer larger than hundreds of nm having dark spots corresponding to smaller MoS$_2$ particles of size from 20–140 nm. High resolution TEM images show the expected hexagonal arrangement characteristic of graphene and MoS$_2$ (Figure 3c). Importantly, selected area electron diffraction at every point of the TEM image shows bright hexagonal spots corresponding to graphene. Coincident with those spots, there were also other points due to MoS$_2$. The coincidence of the electron diffraction patterns between two 2D materials (Figure 3d) is taken as the best evidence for the superlattice configuration.

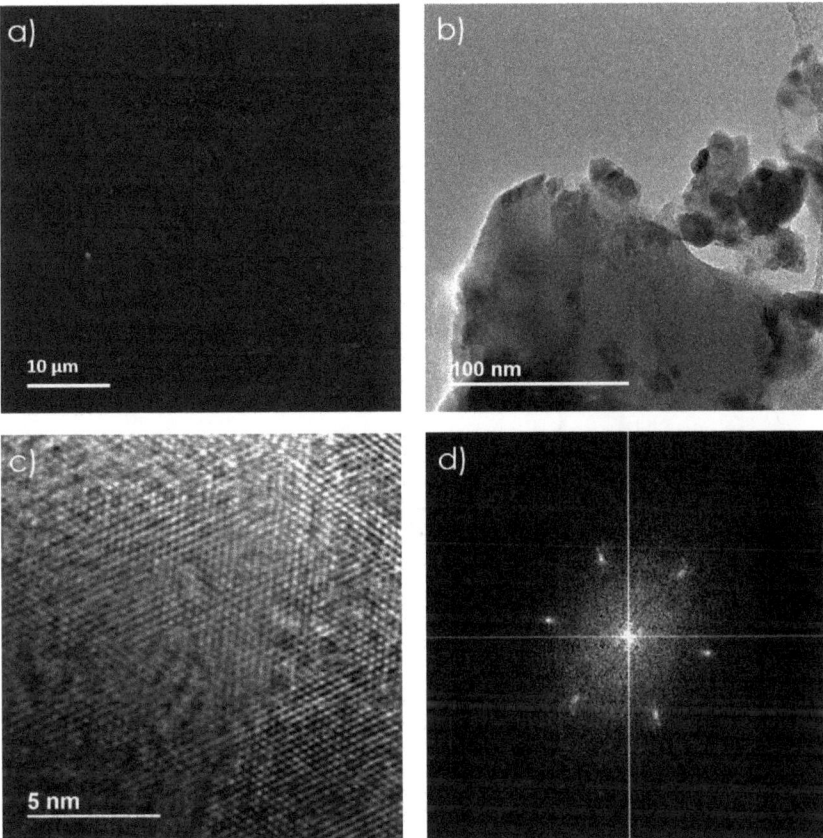

Figure 3. (**a**) SEM image of a MoS$_2$/G film on quartz. (**b**) TEM image of a piece of the MoS$_2$/G film detached from the quartz support. The darker spots in the image correspond to MoS$_2$ particles and the large, low contrast sheet is the graphene layer. (**c**) High-resolution TEM image of MoS$_2$/G visualizing the hexagonal arrangement. (**d**) Selected area diffraction patterns of MoS$_2$/G sample showing two sets of coincident spots corresponding to MoS$_2$ and G in superlattice configuration.

It is similarly proposed here that the existing MoS$_2$ sheets resulting from bulk MoS$_2$ exfoliation are templating during the pyrolysis the formation of the nascent graphene layers in such a way that the growing graphene is replicating the existing MoS$_2$ sheet resulting in lattice matching throughout the heterojunction, as in the precedent reporting the BN/G superlattice heterojunction.

XRD and Raman spectroscopy convincingly evidence the presence of both components, graphene and MoS$_2$, in the heterojunction. Figure S3 in the supporting information shows the XRD pattern recorded for MoS$_2$/G. This XRD shows a sharp peak at 2θ 14.60 corresponding to the 0.0.2 facet of MoS$_2$. Other peaks of much lesser intensity correspond to other MoS$_2$ planes. In addition to the peaks of MoS$_2$, a broad diffraction band at 24° due to few layer defective graphene is also recorded [12].

Figure 4 shows the corresponding Raman for the MoS$_2$/G heterojunction. At high wavenumbers, the expected broad peaks at 2840, 1540, and 1355 cm^{-1} corresponding to the 2D, G, and D peaks characteristic of defective graphenes are recorded. Two very sharp lines at 383 and 408 cm^{-1} that are characteristic of MoS$_2$ are also seen. No changes in the position of the MoS$_2$ or G bands is observed, indicating that the superlattice configuration does not alter the lattice constants of MoS$_2$ or G, in agreement with the lattice match. The difference in wavenumber of these two peaks indicates the single layer (18 cm^{-1}) or few layers structure of MoS$_2$ (25 cm^{-1}). In the present case, the difference between the A$_{1g}$ and E$_{2g}$ vibration modes of MoS$_2$ was 24 cm^{-1}, which agrees with the case of few layers MoS$_2$. This configuration of MoS$_2$ as few layers platelets is also in accordance with the previously commented AFM measurements of MoS$_2$ particles that were supported on graphene.

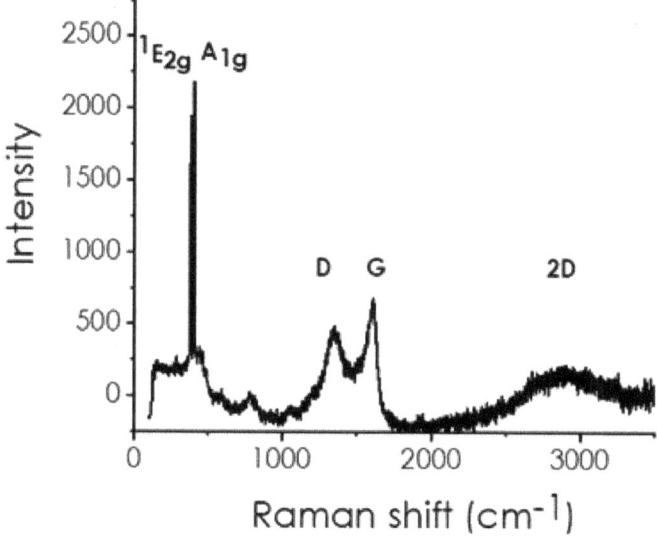

Figure 4. Raman spectrum recorded for MoS$_2$/G. The peaks two-dimensional (2D), G, and D marked on the plot correspond to G, while the sharp intense peaks denoted as ^{1}E$_{2g}$ and A$_{1g}$ are due to MoS$_2$.

Catalytic Activity

The main purpose of this study is to determine the possibility of preparing large area films of superlattice MoS$_2$/graphene (s-MoS$_2$/G, s-meaning superlattice and corresponding to the material prepared, according to Scheme 1) heterojunctions that are suitable for electrocatalytic characterization and compare their electrochemical properties with those of a random MoS$_2$/graphene (r-MoS$_2$/G, r-meaning random) heterojunction, as stated in the introduction. Aimed at this purpose, an additional sample of r-MoS$_2$/G was prepared by introducing commercial preformed graphene during the process of exfoliation of bulk commercially available MoS$_2$ (Aldrich). Supporting information provides characterization data of r-MoS$_2$/G sample used for the comparison, including a set of TEM images and EDS analysis with the corresponding elemental mapping. (Figure S6) Importantly, selected area diffraction patterns show the spots corresponding to hexagonal patterns of MoS$_2$ and graphene not overlapped and clearly distinguishable for each 2D material, as expected for samples lacking the superlattice configuration (Figure S5, frame d).

Electrochemical measurements were performed with a conventional three-electrode single cell with Pt as auxiliary electrode accompanying the MoS$_2$/G electrode and a reference electrode. The quartz plates used for preparation of electrodes were previously modified by subliming carbon before MoS$_2$/G film preparation, thus improving the electrical contacts, in order to increase the electrical conductivity of MoS$_2$/G electrodes. Prior controls with sublimed carbon quartz plates lacking MoS$_2$/G electrocatalyst did not exhibit any peak in cyclic voltammetry. Experiments were carried out for air-saturated, 1.0 M aqueous H_2SO_4 solution as electrolyte. Figure 5 shows the cyclic voltammetry corresponding to s-MoS_2/G and r-MoS_2/G under these conditions n. As it can be seen there, for s-MoS_2/G a cathodic peak at −0.8 V vs. Ag/AgCl corresponding to ORR is recorded before a step peak corresponding to HER with onset being determined by extrapolation of the current density plot at −0.70 V. In the anodic region, the s-MoS_2/G electrode exhibits the expected OER process with an onset potential of 0.87 V. For comparison, Table 1 compiles reported electrochemical data [28,29] for Pt/C and MoS_2 electrodes together with the new data herein obtained for s-MoS_2/G.

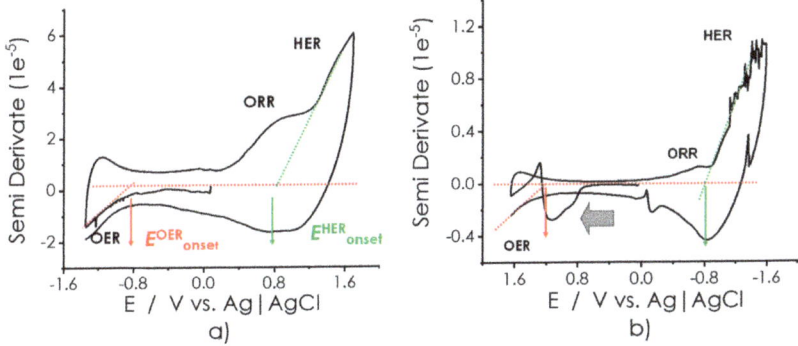

Figure 5. Cyclic voltammograms measured for s-MoS_2/G (**a**) and r-MoS_2/G (**b**) plotted as semiderivative of the current density vs. the applied voltage. Conditions: three electrode configurations; Pt counter electrode; Electrolyte: air-saturated 1.0 M H_2SO_4. The peaks corresponding to hydrogen evolution reaction (HER), oxygen evolution reaction (OER), and oxygen reduction reaction (ORR) have been indicated in the plot, as well as the corresponding onset potentials.

Table 1. Electrochemical data for the catalytic HER process of reference electrodes in comparison to s-MoS_2/G. From linear potential scan voltammograms of material-modified graphene modified glassy carbon (GCE) electrodes in contact with 0.50 M H_2SO_4; potential scan rate 10 mV s^{-1}. The exchange current density (jo) was calculated from Tafel graphs using the current extrapolation method.

Material	E_{onset} (V vs. RHE)	Tafel Slope (mV per Decade)	Jo (mA cm^{-2})
Pt/C	0.0	30	7.1×10^{-1}
MoS_2	−0.24	101	9.1×10^{-4}
s-MoS_2/G	−0.60	54	7.9×10^{-3}

4. Conclusions

In summary, it has been shown that the exfoliation of bulk MoS_2 crystals using polystyrene as promoter, followed by film casting and pyrolysis is a suitable procedure for obtaining few layers MoS_2 deposited on few layers defective graphene in superlattice configuration. The procedure allows for obtaining large area films that are suitable for electrochemical characterization. By comparing the performance of two MoS_2/G electrodes, one with lattice matching and another with random configuration in the heterojunction, it has been observed that the electrocatalytic activity of MoS_2/G improves significantly in the superlattice configuration. This probably reflects the favourable orbital overlapping and electron migration in the MoS_2/G heterojunction when the two lattices match one on top of the other. The present results show the far-reaching potential of superlattice assembly of 2D heterojunctions for application in electrocatalysis.

Supplementary Materials: The following are available online at http://www.mdpi.com/2079-4991/10/5/839/s1, Figure S1: Lateral size (a) and height (b) of MoS_2 nanoparticles present on MoS_2/G, Figure S2: EDX analysis of S, Mo, C and O for the MoS_2/G particles shown in the TEM image (left frame); Figure S3: XRD spectrum of MoS_2/G; Figure S4: Frontal view (a and b) and height profiles (c and d) along the white lines of r-MoS_2/polystyrene (a and c) and r-MoS_2/G (b and d); Figure S5: TEM at different magnifications and selected area electron diffraction of r-MoS_2/G; Figure S6: TEM image (left) and EDS analysis of C, S and Mo for r-MoS_2/G.

Author Contributions: The research was performed with the contribution of all the authors. The concept of the study was developed by A.P., and H.G. Sample preparation and characterization was performed by A.R.-P. and Electrochemical measurements were carried out by A.R.-P and A.D.-C. Drafting of the manuscript was performed by H.G. and A.P. All authors have read and agreed to the published version of the manuscript.

Funding: This research was funded by the Spanish Ministry of Economy and Competitiveness (Severo Ochoa and CTQ2015-68653-CO2-R1) and Generalitat Valenciana (Prometeo 2017-083).

Acknowledgments: A.P. thanks the Spanish Ministry for a Ramón y Cajal research associate contract.

Conflicts of Interest: The authors declare no conflict of interest.

References

1. Feng, J.-X.; Xu, H.; Ye, S.-H.; Ouyang, G.; Tong, Y.-X.; Li, G.-R. Silica-Polypyrrole hybrids as high-performance metal-free Electrocatalysts for the hydrogen evolution reaction in neutral media. *Angew. Chem. Int. Ed.* **2017**, *56*, 8120–8124. [CrossRef] [PubMed]
2. Lai, J.; Li, S.; Wu, F.; Saqib, M.; Luque, R.; Xu, G. Unprecedented metal-free 3D porous carbonaceous electrodes for full water splitting. *Energy Environ. Sci.* **2016**, *9*, 1210–1214. [CrossRef]
3. Zhang, Z.; Yi, Z.; Wang, J.; Tian, X.; Xu, P.; Shi, G.; Wang, S.J. Nitrogen-enriched polydopamine analogue-derived defect-rich porous carbon as a bifunctional metal-free electrocatalyst for highly efficient overall water splitting. *J. Mater. Chem. A* **2017**, *5*, 17064–17072. [CrossRef]
4. Zhong, H.-X.; Zhang, Q.; Wang, J.; Zhang, X.-B.; Wei, X.-L.; Wu, Z.-J.; Li, K.; Meng, F.-L.; Bao, D.; Yan, J.-M.; et al. Engineering ultrathin C3N4 quantum dots on graphene as a metal-free water reduction electrocatalyst. *ACS Catal.* **2018**, *8*, 3965–3970. [CrossRef]
5. Sadighi, Z.; Liu, J.; Zhao, L.; Ciucci, F.; Kim, J.-K. Metallic MoS_2 nanosheets: Multifunctional electrocatalyst for the ORR, OER and Li-O2 batteries. *Nanoscale* **2018**, *10*, 22549–22559. [CrossRef] [PubMed]
6. Yang, L.; Zhang, L.; Xu, G.; Ma, X.; Wang, W.; Song, H.; Jia, D. Metal-organic-framework-derived hollow CoSx@MoS2 microcubes as superior bifunctional electrocatalysts for hydrogen evolution and oxygen evolution reactions. *ACS Sustain. Chem. Eng.* **2018**, *6*, 12961–12968. [CrossRef]
7. Wang, C. Co doped MoS2 as bifunctional electrocatalyst for hydrogen evolution and oxygen reduction reactions. *Int. J. Electrochem. Sci.* **2019**, 9805–9814. [CrossRef]
8. Xue, J.-Y.; Li, F.-L.; Zhao, Z.-Y.; Li, C.; Ni, C.-Y.; Gu, H.-W.; Braunstein, P.; Huang, X.-Q.; Lang, J.-P. A hierarchically-assembled Fe-MoS2/Ni3S2/nickel foam electrocatalyst for efficient water splitting. *Dalton Trans.* **2019**, *48*, 12186–12192. [CrossRef]
9. Zhang, S.; Yang, H.; Gao, H.; Cao, R.; Huang, J.; Xu, X. One-pot Synthesis of CdS Irregular nanospheres hybridized with oxygen-incorporated defect-rich mos2 ultrathin nanosheets for efficient photocatalytic hydrogen evolution. *ACS Appl. Mater. Interfaces* **2017**, *9*, 23635–23646. [CrossRef]
10. Askari, M.B.; Salarizadeh, P.; Seifi, M.; Rozati, S. Electrocatalytic properties of CoS2/MoS2/rGO as a non-noble dual metal electrocatalyst: The investigation of hydrogen evolution and methanol oxidation. *J. Phys. Chem. Solids* **2019**, *135*. [CrossRef]
11. Guruprasad, K.; Maiyalagan, T. Shanmugam, Phosphorus doped MoS2 nanosheet promoted with nitrogen, sulphur dual doped reduced graphene oxide as an effective electrocatalyst for hydrogen evolution reaction. *ACS Appl. Energy Mater.* **2019**, *2*, 6184. [CrossRef]
12. Zhang, X.; Zhang, M.; Tian, Y.; You, J.; Yang, C.; Su, J.; Li, Y.; Gao, Y.; Gu, H. In situ synthesis of MoS 2 /graphene nanosheets as free-standing and flexible electrode paper for high-efficiency hydrogen evolution reaction. *RSC Adv.* **2018**, *8*, 10698. [CrossRef]
13. Adarakatti, P.; Mahanthappa, M.; Hughes, J.; Rowley-Neale, S.; Smith, G.; Siddaramanna, A.; Banks, C. MoS2-graphene-CuNi2S4 nanocomposite an efficient electrocatalyst for the hydrogen evolution reaction. *Int. J. Hydrogen Energy* **2019**, *44*, 16069. [CrossRef]

14. Ge, R.; Li, W.; Huo, J.; Liao, T.; Cheng, N.; Du, Y.; Zhu, M.; Li, Y.; Zhang, J. Metal-ion bridged high conductive RGO-M-MoS2 (M = Fe3+, Co2+, Ni2+, Cu2+ and Zn2+) composite electrocatalysts for photo-assisted hydrogen evolution. *Appl. Catal. B Environ.* **2019**, *246*, 129–139. [CrossRef]
15. Han, X.; Tong, X.; Liu, X.; Chen, A.; Wen, X.; Yang, N.; Guo, X.-Y. Hydrogen evolution reaction on hybrid catalysts of vertical MoS2 nanosheets and hydrogenated graphene. *ACS Catal.* **2018**, *8*, 1828. [CrossRef]
16. He, J.; Fernandez, C.; Primo, A.; Garcia, H. One-step preparation of large area films of oriented MoS2 nanoparticles on multilayer graphene and its electrocatalytic activity for hydrogen evolution. *Materials* **2018**, *11*, 168.
17. Latorre-Sanchez, M.; Esteve-Adell, I.; Primo, A.; Garcia, H. Innovative preparation of MoS2-graphene heterostructures based on alginate containing (NH4)2MoS4 and their photocatalytic activity for H2 generation. *Carbon* **2015**, *81*, 587. [CrossRef]
18. Primo, A.; He, J.; Jurca, B.; Cojocaru, B.; Bucur, C.; Parvulescu, V.I.; Garcia, H. CO2 methanation catalyzed by oriented MoS2 nanoplatelets supported on few layers graphene. *Appl. Catal. B* **2019**, *245*, 351. [CrossRef]
19. Rendon-Patino, A.; Domenech, A.; Garcia, H.; Primo, A. A reliable procedure for the preparation of graphene-boron nitride superlattices as large area (cm × cm) films on arbitrary substrates or powders (gram scale) and unexpected electrocatalytic properties. *Nanoscale* **2019**, *11*, 2981. [CrossRef]
20. Davies, A.; Albar, J.E.; Summerfield, A.; Thomas, J.C.; Cheng, T.S.; Korolkov, V.V.; Stapleton, E.; Wrigley, J.; Goodey, N.L.; Mellor, C.J.; et al. Lattice-matched epitaxial graphene grown on boron nitride. *Nano Lett.* **2018**, *18*, 498. [CrossRef]
21. Zuo, Z.; Xu, Z.; Zheng, R.; Khanaki, A.; Zheng, J.-G.; Liu, J. In-situ epitaxial growth of graphene/h-BN van der Waals heterostructures by molecular beam epitaxy. *Sci. Rep.* **2015**, *5*, 14760. [CrossRef]
22. Hirai, H.; Tsuchiya, H.; Kamakura, Y.; Mori, N.; Ogawa, M. Electron mobility calculation for graphene on substrates. *J. Appl. Phys. (Melville N. Y. USA)* **2014**, *116*. [CrossRef]
23. Lee, K.H.; Shin, H.-J.; Lee, J.; Lee, I.-Y.; Kim, G.-H.; Choi, J.-Y.; Kim, S.-W. Large-scale synthesis of high-quality hexagonal boron nitride nanosheets for large-area graphene electronics. *Nano Lett.* **2018**, *12*, 714–718. [CrossRef] [PubMed]
24. Sakai, Y.; Saito, S.; Cohen, M.L. Lattice matching and electronic structure of finite-layer graphene/h-BN thin films. *Phys. Rev. B Condens. Matter Mater. Phys.* **2014**, *89*, 115424. [CrossRef]
25. Li, X.D.; Yu, S.; Wu, S.Q.; Wen, Y.H.; Zhou, S.; Zhu, Z.Z. Structural and electronic properties of superlattice composed of graphene and monolayer MoS2. *J. Phys. Chem. C* **2013**, *117*, 15347. [CrossRef]
26. Xiong, P.; Ma, R.; Sakai, N.; Nurdiwijayanto, L.; Sasaki, T. Unilamellar Metallic MoS2/graphene superlattice for efficient sodium storage and hydrogen evolution. *ACS Energy Lett.* **2018**, *3*, 997. [CrossRef]
27. Rendón-Patiño, A.; Niu, J.; Doménech-Carbó, A.; García, H.; Primo, A. Polystyrene as graphene film and 3D graphene sponge precursor. *Nanomaterials* **2019**, *9*, 101.
28. Gao, M.-R.; Liang, J.-X.; Zheng, Y.-R.; Xu, Y.-F.; Jiang, J.; Gao, Q.; Li, J.; Yu, S.-H. An efficient molybdenum disulfide/cobalt diselenide hybrid catalyst for electrochemical hydrogen generation. *Nat. Commun.* **2015**, *6*, 5982. [CrossRef]
29. Ding, Q.; Song, B.; Xu, P.; Jin, S. Efficient electrocatalytic and photoelectrochemical hydrogen generation using MoS2 and related compounds. *Chem* **2016**, *1*, 699–726. [CrossRef]

 © 2020 by the authors. Licensee MDPI, Basel, Switzerland. This article is an open access article distributed under the terms and conditions of the Creative Commons Attribution (CC BY) license (http://creativecommons.org/licenses/by/4.0/).

Article

Scalable Fabrication of Modified Graphene Nanoplatelets as an Effective Additive for Engine Lubricant Oil

Duong Duc La [1,*], Tuan Ngoc Truong [1], Thuan Q. Pham [1], Hoang Tung Vo [2], Nam The Tran [2,*], Tuan Anh Nguyen [3], Ashok Kumar Nadda [4], Thanh Tung Nguyen [5], S. Woong Chang [6], W. Jin Chung [6] and D. Duc Nguyen [7,*]

1. Institute of Chemistry and Materials, Nghia Do, Cau Giay, Hanoi 10000, Vietnam; ngoctuan109@gmail.com (T.N.T.); phamquangthuan1982@gmail.com (T.Q.P.)
2. Environmental Institute, Vietnam Maritime University, Haiphong city 180000, Vietnam; tungvh.vmt@vimaru.edu.vn
3. Advanced Nanomaterial Lab, Applied Nano Technology Jsc., Xuan La, Tay Ho, Hanoi 100000, Vietnam; mark@nanoungdung.vn
4. Department of Biotechnology and Bioinformatics, Jaypee University of Information Technology, Waknaghat 173215, India; ashok.nadda@juit.ac.in
5. Institute of Materials Science, Vietnam Academy of Science and Technology, Hanoi 100000, Vietnam; tungnt@ims.vast.ac.vn
6. Department of Environmental Energy Engineering, Kyonggi University, Suwon 16227, Korea; swchang@kyonggi.ac.kr (S.W.C.); cine23@kyonggi.ac.kr (W.J.C.)
7. Institution of Research and Development, Duy Tan University, Da Nang 550000, Vietnam
* Correspondence: duc.duong.la@gmail.com (D.D.L.); thenam@vimaru.edu.vn (N.T.T.); nguyendinhduc2@duytan.edu.vn or nguyensyduc@gmail.com (D.D.N.); Tel.: +84-966-185368 (D.D.L.)

Received: 13 March 2020; Accepted: 11 April 2020; Published: 1 May 2020

Abstract: The use of nano-additives is widely recognized as a cheap and effective pathway to improve the performance of lubrication by minimizing the energy loss from friction and wear, especially in diesel engines. In this work, a simple and scalable protocol was proposed to fabricate a graphene additive to improve the engine lubricant oil. Graphene nanoplates (GNPs) were obtained by a one-step chemical exfoliation of natural graphite and were successfully modified with a surfactant and an organic compound to obtain a modified GNP additive, that can be facilely dispersed in lubricant oil. The GNPs and modified GNP additive were characterized using scanning electron microscopy, X-ray diffraction, atomic force microscopy, Raman spectroscopy, and Fourier-transform infrared spectroscopy. The prepared GNPs had wrinkled and crumpled structures with a diameter of 10–30 μm and a thickness of less than 15 nm. After modification, the GNP surfaces were uniformly covered with the organic compound. The addition of the modified GNP additive to the engine lubricant oil significantly enhanced the friction and antiwear performance. The highest reduction of 35% was determined for the wear scar diameter with a GNP additive concentration of approximately 0.05%. The mechanism for lubrication enhancement by graphene additives was also briefly discussed.

Keywords: modified graphene nanoplates; graphene additives; antifriction; engine lubricant oil additives; antiwear

1. Introduction

The worldwide urgency to minimize the effect of greenhouse gases and climate change requires new measures to improve engine efficiency [1]. The freeload and friction losses of diesel engine vehicles account for approximately 10% of the total energy in fuel [2]. The reduction in these losses is crucial for energy efficiency. Techniques such as system design and handling (reducing the size, electrification, and boosting), the addition of systems for the recovery of heating waste, the reduction in friction in the engine, and the improvement in the combustion efficiency have been successfully utilized to improve engine efficiency. In order to theoretically find suitable methods to enhance engine efficiency, the open-source software framework called PERMIX can be employed [3]. Among these techniques, friction reduction has been receiving significant attention from scientists around the world as a key and cost-effective method to maximize the energy efficiency of diesel fuel. One of the major approaches to reducing friction is the use of lubricants, which can be widely applied in automotive, mechanical, and other parts. Lubricants reduce the friction between the interface of two metal parts in relative motion [4]. Additives are commonly added to the blend of lubricants to improve the lubricating efficiency [5–7].

Since emerging as a technique to fabricate advanced materials, nanotechnology has provided properties superior to those of traditional bulk materials, and nanomaterials have been intensively used as additives for enhancing lubricant performance. Many nanomaterials such as copper [8], MoS_2 [9], PbS [10], WS_2 [11], Zinc borate [12], ZrS_2 [13], boric acid [14,15], and SiO_2 [16], have been employed for this purpose. Carbon nanomaterials with many allotropes have remarkable lubricating properties and have also been utilized as additives to improve the performance of lubricant oils. These carbon nanomaterials consist of carbon nanotubes [17,18], porous carbon [19], fullerence [20,21], and graphene [22–24].

Graphene, a two-dimensional (2D) carbon material with substantial mechanical, electrical, and thermal properties, has been extensively used in a wide range of industrial applications in the fields of engineering, chemistry, and physics [4,25–30]. Additionally, the 2D structures easily slide together, making graphene an effective additive for lowering the friction in mechanical parts and vehicle engines [31–34]. For example, Zhang et al. successfully fabricated graphene nanosheets from graphene oxides and modified them with oleic acid to be used as additives in lubricant oil to reduce the friction coefficient and wear scar diameter by 17% and 14%, respectively [35]. In another study, Azman et al. blended graphene with 95 vol % synthetic based oil (PAO 10) and 15 vol % palm-oil trimethylolpropane (TMP) ester to reduce the wear scar diameter by 15% [36]. Several works have used graphene as an additive in engine lubricant oil; however, these works either used an expensive graphene-fabricating method (hummer methods) or a low dispersion of graphene in the lubricant oil, which hinder the widespread use of graphene as an additive in the lubricating industry. Furthermore, in order to effectively employ the graphene in practical applications, the dispersion and modification of graphene in any solution are crucial factors.

Herein, we adopt a new and facile method, continuing from our previous work, for the mass production of graphene nanoplatelets (GNPs) by the simple one-spot chemical exfoliation of natural graphite. The resultant GNPs are well-dispersed in water with the assistance of a surfactant. The surfaces of the GNPs are modified to easily and homogeneously disperse the GNPs in the targeted engine lubrication oil with a high stability over a long period of storage time. The prepared and modified GNPs are thoroughly characterized. The enhanced lubricating performance of the GNPs additive-containing lubricant is investigated and discussed.

2. Materials and Methods

2.1. Materials

Natural graphite flakes were purchased from VNgraphene. Dried acetone, concentrated sulfuric acid (98%), ethanol, sodium dodecyl persulfate (SDS), sodium persulfate ($Na_2S_2O_8$), and oleic acid were obtained from the Van Minh Company Ltd., Hanoi, Vietnam. The commercial HD-50-based oil was obtained from the petrol station. All chemicals were used as received without further purification.

2.2. Synthesis of Graphene Nanoplatelets

The fabricating protocol for graphene nanoplatelets (GNPs) was adopted from our previous work [37]. Natural graphite flakes were added to a 1000-mL reactor containing concentrated sulfuric acid and stirred for 30 min. Sodium persulfate was gradually added into the reaction mixture and further stirred for 3 h at room temperature. The resultant reaction mixture was directly filtered using a glass sintered filter and thoroughly rinsed three times with dry acetone and water to remove any residual reactants. The GNP powder was dried at 60 °C in air and stored for further processing.

2.3. GNP Modification

Figure 1 illustrates the modification procedures for the graphene nanoplatelets. The GNP powder was first dispersed in an aqueous solution using a combined high shear mixer/probe sonicator system with the assistance of a sodium dodecyl persulfate (SDS) surfactant for 12 h. The homogeneous GNP dispersion in the water with a GNP content of 5% w/w was used for further modification with oleic acid. For the modification, 15 g of oleic acid was gradually added to 300 mL of GNPs in water with a high shear mixer at 7000 rpm and 80 °C for 3 h. The resultant solution was thoroughly dried at 140 °C, and the oleic acid-modified GNP additive was obtained. The modified GNP additive was added to the HD-50 lubricant base oil with various concentrations ranging from 0.005–0.1% w/w to evaluate the effectiveness of the additive for enhancing the properties of the lubricant oil.

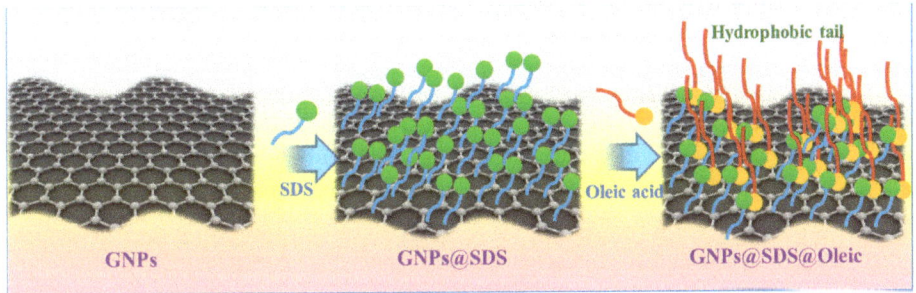

Figure 1. Modification procedure of graphene nanoplatelets with surfactant and organic compound.

2.4. Characterization

Scanning electron microscopy (SEM), FEI Nova NanoSEM (Hillsboro, OR, USA), was utilized to investigate the morphology of the GNPs obtained from the exfoliation of the graphite flakes. The thickness of the prepared GNPs was measured with an AFM (Bruker Multimode 8 with PF TUNA, CA, USA). Fourier transform infrared (FTIR) measurements were performed on a PerkinElmer D100 spectrometer (Ohio, USA) in attenuated total reflectance mode. Raman spectra were obtained with a PerkinElmer Raman Station 200F (Ohio, USA). Bruker AXS D8 Discover instruments (Texas, USA) with a general area detector diffraction system using Cu Kα source were utilized to obtain X-ray diffraction (XRD) patterns of the prepared samples. A tribological test was performed on the four-ball tribometer (MRS-10A, Shandong, China). The test was carried out at room temperature under a load of 400 N with a speed of 1450 rpm.

3. Results

The graphene nanoplatelets were facilely fabricated by employing our reportedly improved approach that obtained GNPs from the direct chemical exfoliation of graphite [37]. This method is environment friendly and can be utilized at industrial scale, which is critical for practical applications. The morphology of the natural graphite and prepared GNPs in this work was observed by SEM (Figure 2). The natural graphite flakes have a thick plate structure with dense stacks of graphene layers

(Figure 2a). After chemical exfoliation with an oxidant, the graphene layers were detached from a thick plate of graphite flakes, as seen in Figure 2b and Figure S1. The GNPs have a wrinkled structure and a diameter of 10–30 μm. The wrinkled and crumpled morphology indicates that the obtained GNPs consist of a few layers of graphene in each stack, as had been demonstrated previously [38,39]. Additionally, the GNPs' sheets are semi-transparent to the electron beam (Figure 2b), which is clear evidence that the GNPs contain less than 30 layers of graphene [40–42].

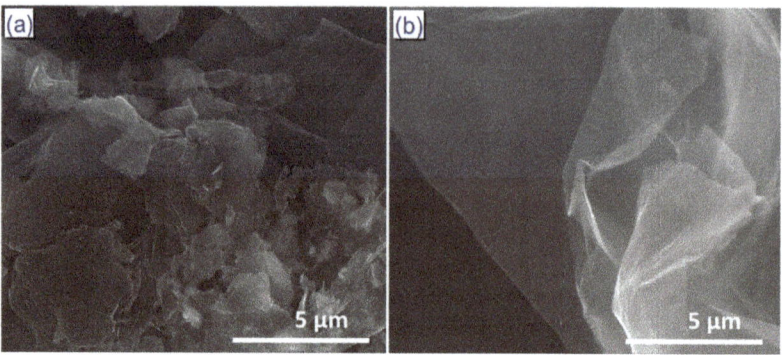

Figure 2. Scanning electron microscopy (SEM) images of (**a**) natural graphite and (**b**) graphene nanoplatelets (GNPs).

The crystalline nature of natural graphite and GNPs was determined, and the X-ray-diffraction (XRD) analysis and the results are shown in Figure 3a. The XRD pattern of graphite showed a sharp characteristic peak at 26.9°, which is a 002-diffraction signal [43]. Interestingly, in the XRD pattern of the GNPs, this peak shifts to 26.4° with a significantly broadened and weakened intensity compared to that of graphite, indicating a less orderly structure with multi-layered graphene [37]. This result is consistent with the aforementioned SEM images. The continuous graphene layers as plate structures in natural graphite flakes no longer exist [37]. The multi-layered nature of the resultant GNPs was further investigated using Raman spectrum excited at the wavelength of 633 nm (Figure 3b). The graphite Raman spectrum shows a characteristic G peak at 1580 cm^{-1} and a band at ~2700 cm^{-1}, which belongs to the graphite samples [44]. The Raman spectrum of the GNPs has two characteristic peaks at 1336 (D band) and 1581 cm^{-1} (G band), which correspond to the defects in carbon networks and sp^2 bonding in carbon elements, respectively [45]. The intensity of the D band peak is significantly lower than that of the G band peak, indicating that the obtained GNPs have fewer defects and a lower oxidant degree when fabricated using this approach. Moreover, the appearance of a peak at 2658 cm^{-1} (assigned to the 2D band), with an intensity significantly lower than that of the G band, indicates that the GNPs are multilayered. The broad photo luminescent band in the Raman spectrum of the GNPs might be due to the amorphous nature of the graphene nanoplatelets. This is consistent with SEM and XRD results. In the Raman spectrum of the oleic-modified GNPs, it can be clearly seen that along with the presence of characteristic bands of GNPs, the CD-stretching vibrations with maximum band intensities around 2100 and 2195 cm^{-1} belong to the oleic acid [46].

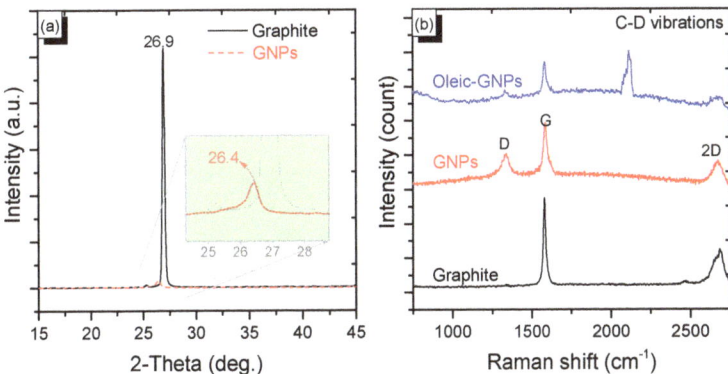

Figure 3. (a) X-ray diffraction (XRD) patterns of natural graphite (black line) and graphene nanoplatelets (red line) and (b) Raman spectrum of graphite (black curve), graphene nanoplatelets (red curve), and oleic-modified graphenenanoplatelets (GNPs, blue curve).

The relative thickness of the GNPs can be calculated by atomic force microscopy (AFM) as shown in Figure 4a. The GNPs are not flat because of their crumpled structure that causes the upper graphene layers to protrude from the surface of the Si wafer (Si wafer is substrate to deposit GNPs for AFM measurement) [37]. Figure 4b and Figure S2 exhibit the topographic AFM image of graphene nanoplatelets the height profile derived from the AFM image. The height profile between the GNPs and the Si substrate is utilized to relatively determine the thickness of the GNPs. The average calculated height is approximately 15 nm, which is approximately less than 30 layers considering the gaps between layers. This result is consistent with the aforementioned SEM, Raman, and XRD results on the multilayer nature of the resultant GNPs.

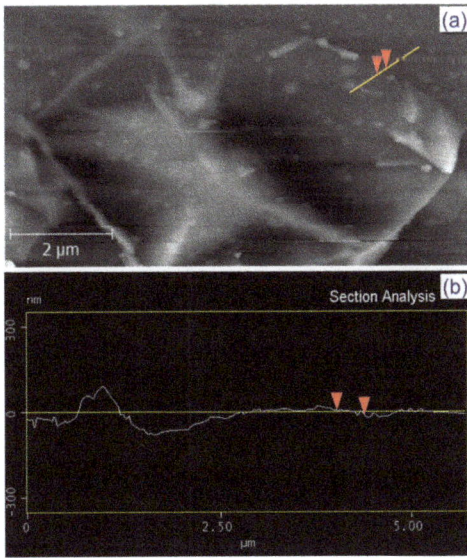

Figure 4. (a) Atomic force microscopy (AFM) images of graphene nanoplatelets and (b) the height profile calculated from AFM imagery.

The successful oleic modification of the GNP surface was investigated by FTIR spectra and XRD patterns (Figure 5). In the IR spectrum of the GNPs, the absorption peaks at 3424 and 1629 cm^{-1} are assigned to the vibration band of the –OH stretching group from moisture, which is physically absorbed on the surface of the GNPs [47]. The remaining absorption peaks at 2369 and 1055 cm^{-1} are attributed to the vibration of COO– and C–O stretching, respectively, which could be ascribed to the absorbed CO_2 [48]. This indicates that the prepared GNPs were virtually not oxidized during the synthesizing process. This is also supported by the X-ray photoelectron spectrometry (XPS) spectrum of C 1s as shown in Figure S3, which shows only one peak of binding energy at 284.5 eV (C–C bonds) indicating that the final product is pure GNPs, and the absence of peaks at 285.5 eV or 286.6 eV is evidence of no oxidizing species. Interestingly, all absorption peaks in the IR spectrum of the GNPs are remarkably weaker, or almost absent, in the IR spectrum of the oleic-modified GNPs, indicating that the surface of the GNPs is uniformly coated by oleic acid. In the IR spectrum of the oleic-modified GNPs, the absorption peaks at 2929 and 2855 cm^{-1} are characteristic of the symmetric and asymmetric vibrations of –CH_2 (which belongs to the long alkyl chains of oleic acid) stretching, respectively [49]. The sharp absorption peaks at 1710 and 1285 cm^{-1} are assigned to the C=O and C–O stretching vibrations of the carboxylic group, respectively [50]. The band at 1461 cm^{-1} is attributed to the bending vibration of (CH2–) [51]. This result confirms that the entire surface of the GNPs was covered by oleic acid. The bonding between the GNPs and the oleic acid is probably due to π–π interactions [52]. The XRD patterns were further employed to confirm the coverage of oleic acid on the graphene surface (Figure 5b). The characteristic peak at 26.4° for graphene nanoplatelets can be clearly seen in the XRD pattern of the GNPs. After modification with oleic acid, this peak virtually disappeared, indicating that the modified GNP surface is uniformly covered with oleic acid.

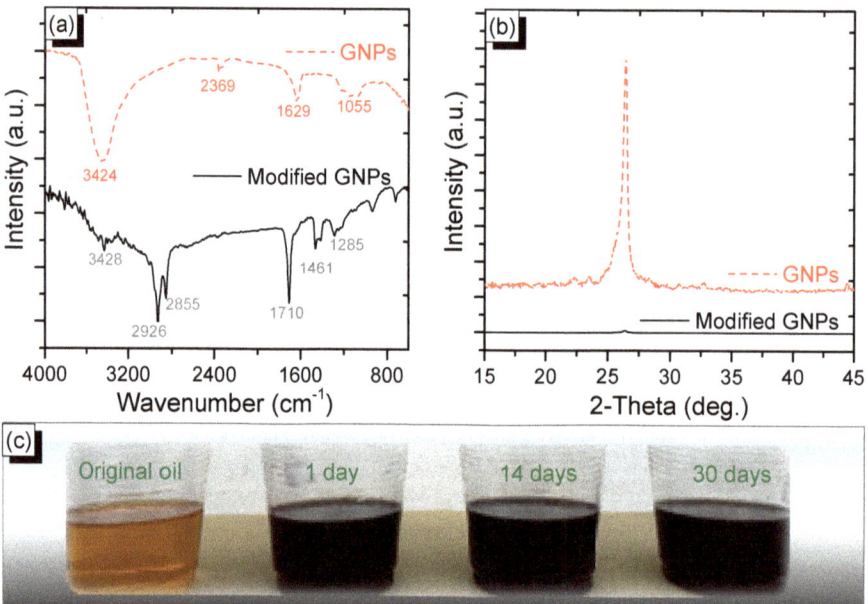

Figure 5. (a) Fourier transform infrared (FTIR) spectra, (b) X-ray diffraction (XRD) patterns of graphene nanoplatelets (red line) and oleic-modified GNPs (black line), and (c) the stability of the oleic-modified GNP additive with 0.01% *w/w* in the lubricant oil.

The homogeneous dispersion stability of modified GNPs in engine lubricant oil (the oil base is HD 50) was evaluated by observation tests to examine the time period in which graphene can remain in the lubricant oil after the mixing process. The concentration of GNPs in the lubricant oil is 0.01% by weight (Figure 5c). Virtually no sediment is observed after 30 days of storage in static conditions. Additionally, clear straight laser beams were employed to further evaluate the stability of the modified GNPs additive in lubricant oil (Figure S4). The Tyndall effect of lubricant oil with the modified GNPs concentration of 0.01% after one day and 30 days of storage clearly indicate that the modified GNPs additive was well-dispersed in the lubricant oil. Therefore, the modified GNP additive is highly stable in blended engine lubricant oil.

The wear scar diameter (WSD) is a critical parameter to determine the antiwear performance of lubricant oil. The WSD was evaluated using a four-ball tribometer (MRS-10A, more information about the instrument). The tribological test was performed at room temperature under a load of 400 N and a speed of 1450 rpm. An optical microscope was utilized to measure the diameter of the wear scar on the ball (Figure 6). The WSD is significantly reduced after the addition of the modified GNP additive with a concentration of 0.005% to 0.01% w/w, indicating that the addition of a small amount of graphene can remarkably enhance the antiwear performance of the lubricant oils. Further increasing the modified GNP concentration from 0.01% to 0.05% w/w reduced the WSD, which reached a minimum diameter of 0.65 mm with an additive concentration of 0.05% w/w, representing a 35% reduction in comparison with the WSD using controlled HD-50 base oil. The WSD of the HD-50 with the addition of the oleic-modified GNPs additive is smaller than that of previous works that used graphene as additives for engine lubricant oils (Table 1), most likely caused by better dispersion of the modified graphene in the lubricant oil. An additional increase in the modified GNP content decreases the antiwear properties of the GNP additive for lubricating oil. Thus, the maximal modified GNPs concentration with the highest antiwear properties is approximately 0.05% by weight. Thus, the GNPs additive concentration of less than 0.05% could be selected as the optimal content inside the lubricant oil. The enhanced antiwear performance upon the addition of small amounts of graphene can be explained by the formation of a protective graphene layer on the steel surface. However, when the graphene content increases, the accumulation of the discontinuous graphene film decreases the antiwear properties and causes friction drying [35].

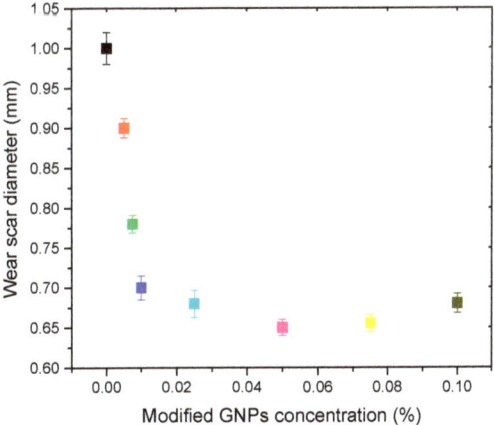

Figure 6. The tribological performance of the engine lubricant oil upon addition of various modified GNPs concentrations.

Table 1. Comparison of the tribological performance between the modified GNPs and those of previous works.

Decreased in Wear Scar Diameter (%)	References
18	[53]
14	[35]
12.6	[32]
Up to 32	[54]
Up to 18.9	[55]
Up to 35	This work

In order to evaluate the thermal stability of lubricant oil upon addition of the modified GNPs additives, the open cup flash points of fabricated oil were determined following ASTM-D92 standard. The results showed that the open cup flash points of the lubricant oil with and without the addition of modified GNPs were 175 °C and 172 °C, respectively, indicating that the GNPs-added lubricant oil was highly stable under the operation condition of diesel oil.

The morphologies of the wear scars' surfaces using lubricants with various modified GNPs contents were investigated by optical microscopy as shown in Figure 7. It can be clearly seen from the figure that when using only base oil, the wear scar is large and the surface is rough with deep narrow trenches. Upon addition of a small amount of the modified GNPs (0.005%), the diameter of the wear scar is reduced and surface becomes smoother, but there still remain deep furrows. However, when the content of the modified GNPs additives was increased to 0.01%, the diameter of the wear scar is significantly reduced to approximately 0.7 mm and the surface becomes much smoother (Figure 7e,f). Further increases in additive contents witness negligible reduction in WSD and smoothness of the wear scar surface. Thus, 0.01 wt % of the modified GNPs additive for lubricant oils was selected as an economically optimized concentration.

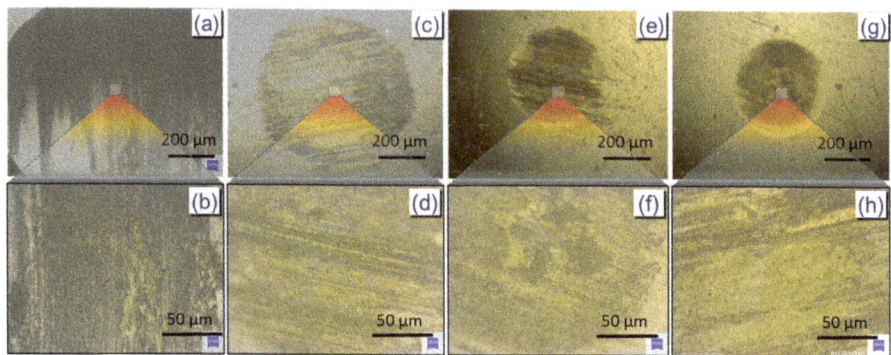

Figure 7. The surface morphologies of the wear scars observed by optical microscopy using different lubricant: (**a,b**) base oil, (**c,d**) 0.005%, (**e,f**) 0.01%, and (**g,h**) 0.05%.

In terms of practical application, the price of materials is essential for commercialization. In the market, the average price of graphene nanoplatelets (analytical and industrial grades) ranges from USD 0.6 to 140 per gram (Figure S5). Meanwhile, the GNPs fabricated from the present approach have a price of around USD 1.2 and 15 per gram for the industrial and analytic grades, respectively, including all the expenditures, which is comparative with available commercial GNPs on the market. When it come to the modified GNPs additives for lubricant oils, the determined price of the additive is approximately USD 0.9 per gram. Considering the significant WSD enhancement with only 0.05 w/w % of the GNPs additives in the lubricant oils, the additional cost calculated for 1 kg of lubricant oil is around USD 0.45, which is reasonable in terms of a 35% lubricating enhancement using the prepared

GNPs additives. Compared with other available nano-additives for lubricant oils in the literature, the modified GNPs content of 0.05 w/w % is much smaller than that of other nanoparticles (Table 2).

Table 2. Optimal concentrations of nano-additives for different lubricant oils.

Nano Additives	Optimum Concentrations, w/w %	References
ZnO	0.5	[56]
CuO	1	[57]
MoS_2	1	[57]
SiO_2	0.05–0.5	[16]
Cu-coated carbon	0.5	[58]
ZrO_2	0.5	[59]
TiO_2	0.3	[60]
GNPs	0.05	This work

4. Conclusions

In conclusion, graphene nanoplatelets were successfully fabricated from natural graphite by direct chemical exfoliation. The resultant GNPs were well-dispersed in an aqueous solution with the assistance of a surfactant and a combination of a high shear mixer and a probe sonicator system. The surface of the graphene was then modified with an organic compound. The as-prepared GNPs were less than 15 nm thick and 10–30 μm in diameter. The results indicate that the modified GNP surface was uniformly covered with oleic acid after modification. The modified GNP additive is facilely dispersed in lubricant oil with remarkable stability, and the GNPs remained in the oil for more than 30 days without settling. The addition of the GNP additive to lubricant oil shows a significant improvement in the tribological performance with a maximal wear scar diameter reduction of 35% at a modified GNP concentration of 0.05% w/w. The formation of a protective graphene layer on the steel surface is responsible for the enhancement of antiwear performance when using the GNP additive in lubricant oil. This remarkable enhancement of the lubricating efficiency (more than 35% enhancement) uses small amounts of the modified GNP additive (approximately 0.05%) that are cost-effectively fabricated and will diversify the practical applications of graphene in the reduction in energy losses from friction and wear in mechanical processing and automotive components.

Supplementary Materials: The following are available online at http://www.mdpi.com/2079-4991/10/5/877/s1, Figure S1: SEM images of prepared graphene nanoplatelets; Figure S2: Topographic AFM image of graphene nanoplatelets and the height profile taken across the white line on the AFM image; Figure S3: XPS spectrum of C 1s; Figure S4: The Tyndall effect of lubricant oil with modified GNPs concentration of 0.01% after 1 day and 30 days; Figure S5: The market price comparisons of graphene nanoplatelets from US, UK, and Chinese companies.

Author Contributions: Conceptualization, D.D.L., D.D.N. and T.A.N.; methodology, D.D.L. and T.N.T.; software, T.Q.P.; validation, D.D.L. and D.D.N.; formal analysis, N.T.T.; investigation, N.T.T.; resources, H.T.V. and S.W.C.; data curation, T.A.N.; writing—original draft preparation, D.D.L., T.A.N., T.T.N., W.J.C. and D.D.N.; writing—review and editing, T.H.V., A.K.N. and S.W.C.; supervision, D.D.N.; project administration, D.D.L.; funding acquisition, D.D.L. and N.T.T. All authors have read and agreed to the published version of the manuscript.

Funding: This work was financially supported in part by Vietnam National Foundation for Science and Technology Development (NAFOSTED) under grant number 104.05-2019.01.

Acknowledgments: We are grateful to the research collaboration among the groups, institutions, and universities of the authors. The authors also would like to thank the reviewers and editors for helpful comments and constructive advice to improve this manuscript.

Conflicts of Interest: The authors declare no conflict of interest.

References

1. Wong, V.W.; Tung, S.C. Overview of automotive engine friction and reduction trends–Effects of surface, material, and lubricant-additive technologies. *Friction* **2016**, *4*, 1–28. [CrossRef]
2. Holmberg, K.; Andersson, P.; Erdemir, A. Global energy consumption due to friction in passenger cars. *Tribol. Int.* **2012**, *47*, 221–234. [CrossRef]
3. Talebi, H.; Silani, M.; Bordas, S.P.; Kerfriden, P.; Rabczuk, T. A computational library for multiscale modeling of material failure. *Comput. Mech.* **2014**, *53*, 1047–1071. [CrossRef]
4. Abdalla, H.; Patel, S. The performance and oxidation stability of sustainable metalworking fluid derived from vegetable extracts. *Proc. Inst. Mech. Eng. Part. B* **2006**, *220*, 2027–2040. [CrossRef]
5. Rizvi, S. Lubricant additives and their functions. *Mater. Park OH ASM Int.* **1992**, 98–112.
6. Guo, J.; Peng, R.; Du, H.; Shen, Y.; Li, Y.; Li, J.; Dong, G. The Application of Nano-MoS_2 Quantum Dots as Liquid Lubricant Additive for Tribological Behavior Improvement. *Nanomaterials* **2020**, *10*, 200. [CrossRef]
7. Li, C.; Li, M.; Wang, X.; Feng, W.; Zhang, Q.; Wu, B.; Hu, X. Novel Carbon Nanoparticles Derived from Biodiesel Soot as Lubricant Additives. *Nanomaterials* **2019**, *9*, 1115. [CrossRef]
8. Borda, F.L.G.; de Oliveira, S.J.R.; Lazaro, L.M.S.M.; Leiróz, A.J.K. Experimental investigation of the tribological behavior of lubricants with additive containing copper nanoparticles. *Tribol. Int.* **2018**, *117*, 52–58. [CrossRef]
9. Grossiord, C.; Varlot, K.; Martin, J.-M.; Le Mogne, T.; Esnouf, C.; Inoue, K. MoS_2 single sheet lubrication by molybdenum dithiocarbamate. *Tribol. Int.* **1998**, *31*, 737–743. [CrossRef]
10. Chen, S.; Liu, W. Oleic acid capped PbS nanoparticles: Synthesis, characterization and tribological properties. *Mater. Chem. Phys.* **2006**, *98*, 183–189. [CrossRef]
11. Zhang, X.; Wang, J.; Xu, H.; Tan, H.; Ye, X. Preparation and Tribological Properties of WS2 Hexagonal Nanoplates and Nanoflowers. *Nanomaterials* **2019**, *9*, 840. [CrossRef] [PubMed]
12. Tian, Y.; Guo, Y.; Jiang, M.; Sheng, Y.; Hari, B.; Zhang, G.; Jiang, Y.; Zhou, B.; Zhu, Y.; Wang, Z. Synthesis of hydrophobic zinc borate nanodiscs for lubrication. *Mater. Lett.* **2006**, *60*, 2511–2515. [CrossRef]
13. Tang, W.; Yu, C.; Zhang, S.; Liu, S.; Wu, X.; Zhu, H. Antifriction and Antiwear Effect of Lamellar ZrS2 Nanobelts as Lubricant Additives. *Nanomaterials* **2019**, *9*, 329. [CrossRef] [PubMed]
14. Deshmukh, P.; Lovell, M.; Sawyer, W.G.; Mobley, A. On the friction and wear performance of boric acid lubricant combinations in extended duration operations. *Wear* **2006**, *260*, 1295–1304. [CrossRef]
15. Lovell, M.R.; Kabir, M.; Menezes, P.L.; Higgs, C.F., III. Influence of boric acid additive size on green lubricant performance. *Philosoph. Trans. R. Soc. A* **2010**, *368*, 4851–4868. [CrossRef] [PubMed]
16. Peng, D.X.; Chen, C.H.; Kang, Y.; Chang, Y.P.; Chang, S.Y. Size effects of SiO_2 nanoparticles as oil additives on tribology of lubricant. *Ind. Lubricat. Tribol.* **2010**, *62*, 111–120. [CrossRef]
17. Moghadam, A.D.; Omrani, E.; Menezes, P.L.; Rohatgi, P.K. Mechanical and tribological properties of self-lubricating metal matrix nanocomposites reinforced by carbon nanotubes (CNTs) and graphene—A review. *Compos. Part B Eng.* **2015**, *77*, 402–420. [CrossRef]
18. Ahmadi, H.; Rashidi, A.; Nouralishahi, A.; Mohtasebi, S.S. Preparation and thermal properties of oil-based nanofluid from multi-walled carbon nanotubes and engine oil as nano-lubricant. *Int. Commun. Heat Mass Transf.* **2013**, *46*, 142–147.
19. Hokkirigawa, K.; Okabe, T.; Saito, K. Friction properties of new porous carbon materials: Woodceramics. *J. Porous Mater.* **1996**, *2*, 237–243. [CrossRef]
20. Rapoport, L.; Fleischer, N.; Tenne, R. Fullerene-like WS_2 nanoparticles: Superior lubricants for harsh conditions. *Adv. Mater.* **2003**, *15*, 651–655. [CrossRef]
21. Lee, K.; Hwang, Y.; Cheong, S.; Kwon, L.; Kim, S.; Lee, J. Performance evaluation of nano-lubricants of fullerene nanoparticles in refrigeration mineral oil. *Curr. Appl. Phys.* **2009**, *9*, e128–e131. [CrossRef]
22. Kinoshita, H.; Nishina, Y.; Alias, A.A.; Fujii, M. Tribological properties of monolayer graphene oxide sheets as water-based lubricant additives. *Carbon* **2014**, *66*, 720–723. [CrossRef]
23. Song, H.-J.; Li, N. Frictional behavior of oxide graphene nanosheets as water-base lubricant additive. *Appl. Phys. A* **2011**, *105*, 827–832. [CrossRef]
24. Svadlakova, T.; Hubatka, F.; Turanek Knotigova, P.; Kulich, P.; Masek, J.; Kotoucek, J.; Macak, J.; Motola, M.; Kalbac, M.; Kolackova, M.; et al. Proinflammatory Effect of Carbon-Based Nanomaterials: In Vitro Study on Stimulation of Inflammasome NLRP3 via Destabilisation of Lysosomes. *Nanomaterials* **2020**, *10*, 418. [CrossRef] [PubMed]

25. Wu, T.; Chen, M.; Zhang, L.; Xu, X.; Liu, Y.; Yan, J.; Wang, W.; Gao, J. Three-dimensional graphene-based aerogels prepared by a self-assembly process and its excellent catalytic and absorbing performance. *J. Mater. Chem. A* **2013**, *1*, 7612–7621. [CrossRef]
26. Kopelevich, Y.; Esquinazi, P. Graphene physics in graphite. *Adv. Mater.* **2007**, *19*, 4559–4563. [CrossRef]
27. Yang, K.; Huang, L.-J.; Wang, Y.-X.; Du, Y.-C.; Zhang, Z.-J.; Wang, Y.; Kipper, M.J.; Belfiore, L.A.; Tang, J.-G. Graphene Oxide Nanofiltration Membranes Containing Silver Nanoparticles: Tuning Separation Efficiency via Nanoparticle Size. *Nanomaterials* **2020**, *10*, 454. [CrossRef]
28. La, D.D.; Hangarge, R.V.; Bhosale, S.; Ninh, H.D.; Jones, L.A.; Bhosale, S.V. Arginine-mediated self-assembly of porphyrin on graphene: A photocatalyst for degradation of dyes. *Appl. Sci.* **2017**, *7*, 643. [CrossRef]
29. La, D.D.; Nguyen, T.A.; Nguyen, T.T.; Ninh, H.D.; Thi, H.P.N.; Nguyen, T.T.; Nguyen, D.A.; Dang, T.D.; Rene, E.R.; Chang, S.W. Absorption Behavior of Graphene Nanoplates toward Oils and Organic Solvents in Contaminated Water. *Sustainability* **2019**, *11*, 7228. [CrossRef]
30. La, D.D.; Patwari, J.M.; Jones, L.A.; Antolasic, F.; Bhosale, S.V. Fabrication of a GNP/Fe–Mg binary oxide composite for effective removal of arsenic from aqueous solution. *ACS Omega* **2017**, *2*, 218–226. [CrossRef]
31. Berman, D.; Erdemir, A.; Sumant, A.V. Graphene: A new emerging lubricant. *Mater. Today* **2014**, *17*, 31–42. [CrossRef]
32. Kiu, S.S.K.; Yusup, S.; Soon, C.V.; Arpin, T.; Samion, S.; Kamil, R.N.M. Tribological investigation of graphene as lubricant additive in vegetable oil. *J. Phys. Sci.* **2017**, *28*, 257.
33. González-Domínguez, J.M.; León, V.; Lucío, M.I.; Prato, M.; Vázquez, E. Production of ready-to-use few-layer graphene in aqueous suspensions. *Nat. Protoc.* **2018**, *13*, 495. [CrossRef] [PubMed]
34. Reina, G.; González-Domínguez, J.M.; Criado, A.; Vázquez, E.; Bianco, A.; Prato, M. Promises, facts and challenges for graphene in biomedical applications. *Chem. Soc. Rev.* **2017**, *46*, 4400–4416. [CrossRef] [PubMed]
35. Zhang, W.; Zhou, M.; Zhu, H.; Tian, Y.; Wang, K.; Wei, J.; Ji, F.; Li, X.; Li, Z.; Zhang, P. Tribological properties of oleic acid-modified graphene as lubricant oil additives. *J. Phys. D Appl. Phys.* **2011**, *44*, 205303. [CrossRef]
36. Azman, S.S.N.; Zulkifli, N.W.M.; Masjuki, H.; Gulzar, M.; Zahid, R. Study of tribological properties of lubricating oil blend added with graphene nanoplatelets. *J. Mater. Res.* **2016**, *31*, 1932–1938. [CrossRef]
37. La, M.D.D.; Bhargava, S.; Bhosale, S.V. Improved and a simple approach for mass production of graphene nanoplatelets material. *ChemistrySelect* **2016**, *1*, 949–952. [CrossRef]
38. Parvez, K.; Wu, Z.-S.; Li, R.; Liu, X.; Graf, R.; Feng, X.; Müllen, K. Exfoliation of graphite into graphene in aqueous solutions of inorganic salts. *J. Am. Chem. Soc.* **2014**, *136*, 6083–6091. [CrossRef]
39. Sheka, E.F.; Hołderna-Natkaniec, K.; Natkaniec, I.; Krawczyk, J.X.; Golubev, Y.A.; Rozhkova, N.N.; Kim, V.V.; Popova, N.A.; Popova, V.A. Computationally Supported Neutron Scattering Study of Natural and Synthetic Amorphous Carbons. *J. Phys. Chem. C* **2019**, *123*, 15841–15850. [CrossRef]
40. Lotya, M.; Hernandez, Y.; King, P.J.; Smith, R.J.; Nicolosi, V.; Karlsson, L.S.; Blighe, F.M.; De, S.; Wang, Z.; McGovern, I. Liquid phase production of graphene by exfoliation of graphite in surfactant/water solutions. *J. Am. Chem. Soc.* **2009**, *131*, 3611–3620. [CrossRef]
41. Dimiev, A.; Kosynkin, D.V.; Sinitskii, A.; Slesarev, A.; Sun, Z.; Tour, J.M. Layer-by-layer removal of graphene for device patterning. *Science* **2011**, *331*, 1168–1172. [CrossRef] [PubMed]
42. Genorio, B.; Lu, W.; Dimiev, A.M.; Zhu, Y.; Raji, A.-R.O.; Novosel, B.; Alemany, L.B.; Tour, J.M. In situ intercalation replacement and selective functionalization of graphene nanoribbon stacks. *ACS Nano* **2012**, *6*, 4231–4240. [CrossRef] [PubMed]
43. Sayah, A.; Habelhames, F.; Bahloul, A.; Nessark, B.; Bonnassieux, Y.; Tendelier, D.; El Jouad, M. Electrochemical synthesis of polyaniline-exfoliated graphene composite films and their capacitance properties. *J. Electroanal. Chem.* **2018**, *818*, 26–34. [CrossRef]
44. Ferrari, A.C. Raman spectroscopy of graphene and graphite: Disorder, electron–phonon coupling, doping and nonadiabatic effects. *Solid State Commun.* **2007**, *143*, 47–57. [CrossRef]
45. Meng, Q.; Jin, J.; Wang, R.; Kuan, H.-C.; Ma, J.; Kawashima, N.; Michelmore, A.; Zhu, S.; Wang, C.H. Processable 3-nm thick graphene platelets of high electrical conductivity and their epoxy composites. *Nanotechnology* **2014**, *25*, 125707. [CrossRef]
46. Matthäus, C.; Chernenko, T.; Quintero, L.; Miljković, M.; Milane, L.; Kale, A.; Amiji, M.; Torchilin, V.; Diem, M. Raman micro-spectral imaging of cells and intracellular drug delivery using nanocarrier systems. In *Confocal Raman Microscopy*; Springer: Berlin/Heidelberg, Germany, 2010; pp. 137–163.

47. Fan, H.-L.; Li, L.; Zhou, S.-F.; Liu, Y.-Z. Continuous preparation of Fe_3O_4 nanoparticles combined with surface modification by L-cysteine and their application in heavy metal adsorption. *Ceram. Int.* **2016**, *42*, 4228–4237. [CrossRef]
48. Coenen, K.; Gallucci, F.; Mezari, B.; Hensen, E.; van Sint Annaland, M. An in-situ IR study on the adsorption of CO_2 and H_2O on hydrotalcites. *J. CO_2 Util.* **2018**, *24*, 228–239. [CrossRef]
49. Hong, R.-Y.; Li, J.-H.; Zhang, S.-Z.; Li, H.-Z.; Zheng, Y.; Ding, J.-M.; Wei, D.-G. Preparation and characterization of silica-coated Fe_3O_4 nanoparticles used as precursor of ferrofluids. *Appl. Surf. Sci.* **2009**, *255*, 3485–3492. [CrossRef]
50. Kooter, I.M.; Pierik, A.J.; Merkx, M.; Averill, B.A.; Moguilevsky, N.; Bollen, A.; Wever, R. Difference Fourier transform infrared evidence for ester bonds linking the heme group in myeloperoxidase, lactoperoxidase, and eosinophil peroxidase. *J. Am. Chem. Soc.* **1997**, *119*, 11542–11545. [CrossRef]
51. Ibarra, J.; Melendres, J.; Almada, M.; Burboa, M.G.; Taboada, P.; Juárez, J.; Valdez, M.A. Synthesis and characterization of magnetite/PLGA/chitosan nanoparticles. *Mater. Res. Express* **2015**, *2*, 095010. [CrossRef]
52. Yang, D.; Gao, S.; Fang, Y.; Lin, X.; Jin, X.; Wang, X.; Ke, L.; Shi, K. The π–π stacking-guided supramolecular self-assembly of nanomedicine for effective delivery of antineoplastic therapies. *Nanomed.* **2018**, *13*, 3159–3177. [CrossRef] [PubMed]
53. Raygoza, E.D.R.; Solorio, C.I.R.; Torres, E.G. Lubricating Oil for Automotive and Industrial Applications, Containing Decorated Graphene. U.S. Patent Application No. 9,679,674, 2019.
54. Lee, G.-J.; Rhee, C.K. Enhanced thermal conductivity of nanofluids containing graphene nanoplatelets prepared by ultrasound irradiation. *J. Mater. Sci.* **2014**, *49*, 1506–1511. [CrossRef]
55. Ma, W.; Yang, F.; Shi, J.; Wang, F.; Zhang, Z.; Wang, S. Silicone based nanofluids containing functionalized graphene nanosheets. *Coll. Surf. A Phys. Chem. Eng. Asp.* **2013**, *431*, 120–126. [CrossRef]
56. Alves, S.; Barros, B.; Trajano, M.; Ribeiro, K.; Moura, E. Tribological behavior of vegetable oil-based lubricants with nanoparticles of oxides in boundary lubrication conditions. *Tribol. Int.* **2013**, *65*, 28–36. [CrossRef]
57. Gulzar, M.; Masjuki, H.; Varman, M.; Kalam, M.; Mufti, R.; Zulkifli, N.; Yunus, R.; Zahid, R. Improving the AW/EP ability of chemically modified palm oil by adding CuO and MoS_2 nanoparticles. *Tribol. Int.* **2015**, *88*, 271–279. [CrossRef]
58. Viesca, J.; Battez, A.H.; González, R.; Chou, R.; Cabello, J. Antiwear properties of carbon-coated copper nanoparticles used as an additive to a polyalphaolefin. *Tribol. Int.* **2011**, *44*, 829–833. [CrossRef]
59. Ma, S.; Zheng, S.; Cao, D.; Guo, H. Antiwear and friction performance of ZrO_2 nanoparticles as lubricant additive. *Particuol.* **2010**, *8*, 468–472. [CrossRef]
60. Laad, M.; Jatti, V.K.S. Titanium oxide nanoparticles as additives in engine oil. *J. King Saud Univ. Eng. Sci.* **2018**, *30*, 116–122. [CrossRef]

© 2020 by the authors. Licensee MDPI, Basel, Switzerland. This article is an open access article distributed under the terms and conditions of the Creative Commons Attribution (CC BY) license (http://creativecommons.org/licenses/by/4.0/).

Article

Capacitance Enhancement of Hydrothermally Reduced Graphene Oxide Nanofibers

Daniel Torres *, Sara Pérez-Rodríguez, David Sebastián, José Luis Pinilla, María Jesús Lázaro and Isabel Suelves

Instituto de Carboquímica, Consejo Superior de Investigaciones Científicas (CSIC), Miguel Luesma Castán 4, 50018 Zaragoza, Spain; sperez@icb.csic.es (S.P.-R.); dsebastian@icb.csic.es (D.S.); jlpinilla@icb.csic.es (J.L.P.); mlazaro@icb.csic.es (M.J.L.); isuelves@icb.csic.es (I.S.)
* Correspondence: dtorres@icb.csic.es; Tel.: +34-976-733-977

Received: 27 April 2020; Accepted: 9 May 2020; Published: 30 May 2020

Abstract: Nanocarbon materials present sp^2-carbon domains skilled for electrochemical energy conversion or storage applications. In this work, we investigate graphene oxide nanofibers (GONFs) as a recent interesting carbon material class. This material combines the filamentous morphology of the starting carbon nanofibers (CNFs) and the interlayer spacing of graphene oxide, and exhibits a domain arrangement accessible for fast transport of electrons and ions. Reduced GONFs (RGONFs) present the partial removal of basal functional groups, resulting in higher mesoporosity, turbostratic stacking, and surface chemistry less restrictive for transport phenomena. Besides, the filament morphology minimizes the severe layer restacking shown in the reduction of conventional graphene oxide sheets. The influence of the reduction temperature (140–220 °C) on the electrochemical behaviour in aqueous 0.5 M H_2SO_4 of RGONFs is reported. RGONFs present an improved capacitance up to 16 times higher than GONFs, ascribed to the unique structure of RGONFs containing accessible turbostratic domains and restored electronic conductivity. Hydrothermal reduction at 140 °C results in the highest capacitance as evidenced by cyclic voltammetry and electrochemical impedance spectroscopy measurements (up to 137 F·g^{-1}). Higher temperatures lead to the removal of sulphur groups and slightly thicker graphite domains, and consequently a decrease of the capacitance.

Keywords: carbon nanofibers; reduced graphene oxide nanofibers; hydrothermal reduction; capacitance

1. Introduction

Carbon nanofibers (CNFs), composed entirely of graphite stacks of small lateral size in a 1D filament morphology, present a unique structure that combines properties such as electronic conduction, thermal and chemical resistance, low mass density and good specific surface area, which are essential for a wide range of potential applications. In this sense, CNFs have been used for many different applications, such as heterogeneous catalysis, conversion or storage of electrochemical energy, sensors, electronic devices, and structural and conductive composite materials [1–7]. In the case of electrochemical energy storage, the high electrical conductivity of CNFs in combination with their accessible porosity allows satisfactory charge accumulation but lower than that offered for graphene materials [8,9]. Graphene is the basic structural unit of the most conductive nanocarbon materials including graphite and carbon nanofibers and nanotubes since it exhibits a free movement of π-electrons along with the 2D network of conjugated sp^2-carbon, allowing it to achieve ultrahigh electrical conductivities (10^5–10^6 S·m^{-1}) [7,10]. Graphene has already been used in all possible electrochemical devices with remarkable results [11], whereas expanded graphite presented a large volumetric capacitance (177 F·cm^{-3}) in aqueous acid media [12]. Likewise, graphite intercalation compounds have even offered conductivities higher than metallic copper, showing notable performance in batteries motivated by rapid diffusion of species between its stacks of distanced graphene layers [13].

In this direction, graphene oxide (GO) has been tentatively investigated for energy conversion and storage applications but oxygen interrupts the free electron movement compromising the electrical conductivity of the graphene layers [14,15]. On the other hand, functionalization enhances the hydrophilicity of nanocarbon materials, which leads to an improved electrochemical capacitance in aqueous electrolytes. GO is conventionally obtained from graphite by chemical oxidation routes, which involve the use of a strong oxidant in acid medium, such as the Hummers method [16], which besides to the creation of oxygen functionalities, including in-plane (hydroxyl and epoxide groups) and edge groups (carbonyl and carboxylic groups) [17,18], results in the generation of sulphur-containing groups [19,20]. As a result of the oxidation, lattice defects (vacancies, holes) are generated in the graphene network which may alter the interaction of the carbon surface with the electrolyte, and thus the electrochemical behaviour.

Graphene oxide nanofibers (GONFs) arise as a new alternative for catalytic and electrochemical applications offering a combination of the best characteristics of CNFs and GO materials: filamentous configuration, accessible microporosity, and tuneable functionalization. Promising approaches in the recent literature related to the combination of GO and CNFs in nanocomposites [21–23] or CNFs produced from GO precursors by electrospinning [24], by templating procedures [25], or by controlled assembly [26], with interesting synergies within phases, have been reported. Only a few recent works are found to report the oxidation and exfoliation of CNFs as a procedure to result in GONFs [27,28]. The innovative aspect of this article is that GONFs are a result of the chemical exfoliation of CNFs at the surface level, representing a novel approach to the best of our knowledge. The main advantage is that the controlled oxidation of CNFs allows keeping the filament morphology. An extended oxidation of CNFs causes its downsizing to few-layer graphene sheets and graphene quantum dots [29].

In the present work, the hydrothermal reduction of GONFs and the effect of the temperature on their physicochemical and electrochemical properties are studied. Stoller and co-workers reported an enhancement of the electrochemical behaviour of GO materials by removal of the large amounts of oxygen groups that diminish the electrical conductivity and hinder the access of species through its porosity [30]. However, oxygen removal gives place to the severe and irreversible restacking of individual graphene sheets (higher L_c and number of layers per stack), which decreases the specific capacitance of RGO (reduced GO) with respect to the expected theoretical values for graphene (550 $F \cdot g^{-1}$) [31]. In that sense, layer restacking causes that a large surface area of RGO becomes inaccessible for charge storage [32]. In case of processing reduced GONFs (RGONFs), the filamentous arrangement of GO stacks largely inhibits this restacking (stacks with a number of layers below 10), offering the possibility of obtaining carbon nanofibers with domains of turbostratic graphite, structure more accessible for fast transport of electrons and reactants than hexagonal graphite. Likewise, particles of nanofilaments simulate the inverse structure of a porous support preserving mesoporosity, which prevents any mass or electron transfer limitations [33]. Finally, the partial reduction of this material barely alters its hydrophilicity, so important for electrochemical applications in aqueous media [34]. The electrochemical capacitance and the electrode/electrolyte interface of the resultant RGONFs are analysed by cyclic voltammetry and impedance electrochemical spectroscopy in acid medium (0.5 M H_2SO_4). Herein, the electrochemical properties of RGONFs are studied for the first time and correlated to their physicochemical features.

2. Materials and Methods

2.1. Preparation of Fishbone GONFs and RGONFs

GONFs were obtained by chemical oxidation, using the modified Hummers method [16,35,36], and ultrasound-assisted exfoliation of fishbone CNFs according to our optimized method [29], where the synthesis method of the starting CNFs is also described. Briefly, 3.0 g of CNFs, 3.0 g of $NaNO_3$ (99%), and 138 mL of H_2SO_4 (96%) were mixed in an ice bath. After that, 26 g of $KMnO_4$ was very slowly added to the solution under vigorous stirring. The temperature was kept below 20 °C during mixing, then at

30 ± 5 °C for 2 h and at room temperature overnight, always under stirring. Then, 240 mL of deionized water was slowly added to prevent the temperature from rising above 70 °C. Subsequently, the solution was stirred for 60 min and diluted with 600 mL of deionized water. Then, 26 mL of H_2O_2 (33%) was added dropwise, turning the solution to yellowish-brown. This solution is finally sonicated in an ultrasounds bath for 60 min to achieve the exfoliation of the oxidized material. This GO suspension contains GONFs (40% of the initial weight of CNFs), completely exfoliated materials like few-layer graphene oxide flakes (FLGOs) and GO quantum dots (GOQDs) and other inorganic salts and thus, separation and posterior washing of phases were necessary. Product was washed by centrifugation at 9500 rpm with HCl (10%) first and deionized water until neutral pH. The clean precipitate was dispersed in deionized water and separated from the rest of the GO products (FLGOs and GOQDs) as the precipitate of its centrifugation at 4500 rpm. More details about the whole process by differential degressive centrifugation can be found elsewhere [29]. Finally, fractions were reduced in suspension by the hydrothermal reduction method [37] in a 40 mL autoclave placed in the oven at 140, 180 and 220 °C for 6 h. In this case, 1 g of GONFs (obtained after drying the GONF solution at 60 °C in a vacuum oven), was dispersed in 25 mL of deionized water. After reduction, products were dried at 60 °C in a vacuum oven overnight. Reduced materials will be indicated hereafter as RGONF-140, RGONF-180, and RGONF-220.

2.2. Physicochemical Characterization

Characterization of GONF and RGONF samples was carried out by X-ray diffraction (XRD), X-ray photoelectron spectroscopy (XPS), elemental analysis (EA), and N_2 and CO_2 physisorption at 77 K and 273 K, respectively. GONF dispersions were dried at 60 °C overnight for its characterization.

XRD patterns of GONFs and RGONFs were acquired in a Bruker D8 Advance Series 2 diffractometer (Bruker AXS, Karlsruhe, Deutschland). The angle range scanned was 3–55° using a counting step of 0.02° and a counting time per step of 4 s. XRD data were fitted using the structure analysis software TOPAS (Bruker AXS, Karlsruhe, Deutschland). The mean interlayer spacings (d-spacing) were evaluated from the position of the corresponding peak applying Bragg's Law [38], while the mean crystallite sizes along c axis (L_c) were calculated using the Scherrer formula, with a value of K = 0.89 [38]. From these, the number of graphene layers (n) was calculated as (L_c/d_{002}) + 1.

The bulk and surface chemistry was analysed by EA and XPS, respectively. The ESCAPlus OMICROM spectrometer (Omicron, Houston, TX, USA), equipped with a hemispherical electron energy analyser, was operated at 18.75 kV and 12 mA (225 W), using a monochromatic AlKα X-ray source (hv = 1486.7 eV) and under vacuum (<5 × 10^{-9} Torr). A survey scan between 1000 and 0 eV was acquired using steps of 0.5 eV and 200 ms of dwell, while C 1s region was acquired each 0.1 eV for 500 ms. Analyser pass energies of 50 and 20 eV were used for survey scans and 20 eV for C1s region, respectively. Elemental analysis of GONFs and RGONFs was performed in a CHNS-O Analyser Thermo FlashEA 1112 (Thermo Fisher Scientific, Waltham, MA, USA).

The textural properties were measured using a Micromeritics ASAP2020 apparatus for N_2 and CO_2 physisorption at 77 and 273 K, respectively. The specific surface area (S_{BET}) was calculated by the BET method applied to the N_2 adsorption isotherm. The total pore volume (V_t) was calculated from the N_2 adsorbed volume at a relative pressure of p/p_0 > 0.994. In addition, micropore surface area (S_{mic}) and the total micropore volume (V_{t_mic}) were calculated by the Dubinin–Radushkevich equation and the adsorbed volume at p/p_0 > 0.031 (pores < 0.8 nm), respectively, using the CO_2 adsorption data.

The morphology of carbon nanofilaments was visualized by transmission electron microscopy (TEM; Tecnai F30, FEI company, Eindhoven, The Netherlands).

2.3. Electrochemical Characterization

Electrochemical measurements were carried out in a three-electrode electrochemical cell and an Autolab PGSTAT302 (Metrohm, Utrecht, Netherlands) potentiostat–galvanostat was used to record the data. A carbon rod and a reversible hydrogen electrode (RHE) were used as counter and reference electrodes, respectively. All potentials in the text are referred to the RHE. Working electrodes were prepared by depositing a layer of a carbon ink on glassy carbon (diameter = 7 mm), resulting in a mass loading of the active material of 1 mg·cm^{-2}. Carbon inks were obtained by dispersing 2 mg of the corresponding material in a deionized water/isopropyl alcohol 50/50 (*v/v*) solution containing 15 wt. % Nafion® (Sigma Aldrich, 5 wt. %). The inks were sonicated for 30 min.

Working electrodes were introduced in the base electrolyte (0.5 M H_2SO_4) saturated with nitrogen. Cyclic voltammetries were performed from 0.2 to 0.8 V vs. RHE at several scan rates: 5, 10, 20, 50, 100, and 200 mV·s^{-1}. Specific capacitances, C (F·g^{-1}), were obtained by integration of the area enclosed in the cyclic voltammograms (CV), according to Equation (1) [39,40]:

$$C\left[F \cdot g^{-1}\right] = \frac{\oint I\, dV}{2\, m\, v\, V} \quad (1)$$

where *I* is the current (A), *V* is the potential (V), ΔV is the potential window (V), *v* is the scan rate (V·s^{-1}), and *m* is the mass of the active material in the working electrode (g).

A frequency response analyser (FRA) was employed to record the electrochemical impedance spectroscopy (EIS) measurements. A frequency range from 100 kHz to 0.01 Hz was always used with a 10 mV r.m.s. AC amplitude around 0.5 V vs RHE.

3. Results and Discussion

3.1. Changes in the Physicochemical Properties of GONFs after Reduction

GONFs, characterized by preserving the filamentous structure of the CNFs used in the synthesis but with a higher degree of interlayer spacing (d > 0.75 nm), presented a higher electrochemical capacitance than the bare CNFs [28]. In this work, GONFs were used as the starting material to study the effect of the hydrothermal reduction temperature (140, 180, and 220 °C) on the physicochemical and electrochemical properties of RGONFs. The 1-D morphology of GONF and RGONFs was revealed by TEM as images in Figure 1 shown. These structures presented fishbone arrangement, inherited from the original CNFs, where the graphite/graphene stacks form an oblique angle to the longitudinal growth axis [41]. GONF (Figure 1a,b) showed disordered arrangements of graphene layers due to the intercalated basal oxygenated groups (hydroxyl and epoxide). In this case, a non-oxidized core preserves the main structure of the nanofilament. As a result of the hydrothermal reduction, RGONFs (Figure 1c–h) presented more compact structures and better definition of layers and edges. Regarding the effect of the temperature used in the hydrothermal reduction, the higher the temperature, the greater the restacking of the graphene layers. In Figure 1h, a height profile of graphene layers in a turbostratic stack of the RGONF-220 sample is shown, with a regular interlayer distance of 0.34 nm.

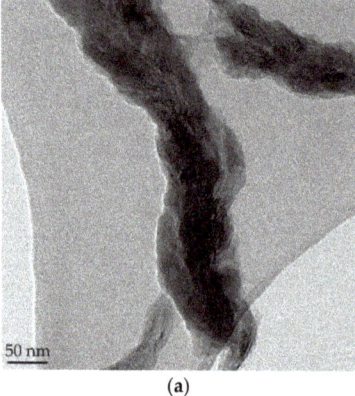
(a)

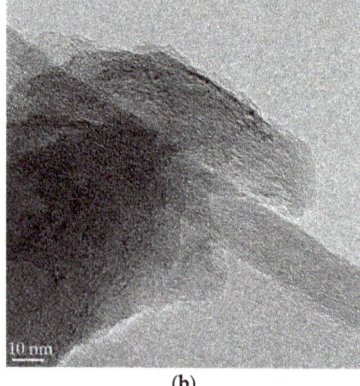
(b)

Figure 1. *Cont.*

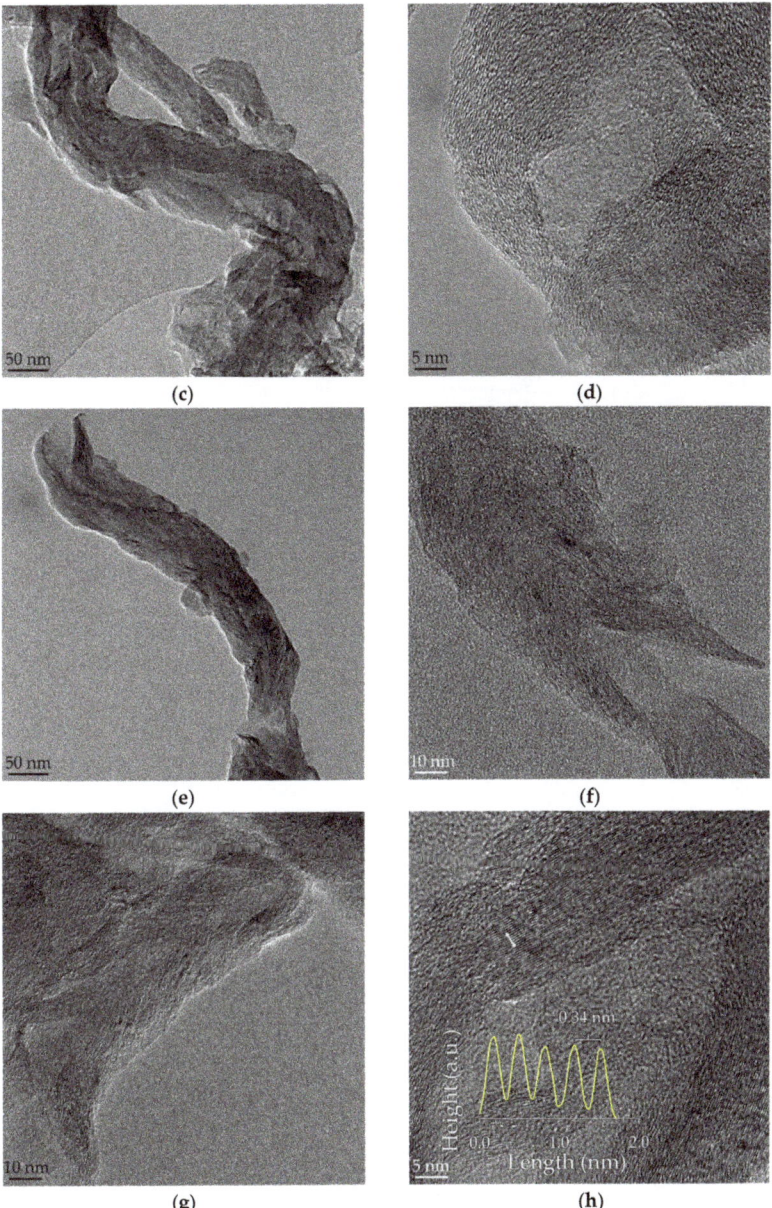

Figure 1. TEM images of: (**a,b**) graphene oxide nanofibers (GONF); and reduced GONFs (RGONFs) obtained at (**c,d**) 140, (**e,f**) 180, and (**g,h**) 220 °C. Height profile of graphene layers in a turbostratic stack is included in (**h**).

XRD patterns of GONF and RGONFs and their associated structural parameters such as d-spacings, L_c and the number of layers are depicted in Figure 2 and listed in Table 1, respectively. GONF showed the presence of two main diffraction peaks corresponding to the basal plane (002) of the graphite stacking at 25.9° (2θ) and the shifted (002) plane (labelled with *) at lower angles (around 10.9°), corresponding to d-spacings of 0.344 and 0.810 nm, respectively. A crystal arrangement of 0.81 nm corresponds to the distance between layers with oxygen intercalated graphite/graphene domains that conform to this fishbone-type nanofilament and is in line with typical values for GO materials [42]. The presence of (002) and (002)* planes is a consequence of the partial oxidation that takes place from the outside towards the inner axis of the starting CNF and is essential in GONF materials as they maintain their tubular structure [29]. After the reduction process, the (002) plane (25.6–25.8°) is recovered at the expense of the (002)* plane motivated by the graphene layers approaching after the removal of oxygenated groups [43]. Only RGONF-140 showed a residual broad band of (002)*. All RGONF samples presented a slightly higher interlayer spacing (about 0.345–0.347 nm) than that in theoretical graphite (0.3354 nm), as confirmed by XRD and in line with TEM images, as is consistent with a non-graphitic or turbostratic arrangement [44,45]. In fact, a deconvolution of this asymmetric peak permits differentiate between turbostratic and hexagonal graphite and to qualitatively calculate the graphitization degree achieved [43,45,46]. At 220 °C, the average number of graphene layers ($L_c/d_{002} + 1$) was increased from 7.0 in GONF (L_c = 2.1 nm) to 9.5 layers in RGONF-220 (L_c = 2.9 nm). These slightly thicker graphite domains are due to the light restacking of the graphene layers by strong π–π interactions and Van der Waals forces between graphene layers and intercalated water molecules [32,42,47,48].

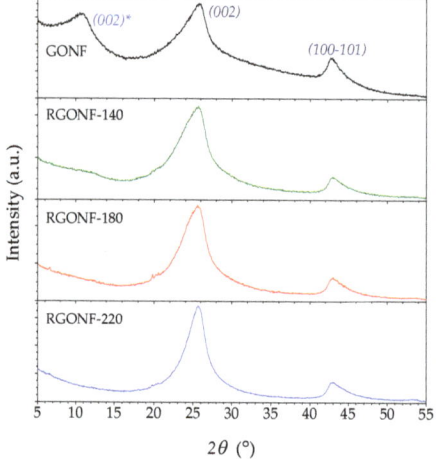

Figure 2. X-ray diffraction (XRD) patterns of GONF and RGONFs obtained at 140, 180, and 220 °C.

Table 1. Structural parameters of GONF and RGONFs obtained at 140, 180, and 220 °C determined by XRD.

Sample	d (nm)		L_c (nm)		n	
	(002) *	(002)	(002) *	(002)	(002) *	(002)
GONF	0.810	0.344	1.5	2.1	2.9	7.0
RGONF-140	-	0.347	-	2.1	-	7.0
RGONF-180	-	0.345	-	2.3	-	7.6
RGONF-220	-	0.346	-	2.9	-	9.5

* For shifted (002) peak.

The bulk and surface compositions of GONF and RGONFs were analysed by elemental analysis and XPS, respectively, as presented in Figure 3 and Table 2. XPS survey spectra of the samples (Figure 3a) showed an inverse behaviour in the trends of C and O which increased and decreased, respectively, as the reduction temperature increased. Surface O content decreased from 18.4 at. % in GONF to a minimum of 12.4 in RGONF-220 (or 32.5 and 17.8 wt. % for the bulk composition, respectively). C/O atomic ratios of RGONFs ranged from 5.3 and 5.4 for RGONF-140 and RGONF-180 to 7.0 for RGONF-220, and they were within typical values for reduced GO obtained by hydrothermal processes [49–51]. N and S contents are derived from the use of $NaNO_3$ and H_2SO_4 during GONF synthesis. The evolution of the different oxygen-containing groups was followed by deconvolution of the C 1s high-resolution regions (Figure 3b and Table 2). Details about C 1s deconvolution can be found elsewhere [9,29]. C 1s was fitted to the components: sp^2 (C=C) and sp^3 hybridized carbon (C–C), C–O bonds in hydroxyls and epoxides (in-plane)s and C–S in sulfonic acid and organosulfates [9], C=O bonds in carbonyls and carboxyls (edge groups), and the π–π* shake-up satellite belonging to aromatic and unsaturated bonds. All samples showed both in-plane and edge oxygen functionalities disrupting the sp^2 graphene network. Reduction removed both, although only RGONF-220 reached a significant variation for edge functional groups. More evident was the restoration of the saturation in RGONFs as incremented the contribution of π–π* transitions.

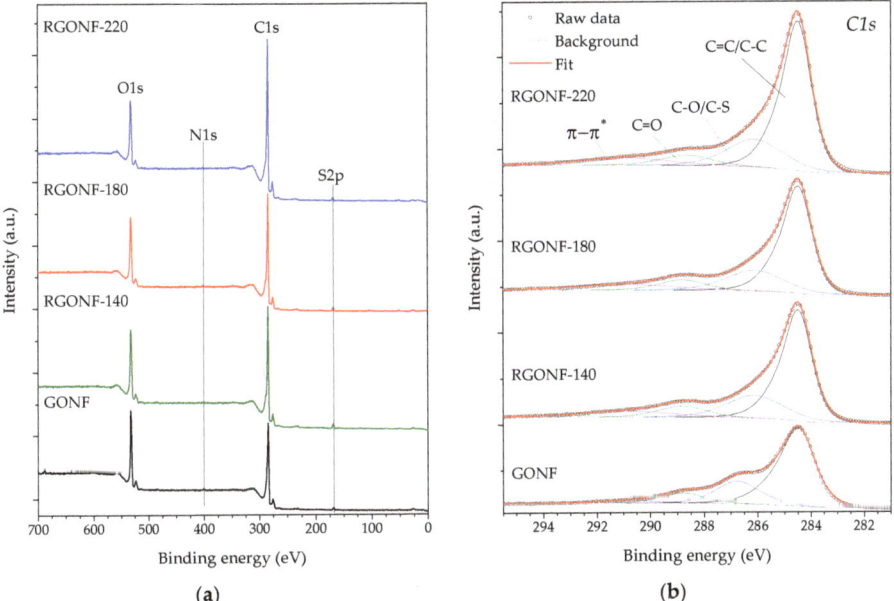

Figure 3. (a) X-ray photoelectron spectroscopy (XPS) Survey; and spectra of the (b) C1s region of GONF and RGONFs obtained at 140, 180, and 220 °C.

Table 2. Bulk and surface composition measured by elemental analysis and XPS, respectively.

Sample	EA (wt. %)					XPS—Survey (at. %)					XPS—C 1s (%)			
	C	O	N	S	H	C	O	N	S	C/O	C sp^2	C–O	C=O	π–π*
GONF	55.9	32.5	0.1	1.6	2.4	80.1	18.4	0.3	1.2	4.4	62.2	23.7	10.4	3.7
RGONF-140	67.4	22.9	0.1	1.8	1.3	82.3	15.5	0.3	1.9	5.3	56.4	23.7	9.8	10.0
RGONF-180	67.7	21.9	0.1	1.9	1.3	82.9	15.4	0.3	1.4	5.4	56.8	23.0	9.4	10.8
RGONF-220	74.2	17.8	0.1	1.1	1.2	86.4	12.4	0.3	1.0	7.0	58.9	22.9	8.0	10.1

Although the hydrothermal reduction only causes the partial oxidation of the starting GONFs, it is enough to modify some physicochemical properties, such as layer restacking or texture.

Figure 4 shows the N_2 and CO_2 isotherms of GONF and RGONF samples. Likewise, textural parameters are listed in Table 3. Based on the results of N_2 physisorption, GONF and RGONFs exhibited type IV isotherms (according to the IUPAC [52]), typical of mesoporous solids, with a H3 hysteresis loop closing at p/p_0 = 0.45. This hysteresis type corresponds to non-rigid aggregates of plate-like particles, where condensation takes place between parallel plates or open slit-shaped capillaries [52]. The sharp step-down located in desorption branch at p/p_0 = 0.43–0.53 is attributed to pore blocking in pore necks [52]. BET surface areas (S_{BET} in Table 3) obtained using the adsorption branch of the N_2 physisorption decreased from 21.7 $m^2 \cdot g^{-1}$ for GONF to 14.3 $m^2 \cdot g^{-1}$ after hydrothermal reduction at the lowest tested temperature (140 °C). Higher temperatures resulted in an increase of S_{BET} (25.1 and 46.3 $m^2 \cdot g^{-1}$ for RGONF-180 and RGONF-220, respectively). Oxygen and sulphur functional groups contained in GONFs hinder the access for N_2 to certain type of pores. After reduction, these groups are partially removed and the graphene layers are approached. Consequently, the development of porosity in RGONFs is mainly due to the mesopores generated in their more difficult compaction as V_t values indicated. RGONF structures, where the layer restacking occurred, are more rigid than GONFs. On the other hand, due to the importance of micropores in GO samples, which are ascribed to the cuneiform pores between the graphenes (slit type pores), CO_2 adsorption measurements were also carried out in order to know the surface area of GONF and RGONFs including narrow micropores (from 0.4 nm). The micropore surface areas (S_{mic}) offered the same evolution as observed for S_{BET} but showing values much higher than those. In this case, the progressive removal of surface functional groups resulted in the clear recovery of the micropore surface area. An even higher development of micro- and mesoporosity could be hydrothermally obtained at more severe conditions [43].

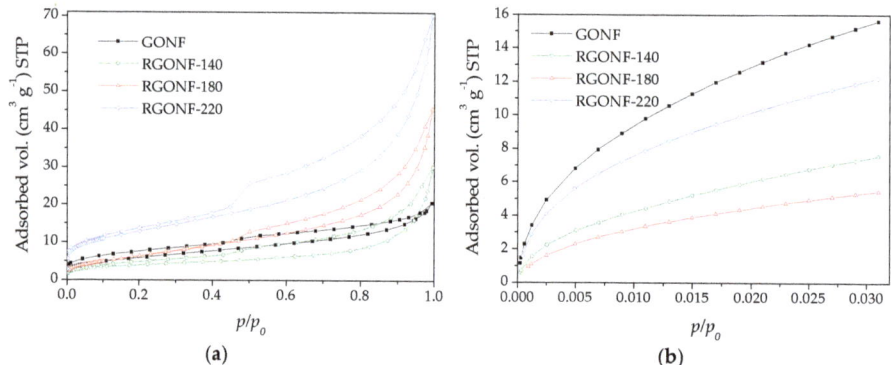

Figure 4. (a) Isotherms of N_2 at 77 K and (b) CO_2 at 273 K for GONF and RGONFs.

Table 3. Textural parameters of GONF and RGONFs obtained at 140, 180, and 220 °C.

Sample	N_2		CO_2	
	S_{BET} ($m^2 \cdot g^{-1}$)	V_t [a] ($cm^3 \cdot g^{-1}$)	S_{mic} [b] ($m^2 \cdot g^{-1}$)	V_{t_mic} [c] ($cm^3 \cdot g^{-1}$)
GONF	21.7	0.032	101.8	0.029
RGONF-140	14.3	0.047	43.6	0.014
RGONF-180	25.1	0.070	36.4	0.010
RGONF-220	46.3	0.109	79.5	0.022

[a] Total pore volume at p/p_0 = 0.994 (maximum pore size of 495 nm); [b] S_{mic} using the Dubinin–Radushkevich equation; [c] Total pore volume at p/p_0 = 0.031 (maximum pore size of 0.8 nm).

3.2. Electrochemical Characterization of GONF and RGONFs

Cyclic voltammograms (CV) were recorded in deaerated 0.5 M H_2SO_4 at six different scan rates: 5, 10, 20, 50, 100, and 200 $mV \cdot s^{-1}$ for GONF and RGONFs. Figure 5 shows the current density-potential behavior for every carbon material at the different scan rates. As expected, the increase of scan rate is associated with an increase of the charge. A quasi rectangular shape is observed for all of them, accounting for the influence of double layer capacitance, together with a small redox contribution with peak current density at about 0.6 V vs. RHE in the positive-going scan and 0.4 V vs. RHE in the negative-going one.

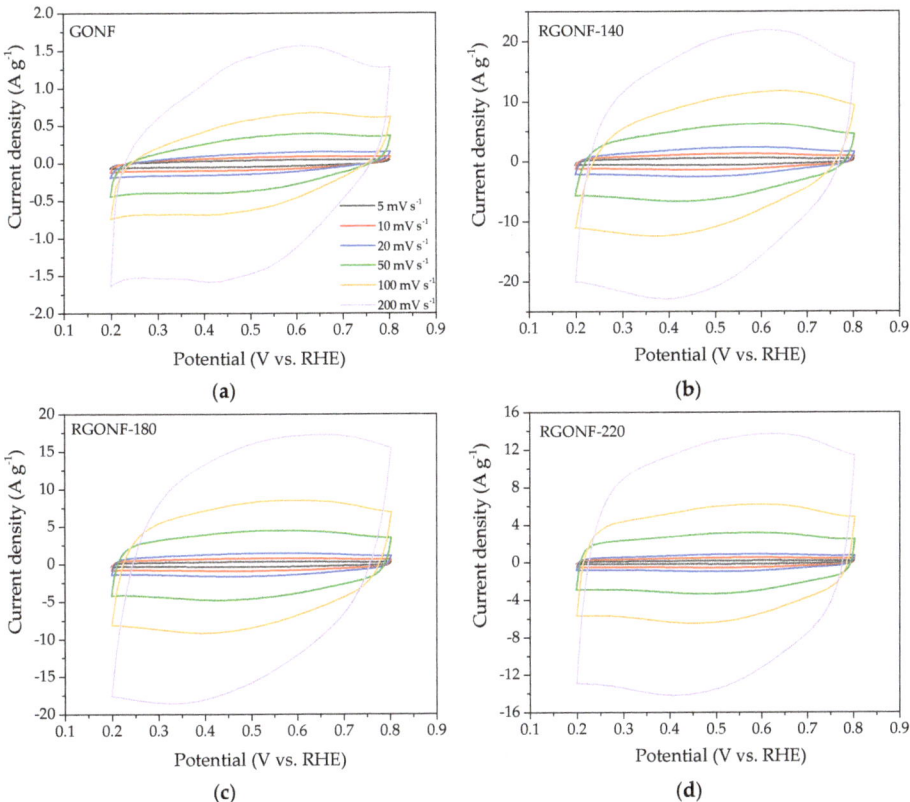

Figure 5. Cyclic voltammograms (CV) from 0.2 to 0.8 V vs. reversible hydrogen electrode (RHE) at several scan rates (5, 10, 20, 50, 100, and 200 mV·s^{-1}) of (**a**) GONF; and RGONFs obtained at (**b**) 140; (**c**) 180; and (**d**) 220 °C. Aqueous electrolyte: 0.5 M H_2SO_4.

Specific capacitances were obtained following the Equation (1). Figure 6a shows an example of the specific capacitance-potential curves obtained at 20 mV·s^{-1} for GONF and RGONFs, while the capacitance values as function of the scan rate are given in Figure 6b. In spite of the partial removal of polar oxygen groups upon hydrothermal reduction treatments (oxygen groups may enhance the hydrophilicity and thus the available electrochemical surface area), an important increase of the double-layer current (see Figure 6a) was observed for RGONFs with specific capacitances from 8- to 16-times higher than that obtained for GONF (~6 F·g^{-1}). RGONFs exhibit a much higher capacitance than GONF, with a good capacitance retention (Figure 6b), despite the relatively low surface area of RGONFs, as evidenced by N_2 and CO_2 physisorption experiments. Although the morphology and structure of RGONFs are different to those reported in the literature regarding reduced graphene oxide, carbon nanofibers decorated with reduced graphene oxide or hybrid materials [53–57], other graphene-based materials have exhibited similar values of specific capacitances in the order of 100 F·g^{-1} with BET surface areas of same order of magnitude than RGONFs reported in this work [21,58].

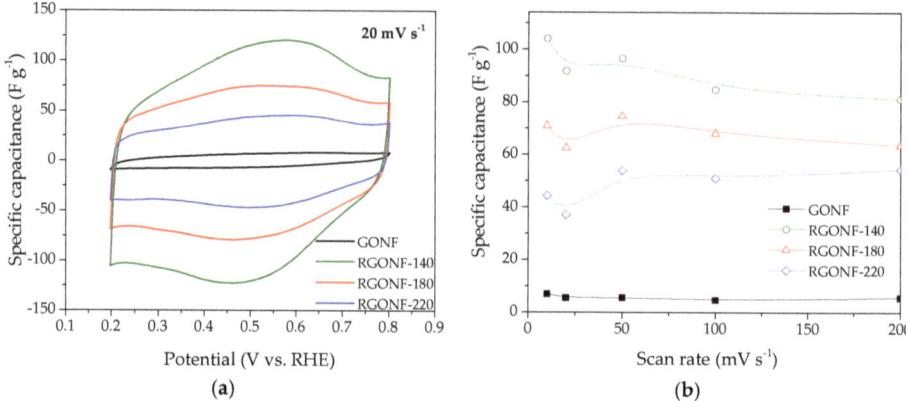

Figure 6. (a) Specific capacitance vs. potential curves at 20 mV·s^{-1} of GONF and RGONF samples. (b) Specific capacitance as a function of the scan rate. Aqueous electrolyte: 0.5 M H_2SO_4.

The samples reduced at 140, 180, and 220 °C present capacitance values up to 104, 75, and 55 F·g^{-1}, respectively. RGONFs present fishbone arrangement (from the original CNFs) with turbostratic stacks forming an oblique angle to the longitudinal growth axis (Figure 1). These highly crystalline structures consisting on few-atom-thick stacks with sp^2-hybridized carbon and lateral dimensions less than 100 nm are expected to exhibit unique properties due to edge effects, quantum confinement and structural defects [58,59]. Consequently, these structures may confine energy band gaps and delocalized charge carriers, resulting in high specific capacitance. On the other hand, other authors have reported an enhanced capacitance of reduced graphene oxide in comparison to graphene oxide [60,61]. The latter can be related to: (i) an increase of the electrical conductivity upon hydrothermal reduction, which favours the electron mobility at RGONF electrodes [61–63], and/or (ii) the particular porous texture and turbostratic structure of RGONFs, which is less restrictive for ion/electrolyte transport. In this context, an important recovery of the π–π* shake-up satellite contribution was observed by XPS (Table 2) upon hydrothermal reduction, as well as a progressive removal of oxygenated species, which may lead to an improved carbon conductivity [7]. On the other hand, a significant decrease of the narrow microporosity was observed after reduction treatments (Table 3, CO_2 physisorption). These narrow micropores may hinder the access of the ions to the surface of carbon electrodes. Additionally, reduction of GONFs is expected to reduce the band gap and to increase the electron-hole pairs, which leads to an enhancement of the quantum confinement [64].

Interestingly, as the reduction temperature decreases, a larger electrochemical capacitance was obtained. These results indicate that the material reduced at 140 °C presents balanced characteristics for charge storage. Higher temperatures resulted in the increase of narrow microporosity as well as some slight restacking of the graphene layers (as confirmed by XRD), which may result in a more restricted quantum confinement effect [32]. Whereas the bare GONF exhibits poor electrical conductivity hindering the electron mobility. Additionally, RGONF-140 presented the highest surface S content (XPS, Table 2) and the largest O concentration among RGONFs. Sulphur and oxygen species have also been reported to present a positive effect on the capacitance of carbon materials in both acid and alkaline electrochemical environments, as a result of Faradaic reactions (pseudocapacitive contribution) and/or a lower affinity to adsorb water by changes in the charge distribution on carbon atoms [34,65–68].

In order to gain further insights on the relative influence of pseudocapacitive contribution to the total specific capacitance, current density was deconvoluted in two components as in Equation (2), as reported in [69]:

$$I(v) = k_1 v + k_2 v^{1/2} \qquad (2)$$

where I is the current (A), k_1 and k_2 are scan rate independent constants and ν is the scan rate (V·s^{-1}). The component associated to k_1 (proportional to ν) encompasses electrochemical double layer charging (non faradaic), while k_2 (proportional to $\nu^{1/2}$) is related to diffusion limited charge from the faradaic contribution of pseudocapacitance, as derived from the Randles–Sevcik equation [39,70,71]. By linearization of Equation (2), a simple linear regression can be applied to the representation of $I/\nu^{1/2}$ against $\nu^{1/2}$ as follows in Equation (3):

$$I(\nu)/\nu^{1/2} = k_1 \nu^{1/2} + k_2 \qquad (3)$$

Figure 7a depicts the variation of $i/\nu^{1/2}$ with the square root of ν for all the materials, including both positive- and negative-going values of current at 0.5 V vs. RHE, together with their linear regression. In RGONF materials, the intercept with the y-axis increases as reduction temperature decreases, indicating a larger pseudocapacitive contribution in the nanofibers treated at lower temperature. Double layer capacitances and pseudocapacitances were calculated from these linear regressions for all the samples, as represented in Figure 7b. The double layer contribution to the total specific capacitance accounts for 87–100% according to this methodology. The largest pseudocapacitance was observed for RGONF-140 (Figure 7b), which is related to the larger extent of O and S groups on the structure of the former material in comparison with the other two RGONFs. The relative contribution of redox processes to the total capacitance is depicted in the inset graph of Figure 7b as a function of the C/O ratio, determined by XPS. As expected, an increase of oxygen surface groups (decrease of C/O ratio) in the form of quinone/hydroquinone (C=O species, XPS, Table 3) results in an increase of the relative pseudocapacitive effect.

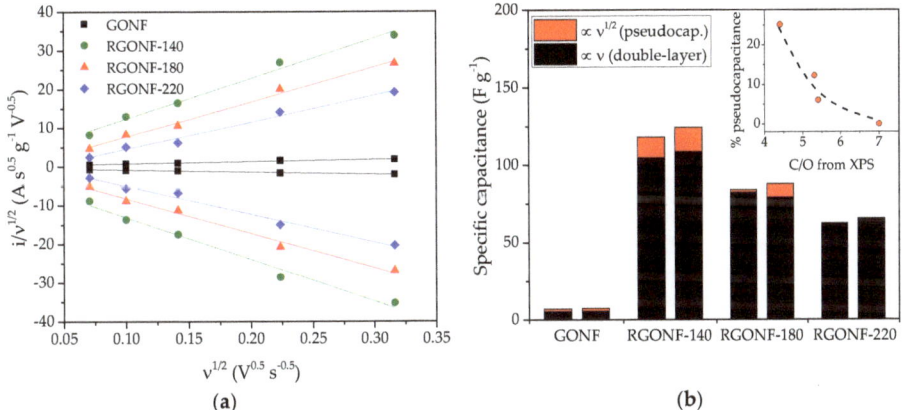

Figure 7. (a) Dependence of $i/\nu^{1/2}$ with $\nu^{1/2}$ for GONF and RGONFs at 0.5 V vs. RHE. (b) Specific capacitances determined from the deconvolution in Figure 7a separated in pseudocapacitance (red, $\propto \nu^{1/2}$) and double layer capacitance (black, $\propto \nu$), for the positive- and negative-going scans being the left and right column, respectively, for each material; the inset represents the percentage of pseudocapacitance as a function of C/O ratio from XPS.

In order to study the overall resistance and the electrolyte ion transport, alternative current EIS measurements were carried out at 0.5 V vs. RHE with 10 mV amplitude (r.m.s.). Figure 8a shows the Nyquist plots, while the Bode-impedance and Bode-phase plots are given in Figure 8b,c, respectively, for all the materials. The Nyquist plots of RGONFs exhibit a sharp increase of the imaginary component of impedance (Z'') with the decrease of frequency, which is indicative of their almost ideal capacitive behaviour, together with a small semicircle at high frequencies [55]. Interestingly, RGONFs present a higher slope at low frequencies than GONF indicating faster ion movement [53,72]. Therefore, RGONF materials present a low charge-transfer resistance with high electrolyte diffusion. The intersection point with the real axis in the Nyquist plot corresponds to the equivalent cell series resistance (R_s), which is in the range 1.5–2.4 $\Omega \cdot cm^{-2}$. The latter is also evident in the Bode-impedance plot at high frequencies (Figure 8b).

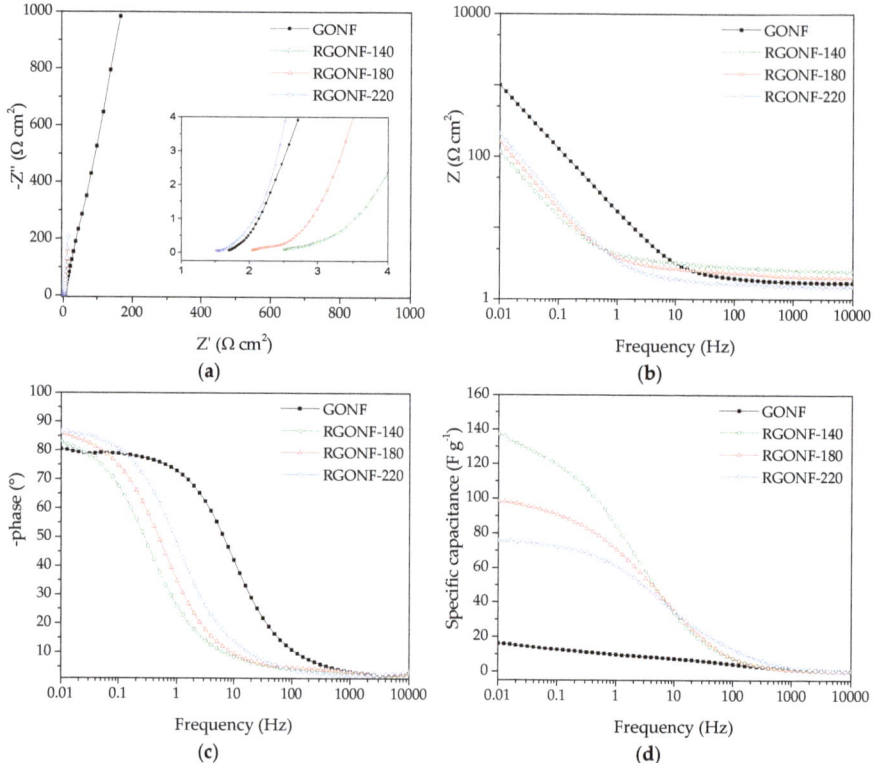

Figure 8. (**a**) Nyquist impedance plots with a 10 mV AC amplitude around 0.5 V of GONF and RGONFs. (**b**) Frequency-dependent impedance modulus Bode plots with a 10 mV AC amplitude around 0.5 V. (**c**) Frequency-dependent phase angle Bode plots with a 10 mV AC amplitude around 0.5 V. (**d**) Specific capacitance vs. frequency curves with a 10 mV AC amplitude around 0.5 V of GONF and RGONFs. Aqueous electrolyte: 0.5 M H_2SO_4.

Figure 8d shows the dependence of the specific capacitance ($F \cdot g^{-1}$) with the frequency, where capacitance values were obtained by Equation (4):

$$C = -\left(\frac{1}{2\pi f Z''}\right) \quad (4)$$

As expected, an increase of the capacitance as the frequency decreases is evident for all materials. At the lowest frequencies (0.01 Hz), the curves reach a plateau at the maximum capacitances. In line with the results obtained by CV, improved capacitances were obtained for RGONFs (76–137 $F \cdot g^{-1}$) in comparison to GONF (16 $F \cdot g^{-1}$). On the other hand, RGONF-140 presented the highest capacitive performance (137 $F \cdot g^{-1}$). As the severity of the reduction conditions increases, a lower capacitance was obtained for RGONF-180 (98 $F \cdot g^{-1}$) and RGONF-220 (76 $F \cdot g^{-1}$), following the same trend already observed by CV experiments. Although superior capacitance values were obtained by EIS measurements in comparison to those calculated by CV, comparable results were observed at 0.5 V vs. RHE (see Figures 8d and 7b). On the other hand, a decrease of the capacitance determined by EIS measurements was evident at 0.2 and 0.8 V vs. RHE. For example, values of 72 and 58 $F \cdot g^{-1}$ were recorded at 0.2 and 0.8 V vs. RHE for the carbon RGONF-180 (Figure 9), whereas a higher capacitance (98 F·g^{-1}) was observed at 0.5 V. This is ascribed to the contribution of the pseudocapacitive component observed in CV experiments at potentials close to 0.5 V vs. RHE, and related to reversible redox processes from oxygen surface groups, most probably quinone-hydroquinone redox pair.

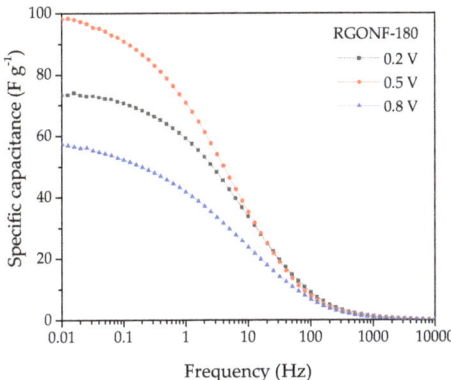

Figure 9. Specific capacitance vs. frequency curves with a 10 mV AC amplitude around 0.2, 0.5, and 0.8 V of RGONF-180. Aqueous electrolyte: 0.5 M H$_2$SO$_4$.

This work reveals that hydrothermal reduction of GONF leads to an enhancement of the specific capacitance, with reduction temperature playing an important role in the morphology and structure of the RGONF materials and also on the electrochemical performance. These results are of interest to design electrodes with potential applications in electrochemical energy conversion and storage, such as supercapacitors, batteries or fuel cells. However, further electrochemical characterization would be necessary to achieve the requirements of commercial devices. For instance, optimization of the mass loading and thickness of the electrode and long-term stability studies.

4. Conclusions

In this work, the influence of hydrothermal reduction at different temperatures (140, 180, and 220 °C) of fishbone graphene oxide nanofibers was carried out, obtaining a novel material with enhanced electrochemical capacitance. After reduction, the resultant carbon presented the partial removal of basal functional groups on graphene/graphite domains, resulting in accessible turbostratic domains and restored electronic conductivity. In particular, RGONFs showed an improved double-layer current with capacitance values from 8- to 16-times higher than GONF, which is ascribed to the unique structure of RGONFs. Hydrothermal reduction at 140 °C led to the highest capacitance as evidenced by cyclic voltammetry and electrochemical impedance spectroscopy (up to 137 F·g^{-1}). A rise in the hydrothermal reduction temperature resulted in a removal of S- and O-containing functional groups and an increase of microporosity, leading consequently to a lower specific capacitance. Additionally, XRD evidenced the presence of slightly thicker graphite domains for the materials treated at 180 and 220 °C, which might decrease the quantum confinement effect, reducing capacitance values.

The proposed approach represents a novel strategy for the production of reduced graphene oxide nanofibers with tuneable properties. The reduced materials exhibit a unique structure and morphology with an enhanced capacitance in comparison to GONFs.

Author Contributions: Conceptualization, D.T. and S.P.-R.; methodology, D.T., D.S. and S.P.-R.; validation, D.T., S.P.-R., D.S. and J.L.P.; investigation, D.T. and S.P.-R.; writing—original draft preparation, D.T. and S.P.-R.; writing—review and editing, D.T., S.P.-R., D.S., J.L.P., M.J.L. and I.S.; supervision, D.T., S.P.-R., D.S. and J.L.P.; project administration, D.T., S.P.-R., D.S., J.L.P., M.J.L. and I.S.; funding acquisition, D.S., J.L.P., M.J.L. and I.S. All authors have read and agreed to the published version of the manuscript.

Funding: This research was funded by FEDER, the Spanish Ministry of Science, Innovation and Universities MICINN (ENE2017-83976-C2-1-R & ENE2017-83854-R) and by the Aragón Government to the Fuel Conversion Group (T06_17R). DS thanks MICINN for his Ramón y Cajal research contract (RyC-2016-20944).

Conflicts of Interest: The authors declare no conflict of interest. The funders had no role in the design of the study; in the collection, analyses, or interpretation of data; in the writing of the manuscript, or in the decision to publish the results.

References

1. De Jong, K.P.; Geus, J.W. Carbon Nanofibers: Catalytic Synthesis and Applications. *Catal. Rev. Sci. Eng.* **2000**, *42*, 481–510. [CrossRef]
2. Bessel, C.A.; Laubernds, K.; Rodriguez, N.M.; Baker, R.T.K. Graphite Nanofibers as an Electrode for Fuel Cell Applications. *J. Phys. Chem. B* **2001**, *105*, 1115–1118. [CrossRef]
3. Serp, P.; Corrias, M.; Kalck, P. Carbon nanotubes and nanofibers in catalysis. *Appl. Catal. A* **2003**, *253*, 337–358. [CrossRef]
4. Pandolfo, A.G.; Hollenkamp, A.F. Carbon properties and their role in supercapacitors. *J. Power Sources* **2006**, *157*, 11–27. [CrossRef]
5. Huang, J.; Liu, Y.; You, T. Carbon nanofiber based electrochemical biosensors: A review. *Anal. Methods* **2010**, *2*, 202–211. [CrossRef]
6. Pampal, E.S.; Stojanovska, E.; Simon, B.; Kilic, A. A review of nanofibrous structures in lithium ion batteries. *J. Power Sources* **2015**, *300*, 199–215. [CrossRef]
7. Pérez-Rodríguez, S.; Torres, D.; Lázaro, M.J. Effect of oxygen and structural properties on the electrical conductivity of powders of nanostructured carbon materials. *Powder Technol.* **2018**, *340*, 380–388. [CrossRef]
8. Ambrosi, A.; Chua, C.K.; Bonanni, A.; Pumera, M. Electrochemistry of Graphene and Related Materials. *Chem. Rev.* **2014**, *114*, 7150–7188. [CrossRef]
9. Torres, D.; Sebastián, D.; Lázaro, M.J.; Pinilla, J.L.; Suelves, I.; Aricò, A.S.; Baglio, V. Performance and stability of counter electrodes based on reduced few-layer graphene oxide sheets and reduced graphene oxide quantum dots for dye-sensitized solar cells. *Electrochim. Acta* **2019**, *306*, 396–406. [CrossRef]
10. Charlier, J.C.; Issi, J.P. Electrical conductivity of novel forms of carbon. *J. Phys. Chem. Solids* **1996**, *57*, 957–965. [CrossRef]
11. Hou, C.; Zhang, M.; Halder, A.; Chi, Q. Graphene directed architecture of fine engineered nanostructures with electrochemical applications. *Electrochim. Acta* **2017**, *242*, 202–218. [CrossRef]
12. Lobato, B.; Wendelbo, R.; Barranco, V.; Centeno, T.A. Graphite Oxide: An Interesting Candidate for Aqueous Supercapacitors. *Electrochim. Acta* **2014**, *149*, 245–251. [CrossRef]
13. Inagaki, M. Applications of graphite intercalation compounds. *J. Mater. Res.* **2011**, *4*, 1560–1568. [CrossRef]
14. Calvillo, L.; Lázaro, M.J.; Suelves, I.; Echegoyen, Y.; Bordejé, E.G.; Moliner, R. Study of the Surface Chemistry of Modified Carbon Nanofibers by Oxidation Treatments in Liquid Phase. *J. Nanosci. Nanotechnol.* **2009**, *9*, 4164–4169. [CrossRef] [PubMed]
15. Sebastián, D.; Suelves, I.; Moliner, R.; Lázaro, M.J. The effect of the functionalization of carbon nanofibers on their electronic conductivity. *Carbon* **2010**, *48*, 4421–4431. [CrossRef]
16. Hummers, W.S.; Offeman, R.E. Preparation of graphitic oxide. *J. Am. Chem. Soc.* **1958**, *80*, 1339. [CrossRef]
17. Lerf, A.; He, H.; Forster, M.; Klinowski, J. Structure of Graphite Oxide Revisited. *J. Phys. Chem. B* **1998**, *102*, 4477–4482. [CrossRef]
18. Szabó, T.; Berkesi, O.; Forgó, P.; Josepovits, K.; Sanakis, Y.; Petridis, D.; Dékány, I. Evolution of Surface Functional Groups in a Series of Progressively Oxidized Graphite Oxides. *Chem. Mater.* **2006**, *18*, 2740–2749. [CrossRef]
19. Eigler, S.; Dotzer, C.; Hof, F.; Bauer, W.; Hirsch, A. Sulfur Species in Graphene Oxide. *Chem. Eur. J.* **2013**, *19*, 9490–9496. [CrossRef]
20. Feicht, P.; Kunz, D.A.; Lerf, A.; Breu, J. Facile and scalable one-step production of organically modified graphene oxide by a two-phase extraction. *Carbon* **2014**, *80*, 229–234. [CrossRef]
21. Zhang, K.; Zhang, L.L.; Zhao, X.S.; Wu, J. Graphene/Polyaniline Nanofiber Composites as Supercapacitor Electrodes. *Chem. Mater.* **2010**, *22*, 1392–1401. [CrossRef]
22. Luo, G.; Wang, Y.; Gao, L.; Zhang, D.; Lin, T. Graphene bonded carbon nanofiber aerogels with high capacitive deionization capability. *Electrochim. Acta* **2018**, *260*, 656–663. [CrossRef]

23. Zhang, Z.; Wang, G.; Lai, Y.; Li, J. A freestanding hollow carbon nanofiber/reduced graphene oxide interlayer for high-performance lithium–sulfur batteries. *J. Alloys Compd.* **2016**, *663*, 501–506. [CrossRef]
24. Gil-Castell, O.; Galindo-Alfaro, D.; Sánchez-Ballester, S.; Teruel-Juanes, R.; Badia, J.D.; Ribes-Greus, A. Crosslinked Sulfonated Poly(vinyl alcohol)/Graphene Oxide Electrospun Nanofibers as Polyelectrolytes. *Nanomaterials* **2019**, *9*, 397. [CrossRef] [PubMed]
25. Cui, C.; Qian, W.; Yu, Y.; Kong, C.; Yu, B.; Xiang, L.; Wei, F. Highly Electroconductive Mesoporous Graphene Nanofibers and Their Capacitance Performance at 4 V. *J. Am. Chem. Soc.* **2014**, *136*, 2256–2259. [CrossRef] [PubMed]
26. Feng, Z.-Q.; Wang, T.; Zhao, B.; Li, J.; Jin, L. Soft Graphene Nanofibers Designed for the Acceleration of Nerve Growth and Development. *Adv. Mater.* **2015**, *27*, 6462–6468. [CrossRef]
27. Zhang, C.; Chen, Q.; Zhan, H. Supercapacitors Based on Reduced Graphene Oxide Nanofibers Supported Ni(OH)2 Nanoplates with Enhanced Electrochemical Performance. *ACS Appl. Mater. Interfaces* **2016**, *8*, 22977–22987. [CrossRef]
28. Torres, D.; Pérez-Rodríguez, S.; Sebastián, D.; Pinilla, J.L.; Lázaro, M.J.; Suelves, I. Graphene oxide nanofibers: A nanocarbon material with tuneable electrochemical properties. *Appl. Surf. Sci.* **2020**, *509*, 144774. [CrossRef]
29. Torres, D.; Pinilla, J.L.; Galvez, E.M.; Suelves, I. Graphene quantum dots from fishbone carbon nanofibers. *RSC Adv.* **2016**, *6*, 48504–48514. [CrossRef]
30. Stoller, M.D.; Park, S.; Zhu, Y.; An, J.; Ruoff, R.S. Graphene-Based Ultracapacitors. *Nano Lett.* **2008**, *8*, 3498–3502. [CrossRef]
31. Xia, J.; Chen, F.; Li, J.; Tao, N. Measurement of the quantum capacitance of graphene. *Nat. Nanotechnol.* **2009**, *4*, 505–509. [CrossRef] [PubMed]
32. Lee, J.H.; Park, N.; Kim, B.G.; Jung, D.S.; Im, K.; Hur, J.; Choi, J.W. Restacking-Inhibited 3D Reduced Graphene Oxide for High Performance Supercapacitor Electrodes. *ACS Nano* **2013**, *7*, 9366–9374. [CrossRef] [PubMed]
33. Chinthaginjala, J.K.; Seshan, K.; Lefferts, L. Preparation and Application of Carbon-Nanofiber Based Microstructured Materials as Catalyst Supports. *Ind. Eng. Chem. Res.* **2007**, *46*, 3968–3978. [CrossRef]
34. Kiciński, W.; Szala, M.; Bystrzejewski, M. Sulfur-doped porous carbons: Synthesis and applications. *Carbon* **2014**, *68*, 1–32. [CrossRef]
35. Stankovich, S.; Piner, R.D.; Chen, X.; Wu, N.; Nguyen, S.T.; Ruoff, R.S. Stable aqueous dispersions of graphitic nanoplatelets via the reduction of exfoliated graphite oxide in the presence of poly(sodium 4-styrenesulfonate). *J. Mater. Chem.* **2006**, *16*, 155–158. [CrossRef]
36. Park, S.; Ruoff, R.S. Chemical methods for the production of graphenes. *Nat. Nanotechnol.* **2009**, *4*, 217–224. [CrossRef]
37. Zhou, Y.; Bao, Q.; Tang, L.A.L.; Zhong, Y.; Loh, K.P. Hydrothermal Dehydration for the "Green" Reduction of Exfoliated Graphene Oxide to Graphene and Demonstration of Tunable Optical Limiting Properties. *Chem. Mater.* **2009**, *21*, 2950–2956. [CrossRef]
38. Biscoe, J. An x-ray study of carbon black. *J. Appl. Phys.* **1942**, *13*, 364–371. [CrossRef]
39. Lee, Y.-H.; Chang, K.-H.; Hu, C.-C. Differentiate the pseudocapacitance and double-layer capacitance contributions for nitrogen-doped reduced graphene oxide in acidic and alkaline electrolytes. *J. Power Sources* **2013**, *227*, 300–308. [CrossRef]
40. Hsiao, C.; Lee, C.; Tai, N. Biomass-derived three-dimensional carbon framework for a flexible fibrous supercapacitor and its application as a wearable smart textile. *RSC Adv.* **2020**, *10*, 6960–6972. [CrossRef]
41. Martin-Gullon, I.; Vera, J.; Conesa, J.A.; González, J.L.; Merino, C. Differences between carbon nanofibers produced using Fe and Ni catalysts in a floating catalyst reactor. *Carbon* **2006**, *44*, 1572–1580. [CrossRef]
42. Dreyer, D.R.; Park, S.; Bielawski, C.W.; Ruoff, R.S. The chemistry of graphene oxide. *Chem. Soc. Rev.* **2010**, *39*, 228–240. [CrossRef] [PubMed]
43. Torres, D.; Arcelus-Arrillaga, P.; Millan, M.; Pinilla, J.; Suelves, I. Enhanced reduction of few-layer graphene oxide via supercritical water gasification of glycerol. *Nanomaterials* **2017**, *7*, 447. [CrossRef] [PubMed]
44. Franklin, R. The structure of graphitic carbons. *Acta Crystallogr.* **1951**, *4*, 253–261. [CrossRef]
45. Li, Z.Q.; Lu, C.J.; Xia, Z.P.; Zhou, Y.; Luo, Z. X-ray diffraction patterns of graphite and turbostratic carbon. *Carbon* **2007**, *45*, 1686–1695. [CrossRef]

46. Feret, F.R. Determination of the crystallinity of calcined and graphitic cokes by X-ray diffraction. *Analyst* **1998**, *123*, 595–600. [CrossRef]
47. Stankovich, S.; Dikin, D.A.; Piner, R.D.; Kohlhaas, K.A.; Kleinhammes, A.; Jia, Y.; Wu, Y.; Nguyen, S.T.; Ruoff, R.S. Synthesis of graphene-based nanosheets via chemical reduction of exfoliated graphite oxide. *Carbon* **2007**, *45*, 1558–1565. [CrossRef]
48. Acik, M.; Mattevi, C.; Gong, C.; Lee, G.; Cho, K.; Chhowalla, M.; Chabal, Y.J. The Role of Intercalated Water in Multilayered Graphene Oxide. *ACS Nano* **2010**, *4*, 5861–5868. [CrossRef]
49. Xu, Y.; Sheng, K.; Li, C.; Shi, G. Self-Assembled Graphene Hydrogel via a One-Step Hydrothermal Process. *ACS Nano* **2010**, *4*, 4324–4330. [CrossRef]
50. Shi, J.L.; Du, W.C.; Yin, Y.X.; Guo, Y.G.; Wan, L.J. Hydrothermal reduction of three-dimensional graphene oxide for binder-free flexible supercapacitors. *J. Mater. Chem. A* **2014**, *2*, 10830–10834. [CrossRef]
51. Diez, N.; Sliwak, A.; Gryglewicz, S.; Grzyb, B.; Gryglewicz, G. Enhanced reduction of graphene oxide by high-pressure hydrothermal treatment. *RSC Adv.* **2015**, *5*, 81831–81837. [CrossRef]
52. Thommes, M.; Kaneko, K.; Neimark Alexander, V.; Olivier James, P.; Rodriguez-Reinoso, F.; Rouquerol, J.; Sing Kenneth, S.W. Physisorption of gases, with special reference to the evaluation of surface area and pore size distribution (IUPAC Technical Report). *Pure Appl. Chem.* **2015**, *87*, 1051–1069. [CrossRef]
53. Bai, Y.; Rakhi, R.B.; Chen, W.; Alshareef, H.N. Effect of pH-induced chemical modification of hydrothermally reduced graphene oxide on supercapacitor performance. *J. Power Sources* **2013**, *233*, 313–319. [CrossRef]
54. Le Fevre, L.W.; Cao, J.; Kinloch, I.A.; Forsyth, A.J.; Dryfe, R.A.W. Systematic Comparison of Graphene Materials for Supercapacitor Electrodes. *ChemistryOpen* **2019**, *8*, 418–428. [CrossRef]
55. Jha, P.K.; Singh, S.K.; Kumar, V.; Rana, S.; Kurungot, S.; Ballav, N. High-Level Supercapacitive Performance of Chemically Reduced Graphene Oxide. *Chem* **2017**, *3*, 846–860. [CrossRef]
56. Gao, Z.; Yang, W.; Wang, J.; Yan, H.; Yao, Y.; Ma, J.; Wang, B.; Zhang, M.; Liu, L. Electrochemical synthesis of layer-by-layer reduced graphene oxide sheets/polyaniline nanofibers composite and its electrochemical performance. *Electrochim. Acta* **2013**, *91*, 185–194. [CrossRef]
57. Jin, Y.; Fang, M.; Jia, M. In situ one-pot synthesis of graphene–polyaniline nanofiber composite for high-performance electrochemical capacitors. *Appl. Surf. Sci.* **2014**, *308*, 333–340. [CrossRef]
58. Zhang, S.; Sui, L.; Dong, H.; He, W.; Dong, L.; Yu, L. High-Performance Supercapacitor of Graphene Quantum Dots with Uniform Sizes. *ACS Appl. Mater. Interfaces* **2018**, *10*, 12983–12991. [CrossRef]
59. Qing, Y.; Jiang, Y.; Lin, H.; Wang, L.; Liu, A.; Cao, Y.; Sheng, R.; Guo, Y.; Fan, C.; Zhang, S.; et al. Boosting the supercapacitor performance of activated carbon by constructing overall conductive networks using graphene quantum dots. *J. Mater. Chem. A* **2019**, *7*, 6021–6027. [CrossRef]
60. Johra, F.T.; Jung, W.-G. Hydrothermally reduced graphene oxide as a supercapacitor. *Appl. Surf. Sci.* **2015**, *357*, 1911–1914. [CrossRef]
61. Gong, Y.; Li, D.; Fu, Q.; Pan, C. Influence of graphene microstructures on electrochemical performance for supercapacitors. *Prog. Nat. Sci. Mater. Int.* **2015**, *25*, 379–385. [CrossRef]
62. Hayes, W.I.; Joseph, P.; Mughal, M.Z.; Papakonstantinou, P. Production of reduced graphene oxide via hydrothermal reduction in an aqueous sulphuric acid suspension and its electrochemical behaviour. *J. Solid State Electrochem.* **2015**, *19*, 361–380. [CrossRef]
63. Li, S.; Chen, Y.; He, X.; Mao, X.; Zhou, Y.; Xu, J.; Yang, Y. Modifying Reduced Graphene Oxide by Conducting Polymer Through a Hydrothermal Polymerization Method and its Application as Energy Storage Electrodes. *Nanoscale Res. Lett.* **2019**, *14*, 226. [CrossRef] [PubMed]
64. Zhang, Y.; Yang, C.; Yang, D.; Shao, Z.; Hu, Y.; Chen, J.; Yuwen, L.; Weng, L.; Luo, Z.; Wang, L. Reduction of graphene oxide quantum dots to enhance the yield of reactive oxygen species for photodynamic therapy. *Phys. Chem. Chem. Phys.* **2018**, *20*, 17262–17267. [CrossRef]
65. Seredych, M.; Bandosz, T.J. S-doped micro/mesoporous carbon–graphene composites as efficient supercapacitors in alkaline media. *J. Mater. Chem. A* **2013**, *1*, 11717–11727. [CrossRef]
66. Seredych, M.; Idrobo, J.-C.; Bandosz, T.J. Effect of confined space reduction of graphite oxide followed by sulfur doping on oxygen reduction reaction in neutral electrolyte. *J. Mater. Chem. A* **2013**, *1*, 7059–7067. [CrossRef]
67. Seredych, M.; Singh, K.; Bandosz, T.J. Insight into the Capacitive Performance of Sulfur-Doped Nanoporous CarbonS·modified by Addition of Graphene Phase. *Electroanalysis* **2014**, *26*, 109–120. [CrossRef]

68. Zhao, X.; Zhang, Q.; Chen, C.-M.; Zhang, B.; Reiche, S.; Wang, A.; Zhang, T.; Schlögl, R.; Sheng Su, D. Aromatic sulfide, sulfoxide, and sulfone mediated mesoporous carbon monolith for use in supercapacitor. *Nano Energy* **2012**, *1*, 624–630. [CrossRef]
69. Augustyn, V.; Simon, P.; Dunn, B. Pseudocapacitive oxide materials for high-rate electrochemical energy storage. *Energy Environ. Sci.* **2014**, *7*, 1597–1614. [CrossRef]
70. Anjos, D.M.; McDonough, J.K.; Perre, E.; Brown, G.M.; Overbury, S.H.; Gogotsi, Y.; Presser, V. Pseudocapacitance and performance stability of quinone-coated carbon onions. *Nano Energy* **2013**, *2*, 702–712. [CrossRef]
71. Lee, J.-S.M.; Briggs, M.E.; Hu, C.-C.; Cooper, A.I. Controlling electric double-layer capacitance and pseudocapacitance in heteroatom-doped carbons derived from hypercrosslinked microporous polymers. *Nano Energy* **2018**, *46*, 277–289. [CrossRef]
72. Zhang, R.; Jing, X.; Chu, Y.; Wang, L.; Kang, W.; Wei, D.; Li, H.; Xiong, S. Nitrogen/oxygen co-doped monolithic carbon electrodes derived from melamine foam for high-performance supercapacitors. *J. Mater. Chem. A* **2018**, *6*, 17730–17739. [CrossRef]

© 2020 by the authors. Licensee MDPI, Basel, Switzerland. This article is an open access article distributed under the terms and conditions of the Creative Commons Attribution (CC BY) license (http://creativecommons.org/licenses/by/4.0/).

Article

Effect of Varying Amine Functionalities on CO_2 Capture of Carboxylated Graphene Oxide-Based Cryogels

Alina I. Pruna [1], Arturo Barjola [1], Alfonso C. Cárcel [1], Beatriz Alonso [2] and Enrique Giménez [1,*]

[1] Instituto de Tecnología de Materiales, Universitat Politècnica de València (UPV), Camino de Vera s/n, 46022 Valencia, Spain; apruna@itm.upv.es (A.I.P.); arbarrui@doctor.upv.es (A.B.); acarcel@upv.es (A.C.C.)
[2] Graphenea S.A., Paseo Mikeletegi 83, 20009 San Sebastián, Spain; b.alonso@graphenea.com
* Correspondence: enrique.gimenez@mcm.upv.es

Received: 12 May 2020; Accepted: 21 July 2020; Published: 24 July 2020

Abstract: Graphene cryogels synthesis is reported by amine modification of carboxylated graphene oxide via aqueous carbodiimide chemistry. The effect of the amine type on the formation of the cryogels and their properties is presented. In this respect, ethylenediamine (EDA), diethylenetriamine (DETA), triethylenetetramine (TETA), were selected. The obtained cryogels were characterized by Fourier Transformed Infrared spectroscopy, thermogravimetric analysis, X-ray spectroscopy, and Scanning electron microscopy. The CO_2 adsorption performance was evaluated as a function of amine modification. The results showed the best CO_2 adsorption performance was exhibited by ethylenediamine modified aerogel, reaching 2 mmol g^{-1} at 1 bar and 298 K. While the total N content of the cryogels increased with increasing amine groups, the nitrogen configuration and contributions were determined to have more important influence on the adsorption properties. It is also revealed that the residual oxygen functionalities in the obtained cryogels represent another paramount factor to take into account for improving the CO_2 capture properties of amine-modified graphene oxide (GO)-based cryogels.

Keywords: graphene oxide; amine; cryogel; CO_2 capture

1. Introduction

The developments in nanotechnology and innovations in graphene research indicated graphene cryogels as a novel class of three-dimensional (3D) architecture with prodigious potential in varying applications including CO_2 capture, energy storage or pollutant adsorption. The myriad of applications arise from the outstanding properties of these cryogels such as their low density, high specific area, mechanical strength and electrical conductivity [1–4].

The most effective approach to integrate graphene into such bulk materials is the self-assembly of graphene oxide (GO) sheets [5–7]. In this respect, one of the most used and preferred methods is the hydrothermal one, due to its low cost and easy implementation. By applying the simultaneous reduction and self-assembly of GO in the presence of amines and a subsequent freeze-drying procedure, aerogels could be synthesized to tackle the urgent environmental matter of highly selective and efficient CO_2 capture [8–14].

The practical implementation of amine bulk sorbent materials generally requires a high surface area, high pore volume, high amine content, amine stability, etc. [15–17]. However, there are large discrepancies in the most affecting parameters towards tailoring the CO_2 capture properties. For example, while the specific surface area of a polyethyleneimine-modified GO was reportedly low, its CO_2 adsorption capacity reached values as high as 1.9 mmol g^{-1} at 298 K and 1 bar [18].

The approaches to improve the structure and adsorption properties of these macroscopic materials mainly refer to adjusting the properties of GO and to the control of the functionalization degree. Concerning GO, approaches such as narrowing of GO size distribution [19] by employing different mesh size of the parent graphite for oxidation or the control of the oxidation degree were studied [20]. Our group has previously reported on the effect of simple hydrothermal synthesis conditions on the gelation and formation, as well as the effect of GO synthesis conditions on the CO_2 capture properties of ethylenediamine (EDA)-impregnated GO-based 3D monoliths [8,9]. Other studies considered the impregnation with amines such as EDA for obtaining 3D monoliths with improved conductivity and mechanical properties [21].

Concerning the functionalization of GO, the incorporation of nitrogen atoms is commonly applied in order to generate various active sites. The nitrogen content was reported to be greatly influenced by the choice of solvent [16,17], which is usually an organic one. Various reports indicated the N configuration has marked effect on CO_2 conversion and uptake, selectivity and activity of doped material [8,9,22–27]. The type of N configuration could be tailored by suitable synthesis conditions. On the other hand, the residual oxygen functional groups in the graphene aerogels were shown to participate, as well, to the interaction with CO_2, influencing their adsorption capacity and selectivity [28,29].

The preferred approaches to modify the GO surface with amine molecules include impregnation or covalent functionalization [30]. Despite its simplicity, time and cost efficiency, the physical immobilization of amines achieved by the simple wet impregnation route mostly rely on weak interactions between the amine and GO which could compromise the stability and lifetime of the sorbent. This approach could also affect the level of amine loading, which is known to be directly linked to the CO_2 adsorption capacity [31]. As amine leaching is an aspect that needs to be considered for long-term viability as a CO_2 capture agent [19,32,33], the synthesis conditions and operating ones, such as high temperatures must be carefully selected. A proposed approach to improve the loading and amine stability is to exploit the reactivity of the oxygen groups on GO to covalently functionalize the GO surface, similarly to other sorbents [20,34–36].

Therefore, control of the GO functionalization results essential for the design of aerogels with improved CO_2 adsorption properties. It should be noted that the different protocols employed for GO preparation were reported to induce inhomogeneity between the studied samples. As the GO is decorated with different oxygen groups, varying simultaneous derivatization reactions, including epoxy ring opening and the amidation of carboxylic acids of GO, may be induced [15] and, as such, the efficiency evaluation of such reactions is a difficult task. Under these aspects, the carboxylation of GO was suggested as a suitable approach to obtain a precursor material with a similar oxidation degree for the functionalization [37]. Moreover, the amine grafting ability of the oxygen groups on GO was shown to be higher for carboxyl ones [38].

On the other hand, a high nitrogen content and desired nitrogen configuration could be achieved by adjusting the chain length and the number of functionalities in the amine employed for functionalization [16]. These parameters were shown to affect the CO_2 uptake performance also in terms of suitable space between the functionalized GO layers for the reaction with the gas [39]. For example, EDA resulted in better uptake reaching 1.46 mmol g^{-1} than its counterparts functionalized with butanediamine and hexanediamine [40]. The type of amine is another important aspect as well, with the secondary types being indicated as a compromise between primary and tertiary ones in terms of sorbent regeneration [20].

In this work, a systematic study on the amine modification of GO cryogels is presented to enhance the understanding on the effect of functional groups in amine-modified graphene cryogels on their CO_2 capture performance, namely residual oxygen functionalities and nitrogen bonding configuration. In general, the literature reports GO covalent modification with amines in harsh conditions of temperature or by complicated processes. Due to this aspect, a more simple approach is highly desired. In this work, the amidation reaction was achieved by simple carbodiimide chemistry as non-toxic alternative and aqueous solvent medium was considered in order to avoid auxiliary procedures, high costs and generation of large chemical wastes. To this purpose, GO was subjected to a carboxylation process and further modified with varying amines, namely ethylenediamine (EDA), diethylenetriamine (DETA), and triethylentetraamine (TETA) in order to introduce varying nitrogen content and configuration into the cryogels. A variation of GO with increased oxidation degree was tested in order to confirm the results on the effect of residual oxygen groups. The results show that the CO_2 capture performance is greatly influenced by the N configuration and residual oxygen functional groups.

2. Materials and Methods

2.1. Materials

Aqueous slurry of GO nanosheets (3.73 mg mL^{-1}) was provided by Graphenea (Donostia, Spain). The improved oxidation degree GO dispersion (1.5 mg mL^{-1}) was supplied as well by Graphenea. Further information on the characteristics of the dispersions is available on the provider website. Chloroacetic acid (ClCH$_2$COOH, ≥99.7%), hydrocloric acid (HCl, 37%), sodium hydroxide (NaOH, ≥98%), ethylenediamine (EDA, 99%), diethylenetriamine (DETA, 99%), triethylenetetramine (TETA, ≥97%), 1-Ethyl-3-(3-dimethylaminopropyl) carbodiimide (EDC, ≥98%) and N-hydroxysulfosuccinimide (S-NHS, ≥98%) reagent grade were purchased from Sigma Aldrich (Valencia, Spain) and used as received.

2.2. Carboxylation of GO

Prior to use, the GO aqueous dispersions (2 mg mL^{-1}) were prepared from the aqueous slurry by ultrasonic bath treatment for 1 h. Carboxylated GO nanosheets (named GO-COOH hereafter) were obtained through reaction with chloroacetic acid under strong basic conditions, to convert oxygen containing groups of GO to carboxylic groups [41]. Briefly, 10 g of chloroacetic acid and 12.8 g of NaOH were added in a 100 ml of a 2 mg mL^{-1} GO dispersion and ultrasonicated in a water bath for 3 h at room temperature. The resulting solution was neutralized with HCl and then purified by repeated rinsing with distilled water and filtration. The product obtained was finally dried in an oven at 50 °C. The improved oxidation degree GO was subjected to the same carboxylation process and employed for further use.

2.3. Functionalization of GO-COOH with Amines

The schematics of the synthesis of the cryogels can be found in Figure S3 in the Supplementary Materials. For the amine functionalization of GO-COOH by carbodiimide chemistry, the EDC was employed to activate the COOH groups and further form the succinimide ester by reacting with S-NHS. First, 40 mg of GO-COOH was suspended in 20 mL of deionized water and ultrasonicated in a bath for 1 h. Then 40 mg of EDC and 40 mg of S-NHS were added to the above suspension and mixed. Next, the amine modifier was incorporated under magnetic stirring in a 1:5 wt. GO-COOH/modifier ratio and was left overnight with continuous stirring in ice bath. Then suspension was heated in an oven at 85 °C for 4 h and hydrogel was obtained. The same preparation procedure was used to functionalize the GO-COOH with EDA, DETA and TETA amine modifiers. Finally, the hydrogel was rinsed thoroughly with distilled water and ethanol.

2.4. Methods

Three-dimensional porous cryogels were obtained from their respective hydrogels by freeze-drying at −80 °C under a high vacuum at 0.05 mbar in a LyoQuest freeze-drier (Telstar, Madrid, Spain) with three-directional cooling with a rate of about 12 degrees min^{-1} followed by sublimation at 20 °C for 48 h at 0.015 mbar.

The apparent density of the cryogels considered their weight and volume. The cryogel volume was measured with a calliper with an accuracy of 0.05 mm (measurement error ±10%.).

The FTIR spectra were acquired on a FT/IR-6200 (Jasco, Madrid, Spain) spectrometer in the spectral window of 4000–400 cm^{-1} in attenuated total reflectance ATR mode, using a resolution of 4 cm^{-1} and 32 scans.

Thermogravimetric analysis (TGA) was performed on a TGA Q50 thermogravimetric analyzer (TA Instruments, Cerdanyola del Valles, Spain). Samples (5–10 mg) were weighed in titanium crucibles and heated under nitrogen atmosphere from 50 to 800 °C at a heating rate of 10 °C min^{-1}.

X-ray photoelectron spectroscopy (XPS) measurements were performed on the powder samples using a spectrometer (VG-Microtech Multilab 3000) equipped with a monochromatic Al X-ray source (1486.6 eV) (Thermo Fisher Scientific Inc., Waltham, MA, USA). The calibration for the surface charging of the binding energy was performed with reference to the C1s peak binding energy. Curve deconvolution for atomic composition was performed using CASAXPS 2.3.17 software (Casa Software Ltd. Wilmslow, Cheshire, UK) by applying a Shirley baseline subtraction and a Gaussian–Lorentzian (70%:30%) peak shape.

The nitrogen and CO_2 adsorption/desorption isotherms were performed on ASAP 2420 analyzer (Micromeritics, Norcross, GA, USA). The samples were outgassed under vacuum at 80 °C for 24 h before adsorption measurement. The Brunauer–Emmett–Teller (BET) approach was employed to obtain the specific surface areas of modified cryogels from the nitrogen adsorption isotherms measured at 77 K. The CO_2 adsorption isotherms of the modified cryogels were recorded up to 1 bar at varying operating temperatures.

3. Results and Discussion

3.1. Synthesis and Characterization of GO-COOH

A carboxylation procedure was applied to the parent GO nanomaterial. This approach was applied to normalize the content of oxygen in the GO subjected to amine modification. Moreover, this process was indicated to improve aqueous dispersion stability with respect to GO [42,43], which is highly desirable for the solvo-thermal synthesis of 3D GO-based aerogel structures. Figure 1A shows the TGA curves recorded for GO-COOH and GO nanomaterials. GO starts its decomposition at about 170 °C and continues up to about 300 °C. A second pronounced weight loss occurs around 500 °C. GO-COOH on the other hand starts its decomposition earlier at about 120 °C and loses around 20% of its mass up to 170 °C. Then, a slight but gradual weight loss appears. Both materials show an initial weight loss associated with some structural water evaporation and thermal decomposition of more labile oxygen-containing functional groups [44–46]. In the case of GO-COOH, this decomposition starts at a lower temperature due to the higher density of carboxylic acid functional groups attached to its surface. The slight gradual weight loss above 200 °C is attributed to the degradation of more stable oxygen functionalities. Furthermore, at about 500 °C a significant weight loss is observed [47,48].

Figure 1B shows FTIR spectra of GO and GO-COOH, in both spectra the characteristic peaks associated with graphene oxide can be seen. The peak for C=O stretching in carboxylic acid appears at 1728 cm^{-1} and the C=C stretching in aromatic rings occurs at 1618 cm^{-1}. Peaks related with C–O–C from epoxy and C–OH from carboxylic groups at 1032 and 1357 cm^{-1} are also showed. Finally, at 3148 cm^{-1} and 2780 cm^{-1} it can be observed a broad band from the stretching vibrations of O–H and C–H bonds [49,50]. From Figure 1B, we can see how the peaks associated with the carboxyl groups at 1728 and 1357 cm^{-1} show a clear increase in intensity. Furthermore, the peak related to the aromatic domains of GO at 1618 cm^{-1} also increases its relative intensity with respect to the rest. The carboxylation process appears to result in the partial removal of some oxygen groups and most probably their transformation. However, the GO-COOH exhibited good dispersion stability, in agreement with other reports [42,43].

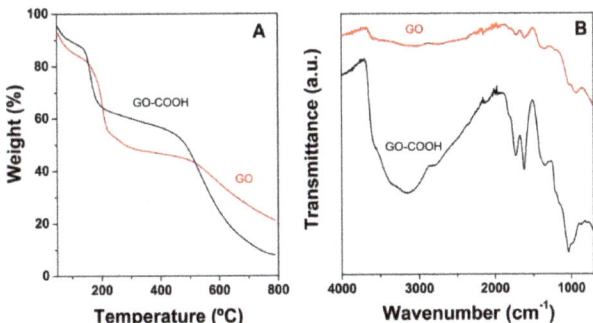

Figure 1. Thermogravimetric curves (**A**) and FTIR spectra (**B**) for graphene oxide (GO) and GO-COOH.

XPS analysis was further employed to determine the species modified by the carboxylation process. The evolution of the survey and C 1s spectra of GO with carboxylation are presented in Figure 2. The fitting curves employed for the deconvolution of the C 1s peak have a binding energy located at about 284.6, 286 and 288.6 eV and are assigned to C=C/C–C, C–OH/C–O–C and O=C–O, respectively [51]. As can be observed from the corresponding contributions of the deconvoluted C 1s peak, the carboxylic carbon contribution increased upon carboxylation of GO from 7.5 at% to 13.1 at% while some oxygen groups got lost, which resulted in about a 23% increase in C/O ratio, from 2.1 to 2.6 for the GO and GO-COOH, respectively. Thus, the conversion of oxygen groups into others is suggested.

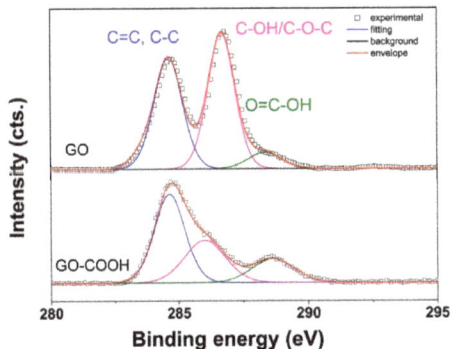

Figure 2. XPS C 1s spectra of GO and GO-COOH.

3.2. Formation of Amine-Modified GO-COOH Cryogels

The effect of functionalization with amines of the GO-COOH was further studied on the volume and density of the modified cryogels and it is depicted in Figure 3. It can be seen that both the volume and the density increase with the molecular mass of the amine incorporated in the cryogel, showing the trend EDA < DETA < TETA for the two properties studied. Digital images of the obtained cryogels are available in Figure S4 in the Supplementary Materials.

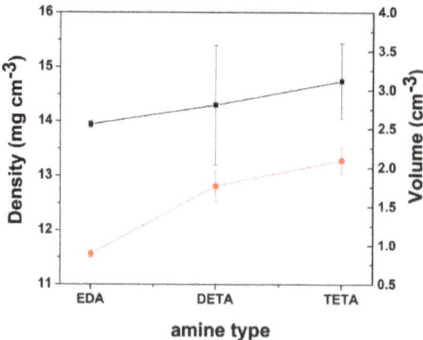

Figure 3. Modified cryogel volume and density with the amine functionalities.

To explain the obtained trend, two effects associated to the amines must be taken into account. On the one hand, there is the thermal and chemical reduction induced by the amine [14,52] that leads to GO-COOH stacking. On the other hand, there is the effect of the amine molecules introduced between the GO-COOH sheets, which prevent their stacking by acting like spacers [53–55]. In this sense, EDA offers the best reducing capacity and, at the same time, is the smallest molecule with the lowest weight.

3.3. Characterization of Amine-Modified GO-COOH Cryogels

The analysis of the effect of the molecular structure of the amines on the morphology of the modified cryogels is further depicted in Figure 4. The SEM measurements indicated that the homogeneity of the distributed sheets decreased, while their stacking increased with the amine functionalities of the modifier, which suggest an improved porosity in the EDA-modified GO-COOH cryogels, while the DETA and TETA-modified ones show enlarged, irregular pores.

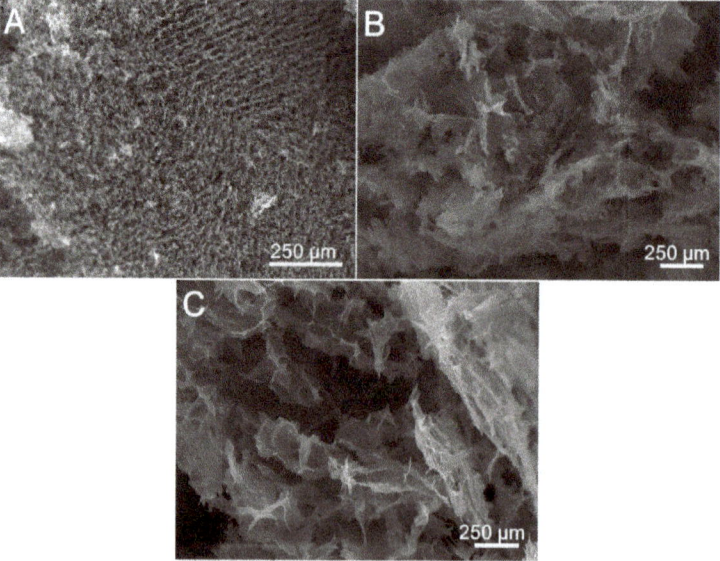

Figure 4. SEM images of GO-COOH cryogels modified with: (**A**) ethylenediamine (EDA) (**B**) diethylenetriamine (DETA) and (**C**) triethylenetetramine (TETA).

Figure 5 shows the TGA curves of amine modified GO-COOH-based cryogels. It can be seen that the amine-functionalized GO-COOH displays a similar profile to that of pristine GO-COOH, while also displaying an improved thermal stability. The amine-functionalized cryogels show a shift towards higher temperatures in the main weight loss compared to pristine GO-COOH, starting decomposition at about 200 °C and going up to 400 °C. However, above 400 °C, only EDA-modified cryogel exhibits enhanced thermal stability, while DETA and TETA-modified GO-COOH show less stability than pristine GO-COOH. The weight loss around 200 °C can be attributed to the degradation of the less stable oxygenated functional groups, as well as to the amine bond [41,42]. Above this temperature, the weight loss is very small and it can be associated with the more stable functional oxygen group decomposition [47,48]. The cryogel functionalized with EDA presents a more improved stability than the rest, indicating a greater number of molecules linked by covalent bonding to the carboxyl groups, forming amide bonds [56]. It was shown that EDA acts as a more efficient reducer for the oxygen functionalities of GO-COOH [49,50]. DETA- and TETA-modified cryogels show a similar and lower thermal stability.

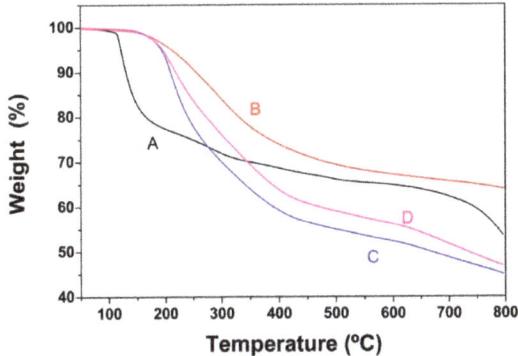

Figure 5. Thermogravimetric curves of GO-COOH before (**A**) and after functionalization with EDA (**B**), DETA (**C**) and TETA (**D**).

The XPS analysis of the modified GO-COOH-based cryogels confirmed the successful functionalization with the amines. Figure 6 depicts the C 1s and N 1s spectra for the GO-COOH upon modification with EDA, DETA and TETA, respectively. The results indicated higher atomic C content and lower O atomic content of the cryogels, thus resulting in increased C/O ratio, reaching the values 3.6, 3.8 and 5.1 for the cryogel modified with EDA, DETA and TETA, respectively. This could be explained by the partial reduction in GO-COOH upon modification with the amines [15] and the increased functionalization degree with the modifier, as the amine functional groups increase. The deconvolution of the C 1s peak presented in Figure 6A indicated that the carbon atoms are present in the form of aromatic rings, with the binding energy of the peak located at 284.6 eV, C–OH/C–O–C compared to the binding energy located at about 286 eV, which is also attributed to the C–N peak due to functionalization, C=O with a binding energy at 287.8 eV and O=C–O with a binding energy located at about 288.6 eV [51,57]. The carboxyl content in the amine-modified cryogels decreased with respect to the GO-COOH material, which is attributed to the functionalization with amines. The EDA-modified cryogel appears to exhibit a peak assigned to C=O which is absent in the other cryogels.

On the other hand, the degree of N-doping and nitrogen configuration were studied by XPS, as depicted in Figure 6B,C. The results showed that both total nitrogen content and configurations were strongly dependent on the modifier's molecular structure. The total N content increased with molecular structure of the modifier, in agreement with the increase in amine functionalities. The N content evaluated by EDAX measurements showed a similar trend, namely it decreased in the order of EDA < DETA < TETA, namely 21.7 < 23.1 < 28.1 (spectra available in the Supplementary Materials).

Moreover, the C/(N+O) ratio increased to 2.5 < 2.7 < 2.9 for the GO-COOH cryogels modified with EDA<DETA<TETA, respectively. It is suggested that, simultaneously with the aqueous functionalization with the amines, the functionalities such as epoxides, hydroxyls and carbonyls on GO-COOH transform into carboxylic acids, which further suffer decarboxylation; thus, a partial reduction takes place in the applied temperature conditions [58].

The N 1s XPS peak was deconvoluted in varying bonding configurations, as shown in Figure 6B,C depicting the evolution of N configuration and their contributions with amine type. The deconvolution of N 1s peak generally exhibits 5 peaks, namely N_a or pyridine-N (398.5 eV), N_b or nitrile (399.5 eV), N_c or pyrrolic–N (400.6 eV), N_d or graphitic–N (401.5 eV) and N_e or pyrridinic oxide–N (403 eV) [59–66].

The modification with EDA was observed to result in a dominant contribution from pyrrolic-N followed by the graphitic one, while the DETA and TETA-modified GO-COOH cryogels exhibited much lower contributions from such configurations. Except for the nitrile contribution, the other ones in the N configuration decreased in the order of pyrrolic–N > graphitic–N > N–oxide and in the order EDA > DETA > TETA in the modified GO-COOH cryogels. The lower pyrrolic and graphitic N contributions are most probably induced by the occurrence of N–H bonds, which are expected given the structure of the corresponding amines that have increased amine functionalities with respect to EDA [67,68].

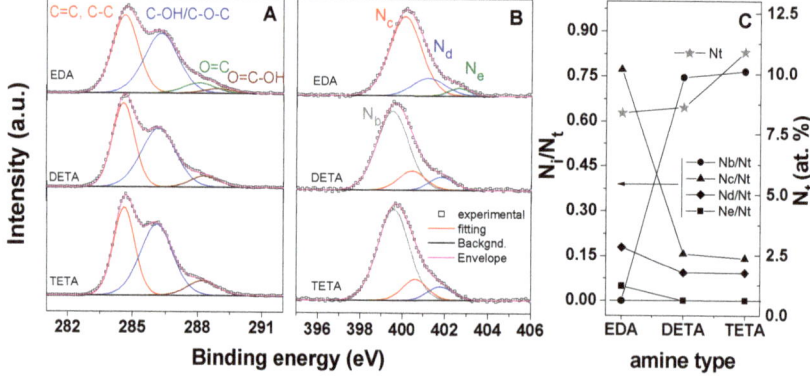

Figure 6. XPS C 1s spectra (**A**), N 1s (**B**) and N configuration contribution (N_i/N_t) and total N content (N_t, at%) for modified cryogels with amine type (**C**).

3.4. CO_2 Adsorption Properties of Amine Modified GO-COOH Cryogels

The cryogels modified with amines by aqueous carbodiimide chemistry were further employed for CO_2 capture measurements. For exemplification, the effect of the molecular structure of the modifier on the CO_2 uptake at 298 K is depicted in Figure 7A. As can be observed, the modified cryogel performs the best when its molecular structure contains less amine functionalities, that is, in the order of EDA > DETA > TETA, as in other reports [69]. Moreover, as Figure 7B shows, the modified cryogels exhibited an increase in the CO_2 capture properties with operating temperature, from 273 K to 298 K. The increase in CO_2 adsorption with the increase in operating temperature from 273 K to 298 K, irrespective of the amine type, could be attributed to enhanced gas molecule mobility, improved pore filling and the activation of active sites in the amine-modified cryogels with the temperature [9].

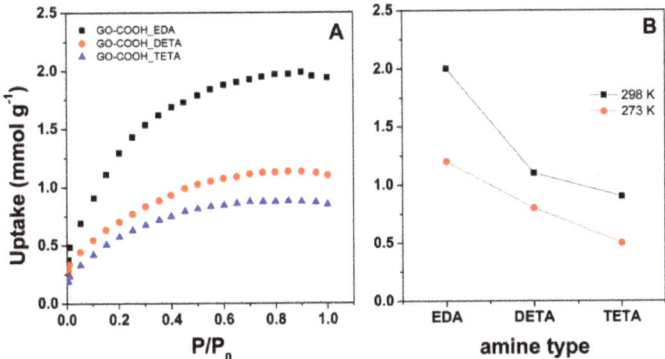

Figure 7. The CO_2 adsorption isotherm at 298 K (**A**) and evolution of CO_2 uptake with temperature (**B**) for the amine-modified cryogels.

The CO_2 uptake evolution with the amine functionalities could be attributed to various factors. On one hand, the homogeneity properties of the cryogel obtained a lower amine group content as a consequence of the improved dispersion and interaction with oxygen functional groups decorating the GO-COOH sheets, as indicated by SEM results. Many reports linked the active surface area to the improved the adsorption performance [70–72]. The surface area obtained from the nitrogen adsorption isotherms at 77 K by the BET theory are presented in Table 1. As can be seen, the surface area increased with the increase in amine functionalities. One may note that the apparent BET surface area values of the cryogels are low and they can be attributed to the low temperature degassing that is used in order to avoid a reduction in GO [73]. Therefore, the BET values could not be directly related to the adsorption properties. Instead, the surface area-normalized uptake (calculated from adsorbed amount at 1bar, 298K divided to BET surface) may be employed to describe the extent of CO_2 uptake [9,74,75]. It is observed that the surface area utilization factor increases with lower amine functional group content in the modifier, reaching an eight-fold increase for EDA compared to TETA.

Table 1. BET surface area and surface area utilization factor at 298 K, 1 bar of GO-COOH cryogels with amine modifier type.

Modifier	BET Surface Area, $m^2\ g^{-1}$	Surface Area Utilization Factor, mmol $CO_2\ m^{-2}$
TETA	42.54	0.012
DETA	25.78	0.029
EDA	21.37	0.094

The N content is known to influence the adsorption properties, not only in terms of configuration type, but also as contribution values to the total N content [76]. In this respect, the pyrrolic or pyridinic-N configurations were identified as the most important in improving the CO_2 adsorption, based on the increase in the basicity character of the aerogel surface due to their presence [77]. However, there are contradictory theories in this regard, pointing either to pyrrolic–N [78] or to pyridinic-N [65] as the most favorable configuration. Moreover, CO_2 adsorption was attributed to take place not only by electrostatic interaction (due to pyrrolic–N and pyridinic–N) but also by dispersion interaction (due to graphitic–N) [79]. In our work, it is shown that the introduction of N atoms into the GO-COOH surface in terms of predominant pyrrolic and graphitic–N greatly enhances the CO_2 adsorption, with the uptake increasing with their contribution, as indicated by XPS results, showing the evolution as pyrrolic–N > graphitic–N > N–oxide with the increase in amine functionalities in the molecular structure of the modifier, namely EDA > DETA > TETA. Moreover, the obtained results show that the CO_2 capture performance decreased with the increase in the reduction degree of the modified cryogel expressed not only as C/O, but also as the C/(N+O) ratio, suggesting the marked influence not only of N content, but also of the residual oxygen functionalities on improving the adsorption properties, in line with other reports on the effect of the extent of the reduction degree on improving the adsorption properties [28,29].

In order to obtain more insight into the effect of oxygen functionalities in improving the adsorption properties, the oxidation conditions of GO were modified so as to introduce more oxygen functionalities. In this respect, the same graphite (previously expanded) was employed as it enhances the accessibility of the oxidizing agents. Table 2 indicates the C/O ratio decreased upon using an expanded graphite, as the oxygen content improved. Although the –COOH% of the higher oxidation degree GO is similar to the previous GO, there is a higher contribution from C–OH/C–O–C that could be successfully exploited to increase the –COOH content upon carboxylation of the new material. As a matter of fact, the carboxylation of the higher oxidation degree GO introduced a 2-fold –COOH contribution with respect to GO-COOH as well as a higher dispersion in the contributions of C–OH/C–O–C and C=O. However, –COOH contribution diminished by functionalization with EDA. A reduction took place, as previously shown, which resulted in a C/O ratio higher than its EDA-modified GO-COOH counterpart. This result could be attributed to the increased functionalization degree, as well as lability of the other oxygen functionalities that underwent transformation and decarboxylation.

Table 2. Evolution of C% and O% atomic composition, –COOH contribution and C/O ratio for higher oxidation degree GO before and upon carboxylation and further EDA modification (XPS based).

Sample	C%	O%	–COOH%	C/O
Higher oxidation degree GO	63.9	34.6	7.6	1.84
Upon carboxylation	65.6	34.4	21.7	1.9
Upon modification with EDA	75.6	17.1	2.1	4.44

Figure 8A depicts the evolution of C 1s spectra for the higher oxidation degree GO and its corresponding carboxylated derivative. The deconvoluted C 1s peak shows similar peaks with previous GO-derivatives, namely the C=C/C–C, the C–OH/C–O–C (286.3 eV, which incorporates contribution from C–N induced by functionalization), C=O (287.6 eV), O=C–O (288.5–288.8 eV) and the π-π * shake-up satellite peak (291.5 eV, due to the sp^2-hybridized C atoms) [58,80,81]. The N 1s spectra of the EDA-modified cryogel obtained from the higher oxidation degree GO is further depicted in Figure 8B. The deconvolution of the N 1s peak indicates the domination of nitriles, as the number of amine groups increased due to enhanced functionalization degree. The occurrence of the nitrile configuration lowers the contributions from pyrrolic and graphitic N to the total N content in comparison to the corresponding counterpart, namely EDA-modified GO-COOH cryogel. The decrease in the pyrrolic and graphitic-N contributions is expected to result in lower CO_2 capture.

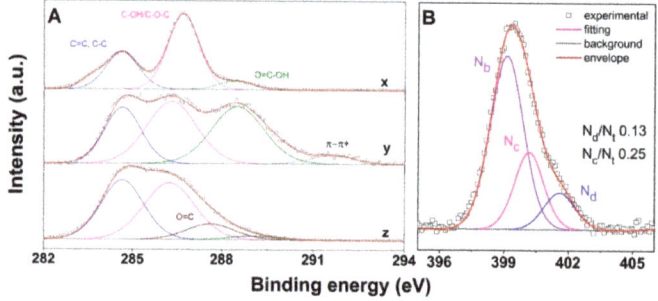

Figure 8. XPS C 1s spectra (**A**) for the higher oxidation degree GO before (x), and after carboxylation (y) and furthermore modification with EDA (z); N 1s spectra upon modification with EDA (**B**).

The CO_2 uptake at 298 K, 1 bar for the EDA-modified new cryogel was obtained as 0.8 mmol g^{-1}, as depicted in Figure 9. The performance is lower with respect to the corresponding GO-COOH counterpart. The evolution of the CO_2 adsorption could be attributed to the lower oxygen functionalities, and decreased contributions from pyrrolic and graphitic-N to the total N content.

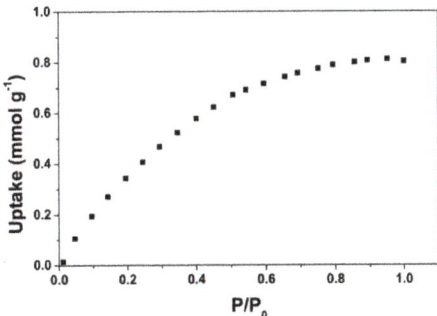

Figure 9. The CO_2 adsorption isotherm of the cryogel obtained from carboxylation and further modification with EDA of the higher degree GO, at 298 K.

4. Conclusions

Varying amine functionalization by aqueous carbodiimide chemistry was employed to modify and introduce N content into the structure of carboxylated GO cryogels in order to study the effect of N bonding configuration on the properties and CO_2 capture performance. The FTIR, TGA and XPS results indicated an increased carboxylic functionality content in the GO-COOH, as the other oxygen groups suffered transformation and partial removal. The functionalization with varying modifiers containing an increased number of amine groups resulted in the further removal of oxygen groups, simultaneous with the introduction of increasing N total content. The XPS analysis revealed a marked influence of the residual oxygen groups and the pyrrolic and graphitic–N bonding configurations of the modified cryogels on their CO_2 uptake—that is, the best performance was obtained by the cryogel with the lowest C/O and C/(N+O) ratios and highest contribution to total N content from the pyrrolic-N as the dominant configuration followed by graphitic-N, namely the EDA-modified GO-COOH based cryogel. The active surface utilization factor confirmed the decrease in CO_2 uptake performance in the order of EDA > DETA > TETA. The results obtained in this work show promising alternatives in addressing the task of improving the CO_2 capture performance of novel amine-modified graphene aerogel.

Supplementary Materials: Supplementary data are available online at http://www.mdpi.com/2079-4991/10/8/1446/s1.

Author Contributions: Conceptualization, A.C.C., A.I.P. and E.G.; methodology, A.C.C., A.I.P. and E.G.; investigation, A.C.C., A.I.P., A.B., B.A. and E.G.; writing—original draft preparation, A.I.P., A.B.; writing—review and editing, A.I.P. A.C.C. and E.G. All authors have read and agreed to the published version of the manuscript.

Funding: This research was supported by the European Commission through the contract no. H2020-LCE-24-2016-727619.

Conflicts of Interest: The authors declare no conflict of interest.

References

1. Li, J.; Li, J.; Meng, H.; Xie, S.; Zhang, B.; Li, L.; Ma, H.; Zhang, J.; Yu, M. Ultra-light, compressible and fire-resistant graphene aerogel as a highly efficient and recyclable absorbent for organic liquids. *J. Mater. Chem. A* **2014**, *2*, 2934–2941. [CrossRef]
2. Nardecchia, S.; Carriazo, D.; Ferrer, M.L.; Gutiérrez, M.C.; Del Monte, F. Three dimensional macroporous architectures and aerogels built of carbon nanotubes and/or graphene: Synthesis and applications. *Chem. Soc. Rev.* **2013**, *42*, 794–830. [CrossRef] [PubMed]

3. Cong, H.-P.; Ren, X.-C.; Wang, P.; Yu, S.-H. Macroscopic Multifunctional Graphene-Based Hydrogels and Aerogels by a Metal Ion Induced Self-Assembly Process. *ACS Nano* **2012**, *6*, 2693–2703. [CrossRef] [PubMed]
4. Sui, Z.-Y.; Bao-Hang, H. Effect of surface chemistry and textural properties on carbon dioxide uptake in hydrothermally reduced graphene oxide. *Carbon* **2015**, *82*, 590–598. [CrossRef]
5. Chen, W.; Yan, L. In situ self-assembly of mild chemical reduction graphene for three-dimensional architectures. *Nanoscale* **2011**, *3*, 3132–3137. [CrossRef]
6. Ai, W.; Du, Z.-Z.; Liu, J.-Q.; Zhao, F.; Yi, M.-D.; Xie, L.-H.; Shi, N.-D.; Ma, Y.-W.; Qian, Y.; Fan, Q.-L.; et al. Formation of graphene oxide gel via the π-stacked supramolecular self-assembly. *RSC Adv.* **2012**, *2*, 12204. [CrossRef]
7. Xeng, K.-X.; Xu, Y.-X.; Li, C.; Shi, G.-Q. High-performance self-assembled graphene hydrogels prepared by chemical reduction of graphene oxide. *New Carbon Mater.* **2011**, *26*, 9–15.
8. Pruna, A.-I.; Cárcel, A.-C.; Benedito, A.; Giménez, E. The Effect of Solvothermal Conditions on the Properties of Three-Dimensional N-Doped Graphene Aerogels. *Nanomaterials* **2019**, *9*, 350. [CrossRef]
9. Pruna, A.; Cárcel, A.C.; Benedito, A.; Giménez, E. Effect of synthesis conditions on CO_2 capture of ethylenediamine-modified graphene aerogels. *Appl. Surf. Sci.* **2019**, *487*, 228–235. [CrossRef]
10. Wang, L.; Park, Y.; Cui, P.; Bak, S.; Lee, H.; Lee, S.-M.; Lee, H. Facile preparation of an n-type reduced graphene oxide field effect transistor at room temperature. *Chem. Commun.* **2014**, *50*, 1224–1226. [CrossRef]
11. Lee, J.-U.; Lee, W.; Yi, J.-W.; Yoon, S.-S.; Lee, S.-S.; Jung, B.-M.; Kim, B.-S.; Byun, J.-H. Preparation of highly stacked graphene papers via site-selective functionalization of graphene oxide. *J. Mater. Chem. A* **2013**, *1*, 12893. [CrossRef]
12. Yu, D.-X.; Wang, A.-J.; He, L.-L.; Yuan, J.; Wu, L.; Chen, J.-R.; Feng, J.-J. Facile synthesis of uniform AuPd@Pd nanocrystals supported on three-dimensional porous N-doped reduced graphene oxide hydrogels as highly active catalyst for methanol oxidation reaction. *Electrochim. Acta* **2016**, *213*, 565–573. [CrossRef]
13. Shu, D.; Feng, F.; Han, H.; Ma, Z. Prominent adsorption performance of amino-functionalized ultra-lightgraphene aerogel for methyl orange and amaranth. *Chem. Eng. J.* **2017**, *324*, 1–9. [CrossRef]
14. Kim, N.H.; Kuila, T.; Lee, J.H. Simultaneous reduction, functionalization and stitching of graphene oxide with ethylenediamine for composites application. *J. Mater. Chem. A* **2013**, *1*, 1349–1358. [CrossRef]
15. Yanga, A.; Li, J.; Zhang, C.; Zhanga, W.; Ma, N. One-step amine modification of graphene oxide to get a green trifunctional metal-free catalyst. *Appl. Surf. Sci.* **2015**, *346*, 443–450. [CrossRef]
16. Dongil, A.B.; Bachiller-Baeza, B.; Rodríguez-Ramos, I.; Guerrero-Ruiz, A. Exploring the insertion of ethylenediamine and bis (3-aminopropyl)amine into graphite oxide. *Nanosci. Methods* **2014**, *3*, 28–39. [CrossRef]
17. Herrera-Alonso, M.; Abdala, A.A.; McAllister, M.J.; Aksay, I.A.; Prud'homme, R.K. Intercalation and stitching of graphite oxide with diaminoalkanes. *Langmuir* **2007**, *23*, 10644–10649. [CrossRef]
18. Shin, G.-J.; Rhee, K.-Y.; Park, S.-J. Improvement of CO_2 capture by graphite oxide in presence of polyethylenimine. *Int. J. Hydrog. Energy* **2016**, *41*, 14351–14359. [CrossRef]
19. Chen, J.; Li, Y.; Huang, L.; Jia, N.; Li, C.; Shi, G. Size fractionation of Graphene Oxide Sheets via Filtration through Track-Etched Membranes. *Adv. Mater.* **2015**, *27*, 1–7. [CrossRef]
20. Wang, J.; Huang, L.; Yang, R.; Zhang, Z.; Wu, J.; Gao, Y.; Wang, Q.; O'Hare, D.; Zhong, Z. Recent advances in solid sorbents for CO_2 capture and new development trends. *Energy Environ. Sci.* **2014**, *7*, 3478–3518. [CrossRef]
21. Hu, H.; Zhao, Z.; Wan, W.; Gogotsi, Y.; Qiu, J. Ultralight and Highly Compressible Graphene Aerogels. *Adv. Mater.* **2013**, *25*, 2219–2223. [CrossRef] [PubMed]
22. Chai, G.-L.; Guo, Z.-X. Highly effective sites and selectivity of nitrogen-doped graphene/CNT catalysts for CO_2 electrochemical reduction. *Chem. Sci.* **2016**, *7*, 1268–1275. [CrossRef] [PubMed]
23. Fiorentin, M.-R.; Gaspari, R.; Quaglio, M.; Massaglia, G.; Saracco, G. Nitrogen doping and CO_2 adsorption on graphene: A thermodynamical study. *Phys. Rev. B* **2018**, *97*, 155428. [CrossRef]
24. Xing, T.; Zheng, Y.; Li, L.-H.; Cowie, B.-C.-C.; Gunzelmann, D.; Qiao, S.-Z.; Huang, S.; Chen, Y. Observation of active sites for oxygen reduction reaction on nitrogen-doped multilayer graphene. *ACS Nano* **2014**, *8*, 6856–6862. [CrossRef] [PubMed]
25. Su, P.; Xiao, H.; Zhao, J.; Yao, Y.; Shao, Z.; Li, C.; Yang, Q. Nitrogen-doped carbon nanotubes derived from Zn–Fe-ZIF nanospheres and their application as efficient oxygen reduction electrocatalysts with in situ generated iron species. *Chem. Sci.* **2013**, *4*, 2941. [CrossRef]

26. Lee, J.-W.; Ko, J.-M.; Kim, J.-D. Hydrothermal preparation of nitrogen-doped graphene sheets via hexamethylenetetramine for application as supercapacitor electrodes. *Electrochim. Acta* **2012**, *85*, 459–466. [CrossRef]
27. Wu, J.; Yadav, R.-M.; Liu, M.; Sharma, P.-P.; Tiwary, C.-S.; Ma, L.; Zou, X.; Zhou, X.-D.; Yakobson, B.-I.; Lou, J.; et al. Achieving highly efficient, selective, and stable CO_2 reduction on nitrogen-doped carbon nanotubes. *ACS Nano* **2015**, *9*, 5364–53718. [CrossRef]
28. Plaza, M.G.; Thurecht, K.J.; Pevida, C.; Rubiera, F.; Drage, T.C. Influence of oxidation upon the CO2 capture performance of a phenolic-resin-derived carbon. *Fuel Process.Technol.* **2013**, *110*, 53–60. [CrossRef]
29. Liu, S.; Peng, W.; Sun, H.; Wang, S. Physical and chemical activation of reduced graphene oxide for enhanced adsorption and catalytic oxidation. *Nanoscale* **2014**, *6*, 766–771. [CrossRef]
30. Li, L.; Song, S.; Maurer, L.; Lin, Z.; Lian, G.; Tuan, C.-C.; Moon, K.-S.; Wong, C.-P. Molecular engineering of aromatic amine spacers for high-performance graphene-based supercapacitors. *Nano Energy* **2016**, *21*, 276–294. [CrossRef]
31. Sayari, A.; Heydari-Gorji, A.; Yang, Y. CO2 -induced degradation of amine-containing adsorbents: Reaction products and pathways. *J. Am. Chem. Soc.* **2012**, *134*, 13834–13842. [CrossRef] [PubMed]
32. Wang, M.; Wang, Z.; Wang, J.; Zhu, Y.; Wang, S. An antioxidative composite membrane with the carboxylate group as a fixed carrier for CO2 separation from flue gas. *Energy Environ. Sci.* **2011**, *4*, 444. [CrossRef]
33. Young, P.-D.; Notestein, J.-M. The Role of Amine Surface Density in Carbon Dioxide Adsorption on Functionalized Mixed Oxide Surfaces. *ChemSusChem* **2011**, *4*, 1671–1678. [CrossRef]
34. Samanta, A.; Zhao, A.; Shimizu, G.-K.-H.; Sarkar, P.; Gupta, R. Post-combustion CO2 capture using solid sorbents: A review. *Ind. Eng. Chem. Res.* **2012**, *51*, 1438–1463. [CrossRef]
35. Georgakilas, V.; Otyepka, M.; Bourlinos, A.-B.; Chandra, V.; Kim, N.; Kemp, C.; Hobza, P.; Zboril, R.; Kim, K.-S. Functionalization of graphene: Covalent and non-covalent approaches, derivatives and applications. *Chem. Rev.* **2012**, *112*, 6156–6214. [CrossRef] [PubMed]
36. Ahmed, M.-S.; Kim, Y.-B. 3D graphene preparation via covalent amide functionalization for efficient metal-free electrocatalysis in oxygen reduction. *Sci. Rep.* **2017**, *7*, 43279. [CrossRef]
37. Xie, B.; Chen, Y.; Yu, M.; Shen, X.; Lei, H.; Xie, T.; Zhang, Y.; Wu, Y. Carboxyl-Assisted Synthesis of Nitrogen-Doped Graphene Sheets for Supercapacitor Applications. *Nanoscale Res. Lett.* **2015**, *10*, 332. [CrossRef] [PubMed]
38. Wen, Z.; Chen, W.; Li, Y.; Xu, J. A Theoretical Mechanism Study on the Ethylenediamine Grafting on Graphene Oxides for CO2 Capture. *Arab. J. Sci. Eng.* **2018**, *43*, 5949–5955. [CrossRef]
39. Xu, J.; Xing, W.; Zhao, L.; Guo, F.; Wu, X.; Xu, W.; Yan, Z. The CO2 Storage Capacity of the Intercalated Diaminoalkane Graphene Oxides: A Combination of Experimental and Simulation Studies. *Nanoscale Res. Lett.* **2015**, *10*, 318. [CrossRef] [PubMed]
40. Cai, J.; Chen, J.; Zeng, P.; Pang, Z.; Kong, X. Molecular Mechanisms of CO2 Adsorption in Diamine-Cross-Linked Graphene Oxide. *Chem. Mater.* **2019**, *31*, 3729–3735. [CrossRef]
41. Ciobotaru, C.C.; Damian, C.M.; Matei, E.; Ionu, H. Covalent functionalization of graphene oxide with cisplatin. *Mater. Plast.* **2014**, *51*, 75–80.
42. Imani, R.; Emami, S.-H.; Faghihi, S. Nano-graphene oxide carboxylation for efficient bioconjugation applications: A quantitative optimization approach. *J. Nanopart Res.* **2015**, *17*, 88. [CrossRef]
43. Yuan, Y.; Gao, X.; Wei, Y.; Wang, X.; Wang, J.; Zhang, Y.; Gao, C. Enhanced desalination performance of carboxyl functionalized graphene oxide nanofiltration membranes. *Desalination* **2017**, *405*, 29–39. [CrossRef]
44. Mallakpour, S.; Abdolmaleki, A.; Borandeh, S. Covalently functionalized graphene sheets with biocompatible natural aminoacids. *Appl. Surf. Sci.* **2014**, *307*, 533–542. [CrossRef]
45. Fang, M.; Wang, K.G.; Lu, H.B.; Yang, Y.L.; Nutt, S. Covalent polymer functionalization of graphene nano-sheets and mechanical properties of composites. *J. Mater. Chem.* **2009**, *19*, 7098. [CrossRef]
46. Fang, M.; Wang, K.G.; Lu, H.B.; Yang, Y.L.; Nutt, S. Single-layer graphene nano-sheets with controlled grafting of polymer chains. *J. Mater. Chem.* **2010**, *20*, 1982. [CrossRef]
47. Shen, J.; Li, T.; Shi, M.; Li, N.; Ye, M. Polyelectrolyte-assisted one-step hydrothermal synthesis of Ag-reduced graphene oxide composite and its antibacterial properties. *Mater. Sci. Eng. C* **2012**, *32*, 2042–2047. [CrossRef]
48. Shen, J.; Hu, Y.; Shi, M.; Lu, X.; Qin, C.; Li, C.; Ye, M. Fast and Facile Preparation of Graphene Oxide and Reduced Graphene Oxide Nanoplatelets. *Chem. Mater.* **2009**, *21*, 3514–3520. [CrossRef]

49. Verma, S.; Dutta, R.K. A facile method of synthesizing ammonia modified graphene oxide for efficient removal of uranyl ions from aqueous medium. *RSC Adv.* **2015**, *5*, 77192–77203. [CrossRef]
50. Song, B.; Li, L.; Lin, Z.; Wu, Z.K.; Moon, K.S.; Wong, C.P. Water-dispersible graphene/polyaniline composites for flexible micro-supercapacitors with high energy densities. *Nano Energy* **2015**, *16*, 470–478. [CrossRef]
51. Shao, L.; Bai, Y.; Huang, X.; Gao, Z.; Meng, L.; Huang, Y.; Ma, J. Multi-walled carbon nanotubes (MWCNTs) functionalized with amino groups by reacting with supercritical ammonia fluids. *J. Mater. Chem. Phys.* **2009**, *116*, 323–326. [CrossRef]
52. Vrettos, K.; Karouta, N.; Loginos, P.; Donthula, S.; Gournis, D.; Georgakilas, C. The role of diamines in the formation of graphene aerogels. *Front. Mater.* **2018**, *5*, 20. [CrossRef]
53. Song, B.; Zhao, J.; Wang, M.; Mullavey, J.; Zhu, Y.; Geng, Z.; Chen, D.; Ding, Y.; Moon, K.S.; Liu, M.; et al. Systematic study on structural and electronic properties of diamine/triamine functionalized graphene networks for supercapacitor application. *Nano Energy* **2017**, *31*, 183–193. [CrossRef]
54. Mungse, H.P.; Singh, R.; Sugimura, H.; Kumar, N.; Khatri, O.P. Molecular pillar supported graphene oxide framework: Conformational heterogeneity and tunable d-spacing. *Phys. Chem. Chem. Phys.* **2015**, *17*, 20822–20829. [CrossRef] [PubMed]
55. Chen, P.; Yang, J.J.; Li, S.S.; Wang, Z.; Xiao, T.Y.; Qian, Y.H.; Yu, S.H. Hydrothermal synthesis of macroscopic nitrogen-doped graphene hydrogels for ultrafast supercapacitor. *Nano Energy* **2013**, *2*, 249–256. [CrossRef]
56. Arrigo, R.; Haevecker, M.; Wrabetz, S.; Blume, R.; Lerch, M.; McGregor, J.; Parrott, E.P.J.; Zeitler, J.A.; Gladden, L.F.; Knop-Gericke, A.; et al. Tuning the acid/base properties of nanocarbons by functionalization via amination. *J. Am. Chem. Soc.* **2010**, *132*, 9616–9630. [CrossRef]
57. Gautam, J.; Thanh, T.-D.; Maiti, K.; Kim, N.-H.; Lee, J.-H. Highly efficient electrocatalyst of N-doped graphene-encapsulated cobalt-iron carbides towards oxygen reduction reaction. *Carbon* **2018**, *137*, 358–367. [CrossRef]
58. Hu, K.; Xie, X.; Szkopek, T.; Cerruti, M. Understanding Hydrothermally Reduced Graphene Oxide Hydrogels: From Reaction Products to Hydrogel Properties. *Chem. Mater.* **2016**, *28*, 1756–1768. [CrossRef]
59. Chen, C.M.; Zhang, Q.; Zhao, X.C.; Zhang, B.; Kong, Q.Q.; Yang, M.G.; Yang, Q.H.; Wang, M.Z.; Yang, Y.G.; Schlogl, R.; et al. Hierarchically aminated graphene honeycombs for electrochemical capacitive energy storage. *J. Mater. Chem.* **2012**, *22*, 14076–14084. [CrossRef]
60. Zhang, C.; Hao, R.; Liao, H.; Hou, Y. Synthesis of amino-functionalized graphene as metal-free catalyst and exploration of the roles of various nitrogen states in oxygen reduction reaction. *Nano Energy* **2013**, *2*, 88–97. [CrossRef]
61. Jiang, Z.; Jiang, Z.J.; Tian, X.; Chen, W. Amine-functionalized holey graphene as a highly active metal-free catalyst for the oxygen reduction reaction. *J. Mater. Chem. A* **2014**, *2*, 441–450. [CrossRef]
62. Wang, B.; Luo, B.; Liang, M.; Wang, A.; Wang, J.; Fang, Y.; Chang, Y.; Zhi, L. Chemical amination of graphene oxides and their extraordinary properties in the detection of lead ions. *Nanoscale* **2011**, *3*, 5059–5066. [CrossRef] [PubMed]
63. Zhang, F.; Jiang, H.; Li, X.; Wu, X.; Li, H. Amine-Functionalized GO as an Active and Reusable Acid–Base Bifunctional Catalyst for One-Pot Cascade Reactions. *ACS Catal.* **2014**, *4*, 394–401. [CrossRef]
64. Yuan, C.; Chen, W.; Yan, L. Amino-grafted graphene as a stable and metal-free solid basic catalyst. *J. Mater. Chem.* **2012**, *22*, 7456–7460. [CrossRef]
65. Tetsuka, H.; Asahi, R.; Nagoya, A.; Okamoto, K.; Tajima, I.; Ohta, R.; Okamoto, A. Optically tunable amino-functionalized graphene quantum dots. *Adv. Mater.* **2012**, *24*, 5333–5338. [CrossRef] [PubMed]
66. Kumar, G.S.; Roy, R.; Sen, D.; Ghorai, U.K.; Thapa, R.; Mazumder, N.; Saha, S.; Chattopadhyay, K.K. Amino-functionalized graphene quantum dots: Origin of tunable heterogeneous photoluminescence. *Nanoscale* **2014**, *6*, 3384–3391. [CrossRef]
67. Navaee, A.; Salimi, A. Efficient amine functionalization of graphene oxide through the Bucherer reaction: An extraordinary metal-free electrocatalyst for the oxygen reduction reaction. *RSC Adv.* **2015**, *5*, 59874–59880. [CrossRef]
68. Caliman, C.C.; Mesquita, A.F.; Cipriano, D.F.; Freitas, J.C.C.; Cotta, A.A.C.; Macedo, W.A.A.; Porto, A.O. One-pot synthesis of amine-functionalized graphene oxide by microwave-assisted reactions: An outstanding alternative for supporting materials in supercapacitors. *RSC Adv.* **2018**, *8*, 6136–6145. [CrossRef]
69. Zhao, Y.; Dinga, H.; Zhong, Q. Preparation and characterization of aminated graphite oxide for CO_2 capture. *Appl. Surf. Sci.* **2012**, *258*, 4301–4307. [CrossRef]

70. Wang, J.; Chen, H.; Liu, X.; Qiao, W.; Long, D.; Ling, L. Carbon dioxide capture using polyethylenimine-loaded mesoporous carbons. *J. Environ. Sci.* **2013**, *25*, 124. [CrossRef]
71. Chabot, V.; Higgins, D.; Yu, A.; Xiao, X.; Chen, Z.; Zhang, J. A review of graphene and graphene oxide sponge: Material synthesis and applications to energy and the environment. *Energy Environ. Sci.* **2014**, *7*, 1564–1596. [CrossRef]
72. Furukawa, H.; Cordova, K.E.; O'Keeffe, M.; Yaghi, O.M. The Chemistry and Applications of Metal-Organic Frameworks. *Science* **2013**, *341*, 974. [CrossRef] [PubMed]
73. Rouquerol, J.; Llewellyn, P.; Navarrete, R.; Rouquerol, F.; Denoyel, R. Assessing microporosity by immersion microcalorimetry into liquid nitrogen or liquid argon. *Stud. Surf. Sci. Catal.* **2002**, *144*, 171–176.
74. Choi, S.; Drese, J.H.; Jones, C.W. Adsorbent materials for carbon dioxide capture from large anthropogenic point sources. *ChemSusChem* **2009**, *2*, 796–854. [CrossRef]
75. Knöfel, C.; Martin, C.; Hornebecq, V.; Llewellyn, P.L. Study of Carbon Dioxide Adsorption on Mesoporous Aminopropylsilane-Functionalized Silica and Titania Combining Microcalorimetry and in Situ Infrared Spectroscopy. *J. Phys. Chem. C* **2009**, *113*, 21726–21734. [CrossRef]
76. Bacsik, Z.; Atluri, R.; Garcia-Bennett, A.E.; Hedin, N. Temperature-Induced Uptake of CO2 and Formation of Carbamates in Mesocaged Silica Modified with n-Propylamines. *Langmuir* **2010**, *26*, 10013–10024. [CrossRef]
77. Hao, G.-P.; Li, W.-C.; Qian, D.; Lu, A.-H. Rapid synthesis of nitrogen-doped porous carbon monolith for CO2 capture. *Adv. Mater.* **2010**, *22*, 853–857. [CrossRef]
78. Sivadas, D.L.; Vijayan, S.; Rajeev, R.; Ninan, K.N.; Prabhakaran, K. Nitrogen-enriched microporous carbon derived from sucrose and urea with superior CO2 capture performance. *Carbon* **2016**, *109*, 7–18. [CrossRef]
79. Sevilla, M.; Valle-Vigón, P.; Fuertes, A.B. N-doped polypyrrole-based porous carbons for CO2 capture. *Adv. Funct. Mater.* **2011**, *21*, 2781–2787. [CrossRef]
80. Khanra, P.; Uddin, M.-D.; Kim, M.-h.; Kuila, T.; Lee, S.-H.; Lee, J.-H. Electrochemical performance of reduced graphene oxide surface-modified with 9-anthracene carboxylic acid. *RSC Adv.* **2015**, *5*, 6443–6451. [CrossRef]
81. Araujo, M.-P.; Soares, O.-S.-G.-P.; Fernandes, A.-J.-S.; Pereira, M.-F.-R.; Freire, C. Tuning the surface chemistry of graphene flakes: New strategies for selective oxidation. *RSC Adv.* **2017**, *7*, 14290–14301. [CrossRef]

© 2020 by the authors. Licensee MDPI, Basel, Switzerland. This article is an open access article distributed under the terms and conditions of the Creative Commons Attribution (CC BY) license (http://creativecommons.org/licenses/by/4.0/).

Article

Graphene Oxide-Based Silico-Phosphate Composite Films for Optical Limiting of Ultrashort Near-Infrared Laser Pulses

Adrian Petris [1], Ileana Cristina Vasiliu [2,*], Petronela Gheorghe [1,*], Ana Maria Iordache [2], Laura Ionel [1], Laurentiu Rusen [1], Stefan Iordache [2], Mihai Elisa [2], Roxana Trusca [3], Dumitru Ulieru [4], Samaneh Etemadi [5], Rune Wendelbo [5], Juan Yang [6] and Knut Thorshaug [6]

[1] National Institute for Laser, Plasma and Radiation Physics, INFLPR, 409 Atomistilor Street, Magurele, 077125 Ilfov, Romania; adrian.petris@inflpr.ro (A.P.); laura.ionel@inflpr.ro (L.I.); laurentiu.rusen@inflpr.ro (L.R.)

[2] National R&D Institute of Optoelectronics-INOE2000, 409 Atomistilor Street, Magurele, 077125 Ilfov, Romania; ana.iordache@inoe.ro (A.M.I.); stefan.iordache@inoe.ro (S.I.); astatin18@yahoo.com (M.E.)

[3] Department of Science and Engineering of Oxide Materials and Nanomaterials, University POLITEHNICA of Bucharest, 313 Independentei Street, 060042 Bucharest, Romania; truscaroxana@yahoo.com

[4] Sitex 45 SRL, 126 A Erou Iancu Nicolae Street, 077190 Voluntari, Romania; ulierud@yahoo.com

[5] Abalonyx AS, Forskningsveien 1, 0373 Oslo, Norway; Samaneh.e@abalonyx.no (S.E.); rw@abalonyx.no (R.W.)

[6] Department of Materials and Nanotechnology, SINTEF AS, Forskningsveien 1, 0343 Oslo, Norway; juan.yang@sintef.no (J.Y.); Knut.Thorshaug@sintef.no (K.T.)

* Correspondence: icvasiliu@inoe.ro (I.C.V.); petronela.doia@inflpr.ro (P.G.)

Received: 7 July 2020; Accepted: 18 August 2020; Published: 20 August 2020

Abstract: The development of graphene-based materials for optical limiting functionality is an active field of research. Optical limiting for femtosecond laser pulses in the infrared-B (IR-B) (1.4–3 µm) spectral domain has been investigated to a lesser extent than that for nanosecond, picosecond and femtosecond laser pulses at wavelengths up to 1.1 µm. Novel nonlinear optical materials, glassy graphene oxide (GO)-based silico-phosphate composites, were prepared, for the first time to our knowledge, by a convenient and low cost sol-gel method, as described in the paper, using tetraethyl orthosilicate (TEOS), H_3PO_4 and GO/reduced GO (rGO) as precursors. The characterisation of the GO/rGO silico-phosphate composite films was performed by spectroscopy (Fourier-transform infrared (FTIR), Ultraviolet–Visible-Near Infrared (UV-VIS-NIR) and Raman) and microscopy (atomic force microscopy (AFM) and scanning electron microscopy (SEM)) techniques. H_3PO_4 was found to reduce the rGO dispersed in the precursor's solution with the formation of vertically agglomerated rGO sheets, uniformly distributed on the substrate surface. The capability of these novel graphene oxide-based materials for the optical limiting of femtosecond laser pulses at 1550 nm wavelength was demonstrated by intensity-scan experiments. The GO or rGO presence in the film, their concentrations, the composite films glassy matrix, and the film substrate influence the optical limiting performance of these novel materials and are discussed accordingly.

Keywords: graphene oxide; sol-gel; silico-phosphate composite films; optical limiting functionality; ultrashort laser pulses

1. Introduction

The rapid progress in high-power laser sources and the numerous civilian and military applications based on them has led to an appropriate development of optical devices for protection of the human eye and sensitive optical systems.

Passive optical limiting (OL) functionality is based on the nonlinear optical (NLO) absorption process specific to certain NLO materials. Such an OL material shows a linear increase in the transmitted intensity/fluence of the laser beam with the incident one below a certain threshold, while above it, the transmitted intensity/fluence remains constant and independent of that of the incident laser beam. The OL functionality is schematically shown in Figure 1 for an ideal optical limiter (red curve) and for a real one (blue curve).

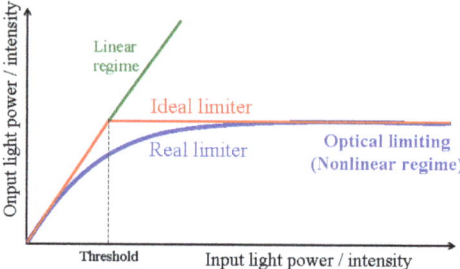

Figure 1. Optical limiting functionality.

In Figure 1, a linear dependence of the transmitted light power/intensity on the same energetic parameters incident on the sample defines a sample characterized by a linear transmittance (green line), with no OL capability. The value of the linear transmittance, T_L, is given by the slope of the corresponding linear dependency.

For a real optical limiter, the desired experimental dependencies of the transmitted light (power/intensity) on the values of the same parameters of the incident beams are not linear. The behaviour is graphically described by the blue line. This type of dependency defines the NLO transmittance, T_{NL}, characterized by a saturation-type curve.

A wide variety of organic and inorganic materials are being studied to achieve efficient OL [1–3]. Graphene has been identified by many industry sectors as a key material that will drive future product development in flexible electronics, smart textiles, biosensors, drug delivery, water filtration, supercapacitors and more, as stated by the Graphene Report 2020 [4]. Lately, graphene has shown great potential as an ideal material for modern photonic, optoelectronic and electronic devices due to its ultrafast carrier relaxation dynamics and ultra-broadband NLO response as a consequence of its extended π-conjugate system and the linear dispersion relation holding for its electronic band structure [5–14].

The optical limiting in carbon-based materials, in particular, in graphene and in its derivatives, has been extensively investigated in the last years [14–17]. The optical limiting functionality of these materials, as suspension, film or bulk, has been mainly studied for visible and near-infrared nanosecond and picosecond laser pulses (for wavelengths shorter than 1100 nm) [18–21] and, to a lesser extent, for femtosecond laser pulses (mostly at 800 nm wavelength) [22–26]. Very few papers have investigated the nonlinear optical absorption and optical limiting of femtosecond laser pulses in the IR-B band (range, 1.4–3 µm), which includes the wavelength of 1550 nm, important for communications [27,28].

In this range of wavelengths, the solvents (water, alcohols and mixture of them) usually used for suspensions of graphene and of its derivatives have larger absorption than in the visible range, favouring the unwanted effect of bubble formation in the cells that optically limit the high-intensity laser beams.

From a practical application point of view, the transformation of the OL properties of graphene-based materials from liquid suspensions to solid-state films with a large NLO effect, low OL threshold, high damage threshold, fast response, broadband spectral response and environmental and mechanical stability is a challenging task [14,29,30].

The progress in graphene-based NLO devices, such as optical limiters, requires the preparation of optically transparent films with controlled thickness and graphene concentration [31,32]. To avoid laser damage to the system, graphene should be embedded into oxide or organic–inorganic matrices because they exhibit a higher damage threshold with respect to organic/polymers [2,33]. Sol-gel chemistry is the most suitable route for preparing homogeneous nanocomposite films from a liquid phase. A significant advantage of the method is the low material synthesis temperature. However, the design of an appropriate synthesis method for doped films via sol-gel is challenging, since the uncontrolled aggregation of the doping moieties often occurs in the precursor sol. A sol-gel synthesis route was reported for the preparation of different graphene-based silica gel glasses with optical limiting properties [29,30,34–36].

H_3PO_4 as a phosphor precursor for silica-phosphate films was reported to form Si–O–P bonds during the sol-gel process [37]. The presence of P_2O_5 in the reduced graphene oxide (rGO)-doped films prepared by sol-gel was reported to yield a more compact graphene-based composite layer [38].

In this paper, we describe the preparation, for the first time to the best of our knowledge, of novel graphene oxide-based silico-phosphate composite glassy materials by the sol-gel method, together with the morphology and structure characterization of the obtained films. We experimentally demonstrate, by intensity-scan experiments, the OL functionality of these NLO materials for ultrashort (~150 fs) laser pulses at the important telecommunication wavelength of 1550 nm, for which there are very few OL reported results. The influence of GO or rGO presence and of their concentrations in the silico-phosphate composite films, of the silico-phosphate matrix as well as of the film substrate, on the linear transmittance and optical limiting performance of these novel materials is discussed. We compare the OL in our samples with several OL results obtained in literature with ns, ps, and fs laser pulses at visible and near infrared wavelengths.

2. Experimental

2.1. Preparation of Silico-Phosphate Films

The sol-gel chemicals for graphene oxide-based silico-phosphate film preparation were as follows: tetraethyl orthosilicate (TEOS, 99% purity, Sigma-Aldrich, Redox Lab Supplies Com S.R.L. Bucharest, Romania) as a precursor for SiO_2, phosphoric acid (H_3PO_4, 85 wt. % in H_2O, Sigma-Aldrich, Redox Lab Supplies Com S.R.L. Bucharest, Romania) as a precursor for P_2O_5, and rGO/GO (as powders, supplied by Abalonyx AS, Oslo, Norway). The compositions of the starting solutions, presented in Table 1, were calculated, aiming to have different concentrations of rGO/GO in the SiO_2-P_2O_5 films.

Table 1. Composition and denomination of samples.

Sample Denomination	Dopant Material	Matrix Composition	
		SiO_2/P_2O_5 (wt. %)	(rGO or GO)/Σ (SiO_2 + P_2O_5) (g/100 g)
1%rGO-SiO_2-P_2O_5	rGO	60/40	1
1.1%rGO-SiO_2	rGO	100/0	1.1
4%rGO-SiO_2-P_2O_5	rGO	60/40	4
1%GO-SiO_2-P_2O_5	GO	60/40	1
1.1%GO-SiO_2	GO	100/0	1.1
4%GO-SiO_2-P_2O_5	GO	60/40	4

The appropriate amount of rGO/GO was dispersed in ethanol and immersed in an ultrasonic bath for 20 min. Then, TEOS was added to the suspension and magnetically stirred for 2 h before adding

H_3PO_4. The reaction mixtures were magnetically stirred for another 24 h and afterwards spin coated onto glass and Indium Tin Oxide (ITO)-coated glass at 2000 rpm for 30 s. The thin films were thermally treated in an oven, first for drying at 200 °C with a heating rate of 5 °C/h and kept for at 200 °C and afterwards sintered at 350 °C with a heating rate of 500 °C/h and kept for 30 min at 350 °C.

A proposed schematic representation of the reactions for the graphene oxide embedment into the SiO_2-P_2O_5 glassy film is presented in Figure 2.

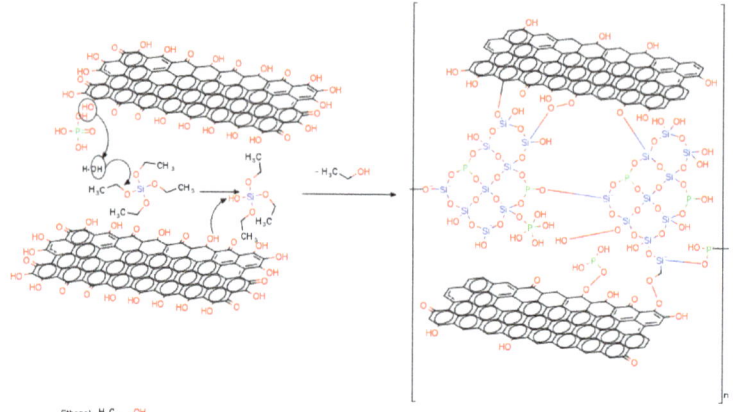

Figure 2. Schematic representation of the process reactions.

The phosphoric acid is involved in more reactions schematically presented as follows:

Hydrolyzation of TEOS: O=P(OH)$_{3-n}$ + RO–Si– $\rightleftharpoons$ =P–(OH)$_{2-n}$(OSi) + R–OH; –Si–O–P– + H_2O $\rightleftharpoons$ –Si–OH + HO–P–

Condensation of reaction' intermediates: –P–OH + –P–OH $\rightleftharpoons$ –P–O–P– + H_2O; –P–OH + –Si–OH $\rightleftharpoons$ –P–O–Si– + H_2O

Reaction with rGO/GO: –C–COOH + H_3PO_4 $\rightleftharpoons$ –C–C=O + H_2O + $H_2PO_4^-$

2.2. Material Characterization

The chemical structure of the samples was investigated using FTIR spectroscopy (with a Spectrum 100 spectrophotometer provided with Universal Attenuated Total Reflectance (UATR) accessory (Perkin Elmer, Llantrisant, UK), in the range 550–4000 cm^{-1}, with a resolution of 4 cm^{-1} and 10 scans, with Atmospheric Vapour Compensation (AVC), and Raman spectroscopy (Nicolet Almega XR, UK-Thermo Fisher Scientific, Oslo, Norway) with an excitation source of λ = 488 nm, a spot of 3 µm diameter and a power of 5 mW at the sample surface. For morphological investigations, atomic force microscopy (AFM) (XE-100 type from Park Systems, Europe GmbH, Mannheim, Germany, non-contact mode) and scanning electron microscopy (SEM) with energy-dispersive X-ray (EDX) analysis using a FEI Inspect F50 system (FEI Europe B.V. Eindhoven, Netherlands) were used. The spectral dependence of the transmittance was investigated using a UV/VIS/NIR spectrophotometer (Perkin Elmer, Lambda 1050, Llantrisant, UK).

The OL capability of the graphene oxide-based silico-phosphate films was investigated by intensity scan (I-scan) experiments [39–43] using ultrashort laser pulses of an Er-doped fibre laser (FemtoFiber Scientific FFS, TOPTICA Photonics AG, Munich, Germany, 1550 nm wavelength, ~150 fs pulse duration). The experimental setup is presented in the Results and Discussion section. In the OL experiments performed on different samples, the transmittance curves (transmitted pulse peak intensity vs. incident one) were determined for a range of average powers incident on the sample increasing up to the maximum value provided by the utilized laser source.

3. Results and Discussion

3.1. FTIR Spectroscopy

The FTIR spectra are presented in Figure 3 and summarized in Table 2.

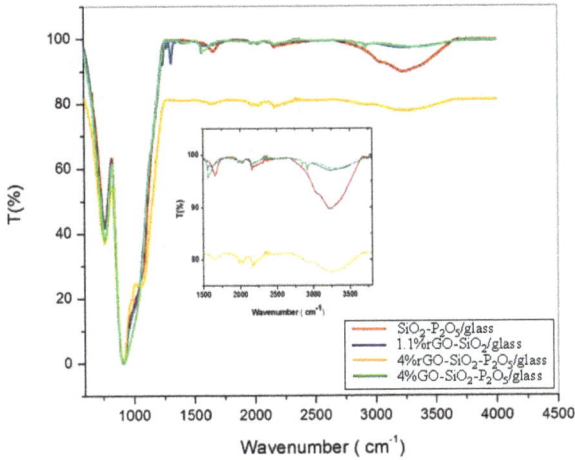

Figure 3. The FTIR spectra of the deposited reduced graphene oxide (rGO)-/GO-containing thin films.

Table 2. Summary of absorption bands in the FTIR spectra of the investigated samples.

Wave Number (cm^{-1})	1.1%rGO-SiO$_2$	4%rGO-SiO$_2$-P$_2$O$_5$	4%GO-SiO$_2$-P$_2$O$_5$	SiO$_2$-P$_2$O$_5$	Assignment
3400	Broad band	Broad band	Broad band	Broad band pronounced	υ(O–H) from H–OH adsorbed, C–OH, CO–OH, Si–OH, P–OH
≈1650	-	Less intense	-	Broad band	(O–H) vibrations of water molecules attached to P–O and GO/rGO bonds, υ (C=C)
~910 Shoulders at: ~1060 ~1100 ~1200	Broad band with shoulders	Broad band with shoulders	Broad band with shoulders	Broad band with shoulders	Si–OH stretching with shoulders: υ_{as} (TO)Si–O–P υ_s (TO)Si–O–Si υ_s (LO) Si–O–Si + υ_s O–P–O
768	Pronounced band	Pronounced band	Pronounced band	Pronounced band	υ_s (Si–O–Si)

The broad peak in the region 2670–3770 cm^{-1} centred around ~3400 cm^{-1} was observed in all the samples, most pronounced for the sample (SiO$_2$-P$_2$O$_5$) without rGO/GO, less pronounced for 1.1%rGO-SiO$_2$ and even less so for samples containing P$_2$O$_5$. This absorption band was assigned to O–H stretching vibrations of hydroxylic, phenolic, carboxylic, P–OH groups and absorbed water molecules. For films containing P$_2$O$_5$, an explanation could be that H$_3$PO$_4$ removes the oxygen-containing functional groups (–OH, C–O and C–OH groups) from the rGO/GO structure, where the main pathway is the protonation of the OH groups followed by H$_2$O elimination. This explanation is in overall agreement with the report from Er and Celikkan [44].

The two small broad peaks near ~2920 and ~2846 cm^{-1} observed in the samples with GO (4%GO-SiO$_2$-P$_2$O$_5$) and in the ones without P–O (1.1%rGO-SiO$_2$) were attributed to the stretching vibrations of C–H in –CH–OH and –CH–COOH belonging to graphene oxide and overlapping with the hydrogen-bonded OH groups of dimeric COOH groups and intra-molecular-bonded O–H stretching of alcohols, respectively [45]. Additionally, small peaks at ~1560 and ~1595 cm^{-1} visible in the FTIR spectrum of the same samples (4%GO-SiO$_2$-P$_2$O$_5$ and 1.1%rGO-SiO$_2$) were attributed to the deformation modes of absorbed water molecules' δ(H–O–H) and O-H groups linked to the –C=O stretching vibration of carboxylic and/or carbonyl moiety functional groups and of skeletal vibrations from un-oxidized graphitic domains from rGO or GO [45–47].

The carboxyl stretching vibrations (C=O) at 1736 cm^{-1} belonging to the rGO/GO were not noticed in any of the samples. However, in the 4%rGO-SiO$_2$-P$_2$O$_5$ sample, the large band at ~1640 cm^{-1} could be due to the shift of the C=O band towards higher wavelengths overlapping with C=C stretching vibrations.

In the 4%rGO-SiO$_2$-P$_2$O$_5$ sample, the vibrational bands at 2920–2846 cm^{-1} were not clearly solved and a broad band centred at ~2900 cm^{-1} was noticed and attributed to the hydrogen inter-layer bonds with water molecules [48].

The large band at ~1650 cm^{-1} in the SiO$_2$-P$_2$O$_5$ sample was attributed to OH vibrations of water molecules attached to P–O bonds.

For all the samples, the characteristic vibration bands for SiO$_2$-P$_2$O$_5$ amorphous films were observed in the FTIR spectra: Si–O–Si symmetric stretching (~760 cm^{-1}), Si–OH stretching (~910 cm^{-1}) with shoulders corresponding to (TO) Si–O–P asymmetric stretching (~1060 cm^{-1}), (TO) Si–O–Si symmetric stretching (~1100 cm^{-1}) and (LO) Si–O–Si symmetric stretching overlapping with O–P–O symmetric stretching (~1200 cm^{-1}) [49].

3.2. Atomic Force Microscopy

A selection of AFM images is presented in Figures 4–9, and the values of the root-mean-squared roughness (Rq) of the films deposited on glass and ITO-coated glass are summarized in Table 3.

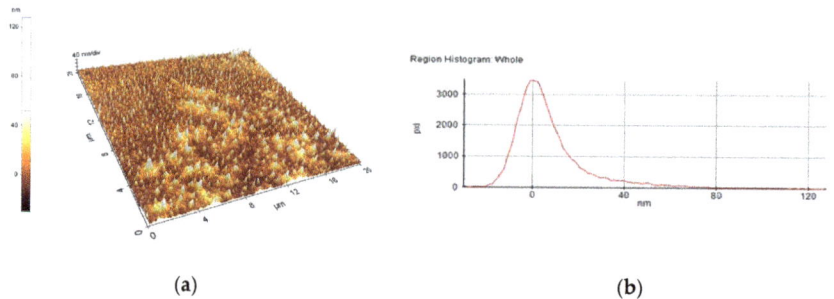

Figure 4. AFM image of 1%rGO-SiO$_2$-P$_2$O$_5$/glass film: (**a**) morphology, (**b**) region histogram.

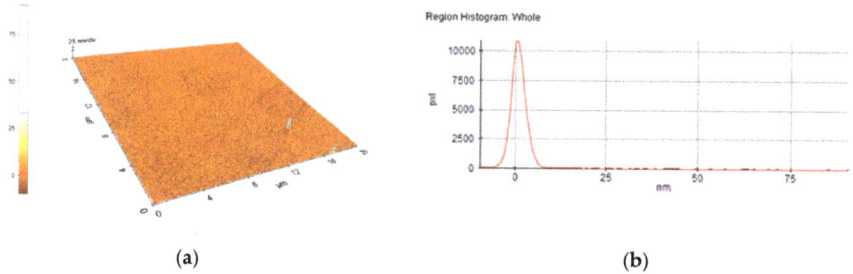

Figure 5. AFM image of 1%rGO-SiO$_2$-P$_2$O$_5$/ITO film: (**a**) morphology, (**b**) region histogram

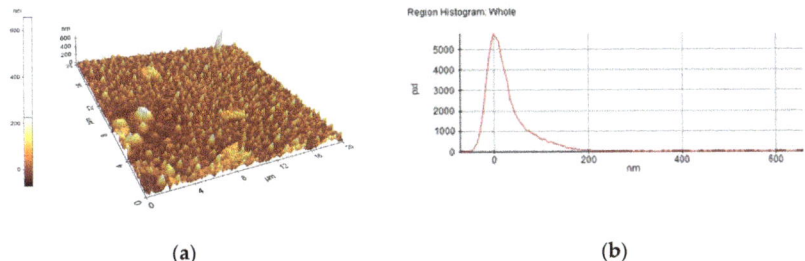

Figure 6. AFM image of 1.1%rGO-SiO$_2$/glass film: (**a**) morphology, (**b**) region histogram.

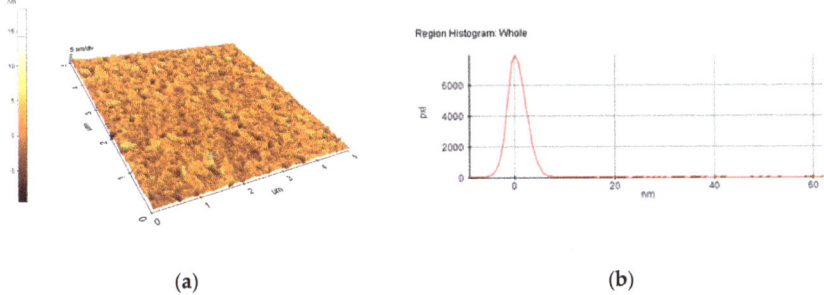

Figure 7. AFM image of 1.1%rGO-SiO$_2$/ITO film: (**a**) morphology, (**b**) region histogram.

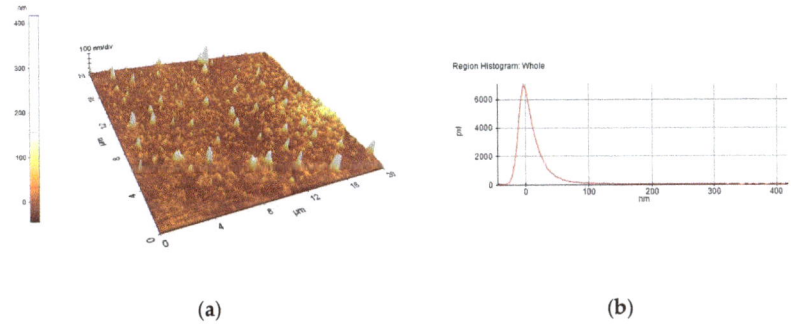

Figure 8. AFM image of 4%GO-SiO$_2$-P$_2$O$_5$/glass film: (**a**) morphology, (**b**) region histogram.

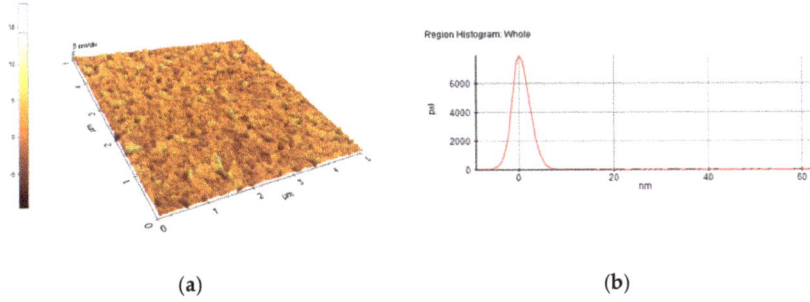

Figure 9. AFM image of 4%GO-SiO$_2$-P$_2$O$_5$/ITO film: (**a**) morphology, (**b**) region histogram.

Table 3. The root-mean-squared roughness (Rq) and peak-to-valley (Rpv) from atomic force microscopy (AFM) investigations.

Sample	Rq–Rpv (nm) for 20 μm Square Surface	
	On glass	On ITO-coated glass
1%rGO-SiO$_2$-P$_2$O$_5$	15–157.5	2.4–100.3
1.1%rGO-SiO$_2$	45–729	2.2–70.5
4%GO-SiO$_2$-P$_2$O$_5$	31–461.2	2.25–70.5

The prepared films were all homogenous, with the standard deviations of the height value (Rq(nm)) being in the intervals of 15–45 nm on the glass substrate and 2.2–2.4 nm on the ITO-coated glass. However, the Rpv values (Rpv is the peak-to-valley of the selected region, that is, the difference between the minimum and maximum values in the selected region) for the samples deposited on glass varied in the interval 157.5–729 nm, meaning that the pores present in the selected regions were deeper and larger for the samples without phosphor content. A more homogenous distribution of pores for the 1%rGO-SiO$_2$-P$_2$O$_5$ samples could be seen from the distribution histograms.

For the films deposited on ITO-coated glass, the roughness was similar for samples with rGO regardless of the presence of phosphor, and the values for Rpv were similar for 1.1%rGO-SiO$_2$ and 4%GO-SiO$_2$-P$_2$O$_5$ and up to 100.3 nm for 1%rGO-SiO$_2$-P$_2$O$_5$. As expected, the films deposited on ITO-coated glass were more compact than the ones on the glass substrate, due to the more homogeneously distributed surface-active sites of the ITO. The glass substrate is a borosilicate glass, which has a surface with low network connectivity, of the glass network formers Si and B, with the coexistence of different types of boron coordination states and with defects (i.e., a nonbridging oxygen, two-membered ring, and three-coordinated silicon) [50,51].

3.3. Scanning Electron Microscopy

A selection of SEM images is presented in Figures 10–12. The 1.1%rGO-SiO$_2$/glass film was less homogeneous than 1%rGO-SiO$_2$-P$_2$O$_5$/glass one, as presented in Figures 10–12, giving evidence of phosphorus pentoxide contribution in the distribution of rGO in the silico-phosphate matrix. This is in agreement with the AFM studies.

The existence of P was noticed in all samples containing P$_2$O$_5$, while C was noticed in the composite films more concentrated in rGO/GO, a lower GO/rGO content being under the detection limit of the equipment. This is demonstrated in Figure 13. The detailed SEM image of the 4%rGO-SiO$_2$-P$_2$O$_5$/ITO sample and the respective EDX spectra are presented in Figures 12 and 13.

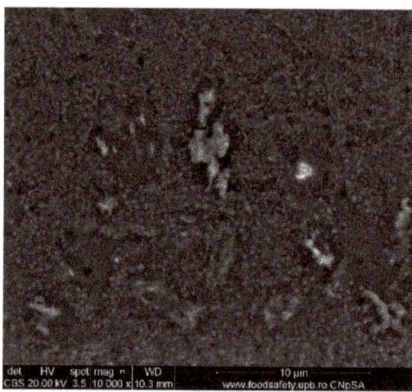

Figure 10. SEM image of 1.1%rGO-SiO$_2$/glass.

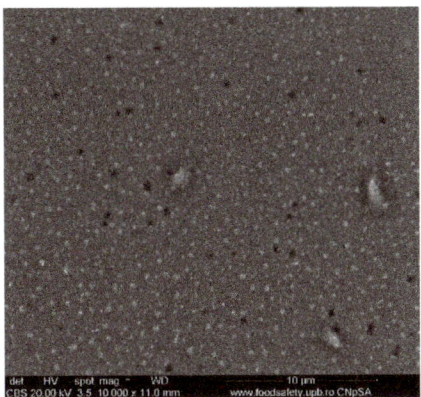

Figure 11. SEM image of 1%rGO-SiO$_2$-P$_2$O$_5$/glass.

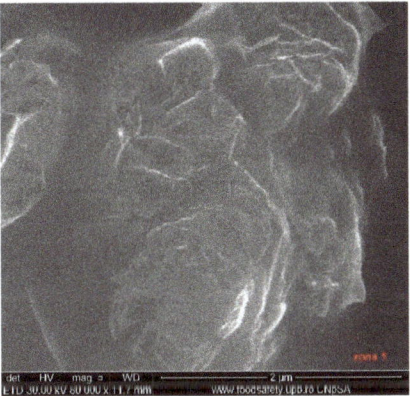

Figure 12. SEM image of 4%rGO-SiO$_2$-P$_2$O$_5$/ITO.

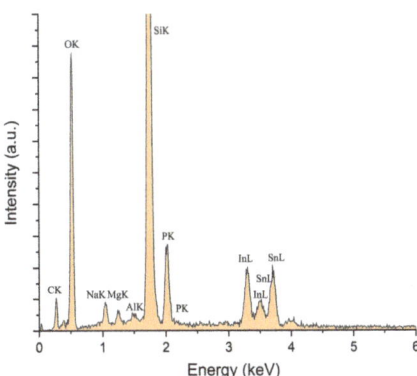

Figure 13. EDX spectra of 4%rGO-SiO$_2$-P$_2$O$_5$/ITO film corresponding to the SEM image from Figure 12.

3.4. UV-VIS-NIR Spectroscopy

The UV-VIS-NIR transmission spectra of the composite sol-gel films prepared on glass and on ITO-coated glass, collected with air as reference, are presented in Figure 14 and in Figure 15, respectively.

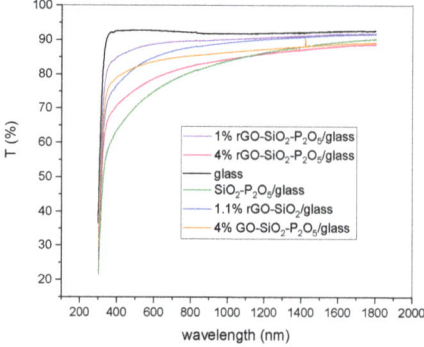

Figure 14. UV-VIS-NIR spectra of the sol-gel films deposited on glass substrate.

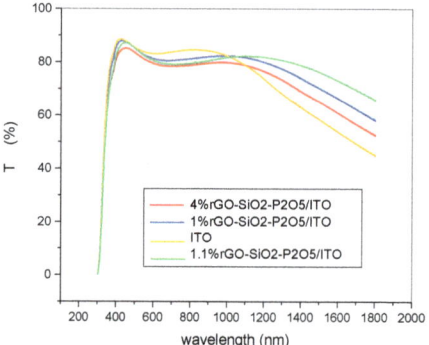

Figure 15. UV-VIS-NIR spectra of the sol-gel films deposited on ITO-coated glass substrate.

The substrate had a strong influence on the UV-VIS-NIR spectra of the prepared samples, as revealed by comparison of the transmission spectra for the films deposited on the glass substrate (Figure 14) with the ones for the films deposited on the ITO-coated glass substrate (Figure 15).

As a general remark, in the visible domain, the transmission of the rGO/GO doped samples on glass was higher than 70%, while on the ITO-coated glass, it was higher than 80%. These films are thus suitable for the protection of sensitive equipment against a NIR laser beam, being transparent enough to see through them.

For films deposited on the glass substrate (Figure 14), the samples containing rGO/GO exhibited a transmission above 85% for λ longer than 1100 nm and followed the general trend of the $SiO_2P_2O_5$-glass sample. For wavelengths longer than this value, the transmission decreased with increasing rGO content. For the same concentration of 4%, the film with the rGO content exhibited a higher absorbance than that of that with GO, as expected. The transmittance of the two samples with quite similar rGO content (1 and 1.1%) was higher for the sample that contained phosphorus pentoxide (P_2O_5) than for the sample without P_2O_5, for all wavelengths. The difference between the two spectra tends to be negligible for wavelengths longer than 1200 nm.

The transmittance of all the samples deposited on the ITO-coated glass substrate (Figure 15) followed the trend line of this particular type of substrate. For longer wavelengths (1100–1800 nm), the increase in rGO content induced a decrease in transmittance, as can be seen from the two spectra of the samples containing P_2O_5, with 1%rGO and with 4%rGO, respectively. In the same spectral range (1100–1800 nm), the transmittance of the two samples with quite similar content of rGO (1% and 1.1%) was higher for the sample without P_2O_5, (1.1%rGO) than for the sample with P_2O_5 (1%rGO), and the difference was larger for longer wavelengths. For both films on ITO-coated glass (1.1%rGO, without

P$_2$O$_5$) and (1%rGO, with P$_2$O$_5$), the transmittance did not decrease under 60% in the considered NIR spectral range.

3.5. Raman Spectroscopy

The Raman spectra of the synthesized films presented in Figure 16 show the characteristic peaks for graphene derivatives, namely, the G band at approximately 1600 cm^{-1} originating from the in-plane vibration of sp^2 carbon atoms and the D band associated with edge planes, defects and disordered structures of carbons found in graphene sheets at 1350 cm^{-1}. The broad D peak suggests a highly disordered regime including the structural imperfections created by the attachment of hydroxyl and epoxide groups on the carbon basal plane [52–55].

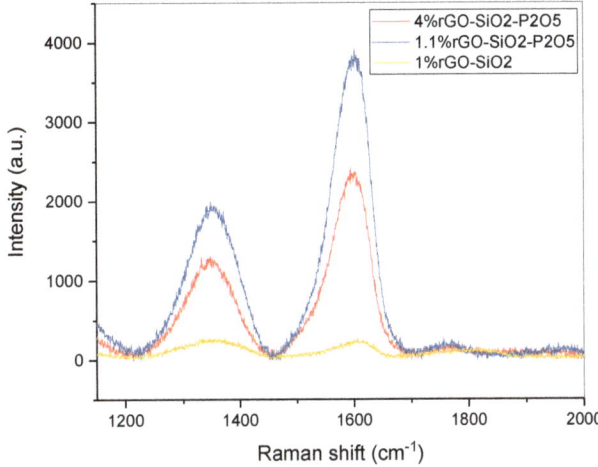

Figure 16. Raman spectra of a selection of the sol-gel films deposited on ITO-coated glass substrate.

Higher D bands were observed in samples 4%rGO-SiO$_2$-P$_2$O$_5$ and 1.1%rGO-SiO$_2$-P$_2$O$_5$ as compared to that in sample 1%rGO-SiO$_2$, indicating a decreased disorder associated with a decreased concentration of oxygen-containing functional groups under the action of H$_3$PO$_4$, which contributes to a further reduction of rGO embedded in the silico-phosphate matrix. The observation is in accordance with the FTIR and AFM results.

3.6. Optical Limiting Capability

The experimental setup used to investigate by intensity scans (I-scan) the OL capability of the synthesized graphene oxide-based silico-phosphate composite films is shown in Figure 17, and it is described below.

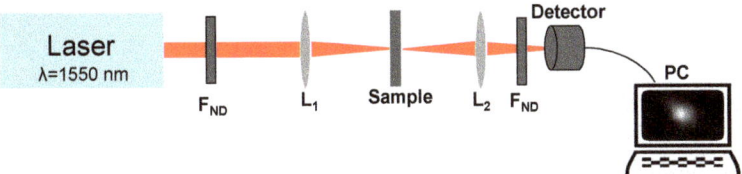

Figure 17. Schematic of the experimental setup for optical limiting (OL) studies.

The laser source is an Er-doped fibre laser (Toptica), which generates ultrashort pulses (~150 fs pulse duration) with a repetition rate of 76 MHz, at the wavelength λ = 1550 nm. The maximum average power is ~230 mW, and the corresponding peak power and the pulse energy of the generated laser pulses are ~19 kW and ~3 nJ, respectively. The lens L_1 (focal length = 2.54 cm) focuses down the laser beam on a spot of 13 μm diameter on the investigated sample, which is placed in its focal plane. The intensity of the laser beam incident on the sample is varied by changing its power only with neutral density filters (F_{ND}) (Thorlabs, Munich, Germany), with the transmission specially calibrated by us at the wavelength of the laser (λ = 1550 nm). Using the lens L_2, the spot size of the transmitted laser beam is adjusted relative to the aperture of the detector used to measure the beam average power. The lenses L_1 and L_2 and the sample are mounted on micrometric translation stages for the fine tuning of their positions. The average powers of the incident and of the transmitted beams were measured using a FieldMax II-TOP power meter (Coherent, Portland, OR, USA) with a type OP-2IR detector (Coherent, Portland, OR, USA). All the measured powers, the incident and the transmitted ones, were corrected for Fresnel reflections. Neutral density filters were also used in front of the detector in order to keep the power of the measured light signal below the maximum power measurable with the detector. The incident average laser powers in the available range are below the values for which the laser-induced damage could appear, as seen in investigations with an optical microscope.

The maximum value of the incident peak intensity in our I-scan experiments was limited by the maximum available average power of the laser source, lower than ~9 GW/cm^2. At these incident peak intensities, no optical damage was observed in the investigated samples.

In order to investigate the OL capability of the GO/rGO composite glassy films, the values of the transmitted peak intensities, $I_{trans}(peak)$, were measured as a function of the incident peak intensities, $I_{inc}(peak)$, for the entire range of $I_{inc}(peak)$ of the available laser source. A deviation from the linear transmittance towards lower values of transmittance is indicative of the optical limiting behaviour [27,28,56].

The values of the experimentally determined transmitted peak intensities, $I_{trans}(peak)$, for incident peak intensities, $I_{inc}(peak)$, lower than ~0.5 GW/cm^2, were fitted with a linear dependence, and from its slope, the linear transmittance, T_L, of each sample was determined.

The entire set of experimentally measured transmitted peak intensities, for each investigated sample, was fitted with a saturation-type function of the form:

$$I_{trans}(peak) = \frac{T_L \cdot I_{inc}(peak)}{1 + [I_{inc}(peak)/I_{sat}]} \quad (1)$$

where I_{sat} is the saturation peak intensity. This equation gives information about the overall optical limiting capability of a sample, without considering the nonlinear optical processes involved in this functionality and their contribution to the overall optical response.

As a general remark regarding the rGO/GO composite films deposited on the two considered substrates (glass and ITO-coated glass, respectively), we mention that the samples that consist of composite films deposited by spin coating on glass substrates have a higher linear transmittance in NIR than those deposited in the same conditions on ITO-coated glass substrates.

This is possible to see from the experimental transmittance results, considering the incident and the transmitted peak intensities of the laser pulses, shown in Figure 18 a,b (samples with rGO) and in Figure 18c,d (samples with GO), by comparing the linear transmittances and T_L (slopes of linear dependencies) of the corresponding sample sets. This is consistent with the transmittances at 1550 nm from the transmission UV-VIS-NIR spectra shown in Figures 14 and 15, respectively. The small differences could be due to the different incident powers (very low in the case of the spectrometer's light source) and to the very different sizes of the illuminated areas in the two cases.

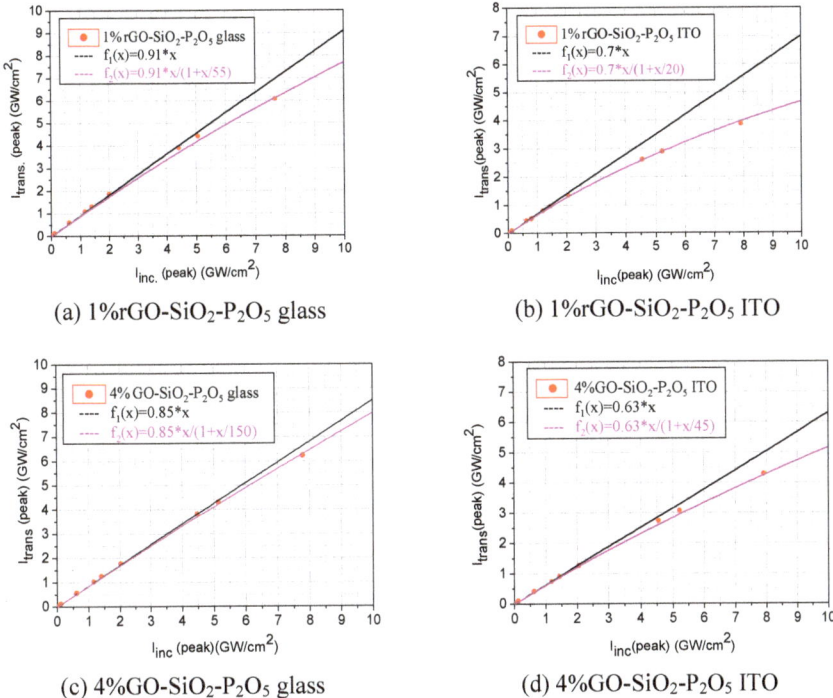

Figure 18. The transmitted peak intensities vs. incident ones for two sets of corresponding composite samples: with rGO deposited on glass (**a**) and on ITO-coated glass (**b**); with GO deposited on glass (**c**) and on ITO-coated glass (**d**).

For both considered substrates, the linear transmittance is appropriate for OL functionality. OL in samples of different materials with comparable [57,58] or even much lower [56] linear transmittance has been reported.

The maximum incident peak intensities are slightly varying in the graphics shown in Figure 18a–d due to the fact that the individual optical limiting experiments were performed at different moments in time with a slightly readjusted internal optical alignment of the fs laser system, which influenced its maximum average power.

On the other hand, the OL experiments revealed the fact that the silico-phosphate composite films, based on rGO or on GO, deposited on ITO-coated glass substrates limit better (a larger deviation of the saturation-type fit of the experimental data from the straight line corresponding to linear transmittance) than those deposited on glass substrates, as observed from Figure 18a,b (samples with rGO) and in Figure 18c,d (samples with GO) of the corresponding sample sets.

The better limiting in films deposited on the ITO-coated glass substrate is probably due to the already-mentioned fact that these films are more compact than the ones deposited on the glass substrate since the active sites on the ITO surface are more uniformly distributed. For the glass substrate, the phosphorus pentoxide content in the films contributes to the decrease in the pore size of the deposited composite films.

In order to compare the influence of the rGO and GO, respectively, and of their concentrations, and to assess the influence of the silico-phosphate matrix on the OL performance of the composite films, in the following, we will show and analyse the results obtained on the composite films deposited on ITO-coated glass substrates only.

The experimental transmittance results, considering the incident, $I_{inc}(peak)$, and the transmitted, $I_{trans}(peak)$, peak intensities of the laser pulses, are shown in Figure 19a–f, for the films deposited on ITO-coated glass substrates for the two categories of the investigated silico-phosphate composite samples, with rGO (Figure 19a–c) and with GO (Figure 19d–f), respectively.

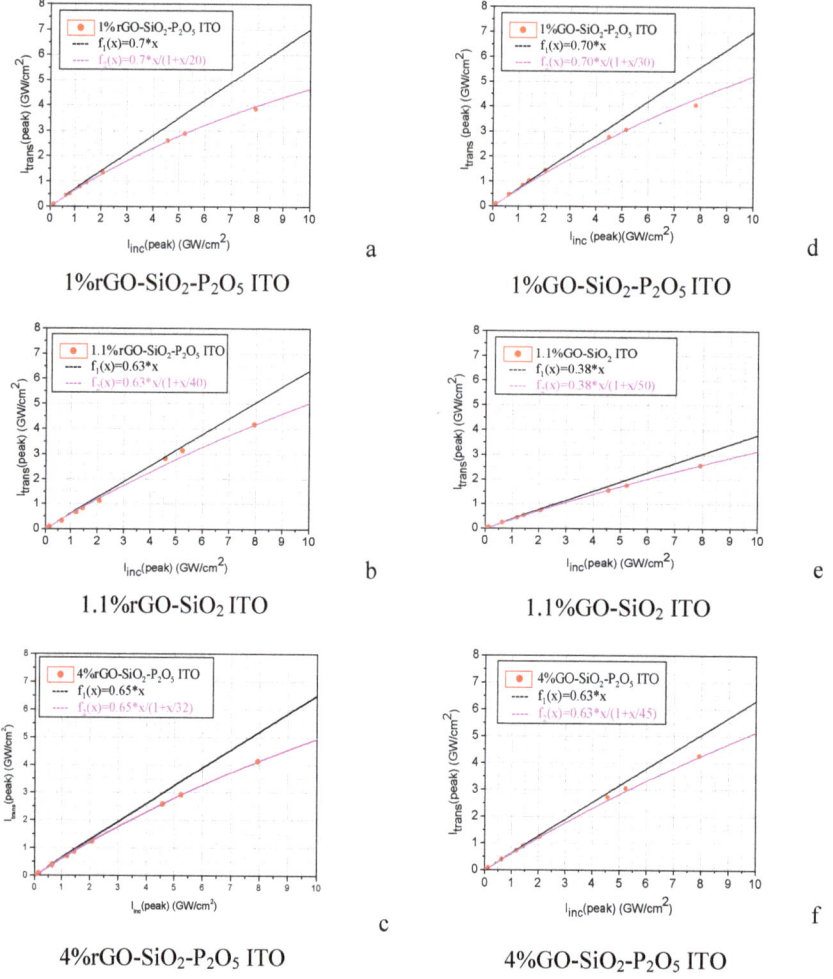

Figure 19. The experimental transmittance results for the films deposited on ITO-coated glass substrates, for the two categories of the investigated silico-phosphate composite samples: with rGO (**a**–**c**) and with GO (**d**–**f**).

Both the linear fit (considering the experimental points at low incident intensity; linear transmittance) and the fit with the saturation-type function from Equation (1) (considering the entire set of experimental points for each sample; nonlinear transmittance) are shown in Figure 19a–f.

In Table 4 are summarized the linear transmittances, T_L, and the saturation intensities, I_{sat}, obtained by the fitting of the experimental points. T_L is given by the slope of the linear fit of the experimental points corresponding to low incident peak intensities, and I_{sat} is obtained from the nonlinear fit of all experimental points with the saturation-type function from Equation (1), respectively.

Table 4. The linear transmittance and the saturation intensity derived from Figure 19.

Sample Denomination	Linear Transmittance (%)	I_{sat} (GW/cm^2)
1%rGO-SiO$_2$-P$_2$O$_5$ ITO (Figure 19a)	70	20
1.1%rGO-SiO$_2$ ITO (Figure 19b)	63	40
4%rGO-SiO$_2$-P$_2$O$_5$ ITO (Figure 19c)	65	32
1%GO-SiO$_2$-P$_2$O$_5$ ITO (Figure 19d)	70	30
1.1%GO-SiO$_2$ ITO (Figure 19e)	38	50
4%GO-SiO$_2$-P$_2$O$_5$ ITO (Figure 19f)	63	45

The comparative analysis of the OL results shown in Figure 19 and in Table 4 reveals that the samples with rGO (Figure 19a–c) limit better (lower I_{sat}) than the corresponding (the same concentration) samples with GO (Figure 19d–f), for all considered concentrations (1, 1.1 and 4%) of rGO/GO.

From the sets of figures Figure 19a,c and Figure 19d,f, respectively, it is possible to see that the samples with P$_2$O$_5$ with a lower concentration of rGO/GO (1%) limit better than the samples with higher concentration of rGO/GO (4%). Additionally, the linear transmittance of the samples with a lower content of rGO/GO is higher than that of the samples with a higher content of rGO/GO. Regarding the linear transmittance of the samples with the same concentration of rGO/GO, it is practically similar, excepting the samples without P$_2$O$_5$ (Figure 19b,e). In this last-mentioned case, the lower linear transmittance of the sample with GO compared to that of the one with rGO could be attributed to the larger content of H$_2$O that can be hydrogen bonded to the OH moieties from the surface of GO.

The comparative analysis of all the results shown in Figure 19a–f also reveals the favourable effect on the OL capability of the phosphoric acid used in the preparation of the sol-gel matrix. H$_3$PO$_4$ contributed to the properties of the films in several ways: (i) the graphenization of GO as a result of the water elimination reaction in the presence of H$_3$PO$_4$ [14]; (ii) the modification of the interlinkage of the rGO sheets (collective strength of interlayers) due to the modification of the network of hydrogen bonds mediated by oxygen-containing functional groups and water molecules and, accordingly, the change in the materials' properties [48]; (iii) a more homogeneous distribution of the rGO sheets in the precursor mixtures and in the obtained films.

The OL trend in the silico-phosphate glassy films with rGO on the ITO-coated glass substrate is similar to that reported in isolated fullerene-rich thin films at the wavelength of 532 nm [57]. In our case, the onset of OL (defined as the point on the transmittance curve at which it starts to diverge from the linear transmittance [19]) is much lower (<0.5 mJ/cm^2) than the values (60–140 mJ/cm^2, for different samples) reported in [57].

The OL trend in our samples is also comparable to those reported in aqueous GO suspensions at 1064 nm, for 35 ps and 4 ns laser pulses [19]. The onset of OL is again much lower in our case than the ones (~0.50 J/cm^2) reported for the two pulse durations.

The minimum value of the normalized transmittance, obtained by dividing the transmittance ($I_{trans}(peak)/I_{inc}(peak)$) corresponding to the maximum incident intensity by the lowest, linear transmittance is in the case of the sample 1%rGO-SiO$_2$-P$_2$O$_5$ ITO equal to ~0.67, which is similar to that obtained in non-covalent functionalized rGO and rGO functionalized with various concentrations of Ag nanoparticles (~0.65), at 800 nm wavelength with 100 fs laser pulses [25]. The minimum normalized transmittances reported in [25] were obtained at incident intensities of ~2×10^{17} W/m^2, a value that is orders of magnitude higher than our maximum incident peak intensity, 8×10^{13} W/m^2.

The normalized transmittance reported by Ren et al. [24] in electrochemical GO samples, at 800 nm wavelength, with 85 fs laser pulses, reached a minimum value of 0.88 at the incident fluence of 200 mJ/cm^2 and of 0.67 at the incident fluence of 400 mJ/cm^2. In our sample 1% rGO-SiO$_2$-P$_2$O$_5$ ITO, the minimum normalized transmittance ~0.67 was obtained at a much lower incident fluence of ~1.3 mJ/cm^2.

In the range of the incident peak intensities from our I-scan experiments, no optical damage was observed in the investigated samples, of which surface was visualized by microscopy before and after exposure to laser radiation (Figure 20).

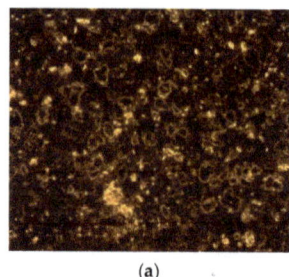
(a)

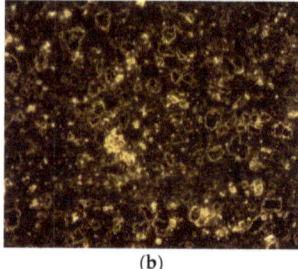
(b)

Figure 20. The images of the 1%rGO-SiO$_2$-P$_2$O$_5$ sample, deposited on glass, obtained by differential interference contrast (DIC) microscopy (a) before exposure to laser radiation and (b) after exposure to laser radiation with $I_{inc}(peak) = 7.7$ GW/cm^2.

The surface of the investigated samples was analysed with an optical microscope (Axiotech Vario microscope, Carl Zeiss, Jena, Germany) using DIC microscopy.

4. Conclusions

We have demonstrated the preparation of graphene oxide-based silico-phosphate composite films for the optical limiting of ultrashort (fs) laser pulses in the NIR spectral domain, by a low cost and environmentally friendly sol-gel method.

The FTIR spectra revealed the H$_3$PO$_4$ action to lower the concentration of oxygen-containing functional groups, mainly –OH, from the rGO/GO structure, presumably via the formation and elimination of H$_2$O, followed by a vertical re-agglomeration of the rGO sheets as a result of the change from hydrophilic to hydrophobic character.

A more homogenous distribution of pores for the 1%rGO-SiO$_2$-P$_2$O$_5$ films in the AFM distribution histograms was noticed. For the films deposited on ITO, the roughness was almost similar for the samples with GO and rGO regardless of the presence of phosphorus pentoxide. H$_3$PO$_4$ contributes to the homogeneity distribution of rGO as noticed from SEM/EDX investigations.

Samples containing rGO/GO deposited on the glass substrate exhibited a transmission as high as 80% for λ higher than 800 nm, and the transmission decreased with increasing rGO/GO content. The phosphorus pentoxide content in the films did not induce a higher absorption. The transmittance of all the samples followed the trend line of the ITO/glass substrate with T = 80% up to 1100 nm. For 1100–1800 nm, the increase in rGO content and the presence of phosphorus pentoxide induced a decrease in the transmittance. For the films on ITO/glass, containing 1%rGO, the transmittance did not decrease under 60%.

The Raman spectra revealed an increase in the D band intensity in samples containing P$_2$O$_5$ as compared with that for the films without P$_2$O$_5$, meaning a decreased concentration of the groups responsible for the intensity of the D band under the action of H$_3$PO$_4$. These are consistent with the FTIR and AFM investigations and support the fact that H$_3$PO$_4$ contributes to a further reduction of rGO, with the rGO sheets uniformly distributed on the substrate.

The optical limiting capability of the graphene-based silico-phosphate composite films was revealed by intensity-scan type experiments. A comparative analysis of the presence and of the concentration of rGO or GO in the structure of the composite glassy films, of the influence of the silico-phosphate matrix, and of the substrate on which the films are deposited on the optical limiting performance of the novel graphene composite films was performed. A comparison of the OL in our samples with several OL results obtained in the literature with ns, ps and fs laser pulses at visible and near infrared wavelengths was performed.

Author Contributions: Conceptualization, supervision, validation: A.P. and I.C.V.; methodology: A.P., I.C.V., P.G., A.M.I. and D.U.; resources: I.C.V. and R.W.; investigation: A.P., I.C.V., P.G., A.M.I., L.I., L.R., S.I., M.E., R.T., S.E. and K.T.; data curation: P.G. and A.M.I.; formal analysis, writing—original draft, writing—review & editing: A.P., I.C.V. and P.G.; visualization: A.P., I.C.V., P.G., A.M.I. and S.I.; funding acquisition, project administration: A.P., I.C.V., D.U., R.W. and J.Y. All authors have read and agreed to the published version of the manuscript.

Funding: Contract 32/2018 funded by the Romanian Ministry of Education and Research, Unitatea Executiva pentru Finantarea Invatamantului Superior, a Cercetarii, Dezvoltarii si Inovarii (UEFISCDI) and a Grant of Norges Forskningsråd (Research Council of Norway) through the program of Nano2021, both under the MANUNET Program (Project MNET17/NMCS-0114, OLIDIGRAPH); Contract 42PCCDI/2018 funded by the Romanian Ministry of Education and Research, UEFISCDI.

Acknowledgments: The authors acknowledge the financial support provided by the Romanian Ministry of Education and Research, UEFISCDI (project number 32/2018), and by the Research Council of Norway (a grant of Nano2021 program), both under the MANUNET Program, Project MNET17/NMCS-0114, OLIDIGRAPH. In addition, the authors acknowledge UEFISCDI for the financial support and APC provided in the frame of the project 42PCCDI/2018.

Conflicts of Interest: The authors declare no conflict of interest.

References

1. Sun, Y.P.; Riggs, J.E. Organic and inorganic optical limiting materials. From fullerenes to nanoparticles. *Int. Rev. Phys. Chem.* **1999**, *18*, 43–90. [CrossRef]
2. Wang, J.; Werner, J.B. Inorganic and hybrid nanostructures for optical limiting. *J. Opt. A Pure Appl. Opt.* **2009**, *11*, 024001. [CrossRef]
3. Parola, S.; Julián-López, B.; Carlos, L.D.; Sanchez, C. Optical Properties of Hybrid Organic–Inorganic Materials and their Applications—Part II: Nonlinear Optics and Plasmonics. In *Handbook of Solid State Chemistry, Part 4 Nano and Hybrid Materials*; Wiley: Hoboken, NJ, USA, 2017.
4. Graphene Report 2020, Description. Available online: https://www.researchandmarkets.com/reports/4901148/the-graphene-report-2020 (accessed on 1 June 2020).
5. Bonaccorso, F.; Sun, Z.; Hasan, T.A.; Ferrari, A.C. Graphene photonics and optoelectronics. *Nat. Photon* **2010**, *4*, 611–622. [CrossRef]
6. Low, T.; Avouris, P. Graphene Plasmonics for Terahertz to Mid-Infrared Applications. *ACS Nano* **2014**, *8*, 1086–1101. [CrossRef]
7. De Abajo, F.J.G. Graphene Plasmonics: Challenges and Opportunities. *ACS Photonics* **2014**, *1*, 135–152. [CrossRef]
8. Ooi, K.J.; Tan, D.T. Nonlinear graphene plasmonics. *Proc. R. Soc. A* **2017**, *473*, 20170433. [CrossRef]
9. Glazov, M.M.; Ganichev, S.D. High frequency electric field induced nonlinear effects in graphene. *Phys. Rep.* **2014**, *535*, 101–138. [CrossRef]
10. Cheng, J.L.; Sipe, E.; Vermeulen, N.; Guo, C. Nonlinear optics of graphene and other 2D materials in layered structures. *J. Phys. Photonics* **2019**, *1*, 015002. [CrossRef]
11. Li, P.; Zhu, B.; Li, P.; Zhang, Z.; Li, L.; Gu, Y. A Facile Method to Synthesize CdSe-Reduced Graphene Oxide Composite with Good Dispersion and High Nonlinear Optical Properties. *Nanomaterials* **2019**, *9*, 957. [CrossRef]
12. Ren, Y.; Zhao, L.; Zou, Y.; Song, L.; Dong, N.; Wang, J. Effects of Different TiO_2 Particle Sizes on the Microstructure and Optical Limiting Properties of TiO_2/Reduced Graphene Oxide Nanocomposites. *Nanomaterials* **2019**, *9*, 730. [CrossRef]
13. Wang, J.; Wang, Y.; Wang, T.; Li, G.; Lou, R.; Cheng, G.; Bai, J. Nonlinear Optical Response of Graphene Oxide Langmuir-Blodgett Film as Saturable Absorbers. *Nanomaterials* **2019**, *9*, 640. [CrossRef] [PubMed]
14. Chen, Y.; Bai, T.; Dong, N.; Fan, F.; Zhang, S.; Zhuang, X.; Sun, J.; Zhang, B.; Zhang, X.; Wang, J.; et al. Graphene and its derivatives for laser protection. *Prog. Mater. Sci.* **2016**, *84*, 118–157. [CrossRef]
15. Agrawal, A.; Park, J.Y.; Sen, P.; Yi, G.-C. Unraveling absorptive and refractive optical nonlinearities in CVD grown graphene layers transferred onto a foreign quartz substrate. *Appl. Surf. Sci.* **2020**, *505*, 144392. [CrossRef]

16. Wang, A.J.; Yu, W.; Fang, Y.; Song, Y.L.; Jia, D.; Long, L.L.; Cifuentes, M.P.; Humphrey, M.G.; Zhang, C. Facile Hydrothermal Synthesis and Optical Limiting Properties of TiO_2-Reduced Graphene Oxide Nanocomposites. *Carbon* **2015**, *89*, 130–141. [CrossRef]
17. Zhao, M.; Peng, R.; Zheng, Q.; Wang, Q.; Chang, M.J.; Liu, Y.; Song, Y.L.; Zhang, H.L. Broadband optical limiting response of a graphene–PbS nanohybrid. *Nanoscale* **2015**, *7*, 9268–9274. [CrossRef] [PubMed]
18. Liu, Z.; Wang, Y.; Zhang, X.; Xu, Y.; Chen, Y.; Tian, J. Nonlinear optical properties of graphene oxide in nanosecond and picosecond regimes. *Appl. Phys. Lett.* **2009**, *94*, 021902. [CrossRef]
19. Liaros, N.; Aloukos, P.; Kolokithas-Ntoukas, A.; Bakandritsos, A.; Szabo, T.; Zboril, R.; Couris, S. Nonlinear Optical Properties and Broadband Optical Power Limiting Action of Graphene Oxide Colloids. *J. Phys. Chem. C* **2013**, *117*, 6842–6850. [CrossRef]
20. Xu, Y.; Liu, Z.; Zhang, X.; Wang, Y.; Tian, J.; Huang, Y.; Ma, Y.; Zhang, X.; Chen, Y. A Graphene Hybrid Material Covalently Functionalized with Porphyrin: Synthesis and Optical Limiting Property. *Adv. Mater.* **2009**, *21*, 1275–1279. [CrossRef]
21. Liaros, N.; Orfanos, I.; Papadakis, I.; Couris, S. Nonlinear optical response of some Graphene oxide and Graphene fluoride derivatives. *Optofluid. Microfluid. Nanofluid.* **2016**, *3*, 53–58. [CrossRef]
22. Jiang, X.F.; Polavarapu, L.; Neo, S.T.; Venkatesan, T.; Xu, Q. Graphene Oxides as Tunable Broadband Nonlinear Optical Materials for Femtosecond Laser Pulses. *J. Phys. Chem. Lett.* **2012**, *3*, 785–790. [CrossRef]
23. Zheng, Z.; Zhu, L.; Zhao, F. Nonlinear Optical and Optical Limiting Properties of Graphene Oxide Dispersion in Femtosecond Regime. *Proc. SPIE* **2014**, *9283*, 92830V-1.
24. Ren, J.; Zheng, X.; Tian, Z.; Li, D.; Wang, P.; Jia, B. Giant third-order nonlinearity from low-loss electrochemical graphene oxide film with a high power stability. *Appl. Phys. Lett.* **2016**, *109*, 221105. [CrossRef]
25. Oluwafemi, O.S.; Sreekanth, P.; Philip, R.; Thomas, S.; Kalarikkal, N. Improved nonlinear optical and optical limiting properties in non-covalent functionalized reduced graphene oxide/silver nanoparticle (NF-RGO/Ag-NPs) hybrid. *Opt. Mater.* **2016**, *58*, 476–483.
26. Chen, W.; Wang, Y.; Ji, W. Two-Photon Absorption in Graphene Enhanced by the Excitonic Fano Resonance. *J. Phys. Chem. C* **2015**, *119*, 16954–16961. [CrossRef]
27. Demetriou, G.; Bookey, H.T.; Biancalana, F.; Abraham, E.; Wang, Y.; Ji, W.; Kar, A.K. Nonlinear optical properties of multilayer graphene in the infrared. *Opt. Express* **2016**, *24*, 13033–13043. [CrossRef]
28. Xu, X.; Zheng, X.; He, F.; Wang, Z.; Subbaraman, H.; Wang, Y.; Jia, B.; Chen, R.T. Observation of Third-order Nonlinearities in Graphene Oxide Film at Telecommunication Wavelengths. *Sci. Rep.* **2017**, *7*, 1–7. [CrossRef]
29. Zheng, C.; Zheng, Y.; Chen, W.; Wei, L. Encapsulation of graphene oxide/metal hybrids in nanostructured sol–gel silica ORMOSIL matrices and its applications in optical limiting. *Opt. Laser Technol.* **2016**, *68*, 52–59. [CrossRef]
30. Monisha, M.; Priyadarshani, N.; Durairaj, M.; Girisun, T.C.S. 2PA induced optical limiting behaviour of metal (Ni, Cu, Zn) niobate decorated reduced graphene oxide. *Opt. Mater.* **2020**, *101*, 109775. [CrossRef]
31. Feng, M.; Zhan, H.B.; Chen, Y. Nonlinear optical and optical limiting properties of graphene families. *Appl. Phys. Lett.* **2010**, *96*, 033107. [CrossRef]
32. Saravanan, M.; Girisun, T.C.S. Enhanced nonlinear optical absorption and optical limiting properties of superparamagnetic spinel zinc ferrite decorated reduced graphene oxide nanostructures. *Appl. Surf. Sci.* **2017**, *392*, 904–911.
33. Loh, V.K.P.; Bao, Q.L.; Eda, G.; Chhowalla, M. Graphene oxide as a chemically tunable platform for optical applications. *Nat. Chem.* **2010**, *2*, 10151024. [CrossRef] [PubMed]
34. Zheng, X.; Feng, M.; Zhan, H. Giant optical limiting effect in Ormosil gel glasses doped with graphene oxide materials. *J. Mater. Chem. C* **2013**, *1*, 6759–6766. [CrossRef]
35. Liu, Z.; Zhang, X.; Yan, X.; Chen, Y.; Tian, J. Nonlinear optical properties of graphene-based materials. *Chin. Sci. Bull.* **2012**, *57*, 2971–2982. [CrossRef]
36. Innocenzi, P.; Malfatti, L.; Lasio, B.; Pinna, A.; Loche, D.; Casula, M.F.; Alzari, V.; Mariani, A. Sol–gel chemistry for graphene–silica nanocomposite films. *New J. Chem.* **2014**, *38*, 3777. [CrossRef]
37. Anastasescu, M.; Gartner, M.; Ghita, A.; Predoana, L.; Todan, L.; Zaharescu, M.; Vasiliu, C.; Grigorescu, C.; Negrila, C. Loss of phosphorous in silica-phosphate sol-gel film. *J. Sol-Gel Sci. Technol.* **2006**, *40*, 325–333. [CrossRef]

38. Baschir, L.; Savastru, D.; Popescu, A.A.; Vasiliu, I.C.; Filipescu, M.; Iordache, A.M.; Elisa, M.; Iordache, S.M.; Buiu, O.; Obreja, C. Morphologic and optical characterization studies of the influence of reduced graphene oxide concentration on optical properties of ZnO-P$_2$O$_5$ composite sol-gel films. *J. Optoelectron. Adv. M.* **2019**, *21*, 524–529.
39. Taheri, A.; Liu, H.; Jassemnejad, B.; Appling, D.; Powell, R.C.; Song, J.J. Intensity scan and two photon absorption and nonlinear refraction of C$_{60}$ in toluene. *Appl. Phys. Lett.* **1996**, *68*, 1317. [CrossRef]
40. Dancus, I.; Vlad, V.I.; Petris, A.; Rujoiu, T.B.; Rau, I.; Kajzar, F.; Meghea, A.; Tane, A. Z-scan and I-scan methods for characterization of DNA optical nonlinearities. *Rom. Rep. Phys.* **2013**, *65*, 966.
41. Dancus, I.; Vlad, V.I.; Petris, A.; Gaponik, N.; Lesnyak, V.; Eychmüller, A. Optical limiting and phase modulation in CdTe nanocrystal devices. *J. Optoelectron. Adv. M.* **2010**, *12*, 119.
42. Dancus, I.; Vlad, V.I.; Petris, A.; Gaponik, N.; Lesnyak, V.; Eychmüller, A. Saturated near-resonant refractive optical nonlinearity in CdTe quantum dots. *Opt. Lett.* **2010**, *35*, 1079. [CrossRef]
43. Bazaru, T.; Vlad, V.I.; Petris, A.; Gheorghe, P.S. Study of the third-order nonlinear optical properties of nano-crystalline porous silicon using a simplified Bruggeman formalism. *J. Optoelectron. Adv. M.* **2009**, *11*, 820–825.
44. Er, E.; Çelikkan, H. An efficient way to reduce graphene oxide by water elimination using phosphoric acid. *RSC Adv.* **2014**, *4*, 29173–29179. [CrossRef]
45. Manoratne, C.H.; Rosa, S.R.D.; Kottegoda, I.R.M. XRD-HTA, UV Visible, FTIR and SEM Interpretation of Reduced Graphene Oxide Synthesized from High Purity Vein Graphite. *Mat. Sci. Res. India* **2017**, *14*, 19–30. [CrossRef]
46. Max, J.J.; Chapados, C. Infrared Spectroscopy of Aqueous Carboxylic Acids: Comparison between Different Acids and Their Salts. *J. Phys. Chem. A* **2004**, *108*, 3324–3337. [CrossRef]
47. Parhizkar, N.; Ramezanzadeh, B.; Shahrabi, T. Corrosion protection and adhesion properties of the epoxy coating applied on the steel substrate pre-treated by a sol-gel based silane coating filled with amino and isocyanate silane functionalized graphene oxide nanosheets. *Appl. Surf. Sci.* **2018**, *439*, 45–59. [CrossRef]
48. Medhekar, N.V.; Ramasubramaniam, A.; Ruoff, R.S.; Shenoy, V.B. Hydrogen Bond Networks in Graphene Oxide Composite Paper: Structure and Mechanical Properties. *ACS Nano* **2010**, *4*, 2300–2306. [CrossRef]
49. Vasiliu, I.; Gartner, M.; Anastasescu, M.; Todan, L.; Predoana, L.; Elisa, M.; Grigorescu, C.; Negrila, C.; Logofatu, C.; Enculescu, M.; et al. SiO$_x$–P$_2$O$_5$ films: Promising components in photonic structure. *Opt. Quant. Electron.* **2007**, *39*, 511–521. [CrossRef]
50. Ren, M.; Deng, L.; Du, J. Surface structures of sodium borosilicate glasses from molecular dynamics simulations. *J. Am. Ceram. Soc.* **2017**, *100*, 2516–2524. [CrossRef]
51. Mason, M.G.; Hung, L.S.; Tang, C.W.; Lee, S.T.; Wong, K.W.; Wang, M. Characterization of treated indium tin oxide surfaces used in electroluminescent devices. *J. Appl. Phys.* **1999**, *86*, 1688–1692. [CrossRef]
52. Yang, D.; Velamakanni, A.; Bozoklu, G.; Park, S.; Stoller, M.; Piner, R.D.; Stankovich, S.; Jung, I.; Field, D.A.; Ventrice, C.A., Jr.; et al. Chemical analysis of graphene oxide films after heat and chemical treatments by X-ray photoelectron and Micro-Raman spectroscopy. *Carbon* **2009**, *47*, 145–152. [CrossRef]
53. Kaniyoor, A.; Ramaprabhua, S. A Raman spectroscopic investigation of graphite oxide derived graphene. *AIP Adv.* **2012**, *2*, 032183. [CrossRef]
54. Lucches, M.M.; Stavale, F.; Ferreira, E.H.M.; Vilani, C.; Moutinho, M.V.O.; Capaz, B.R.; Achete, C.A.; Jorio, A. Quantifying ion-induced defects and Raman relaxation length in graphene. *Carbon* **2010**, *48*, 1592–1597. [CrossRef]
55. Buasri, A.; Ananganjanakit, T.; Peangkom, N.; Khantasema, P.; Pleeram, K.; Lakaeo, A.; Arthnukarn, J.; Loryuenyong, V. A facile route for the synthesis of reduced graphene oxide (RGO) by DVD laser scribing and its applications in the environment-friendly electrochromic devices (ECD). *J. Optoelectron. Adv. M.* **2017**, *19*, 492–500.
56. Wang, A.; Cheng, L.; Chen, X.; Zhao, W.; Li, C.; Zhu, W.; Shang, D. Reduced graphene oxide covalently functionalized with polyaniline for efficient optical nonlinearities at 532 and 1064 nm. *Dyes Pigments* **2019**, *160*, 344–352. [CrossRef]

57. Kang, S.; Zhang, J.; Sang, L.; Shrestha, L.K.; Zhang, Z.; Lu, P.; Li, F.; Li, M.; Ariga, K. "Electrochemically Organized Isolated Fullerene-Rich Thin Films with Optical Limiting Properties" and Supporting Information. *ACS Appl. Mater. Interfaces* **2016**, *8*, 24295–24299. [CrossRef] [PubMed]
58. Bai, T.; Li, C.-Q.; Sun, J.; Song, Y.; Wang, J.; Blau, W.J.; Zhang, B.; Chen, Y. Covalent Modification of Graphene Oxide with Carbazole Groups for Laser Protection. *Chem. Eur. J.* **2015**, *21*, 4622–4627. [CrossRef]

© 2020 by the authors. Licensee MDPI, Basel, Switzerland. This article is an open access article distributed under the terms and conditions of the Creative Commons Attribution (CC BY) license (http://creativecommons.org/licenses/by/4.0/).

Article

Improved Field Emission Properties of Carbon Nanostructures by Laser Surface Engineering

Minh Nhat Dang [1],*, Minh Dang Nguyen [2,3], Nguyen Khac Hiep [4], Phan Ngoc Hong [4], In Hyung Baek [5] and Nguyen Tuan Hong [4],*

1. The Australian Research Council (ARC) Industrial Transformation Training Centre in Surface Engineering for Advanced Materials (SEAM), Faculty of Science, Engineering and Technology, Swinburne University of Technology, P.O. Box 218, Hawthorn, VIC 3122, Australia
2. Department of Chemistry, University of Houston, Houston, TX 77204-5003, USA; dangminh27498@gmail.com
3. Vietnam Academy of Science and Technology (VAST), University of Science and Technology of Hanoi, 18 Hoang Quoc Viet, Hanoi 100000, Vietnam
4. Centre for High Technology Development, VAST, 18 Hoang Quoc Viet, Hanoi 100000, Vietnam; hiephulk@gmail.com (N.K.H.); hongpn@htd.vast.vn (P.N.H.)
5. Korea Atomic Energy Research Institute, Daeduk-Daero 989-111, Daejeon, Korea; ihbaek@kaeri.re.kr
* Correspondence: nhatminh@swin.edu.au (M.N.D.); hongnt@htd.vast.vn (N.T.H.)

Received: 3 August 2020; Accepted: 23 September 2020; Published: 27 September 2020

Abstract: We herein present an alternative geometry of nanostructured carbon cathode capable of obtaining a low turn-on field, and both stable and high current densities. This cathode geometry consisted of a micro-hollow array on planar carbon nanostructures engineered by femtosecond laser. The micro-hollow geometry provides a larger edge area for achieving a lower turn-on field of 0.70 V/µm, a sustainable current of approximately 2 mA (about 112 mA/cm^2) at an applied field of less than 2 V/µm. The electric field in the vicinity of the hollow array (rim edge) is enhanced due to the edge effect, that is key to improving field emission performance. The edge effect of the micro-hollow cathode is confirmed by numerical calculation. This new type of nanostructured carbon cathode geometry can be promisingly applied for high intensity and compact electron sources.

Keywords: carbon nanotubes; hot-filament CVD; graphene; field electron emission; laser machining

1. Introduction

Electron sources have been utilized in X-ray tubes, vacuum microwave amplifiers, and electron guns [1]. In these instruments, thermionic emitters are widely used due to their proven operation. However, the thermionic emitters are bulky and not efficient enough for low-power applications due to the waste heat radiated from the filament and the high applied field. An alternate means to generate electrons is field emission (FE), which exhibits many advantages in comparison with thermionic emission. For instance, as field emitters do not require heat to generate electrons, they are more energy-efficient and may be rapidly switched on and off. Among the family of field emission materials, carbon nanostructures are interesting due to low applied fields and elevated electron currents [2].

Electron field emission from the nanostructured carbon is found to involve the phases showing sp^2 hybridization. The sp^2 content, orientation, and size of these phases, which have a positive electron affinity, can play a considerable role in emissions at a low field [3,4]. Graphene and carbon nanotubes are the types of cold cathodes that extract electrons at the lowest applied field.

Low applied fields of the nanostructured carbon cathodes are also facilitated by the large aspect ratio. For nearly two decades, graphene has drawn significant attention from scholars owing to its extraordinary properties, which are applicable in diverse industries such as aerospace, energy, and infrastructure construction [5–7]. Graphene and its relative composites can be synthesized via bottom-up and top-down methodologies including chemical vapor deposition (CVD), liquid exfoliation, and electrochemical co-synthesis [8–12]. For graphene materials, the existence of abundant sharp-edged structures can act as efficient field emission sites [13,14]. Nevertheless, field emission from graphene is still a challenge because the fabrication methods such as exfoliation or screen-printing will create graphene sheets laterally on the support substrate, which is unfavorable for electron tunneling.

Carbon nanotubes (CNTs) are another novel carbon allotrope, which can be seen as cylindrical graphene sheets [15]. Because of their long, tubular geometry and excellent physicochemical properties, carbon nanotubes are also attractive materials for cold cathodes [1–3]. Planar and point structures are the two predominant types of carbon nanotube cathodes. The CNT point emitter is favorable in devices requiring a small, bright electron source [16], whereas the CNT planar emitter is suited to applications requiring uniform and relatively large emitting sites. Such applications widely use dense carbon nanotubes as the cathode. The dense carbon nanotubes are deposited on the substrate by either CVD, or printing [17]. CVD carbon nanotubes are selectively chosen for devices requiring low turn-on field and high emitting current. Individual carbon nanotubes demonstrate excellent field emission properties; however, vertically-aligned carbon nanotubes (VACNTs) do not present such great performance. In this report, the VACNTs are defined as dense carbon nanotubes grown perpendicular to the substrate platform. This disadvantage is acceptably ascribed to the screen effect, which is mutual shielding of the electric field between adjacent nanotubes [18]. Some studies have reported mitigation of the screen effect by photolithography patterning [19]. One can obtain the high aspect-ratio geometry of patterned VACNTs purposely, which is favorable to local electric field enhancement. The patterned VACNTs such as VACNT-column arrays could not only negate the screen effect but also introduce isolated CNT extrusions and edges, both of which result in the enhanced electric field [20].

Surface engineering methods are known to enhance the pristine properties of bulk materials [21–24]. For cold cathodes, geometry engineering and surface modification are alternative methods to improve field emission performance [25–27]. In this work, we use a femtosecond laser to engineer and hence enhance the electric field of the VACNT cathode. The femtosecond laser is used to selectively trim out carbon nanotubes and create a micro-hollow (MH) array on the top surface. Since the MH-VACNTs cathode is based on the as-grown VACNTs, our work included the pristine and planar VACNTs to compare their field emission properties. The field emission characteristics of the MH-VACNTs cathode showed a low turn-on field, and both stable and high emission current. Initial results demonstrated the 40 μm diameter hollow cathodes with a pitch of 150 μm can achieve a turn-on field of 0.70 V/μm, lower than the 0.89 V/μm of the planar VANCTs (without laser modification). Numerical calculation confirmed the local electric field is increased at the rim edge of the micro hollow, a key factor to improving field emission characteristics.

2. Experimental Procedure

2.1. Materials Preparation

The MH-VACNTs cathode was produced by a two-step process consisting of (i) the VACNTs synthesis and (ii) the femtosecond laser engineering. The VACNTs were grown on silicon (Si) by using the hot-filament CVD method [20,28]. The highly doped, 500 μm thick silicon substrate, following the microelectronic industry standard, was used for the growth of carbon nanotubes. The catalysts are selectively deposited using a routine photolithography process. In brief, the VACNT film was obtained at 750 °C, CH_4/H_2 mixture (20/30 SCCM-standard cubic centimetre per minute at Standard Temperature and Pressure-STP), and reactor pressure of 30 Torr. It is noted that the hot-filament CVD has the advantage of decomposing CH_4/H_2 at relatively high temperatures (2200–2500 °C), which is favourable to a high purity of the as-grown carbon nanotubes [28].

The 1.5 mm diameter VACNTs with a thickness of 250–300 μm was subject to femtosecond laser machining in an ambient atmosphere. A femtosecond laser is a tool that is widely used to fabricate microstructures from several materials, including carbon nanotubes [29]. The laser system consisted of a mode-locked Ti:Sapphire laser (repetition rate of 92 MHz, 80-fs pulse width, average laser power of 200 mW) and the translation stage [30]. Once the laser beam was incident on the sample, it selectively ablates carbon nanotubes. By translating the VACNT substrate stage relative to the laser shots, a periodic array of the micro hollows is formed. Using good selection of laser power, exposure time, and depth of focus, we can achieve the well-shaped hollow geometry on the VACNT film.

2.2. Characterization Preparation

Carbon nanotubes were examined by scanning electron microscope SEM (FESEM S-4800, Hitachi, Tokyo, Japan) and Raman spectroscope (IFS/66 Bruker system, Billerica, MA, USA). Raman signal was recorded at room temperature using an excitation laser wavelength of 1064 nm. The 1.5-millimetre-diameter and about 250-μm-thick VACNTs on Si with and without the laser engineering are field emission (FE) tested. The FE measurement is carried out in a vacuum chamber (KVM-T4060 system, Korea Vacuum, Daegu, Korea) with a working pressure of 5×10^{-6} Torr and a ballast resistor of 200 kΩ. Keithley-248 (Tektronix, Beaverton, OR, USA) is a DC anode voltage source. The emission electron current is measured by Keithley-2001 (Multi-meter, Tektronix, USA). Measured FE data are acquired using LABVIEW software and a personal computer through a general-purpose interface bus (GPIB) card. The diode configuration is used with the anode (aluminium plate) and the carbon nanotube cathode. The anode and the cathode were electrically isolated by Kapton sheets (spacer); the electric field was calculated to be the anode voltage divided by the anode-to-CNT-surface distance. Current densities were estimated from the net currents and the VACNT area. Turn-on field and threshold field are defined as the electric field necessary to extract 100-nA emission current and 1-mA/cm^2 current density, respectively. For current-voltage (I-V) curves, measurements were repeated several times until the curve became stable. The carbon nanotube cathode was also tested in a continuous operating condition for a stability test. The operation stability was evaluated by emission current values at anode voltages with a 50% duty cycle. In practice, each cycle is 60 s; hence, the anode voltage was ON for only 30 s.

To investigate the carbon nanotube cathode geometry, we carried out numerical calculation (COMSOL Multiphysics, 3D simulation) in a diode configuration. Due to periodicity in geometry and hence, electric field, a diode with a 3 × 3 micro-hollow array was used as a stand in for the whole micro-hollow array cathode. A cathode structure that modeled a practical MH-VACNTs (40 μm diameter hollow, depth of about 200 μm) is simulated. The electric field is computed by the Laplace equation solution using finite element analysis (a bias voltage of 3000 V, equivalent to an applied field of 2.40 V/μm). Pitch d (hollow-to-hollow distance) is a variant in the simulation. The local electric field at a distance immediately above the cathode was calculated and studied.

3. Results and Discussions

3.1. Characterization Data

Figure 1a demonstrates the femtosecond-laser process to modify the VACNTs and the obtained results. In practice, the femtosecond laser beam is synchronized with the substrate translation, capable of creating a periodic micro-hollow array. Figure 1b,c presents typical SEM images of the micro hollow on the surface and along the VACNT bulk. The micro hollows were well-shaped and smooth. The femtosecond laser process was able to form a long micro-hollow from the VACNT top to the bottom substrate. A confocal microscope was used to draw the hollow profile and measure the depth (Figure 1d,e). Figure 1f shows typical micro-hollow arrays on the 1.5 mm diameter specimens; the hollow diameter of 40–150 μm and depth of about 200 μm were controlled by periods of the laser exposure time.

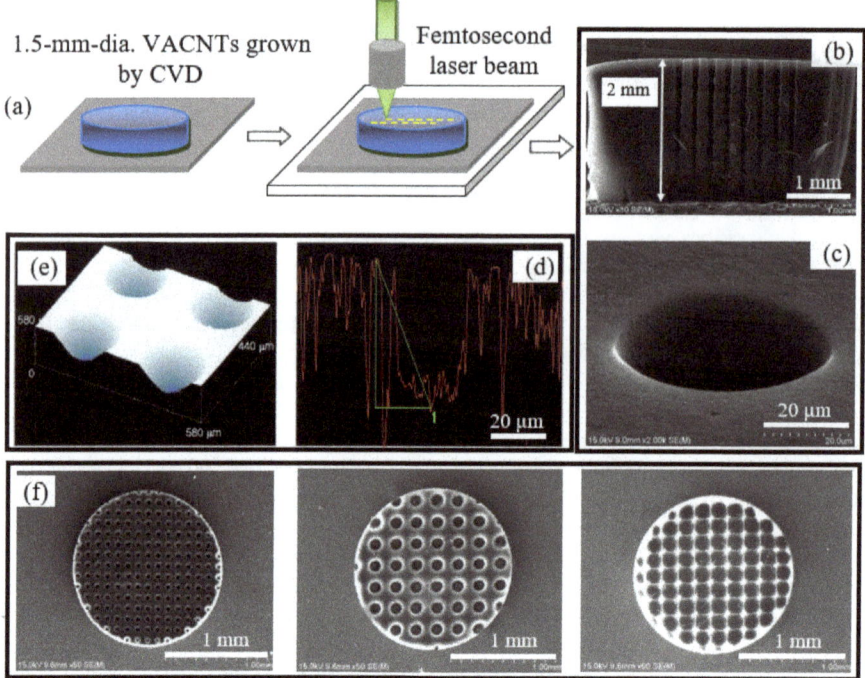

Figure 1. (a) Sketch of femtosecond laser process (repetition rate of 92 MHz, 80-fs pulse width, an average power of 200 mW); (b,c) SEM images of about 2-mm depth hollow cross-section (from the VACNT top to the bottom substrate and hollow shape on the top surface; (d,e) 3D confocal image and profile of the micro hollow with depth of ~200 μm, (f) typical micro-hollow arrays on 1.5 mm diameter, VACNT specimens, with depth of ~200 μm, and hollow diameters of 40, 100, and 150 μm, respectively (from left to right).

Raman spectra of carbon nanotubes in this work are presented in Figure 2a. Raman signals were directly recorded from the MH-VACNT (hollow diameter of 40 μm, pitch $d \sim 100$ μm) cathode at three spots, named site A, B, and C. Site A was on the rim edge, site B was 25 μm away the edge, and site C was a middle spot between two consecutive hollows. In practice, it was found that the Raman spectrum of carbon nanotubes at site C was comparable to those of the as-grown, pristine carbon nanotubes; we assumed those nanotubes (site C) were insignificantly affected by the femtosecond laser process. In Figure 2a, there occurs a graphitic G band at 1560–1580 cm^{-1} and a defective D band at ~1324 cm^{-1} [31]. The disordered peak (D-band) accounts for amorphous carbon, contaminants, and defective nanotubes as well. It is widely accepted that the intensity ratio of I_D to I_G (I_D/I_G) is an indicator of carbon nanotube crystallinity [32]. The I_D/I_G ratio at sites A, B, and C (shown in Figure 2b–d) are 1.48, 1.90, and 1.96, respectively. The G-band of type-A CNTs increases due to the graphene zone-folding effect at the rim edge. Because of the nature of our rim-edge CNTs, which possibly have active phonon coupling to the continuum of electronic states, the red-shift of the optical phonon is associated with the increase in the temperature due to the combinational effect of thermal expansion and temperature contribution, which resulted in the finer crystal lattice of type-A CNTs. Apparently, when the temperature increases at the spot femtosecond laser hit, the G-band is also noticeably broader together with red-shifting. The nanotubes at the rim edge were also observed to be joined together as multi-junctions after the laser process (Figure 2b) which was previously reported [29,33]. Conversely, carbon nanotubes at site B may be less affected by the femtosecond laser energy; hence, their I_D/I_G ratio was insignificantly reduced and mostly comparable with the pristine carbon nanotubes. However, their morphologies were more or less restructured, as can be witnessed in Figure 2c.

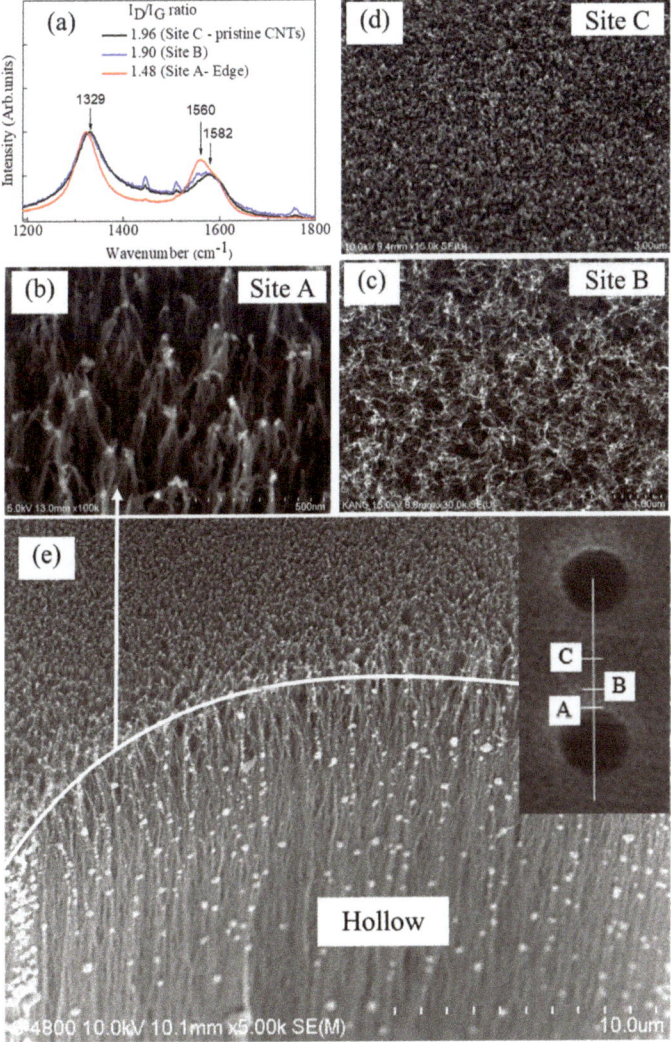

Figure 2. (a) Raman spectra of carbon nanotubes recorded at three different sites (**A**–**C**), which are shown in SEM images (**b**–**d**); (**e**) Hollow edge of CNTs, Inset: Site A, B, and C positions: (**A**) rim edge of hollow; (**B**) 25 µm away from site A; (**C**) middle between two consecutive hollows.

3.2. Field Emission Properties

To examine the operation of both the MH-VACNTs and planar VACNTs cathodes, field emission characteristics were repeatedly measured. The MH-VACNTs, the same 40-µm-dia. hollow, depth of ~200 µm, with pitch of 75, 100, and 150 µm, respectively were measured. The measuring setup is demonstrated in Figure 3a. The anode–cathode spacing was maintained at 1500 µm by a Kapton spacer. The current density–applied field (J–F) curves of both cathodes are shown in Figure 3b. The turn-on field, corresponding to an emission current of 0.1 µA, was 0.70 V/µm and 0.89 V/µm for the MH-VACNTs (40-μm-dia. hollow, depth of ~200 μm, and pitch of 150 μm) and the planar VACNTs, respectively. We extended further field emission measurements to a higher current range by gradually increasing applied fields. The current density reached 10 mA/cm^2 at the threshold field of less than 1.50 V/μm (the MH-VACNTs cathodes). The stability of field emission current was also measured. Before the stability test, an aging process was carried out. Anode voltages of 100–3000 V were swept up/down several times. The aging experiment is aimed to exercise field emission of carbon nanotubes in a wide range of applied fields. The experiment may also remove loosely-entangled carbon nanotubes due to laser machining. The stability test of both cathodes was started at an emitting current of 1 mA. Figure 3c demonstrates emitting current over 15 h. The constant bias voltage, 50% duty cycle, equivalent to an electric field of 1.60 V/μm, and 1.92 V/μm for the MH-VACNTs (40-μm-dia. hollow, depth of ~200 μm, and pitch of 150 μm) and the planar VACNTs, was applied respectively. Although the electrical aging has been carried out, in the onset both cathodes showed a gradual decrease and large fluctuation of the emission current before the stable electron emission was established. It was widely accepted that field emission is a highly selective process, which occurs at a relatively small number of emitting sites with the largest field enhancement factor β. Due to localized high currents, the best-first emitting sites may fail first or collapse abruptly, and the emitting currents are sharply decreased. Then, many other sites that have a more average field enhancement factor start to emit electrons. Because of large numbers and even distribution, these new emitting sites operate reliably, and hence, total electron currents become more stable. For the MH-VACNTs, the standard deviation of the observed emitting currents within two hours (from fifth hour) was about 5.15 μA (the mean current value of 700 μA) and smaller than that of the planar VACNTs (10.5 μA with the mean current value of 615 μA). Therefore, the MH-VACNT cathode operated in a more reproducible manner than the planar VACNTs.

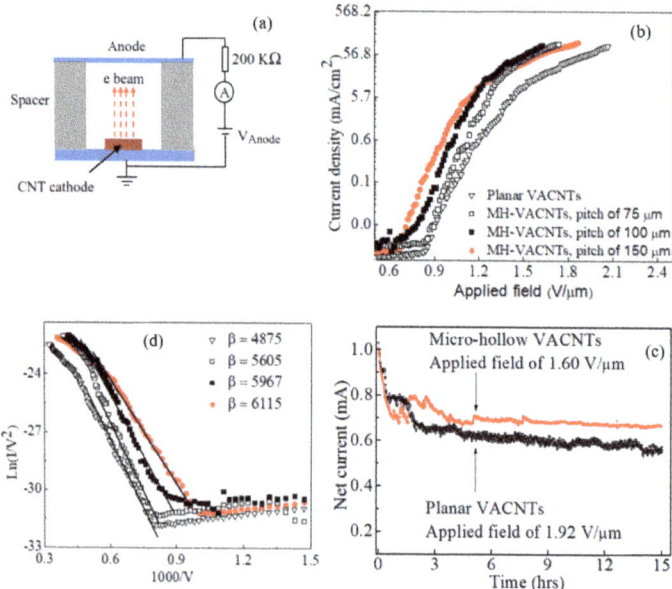

Figure 3. (**a**) Schematic of the field emission setup; (**b**) Current density–applied field (J–F) curves of 1.5 mm dia. MH-VACNTs (40-μm-dia. hollow, depth of ~ 200 μm) with pitch of 75, 100, and 150 μm, and planar VACNTs; (**c**) Stability test of two cathodes: MH-VACNTs (40-μm-dia. hollow, depth of ~200 μm, and pitch of 150 μm) and planar VACNTs, constant bias voltage, 50% duty cycle; (**d**) Corresponding F-N plots.

Generally, electron emission of carbon nanotubes, driven by quantum mechanical tunneling, follows the simplified Folwer–Nordheim (F-N) equation [34]

$$I = A \frac{1.42 \times 10^{-6}}{\phi} F^2 \exp\left(\frac{10.4}{\sqrt{\phi}}\right) \exp\left(-\frac{B\phi^{1.5}}{F}\right) \quad (1)$$

where A is the emitting area (m^2), $B = 6.44 \times 10^9$ VeV$^{-1.5}$m^{-1} (constant); ϕ is work function (eV), and F represents local electric field. The local electric field F is estimated approximately by $F = \beta\frac{V}{d}$, in which V is the applied bias voltage, d is the anode to CNT-top distance, and β is the collective field enhancement factor. Electron emission at low applied fields is possible because of the significant amplification at of the local field. Therefore, β depends on the distribution density and geometry of the carbon nanotube cathode. F–N equation can be simplified as

$$I = A \frac{1.42 \times 10^{-6}}{\varphi} \beta^2 \left(\frac{V}{d}\right)^2 \exp\left(\frac{10.4}{\sqrt{\phi}}\right) \exp\left(-\frac{B\phi^{1.5}d}{\beta V}\right) \quad (2)$$

Figure 3d demonstrates F–N plots of the MH-VACNTs and the planar VACNTs. In a lower field, both F–N plots were well fitted to a straight line, and linear regression analysis using the equation $\ln(I/V^2) = a - b(1/V)$ was performed. This suggests that the observed electron emission from both cathodes follows the quantum tunnelling theory. In the high field, F-N plots show a saturation behavior that may be due to thermionic emission, series resistance, and cathode geometry [35,36]. The field enhancement factor β (given the known fitting value of $b = -B\phi^{1.5}d/\beta$, and assuming $\Phi = 5$ eV for carbon nanotubes, ref. [37] was 4875 for the planar VACNTs. As for the MH-VACNTs (all three samples had the same 40-µm-dia. hollow, depth of ~ 200 µm), β values were 5605, 5967, and 6115 in responses to the pitch of 75, 100, and 150 µm, respectively. In other words, the local fields are increased 14%, 22%, and 25% in comparison with the planar VACNTs (Figure 4c). An increase of β (MH-VACNTs) is attributed to the micro hollow geometry, consistent with the simplified picture that higher emission currents appear at the vicinity of the hollow edges.

To investigate the effects of the micro hollow array on the cathode electric field, we used a 3D diode model and numerically calculated the field by using the COMSOL Multiphysics program. Cathode geometry was based on the MH-VANCT cathode; a 40 µm diameter hollow, with a depth of 200 µm, and a VANCT-layer thickness of 250 µm. Pitch d (hollow-to-hollow distance) is a variant in the simulation. It is noted that our model considered the cathode as a monolithic surface rather than as porous nanotubes. In practice, to correctly calculate the electric field of the CNT cathode, it requires a meshing resolution of a few nanometers, or at least less than a nanotube diameter (~5 nm). Such resolution exceeds the number of degrees of freedom, and the available computer memory; hence, in this work we used an alternative approach to semi-quantitatively estimate field enhancement which is based on the multistage field effects [36,38]. The total field enhancement factor (β_Σ) is a product of the multistage cathode geometry ($\beta_\Sigma = \beta_{\text{hollow-array}} \times \beta_{\text{CNT-film}}$); the simulation is to estimate $\beta_{\text{hollow-array}}$. Assuming even two cathodes possessing the same $\beta_{\text{CNT-film}}$, the cathode with larger $\beta_{\text{hollow-array}}$ will have larger β_Σ in general. Figure 4a,b shows the typical electric field distribution on the micro hollow with pitch $d = 100$ µm (along the centrepiece cross-section line and right on the cathode surface) and contour plots of the 3 × 3 hollow-array electric field. In the contour plots (at the X-Y plane right on the cathode surface), brighter spots are corresponding to stronger fields. In the vicinity of the hollow edge, the electric fields are strongest, and then reduce moving away from the rim edge. There was an increase of electric field moving toward the rim edge of the micro hollow, from 2.40 V/µm in the planar cathode, to a maximum of 3.09 V/µm in the vicinity of the rim edge; i.e., $\beta_{\text{hollow-array}} = 1.28$. From the SEM image of Site A in Figure 2, it can be seen that the carbon nanotube tips at the rim edge were joined together to a greater extent than site B (25 µm away from the edge). Such CNT bundles may also work as efficient emitters [39]. The laser-induced modification of carbon nanotubes, and subsequent effects on both geometry and a work function, would therefore be an interesting topic for future investigation.

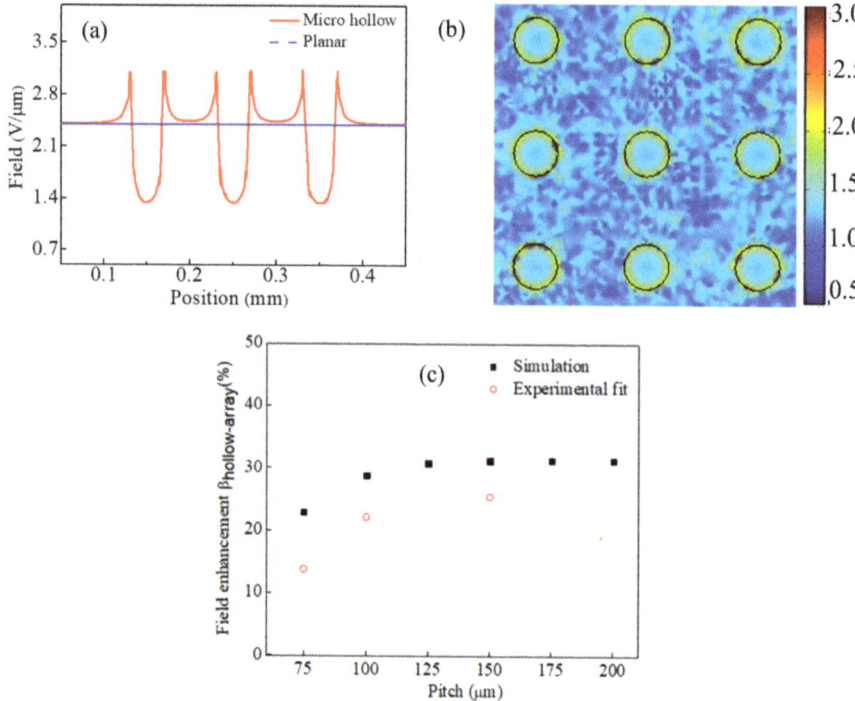

Figure 4. (a) Electric field distribution on three consecutive hollows (along centerpiece cross-section line), a maximum field at rim edges. (b) Contour plots of a local electric field (false color plots in V/μm scale) of 3 × 3 hollow (diameter of 40 μm, a pitch of 100 μm, depth of 200 μm), demonstrates a variation of electric field for the (c) simulated cathode field enhancement as a function of pitch.

A screen effect is an important factor in a cold cathode. In the case of the micro hollow cathode, too many hollows will flatten the field enhancement at the rim edge. Figure 4c plots the field enhancement ($\beta_{hollow-array}$) as a function of pitch d in the theoretical simulation and the experimental fitting. In the numerical calculation (40-μm-dia. hollow, depth of 200 μm), the field enhancement is proportionate to d and saturated when $d > 150$ μm. The MH-VACNT cathode with 40 μm dia. hollow and $d > 150$ μm can enhance the field by about 31%. The field enhancement of the MH-VACNT cathodes with 40-μm-dia. hollow, and pitch d of 75, 100, and 150 μm were 23%, 29%, and 31%. These numerical calculations are reasonably consistent with the experimental data fit.

4. Conclusion

In this report, we presented an innovative geometry of carbon nanotubes with a micro-hollow array on the top surface, created by femtosecond laser machining. Field enhancement effects due to aspect ratios were modeled using COMSOL multiphysics. The increased geometrical field enhancement factor ultimately gives rise to the lowering of the threshold field and improved field emission current. A typical 1.5 mm dia. MH-VACNTs cathode with a hollow diameter of 40 μm, a pitch of 100 μm showed a turn-on field of 0.70 V/μm, a sustainable current of about 2 mA (~112 mA/cm^2) at an applied field of less than 2 V/μm. Compared to the pristine and planar VACNTs, the laser-based MH-VACNT cathode can increase the local electric field by 25%. The initial characterization of VACNTs indicates that the laser process likely restructures (self-assembles) and graphitizes the carbon nanotubes, cleaning defects, and amorphous regions as well. It also improves field emission properties through factors such as geometry, work function, and stability. Finally, it can be stated the micro-hollow cathode geometry of carbon nanotubes has potential applications in high intensity, compact electron sources, and further investigation is currently underway.

Author Contributions: Conceptualization, M.N.D. and N.T.H.; investigation, M.D.N., N.T.H., N.K.H., and I.H.B.; resources, N.T.H. and I.H.B.; writing—original draft preparation, M.N.D. and N.T.H.; writing—review and editing, M.N.D. and N.T.H.; project administration, P.N.H.; funding acquisition, N.T.H. All authors have read and agreed to the published version of the manuscript.

Funding: This research was fully funded by the Vietnam National Foundation for Science and Technology Development (NAFOSTED) under grant number 103.99-2016.58.

Acknowledgments: The first author sincerely thanks James Wang laoshi and Arne Biesiekierski senpai for their fruitful supervision and Vesna Stefanovski for her fantastic assistance at ARC Training Centre in Surface Engineering for Advanced Materials.

Conflicts of Interest: The authors declare no conflict of interest.

References

1. Yamamoto, S. Fundamental physics of vacuum electron sources. *Rep. Prog. Phys.* **2005**, *69*, 181–232. [CrossRef]
2. Hong, X.; Shi, W.; Zheng, H.; Liang, D. Effective carbon nanotubes/graphene hybrid films for electron field emission application. *Vacuum* **2019**, *169*, 108917. [CrossRef]
3. Ilie, A.; Ferrari, A.; Yagi, T.; Rodil, S.E.; Robertson, J.; Barborini, E.; Milani, P. Role of sp2 phase in field emission from nanostructured carbons. *J. Appl. Phys.* **2001**, *90*, 2024–2032. [CrossRef]
4. Küttel, O.M.; Gröning, O.; Emmenegger, C.; Nilsson, L.; Maillard, E.; Diederich, L.; Schlapbach, L. Field emission from diamond, diamond-like and nanostructured carbon films. *Carbon* **1999**, *37*, 745–752. [CrossRef]
5. Novoselov, K.S.; Fal'Ko, V.I.; Colombo, L.; Gellert, P.R.; Schwab, M.G.; Kim, K. A roadmap for graphene. *Nature* **2012**, *490*, 192–200. [CrossRef]
6. Chan, K.-Y.; Pham, D.Q.; Demir, B.; Yang, D.; Mayes, E.L.; Mouritz, A.P.; Ang, A.S.; Fox, B.; Lin, H.; Jia, B.; et al. Graphene oxide thin film structural dielectric capacitors for aviation static electricity harvesting and storage. *Compos. Part B Eng.* **2020**, *201*, 108375. [CrossRef]
7. Minh, D.N.; Duong, H.P.; Hoang, L.; Nguyen, D.; Tran, P.D.; Honf, P.N. Plasma-Assisted Preparation of MoS2/Graphene/MOF Hybrid Materials and Their Electrochemical Behaviors. *Mater. Trans.* **2020**, *61*, 1535–1539. [CrossRef]
8. Dang, M.N.; Nguyen, T.H.; Van Nguyen, T.; Thu, T.V.; Le, H.; Akabori, M.; Ito, N.; Nguyen, H.Y.; Le, L.T.; Nguyen, T.H.; et al. One-pot synthesis of manganese oxide/graphene composites via a plasma-enhanced electrochemical exfoliation process for supercapacitors. *Nanotechnology* **2020**, *31*, 345401. [CrossRef]
9. Van Hau, T.; Trinh, P.; Nam, N.P.H.; Van Tu, N.; Lam, V.D.; Phuong, D.D.; Minh, P.N.; Thang, B.H. Electrodeposited nickel–graphene nanocomposite coating: Effect of graphene nanoplatelet size on its microstructure and hardness. *RSC Adv.* **2020**, *10*, 22080–22090. [CrossRef]
10. Van Chuc, N.; Thanh, C.T.; Van Tu, N.; Phuong, V.T.; Thang, P.V.; Tam, N.T.T. A Simple Approach to the Fabrication of Graphene-Carbon Nanotube Hybrid Films on Copper Substrate by Chemical Vapor Deposition. *J. Mater. Sci. Technol.* **2015**, *31*, 479–483. [CrossRef]
11. Hernandez, Y.; Nicolosi, V.; Lotya, M.; Blighe, F.M.; Sun, Z.; De, S.; McGovern, I.T.; Holland, B.; Byrne, M.; Donnelly, F.C.; et al. High-yield production of graphene by liquid-phase exfoliation of graphite. *Nat. Nanotechnol.* **2008**, *3*, 563–568. [CrossRef] [PubMed]
12. Duoc, P.N.D.; Binh, N.H.; Van Hau, T.; Thanh, C.T.; Van Trinh, P.; Tuyen, N.V.; Van Quynh, N.; Van Tu, N.; Chinh, V.D.; Thu, V.T.; et al. A novel electrochemical sensor based on double-walled carbon nanotubes and graphene hybrid thin film for arsenic(V) detection. *J. Hazard. Mater.* **2020**, *400*, 123185. [CrossRef] [PubMed]
13. Sankaran, K.J.; Bikkarolla, S.K.; Desta, D.; Roy, S.S.; Boyen, H.-G.; Lin, I.-N.; McLaughlin, J.; Haenen, K. Laser-Patternable Graphene Field Emitters for Plasma Displays. *Nanomaterials* **2019**, *9*, 1493. [CrossRef]
14. Chen, L.; Yu, H.; Zhong, J.; Song, L.; Wu, J.; Su, W. Graphene field emitters: A review of fabrication, characterization and properties. *Mater. Sci. Eng. B* **2017**, *220*, 44–58. [CrossRef]
15. Hong, P.N.; Minh, D.N.; Van Hung, N.; Minh, P.N.; Khoi, P.H. Carbon Nanotube and Graphene Aerogels—The World's 3D Lightest Materials for Environment Applications: A Review. *Int. J. Mater. Sci. Appl.* **2017**, *6*, 277.

16. Wang, M.-S.; Chen, Q.; Peng, L.-M. Field-Emission Characteristics of Individual Carbon Nanotubes with a Conical Tip: The Validity of the Fowler-Nordheim Theory and Maximum Emission Current. *Small* **2008**, *4*, 1907–1912. [CrossRef]
17. Li, J.; Lei, W.; Zhang, X.; Zhou, X.; Wang, Q.; Zhang, Y.; Wang, B. Field emission characteristic of screen-printed carbon nanotube cathode. *Appl. Surf. Sci.* **2003**, *220*, 96–104. [CrossRef]
18. Nilsson, L.; Groening, O.; Emmenegger, C.; Kuettel, O.; Schaller, E.; Schlapbach, L.; Kind, H.; Bonard, J.-M.; Kern, K. Scanning field emission from patterned carbon nanotube films. *Appl. Phys. Lett.* **2000**, *76*, 2071–2073. [CrossRef]
19. Silan, J.L.; Niemann, D.L.; Ribaya, B.P.; Rahman, M.; Meyyappan, M.; Nguyen, C.V. Carbon nanotube pillar arrays for achieving high emission current densities. *Appl. Phys. Lett.* **2009**, *95*, 133111. [CrossRef]
20. Hai, N.T.; Minh, D.N.; Minh, D.N.; Dung, N.D.; Hai, L.N.; Hong, P.N.; Hong, N.T. Hot-filament CVD Growth of Vertically-aligned Carbon Nanotubes on Support Materials for Field Electron Emitters. *VNU J. Sci. Math. Phys.* **2020**, *36*, 98–105.
21. Wang, J.; Ghantasala, M.; McLean, R. Bias sputtering effect on ultra-thin SmCo5 films exhibiting large perpendicular coercivity. *Thin Solid Films* **2008**, *517*, 656–660. [CrossRef]
22. Biesiekierski, A.; Wang, J.; Gepreel, M.A.-H.; Wen, C. A new look at biomedical Ti-based shape memory alloys. *Acta Biomater.* **2012**, *8*, 1661–1669. [CrossRef] [PubMed]
23. Wang, J.; Sood, D.K.; Ghantasala, M.K.; Dytlewski, N. Characterization of sputtered SmCo thin films for light element contamination using RBS and HIERDA techniques. *Vacuums* **2004**, *75*, 17–23. [CrossRef]
24. Pham, D.; Berndt, C.; Gbureck, U.; Zreiqat, H.; Truong, V.; Ang, A. Mechanical and chemical properties of Baghdadite coatings manufactured by atmospheric plasma spraying. *Surf. Coat. Technol.* **2019**, *378*, 124945. [CrossRef]
25. Chen, Y.M.; Chen, C.A.; Huang, Y.S.; Lee, K.Y.; Tiong, K.-K. Characterization and enhanced field emission properties of IrO2-coated carbon nanotube bundle arrays. *Nanotechnology* **2009**, *21*, 035702. [CrossRef]
26. Rai, P.; Mohapatra, D.R.; Hazra, K.S.; Misra, D.S.; Tiwari, S.P. Nanotip formation on a carbon nanotube pillar array for field emission application. *Appl. Phys. Lett.* **2008**, *93*, 131921. [CrossRef]
27. Zhang, Y.; Liao, M.; Deng, S.; Chen, J.; Xu, N. In situ oxygen-assisted field emission treatment for improving the uniformity of carbon nanotube pixel arrays and the underlying mechanism. *Carbon* **2011**, *49*, 3299–3306. [CrossRef]
28. Hong, N.T.; Kim, S.Y.; Koh, K.H.; Lee, D. Quantitative elucidation of the rapid growth and growth saturation of millimeter-scale vertically aligned carbon nanotubes by hot-filament chemical vapor deposition. *Thin Solid Films* **2011**, *519*, 4432–4436. [CrossRef]
29. Lim, Z.H.; Lee, A.; Lim, K.Y.Y.; Zhu, Y.; Sow, C.-H. Systematic investigation of sustained laser-induced incandescence in carbon nanotubes. *J. Appl. Phys.* **2010**, *107*, 64319. [CrossRef]
30. Hong, N.T.; Baek, I.H.; Rotermund, F.; Koh, K.H.; Lee, D. Femtosecond laser machining: A new technique to fabricate carbon nanotube based emitters. *J. Vac. Sci. Technol. B* **2010**, *28*, C2B38–C2B42. [CrossRef]
31. Santidrián, A.; González-Domínguez, J.M.; Diez-Cabanes, V.; Hernández-Ferrer, J.; Maser, W.K.; Benito, A.M.; Ansón-Casaos, A.; Cornil, J.; Da Ros, T.; Kalbac, M.; et al. A tool box to ascertain the nature of doping and photoresponse in single-walled carbon nanotubes. *Phys. Chem. Chem. Phys.* **2019**, *21*, 4063–4071. [CrossRef]
32. Flygare, M.; Svensson, K. Quantifying crystallinity in carbon nanotubes and its influence on mechanical behavior. *Mater. Today Commun.* **2019**, *18*, 39–45. [CrossRef]
33. Labunov, V.; Prudnikava, A.; Bushuk, S.; Filatov, S.; Shulitski, B.; Tay, B.K.; Shaman, Y.; Basaev, A. Femtosecond laser modification of an array of vertically aligned carbon nanotubes intercalated with Fe phase nanoparticles. *Nanoscale Res. Lett.* **2013**, *8*, 1–10. [CrossRef]
34. Bonard, J.-M.; Klinke, C.; Dean, K.A.; Coll, B.F. Degradation and failure of carbon nanotube field emitters. *Phys. Rev. B* **2003**, *67*, 115406. [CrossRef]
35. Chen, L.-F.; Ji, Z.; Mi, Y.-H.; Ni, H.-L.; Zhao, H.-F. Nonlinear characteristics of the Fowler–Nordheim plots of carbon nanotube field emission. *Phys. Scr.* **2010**, *82*, 35602. [CrossRef]
36. Gautier, L.-A.; Le Borgne, V.; Al Moussalami, S.; El Khakani, M.A. Enhanced field electron emission properties of hierarchically structured MWCNT-based cold cathodes. *Nanoscale Res. Lett.* **2014**, *9*, 55. [CrossRef] [PubMed]
37. Gröning, O.; Küttel, O.M.; Emmenegger, C.; Gröning, P.; Schlapbach, L. Field emission properties of carbon nanotubes. *J. Vac. Sci.* **2000**, *18*, 665. [CrossRef]

38. Huang, J.Y.; Kempa, K.; Jo, S.H.; Chen, S.; Ren, Z.F. Giant field enhancement at carbon nanotube tips induced by multistage effect. *Appl. Phys. Lett.* **2005**, *87*, 053110. [CrossRef]
39. Pandey, A.; Prasad, A.; Moscatello, J.; Ulmen, B.; Yap, Y.K. Enhanced field emission stability and density produced by conical bundles of catalyst-free carbon nanotubes. *Carbon* **2010**, *48*, 287–292. [CrossRef]

© 2020 by the authors. Licensee MDPI, Basel, Switzerland. This article is an open access article distributed under the terms and conditions of the Creative Commons Attribution (CC BY) license (http://creativecommons.org/licenses/by/4.0/).

Review

Recent Advances of Graphene-Derived Nanocomposites in Water-Based Drilling Fluids

Rabia Ikram [1,*], Badrul Mohamed Jan [1], Jana Vejpravova [2], M. Iqbal Choudhary [3,4] and Zaira Zaman Chowdhury [5]

1. Department of Chemical Engineering, University of Malaya, Kuala Lumpur 50603, Malaysia; badrules@um.edu.my
2. Department of Condensed Matter Physics, Faculty of Mathematics and Physics, Charles University, Ke Karlovu 5, 121 16 Prague 2, Czech Republic; jana@mag.mff.cuni.cz
3. HEJ, Research Institute of Chemistry, International Center for Chemical and Biological Sciences, University of Karachi, Karachi 75270, Pakistan; iqbal.choudhary@iccs.edu
4. Panjwani Center for Molecular Medicine and Drug Research, International Center for Chemical and Biological Sciences, University of Karachi, Karachi 75270, Pakistan
5. Nanotechnology & Catalysis Research Centre, Deputy Vice Chancellor (Research & Innovation) Office, University of Malaya, Kuala Lumpur 50603, Malaysia; dr.zaira.chowdhury@um.edu.my
* Correspondence: raab@um.edu.my

Received: 4 July 2020; Accepted: 24 July 2020; Published: 11 October 2020

Abstract: Nanocomposite materials have distinctive potential for various types of captivating usage in drilling fluids as a well-designed solution for the petroleum industry. Owing to the improvement of drilling fluids, it is of great importance to fabricate unique nanocomposites and advance their functionalities for amplification in base fluids. There is a rising interest in assembling nanocomposites for the progress of rheological and filtration properties. A series of drilling fluid formulations have been reported for graphene-derived nanocomposites as additives. Over the years, the emergence of these graphene-derived nanocomposites has been employed as a paradigm to formulate water-based drilling fluids (WBDF). Herein, we provide an overview of nanocomposites evolution as engineered materials for enhanced rheological attributes in drilling operations. We also demonstrate the state-of-the-art potential graphene-derived nanocomposites for enriched rheology and other significant properties in WBDF. This review could conceivably deliver the inspiration and pathways to produce novel fabrication of nanocomposites and the production of other graphenaceous materials grafted nanocomposites for the variety of drilling fluids.

Keywords: nanotechnology; graphene-derived materials; mud cake; rheology; effect of nanocomposites; fluid loss; water-based drilling fluids

1. Introduction

Over the years, the influential production of nanomaterials through nanotechnology has prominently contributed to the advancement of expertise in many industries. Numerous studies have been performed to address the impact of significant applications of nanomaterials in drilling fluids [1]. Lately, materials sciences and engineering accomplished remarkable progress in the field of nanocomposites fabrication with enriched physical, chemical, and mechanical properties [2]. A widespread range of studies were directed at processing these nanocomposites [3]. The combination of graphene in nanocomposites has resulted in their enrichment of mechanical strength, thermal stability, electrochemical activity, electrical conductivity as well as gas barrier properties [4,5]. Manifestation of graphene in these composites has established unique properties via increasing functional groups on the material's surface for a variety of drilling fluids [6]. A smooth process of drilling requires an appropriate well control system, suitable usage of a blowout preventer, and proper formulation of drilling fluids [7]. To motivate drilling of wellbore, drilling fluids are utilized to circulate in the borehole to ensure an efficient drilling process [8]. Typically, three substantial types of drilling fluids or drilling muds have been reported: water-based drilling fluids (WBDF), oil-based drilling fluids (OBDF), and synthetic-based drilling fluids (SBDF) [9,10].

These drilling fluids motivate the drilling process by transporting suspended cuttings back to the surface, cooling the drill bit and providing stabilization for the rock formation, and also controlling the pressure inside the well [11]. In addition, drilling fluids are reported to prevent corrosion of the equipment and mud cake formation on the wellbore wall as well preserving the cuttings to settle down if the circulation stops abruptly [12]. Due to the inefficient methods of well cleaning, the cuttings inclined to deposit when circulation stops causes the bit to become incapable of operating properly and the drills fresh formation becomes buried under those deposited cuttings, ultimately leading to a delay in the drilling process [13]. Conjointly, drilling fluids are required to remove heat from the bit and transfer the heat to the surface to cool the drill bit [14]. In the case of improper handling, the bits performance is diminished and eventually become damaged [15].

One of the major advantages of nanocomposites include affirmative changes by adding even a small amount of graphene filler in the presence of miscellaneous polymeric matrices [16]. Therefore, processing of nanocomposites is critical for environmental and cost-friendly rheological behavior. The most promising features, such as plastic viscosity, yield point, shear rates, and gel strength along with filtrate loss and mud cake thickness, are frequently assessed for the rheological performance of drilling fluids [17,18]. This review highlights the importance of graphene-derived nanocomposites with polymer, active carbon, metal, metal oxide, carbon fiber, and their applications in WBDF. A summarized effect of nanocomposites is abridged that incorporates nano-sized particles into a matrix of a standard material for drilling fluids. Based on the functionality of these nanocomposites, proper treatment of drilling fluids rheology is indeed imperative, so they tend to deliver their functions for a smooth drilling process.

2. Role of Significant Nanocomposites in Drilling Fluids

There have been advances in fabricating nanocomposites composed of a multiphase solid materials, where one of the dispersed phases is in the nanometer-scale dimension [19] and the other is a major phase, such as ceramic, metals, or polymers [20]; carbon-carbon composites [21] and nanocarbon [22] are combined as a matrix material.

For the past years, several studies have been conducted on the applications of numerous nanocomposites in drilling fluid such as polyacrylamide/clay nanocomposite [23], nanocarboxymethyl cellulose/polystyrene core-shell nanocomposite [24], polymer nanocomposite [25], nanosilica polymer composite [17], TiO_2-polyacrylamide [26], clay nanocomposite [27], ZnO-clay composite, and ZnO-Am nanocomposite [27,28]. The outcome of these studies presented a homogenous dispersion of nanocomposites in the drilling fluids to serve multiple functions simultaneously, for instance, fluid loss control, high thermal stability, enhanced rheological performance and a reduction in mud cake thickness. Furthermore, common methods, such as injection molding, solvent processing, chemical and vapor techniques, in situ polymerization, melt blending, template synthesis, spray pyrolysis, and sol-gel methods have been utilized to improve nanophase dispersion while processing graphene-derived nanocomposites [2,3,6,16]. The application of nanocomposites as additives, such as stabilizers, surfactants or ionic liquids, have had a productive effect on the rheological properties as compared to drilling fluids with single nanoparticles [5,28].

2.1. Effect of Silica Nanocomposites

Cheraghian et al. [29] summarized their research on Clay/SiO_2 nanocomposite as compared to SiO_2 nanoparticles to determine the rheological properties of drilling fluids. Their work fabricated Clay/SiO_2 nanocomposite consisting of nano-fumed silica and sodium bentonite synthesized through an effective hydrothermal method. The effects of SiO_2 nanocomposites in rheological tests at low and high temperatures showed that properties, such as apparent viscosity, plastic viscosity, yield point, and gel strength, increase as the concentration of nanocomposite is augmented. The measured values were higher than the drilling fluid with SiO$_2$ nanoparticles and base fluid. Due to the small size of the SiO$_2$ nanocomposites, the plugged pores were more effective than the SiO$_2$ nanoparticles which then formed a thin and impermeable mud cake to reduce filtrate loss control. Therefore, the SiO$_2$ nanocomposites also showed remarkable filtration control at high temperatures. By the additivation of a minute amount of SiO$_2$ nanocomposites (0.1 wt.%), a sufficient and lower fluid loss control at a recommended value of 15 mL by the American Petroleum Institute (API) was displayed [30]. Consequently, it was established that SiO$_2$ nanocomposites are efficient at high pressure–high temperature (HPHT) wellbore conditions of drilling fluids.

2.2. Effect of Copper Oxide Nanocomposites

Several studies have been implemented on the application of copper oxide nanocomposites in drilling fluids with better results in water-based muds [23,26]. Saboori et al. [31] studied the influence of copper oxide/polyacrylamide (CuO/PAM) nanocomposite synthesized through a solution polymerization method as shown in Figure 1. The improvement in rheology was observed together with the thermal conductivity of bentonite drilling fluids. The outcome of this study represented high viscosity by increasing the concentration of CuO/PAM nanocomposite due to the usage of PAM, which is known as a viscosifier in drilling fluids. Meanwhile, an increase in CuO/PAM nanocomposite concentration has been resulted in reduction of fluid loss volume and mud cake thickness due to the appropriate pore sealing by nanocomposites with an average size of 55.4 nm. Therefore, the application of CuO/PAM nanocomposite has enhanced the thermal conductivity due to the heat transfer between the drill bit and the drilling fluid. It has been proved that instantaneous function of CuO/PAM nanocomposite was achieved as an additive in the drilling fluids [32,33].

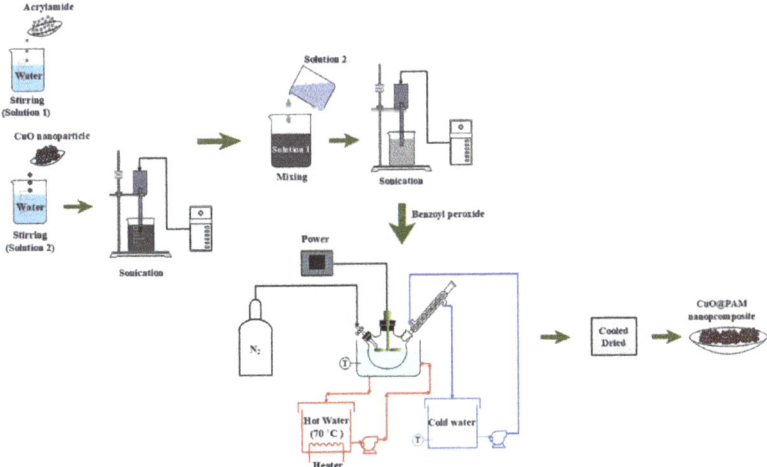

Figure 1. Synthesis of copper oxide/polyacrylamide (CuO/PAM) nanocomposite through the solution polymerization process. Reprinted with permission from [31]. Copyright Elsevier, 2019.

2.3. Role of Titanium Dioxide Nanocomposites

Titanium dioxide (TiO$_2$) as metal oxide nanoparticles have been evidenced as chemically and physically stable. Also, they are exposed as nontoxic by providing high thermal conductivities [34]. Based on the unique characteristics of TiO$_2$, Sadeghalvaad et al. [26] produced TiO$_2$/polyacrylamide (PAM) nanocomposites through the solution polymerization method in order to increase the properties of nanocomposites which acted as viscosifier and thermal conductivity enhancers. The addition of this nanocomposite is exhibited as a nanofluid that stayed in liquid form, regardless of ambient temperature as opposed to changing into solid form in drilling fluid [35]. This property was observed for the reason that capability of TiO_2 nanoparticles in transferring heat efficiently occurred due to the Brownian motion [36]. Besides, TiO_2 nanocomposites also contributed to enhance viscosity and reduced the amount of fluid loss and mud cake thickness. By increasing TiO_2 nanocomposites concentration, the rheological properties of drilling fluids were modified excessively [26].

2.4. Effect of Grass and Other Additives to Improve Rheological Properties

Drilling wastes are considered as the second largest volume of waste generated in the oil and gas industry [37]. Two major generated wastes include drilling cuttings and drilling fluids. These wastes are required pre-treatment before being disposed of in order to protect humans and the environment. Correspondingly, researchers developed an idea of producing an eco-friendly drilling fluid system which entails the same striking rheological and filtration properties among drilling fluids without non-toxic chemical additives [38].

In recent times, researchers have inspected the application of powdered grass in drilling fluids for refining the rheological and filtration performance. However, some studies are limited to the application at low pressure-low temperature (LPLT) conditions. Investigation at HPHT conditions was declared to be important in addressing the ability of this eco-friendly drilling fluid as compared to bottom hole conditions which were found likely to be at HPHT. Therefore, the application of different powdered grass concentrations in drilling fluids under LPLT and HPHT conditions were conducted by Al-Hameedi et al. [39]. Effective concentrations of 0.5%, 1.0%, and 1.5% of powdered grass were added separately to the base fluid which only consisted of bentonite, NaOH, and water. It showed that the addition of powdered grass in each concentration increased the plastic viscosity, yield point, and gel strength of the drilling fluids as compared to the base fluid. In addition, the powdered grass drilling fluid also showed impressive filtration properties at LPLT and HPHT conditions. During LPLT condition, 0.5% of powdered grass showed better fluid loss control and equated to 1.0% and 1.5% of powdered grass. Whereas at HPHT conditions, a 43% reduction in fluid loss volume was recorded by using a 1.5% concentration of powdered grass as compared to the base fluid. Furthermore, an impermeable and thin mud cake was formed in both conditions. Hence, powdered grass additive has revealed great potential for use as a fluid loss control agent.

2.5. Applications of Various Nanocomposites in Drilling Fluids

The depletion of hydrocarbons in conventional shale has caused a surge in investigation of reservoirs that exhibited severe conditions such as HPHT, high salinity [40], and widely distributed nano-sized pores. Numerous mathematical models such as Bingham plastic, Newtonian fluid, and Herschel-Bulkley fluid have been reported as reliable sources to evaluate hydraulic parameters of drilling fluids [41]. This exploration driven to meet the growing demand for energy consumption by consumers.

In order to retrieve oil beneath a layer of earth, drilling should be completed through a successful drilling process. However, drilling through conventional drilling fluids has resulted in several problems, such as excessive fluid loss [42], thick mud cake [43], and poor rheological properties at HPHT, which has led to the intervention of nanomaterials for transformation of conventional drilling fluids. By captivating into the progress of nanotechnology, nanocomposites application has improved rheological and filtration properties in drilling fluids [25,44]. Several types of nanocomposites as additives have helped to increase wellbore stability, forming impermeable mud cake [45], reduced fluid loss into formation [46], and enhanced the viscosity of the drilling fluids [31]. Though nanoparticles were examined due to the fact of their contribution in remodeling the rheological properties by plugging nano-sized pores which helped to control the fluid loss [47]. Numerous studies have indicated that the application of nanocomposites produced better drilling fluid properties. A summary of recent studies of drilling fluids is abridged along with several nanocomposites as presented in Table 1.

Table 1. Summary of various nanocomposites in rheological properties of WBDF.

Nanocomposites	Experimental Conditions	Rheological Properties	Outcomes	References
Nanosilica polymer composite	LPLT and high temperature up to 446 °F	PV, YP, GS, FL, MCT	Under LPLT conditions, the usage of nanosilica (1.0 wt.%) has greatly enhanced the rheological and filtration properties. Nanocomposites have shown no decomposition under high temperatures at 392, 410, 428, and 446 °F, proving nanocomposites to be suitable under HPHT conditions.	[17]
ZnO nanocomposite	HPHT at 109 to 370 °F and 150 to 18,500 psi	PV, YP	2.3 wt.% of ZnO nanocomposite (5 to 50 nm) resulted in upgrading the rheological properties under HPHT conditions.	[27]
TiO$_2$-polyacrylamide	LPLT	PV, YP, GS, FL	Development in rheological properties and filtration behavior under LPLT condition was observed by using 1–14 g of TiO$_2$-PAM nanocomposite.	[26,48]
ZnO-polyacrylamide	At 80 and 150 °F	PV, AV, YP, GS, FL	By adding 0.8 g of ZnO-PAM nanocomposite in the drilling fluid, PV and YP increased by 18.8% and 16.7%, respectively. Fluid loss was reduced by 12.7% and 23% under LPLT and HPHT conditions, respectively, when using 1 g of nanocomposite.	[28]
Sepiolite	LPLT and HPHT at 122 to 356 °F and 500 to 6000 psi	-	The experiment showed that WBDF samples with 1.4 wt.% of sepiolite enriched the rheological properties at 6000 psi and temperatures up to 356 °F conditions.	[49]
Polyacrylamide-grafted polyethylene glycol nanosilica	High temperature up to 203 °F	PV, YP, GS, FL	Enhancement of rheological and filtration properties was observed with 0.7 wt.% of nanocomposite and the values remained stable under a temperature of 203 °F.	[50]
Hydrophobic modified polymer-based silica	LPLT	PV, YP, GS, FL	Rheological and filtration properties were improved by adding 2.0 wt.% before and after hot rolling under 250 °F for 16 h.	[51]
Polyethylene glycol grafted nanosilica	LPLT	PV, AV, YP, GS, FL	Results have showed that PV, YP, and AV values were increased while fluid loss volume was decreased to 15.2% by adding 1 g of nanocomposite in drilling fluid.	[52]
Amphiphilic polymer/nano-silica	LPLT	PV, AV, YP, GS, FL	PV was enhanced by the addition of 7.1% nanocomposite. Addition of 2 wt.% nanocomposite reduced the fluid loss volume to 6.4 mL.	[53]
Nanocarboxymethyl cellulose/polystyrene core-shell nanocomposite	LPLT	PV, AV, YP, GS, FL	PV and AV increased by up surging the concentration of three additives. YP values were the highest for bulk CMC while core-shell nanocomposites recorded the lowest amount of fluid loss volume.	[24]
CuO/ZnO/synthetic polymer nanocomposite	LPLT and high temperature up to 400 °F	PV, YP, GS, FL, MCT	The drilling fluid exhibited stable rheological and filtration properties at 400 °F. Under LPLT conditions, low fluid loss volume was recorded. while the mud cake formed was thin and impermeable.	[54]
Lignosulfonate/Acrylamide graft copolymers	78 °F and 250 °F	PV, YP, GS, FL, MCT	At a temperature of 78 °F and 250 °F, rheological and filtration properties of the drilling fluid were enhanced with the inclusion of nanocomposite (2.4–3.5 g/350 mL water).	[55]
Hybrid polymer nanocomposite poly(styrene-methylmethacrylate-acrylic acid)/nanoclay	High temperature up to 250 °F	PV, YP, GS, FL	The nanocomposite presented stable rheology at temperatures up to 250 °F, and the combination of nanocomposite in nanoclay-based drilling fluid was reduced by up to 22% fluid loss under LPLT conditions, and a 65% reduction in the polymer-based drilling fluid.	[56]
Novel synthetic based acrylamide-styrene copolymer	High temperature up to 250 °F	PV, YP, GS, FL, MCT	Rheological and filtration properties proved a progressive fluid loss control. An ideal filtration performance at LPLT and HPHT conditions was achieved with the addition of 3 g of nanocomposite into the drilling fluid.	[57,58]

Table 1 includes the following designations: WBDF: water-based drilling fluids, LPLT: low pressure-low temperature, HPHT: high pressure-high temperature, PAM: polyacrylamide, CMC: carboxymethyl cellulose, PV: plastic viscosity, YP: yield point, GS: gel strength, FL: fluid loss, MCT: mud cake thickness.

3. Variation of Rheological Properties by Nanocomposites

Combinations of nano-sized particles are triggered by joining with other solid particles present in the mud system, such as bentonite, either directly or via intermediate chemical linkages [59]. Since nanomaterials have revealed a high surface area which greatly enhances interaction between nanocomposites and the matrix of the mud system [60], recent studies have shown high concentrations of nanocomposites and their effects at different temperatures in WBDF [61]. By increasing the number of nanomaterials in drilling fluids, sufficient hole cleaning has improved which substantially reduces many drilling problems. Subsequently, adequate amounts of nanocomposites were evaluated as an important factor to ensure drilling operations efficiently and without adding extra pressure on the drilling pump [62]. Also, the existence of nanocomposites in the drilling mud system has produced a strong repulsion force among negatively charged bentonite particles to improve the rheology of drilling fluids. Correspondingly, the increase in strong repulsion prevented agglomeration. Hence, a strong clay platelet network was formed [63].

3.1. Gel Strength

Gel strength allows the drill cuttings to suspend as soon as the circulation of the mud stops unexpectedly [64]. It prevented a critical mechanism of the wellbore to collapse. By the addition of nanocomposites in the drilling fluid, it facilitated the drilling fluid system to create a gel structure quicker than the base fluid. Moreover, studies have displayed an increase in gel strength by increasing the concentration of nanocomposites in WBDF as disclosed in Figures 2 and 3 [61].

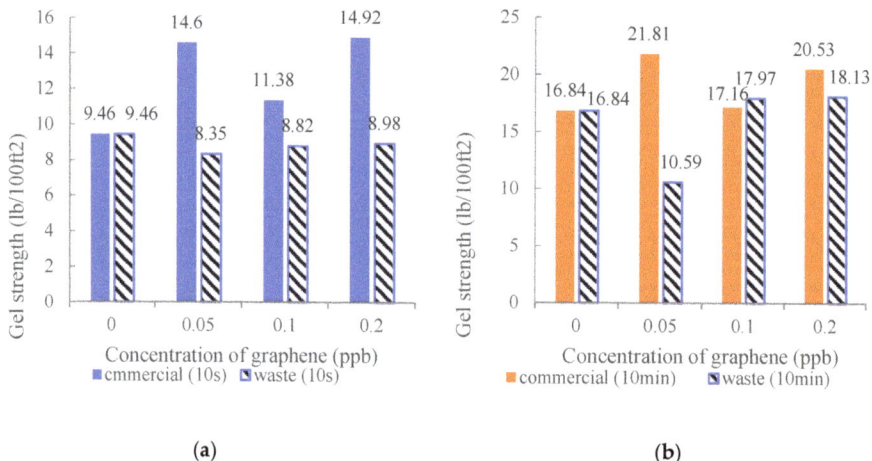

Figure 2. Gel strengths of different concentration of graphene before aging: (**a**) 10 s and (**b**) 10 min. Reprinted with permission from [61]. Copyright IOP Publishing, 2020.

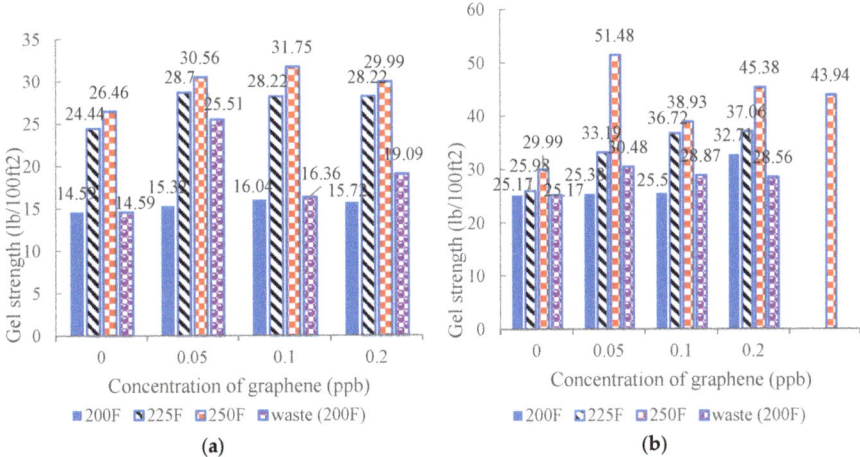

Figure 3. Gel strengths of different concentrations of graphene after aging: (**a**) 10 s and (**b**) 10 min. Reprinted with permission from [61]. Copyright IOP Publishing, 2020.

3.2. Filtrate Loss

Filtrate loss of drilling fluids was measured under LPLT or HPHT using a filter press following the procedure stated in the recommending practice API 13B-1 [65]. Nanocomposites lower than 100 nm in size were established to mix with nanoporous medium and were found challenging to be tailored into conventional drilling fluids performance [66]. Later, by increasing the concentration of nanocomposites, a thin, stiff, and impermeable mud cake was formed. Resultantly, the filtrate loss was greatly reduced into the rock formation. Barry et al. [67] revealed a decrease in filtrate loss volume which contributed by intensification of electrostatic forces between negative ions of nanocomposites to avoid flocculation with other particles as evidenced in Figure 4. Hereafter, it produced a thin and impermeable mud cake on the surface of the wall.

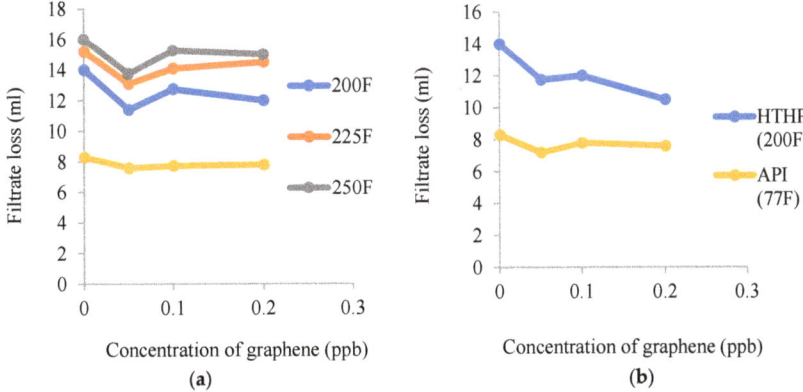

Figure 4. High pressure-high temperature (HPHT) filtrate loss at different concentrations of (**a**) commercial graphene and (**b**) waste graphene. Reprinted with permission from [61]. Copyright IOP Publishing, 2020.

3.3. Shear Rate

Since rheology describes the deformation of an ideal fluid under the influence of stress, it is equally important to know the shear rates of drilling fluids to understand their rheological performance. Researchers have examined carbonaceous materials effects for improved rheology of WBDF. For example, carbon nanotubes were considered to increase the shear stress which found to be proportional to the shearing rate. It was due to the better dispersion of base-fluid mud combined with carbon nanotubes at high shear rates [68].

Similarly, the yield stress and viscosity were up surged due to the resistant of fluid structure malformation as compared to the conventional drilling muds. Also, studies have reported low shear rate viscosity profiles of WBDF by adding nanocomposites. For instance, Vallejo et al. [69,70] conducted studies of loaded dispersions of carbon black (0.25%), nano-diamonds (0.50%), graphite/diamond (1%), and graphene nanoplates (1.5%) to analyze shear rates of 24 nanofluids. By increasing temperature, the samples presented a decrease in viscosity of 76% to 84% at a shear rate of 57.4 s^{-1} and 79% to 83% at a shear rate of 489 s^{-1}.

Likewise, proportional carbon-based nanofluids presented a decline of dynamic viscosity at 70 K for 84% and a shear rate increase of 57.4 s^{-1}, hence, they examined the shear rates between 10 to 100 s^{-1} for elevated viscid conditions [71]. In addition, Sayindla et al. [72] evaluated improved rheological properties in field conditions such as shear rates below 400 s^{-1} for WBDF viscosity profiles. The encountered low shear rate viscosity profile was observed below 400 s^{-1} along with rheological performance comparison between WBDF and OBDF.

Studies have inspected the temperature effect on viscosity of nanoclay/SiO_2 water-based muds (S2-S5), their comparison with base mud (S1) and SiO_2 water-based muds (S6-S9) at 25 and 90 °C. It was detected that nanoparticles enhanced the viscosity of the muds. Conversely, viscosity of nano clay/SiO_2 water-based fluids was decreased with accelerating shear rate. The viscosity of 1000 cP at shear rate of 10 s^{-1} was observed at 25 °C, while viscosity decreased at 90 °C to 200 cP for base fluid (S1). Moreover, clay/SiO_2 water-based mud presented the viscosity of 1200 cP at 25 °C, though it dropped to 600 cP at a higher temperature of 90 °C. By increasing the concentration of these nanomaterials in WBDF, observed viscosity was enhanced at both temperatures which can be seen in Figure 5 [29].

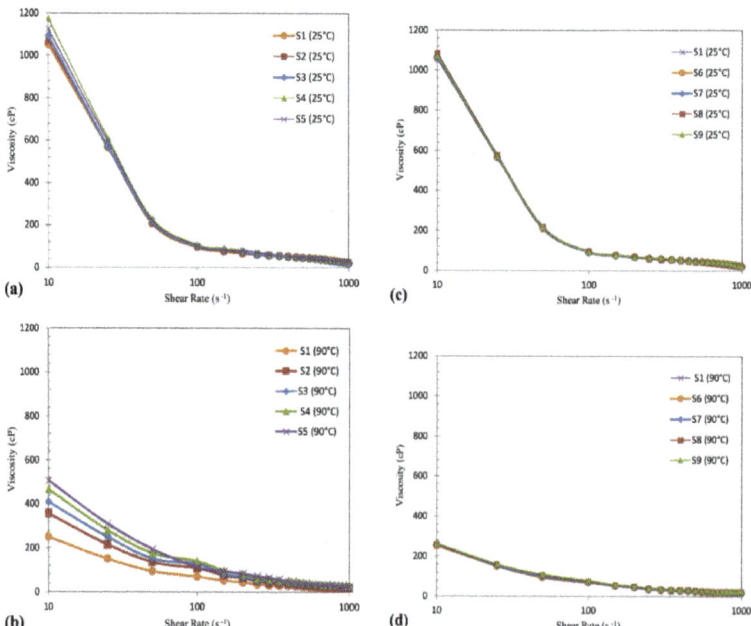

Figure 5. Temperature effects on viscosity: (**a**,**b**) nano clay/SiO_2 WBDF (S2-S5) and (**c**,**d**) SiO_2 WBDF (S6-S9) at 25 and 90 °C compared to base fluid (S1). Reprinted with permission from [29]. Copyright Elsevier, 2018.

4. Graphene Impacts in Drilling Operations

Graphene as a unique material with a one atom thick sheet of sp^2 hybridized carbon atoms is well known due to the fact of its remarkable properties such as physical, electrical, optical, electrochemical, large surface to volume area, thermal stability, and high mechanical strength. These properties make graphene a unique material for a wide range of industrial applications [73]. For example, high electrical conductivity and electrocatalytic efficiency have made graphene an exceptional tool for electrochemical applications. Likewise, other properties, such as high hydrophobicity, prominent electric conductivity, and mechanical strength, have testified to graphene's incorporation as a sensing element for biosensors. Despite these wonderful properties, synthesis of graphene sheets is difficult to produce at a large scale. Due to the poor dispersion of graphene in organic solvents [74], graphene oxide (GO) and reduced graphene oxide (rGO) have been found way more favorable as compared to graphene in drilling fluids.

Earlier studies have been produced GO either by modified Hummers method through oxidation of graphite by using sulfuric acid, nitric acid, and potassium manganate [75] or by dispersion of GO precursors in water and aqueous KOH solution [76]. Li et al. [77] proposed a method without any addition of polymeric or surfactant stabilizers to ensure notable dispersion of graphene and concluded a stress-free way to produce aqueous graphene dispersion for large-scale production. A significant role of graphene-derived structures is represented in Figure 6.

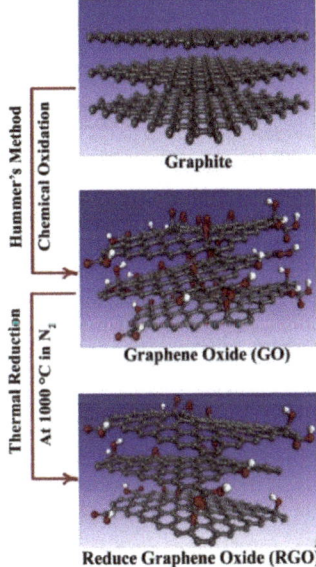

Figure 6. Schematic formation from graphite, GO to rGO. Reprinted with permission from [78]. Copyright Elsevier, 2019.

The presence of oxygenated groups in a GO structure was produced with higher solubility and well-dispersed GO nanosheets in water and organic solvents [79]. Simultaneously, these oxygenated groups distressed the electrical, mechanical, and electrochemical properties of GO which made it slightly different from graphene [73].

Likewise, rGO was produced through thermal and electrochemical reduction of GO [80]. William et al. [81] synthesized rGO through photocatalyzed reactions using TiO_2 as a catalyst stimulated with UV light. This conversion of rGO was testified to minimize the number of oxygenated groups presented in the structure of GO to gain attributes relatively similar to graphene [82]. As compared to the extraordinary properties of graphene, GO and rGO have affirmed advantageous roles to vital industries. Keeping up to date with current technology advancements, novel perspectives have been invented in favor of enhancing the quality of graphene-derived materials [83]. These approaches have been involved the combination of composite materials to counterbalance merging of conventional ceramics, metal alloys, and many polymeric materials. On the contrary, grouping of two or more materials is found to be a superlative classification of composites in the form of fibers, sheets or particles for their fabrication into the matrix phase [84]. While for nanocomposites, one of the materials is composed of dimension less than 100 nm [19].

Although graphene has attracted substantial attention due to the fact of its distinctive properties such as extraordinary mobility and conductivity, contrarily, impurities have been found in order to recover its functionality and electrochemical activity [85]. Therefore, graphene-based nanocomposites including inorganic nanostructures, conducting polymers, and organic materials have been combined for enriched mechanical strength, electrical conductivity, and thermal stability [86]. Lawal et al. [78] stated that a small quantity of graphene filler was needed to incorporate into the polymer matrix in order to enhance the properties and characteristics of nanocomposites. While addressing the drilling operations in extreme conditions, suitable additives are required to use in WBDF to avoid decomposition and better performance of the drilling fluids during drilling process. Nevertheless, due to the current advances in technology, novel nanocomposites have been selected to utilize as additives for modification of rheological properties as well as the filtration behavior of drilling fluids [87].

In 2004, extraction of graphene was successfully done by Andre K. Geim and Konstantin Novoselov which later won them the Noble Prize in 2010 [88]. The extraction method involved peeling off graphite flakes from substances with the least defects [89]. Graphene is considered as a wonderful material due to the fact of its unique attributes and thickness of an atom with high surface-area-to-volume ratio [90] which has made it suitable for many oil and gas merged industrial applications. Many studies have been conducted on refining the rheological and filtration properties of drilling fluids through incredible deployment of graphene-grafted nanoparticles [23]. The application of graphene and its derivatives was reported for rheological performance augmentation and stability of the shale [11]. Due to the fact of its nano-sized particles, it is a favorable fluid loss control additive in contrast to bentonite that is usually offered in base fluids [91]. These nano-sized particles are accomplished to plug the nano-sized pores that are present in the fluid. As a result, enriched mud cake with thin and low permeability characteristics is formed. However, some studies have exposed that graphene tends to flocculate and cause poor dispersion of nanoparticles in the drilling fluid system. This problem is resolved through the application of graphene derivative such as GO and its dispersion in an aqueous solution due to the fact of its highly hydrophilic possessions [92]. Aftab et al. [28] categorized rheological properties using graphene nanoparticles for noteworthy filtrate volume which was remained stable at HPHT conditions as compared to the base fluid. Since graphene contained a high-surface-to-volume ratio, considering a small amount of graphene was found to be sufficient enough to increase the thermal conductivity, heat tolerability, and the effectiveness of interaction among rock surfaces [89]. Hence, the drill bit was able to transfer the heat generated during the drilling process to cool off the bit. Consistently, in a research by Friedheim et al. [93], GO was proved as a viable shale inhibitor option at HPHT conditions due to the fact of its ability to mitigate the swelling effect in shale formation It was triggered by the interaction of water between clay minerals which existed in the shale and enhanced wellbore stability to prevent from collapse.

4.1. Graphene-Derived Nanocomposites in WBDF

The combination of graphene with composites have been utilized to develop superlative properties as a filling agent to advance the applications of nanocomposites in WBDF. A symbolic role of graphene-derived nanocomposites was tested to minimize fluid loss and notable effects on lubricity, viscosity, yield stress, shear rate, etc. Efforts of graphene flakes dispersion in WBDF resulted as unideal remediation for drilling fluids. In contrast, dispersion of graphene-derived nanocomposites compacted the interlocking of diverse materials to maintain desired pore-plugging through mud cake formation. It allows nanocomposites to serve multiple rheological functions with minute quantities due to the fact of their well-exfoliation and enriched functional characteristics in any system as presented in Figure 7 [35].

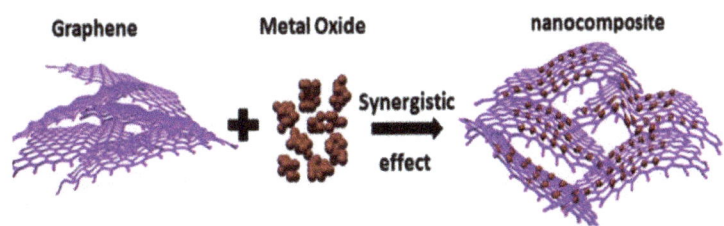

Figure 7. Fabrication of graphene-derived nanocomposite from metal oxide. Reprinted with permission from [78]. Copyright Elsevier, 2019.

A summary of studies has displayed the rheology of drilling fluids by utilizing nanocomposites as filtration reducing agent [24], heat resistant [34], viscosifier [42], shale inhibitor [51], weighing additives through the magnetic field into an environmental responsive product [56], nano-emulsion lubricant for strong inhibition [23], and desulfurizing agent to remove H_2S from drilling fluids due to the fact of their high porosity and surface area [94]. The fabrication of graphene nanocomposites with polymers, organic, inorganic, and carbon materials as well as summarization of their progress for rheological and filtration properties of WBDF is displayed in Table 2.

4.1.1. Graphene-Polymer Nanocomposites

Various studies for graphene–polymer nanocomposites have been developed by chemical or electrochemical polymerization of the monomers in the presence of graphene. Considerable polymer nanocomposites research has motivated on uncovering synthesis routes, for instance, in situ polymerization, solvent blending, melt compounding to fabricate graphene-based materials. Quantitative dispersion of these materials determines the structural deformation to stabilize their properties and unnecessary functionalization into polymer matrix. Concerning this, untangling of sheets during polymer dispersion into other materials incapacitates their unique properties [95]. Furthermore, ultrasonication has deployed better graphene dispersion in the polymer matrix as a nanofiller. On the other hand, consideration was taken for size and wt.% of desired nanomaterial. Individual types of polymer clay nanocomposites and their interaction among the fillers are demonstrated in Figure 8 [96]. Several examples of graphene–polymer nanocomposite in WBDF are enumerated in Table 2.

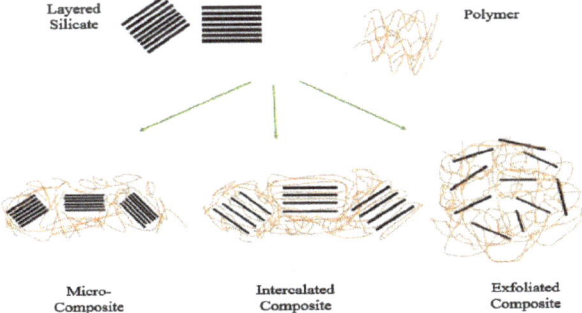

Figure 8. Assembling of polymer–clay nanocomposites through in situ intercalation, melt intercalation, and exfoliation techniques. Reprinted with permission from [96]. Copyright Elsevier, 2018.

4.1.2. Graphene-Activated Carbon Nanocomposites

Fabrication of graphene on activated carbon is employed for commercial usage due to the low cost and availability of activated carbon [97]. Besides, activated carbon has intensified performance due to the fact of its pore structure and large surface area. The resulting nanocomposite presented high performance, suitable yield point, plastic velocity as well as thin mud cake formation [98]. Examples of graphene-activated carbon nanocomposites as a high enactment material in WBDF are listed in Table 2.

4.1.3. Graphene-Metal Nanocomposites

The incorporation of metals as composite, such as copper, gold, and iron, into graphene are considered as the next generation conductors due to the fact of their room temperature tolerance and high resistivity as compared to conventional metals. The effect of graphene-metal nanocomposites has upgraded rheological properties using different concentrations of particles sizes less than 50 nm [99]. A great number of functional groups have been incorporated due to the metal nanocomposites which resulted in low shear thinning and a decrease in mud filtrate. Due to the fact of their high thermal conductivities that dissolved heat efficiently, they have added benefits for upraised electrochemical properties and analytical performances in drilling operations [100]. Influence of these graphene-metal nanocomposites for WBDF are briefly tabulated in Table 2.

4.1.4. Graphene-Metal Oxide Nanocomposites

Lately, graphene-metal oxide nanocomposites have been widely used as an alternative with cost-friendly results in drilling fluid applications [101]. Due to the combination of graphene into pores of metal oxygenated groups, these nanocomposites exhibited a great tendency to tolerate HPHT conditions of drilling operations. Recent studies have contributed to an enhanced usage of metal oxides for the fabrication of graphene nanocomposite for testified high-energy density. Rheological behavior of these nanocomposites has immensely influenced WBDF [102,103]. A significant role of graphene-metal oxide nanocomposites for value-added properties of WBDF is given in Table 2.

4.1.5. Graphene-Fiber Nanocomposites

Graphene–fiber nanocomposites are formed through direct covalent bonding of carbon fiber with graphene to advance WBDF as compared to bentonite-formulated base fluids. Studies have uncovered a great effect of these nanocomposites by novel incorporated fibers for minimizing fluid loss, reduced mud cake damage, and enhanced performance of drilling fluids. Potential studies were presented for sealing of wellbores and fluid production to prevent leakage to the surface resulting in low costs, low environmental risks, persistent wellbore reliability, and well cementing modern technologies [104]. In addition, hybridized nanocomposites enhance the thermal stability of conventional drilling fluids which shows a reduction in mud filtrate and modification of nano-additives in drilling fluids. An example of polyaniline (PANI)-GO nanocomposite dispersion to avoid self-aggregation is presented in Figure 9 [105,106].

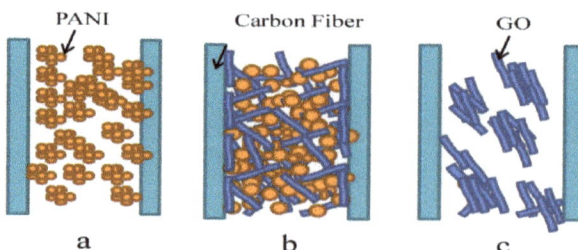

Figure 9. Representative dispersion of matrix: (**a**) random distribution of polyaniline (PANI), (**b**) enhanced dispersion of hybrid fillers due to the electrostatic interaction, (**c**) poor dispersion of GO due to the large particles agglomeration. Reprinted with permission from [105]. Copyright Elsevier, 2016.

Several researchers have investigated the role of graphene-fiber nanocomposites in WBDF which are briefly entailed in Table 2.

Table 2. Summary of the graphene nanocomposites in WBDF.

Graphene-Derived Nanocomposites	Synthesis Routes	Conditions & Outcomes	References
Graphene-polypropylene (PP) nanocomposite	Melt mixing	Enhanced PV versus SR, 20–5000 s^{-1} for nanocomposites at 200 °C. The PV of PP was observed 289 Pa s at 300 s^{-1} SR, which increased up to 513 Pa s due to the stronger interaction of the PP matrix with GO nanocomposite.	[107]
Graphene-acrylonitrile butadiene styrene resin (ABC) nanocomposite	Facile coagulation method	An increase of PV and mechanical modulus was observed due to the graphene nanocomposite.	[108]
Graphene-polyester nanocomposite	Partial pyrolysis	An enriched RP was observed due to the nanocomposite as compared to graphite.	[109]
Graphene-polyurethane nanocomposite	Solution mixing method	0.5–3 wt.% qualitative expansion was presented in the frequency of RP.	[110]
Graphene-low density polyethylene nanocomposite	Melt extrusion and film casting	Established PV, ST, viscoelasticity at 140 °C.	[111]
GO-Fe_2O_3/Al_2O_3 nanocomposite	Vertical bed method	Nanocomposite reduced the FL from 20 mL to 15 mL and MCT from 0.3 mm to 0.1 mm of WBDF with 0.02%.	[112]
GO-ZrO nanocomposite	Microwave synthesis	Enriched HPHT applications using a high-temperature range of 330 °F.	[113]
rGO-SnO_2 nanocomposite	Ultrasonic synthesis	Improved RPs were reported with the effect of vol% of rGO-SnO_2 nanocomposite (three different ratios: 1:7, 1:8, 1:10) in base fluid for PV, ST, ranging from 0 to 10,000 s^{-1} at 25 °C.	[114]
GO-ZnO nanocomposite	Chemical synthesis	A desirable increase of PV (5–28%), YP (25–42%), GS (25–33%), and a considerably reduced FL were examined.	[66]
rGO-thermally polypyrrole nanocomposite	In situ polymerization	RP of rGO-thermally polypyrrole nanocomposite was determined using a cone-plate method with ratios (100:1, 100:3, and 100:5%) and temperature (25–180 °C), and represented an increase of ST and PV due to the addition of thermally reduced GO sheets into polypyrrole.	[115]
GO-polyacrylamide (PAM) nanocomposite	Chemical synthesis (polymerization)	Nanocomposite influenced FL at LPLT and HPHT which was reduced up to 38.96% and 34.36, respectively. A noteworthy decrease in FL, MCT treated with 1.5 wt.% nanocomposite.	[116]
GO-hydrolyzed polyacrylamide nanocomposite	Chemical synthesis	Addition of GO increased PV, the effect was notable at elevated temperatures. Addition of 0.1 wt.% of GO enhanced PV by 47% and 36%, respectively, at 85 °C and 25 °C. GO increased the thermal stability due to the electrostatic hydrogen bonding among nanocomposite functional groups. After ageing for 30 days at 80 °C, PV of the composite's solution decreased very slightly, while a 59% reduction was observed for pure polymer solution.	[117]
GO nanocomposite	Chemical synthesis	Reduced FL was observed using low concentration of GO nanocomposite.	[118]

Table 2. Cont.

Graphene-Derived Nanocomposites	Synthesis Routes	Conditions & Outcomes	References
Applications of Other Graphenaceous Materials in WBDF			
GO/polyanionic cellulose polymers	Hummers method	FL of 6.1 mL over 30 min, MCT ~20μm/FL of 7.2 mL, MCT ~280 μm, high-temperature stability with better-quality RP.	[16]
GO	Hummers method	Concentration of GO increased from 0.2 wt.% to 0.6 wt.%, PV of GO aqueous dispersion noticeably increased, whereas there was no obvious change of YP and GS.	[119]
GNP	Hydrothermal technique	Amended RP was presented at HPHT due to the low friction between nanoplates.	[28]
Graphene/MgO/TiO$_2$	Hydrothermal technique	An increase of GS (92%) and PV (253%) by adding MgO (2%) and graphene (75%) was observed.	[120]
Graphite–Al$_2$O$_3$	-	Upgraded drilling mud properties were revealed; thermal conductivity (10%) and zeta potential (13%) in the presence of 0.8 wt.% of graphite–Al$_2$O$_3$.	[121]
Graphene	-	Graphene with a concentration of 17.5 mL was reduced polymer usage up to 40% for mud cake formula. Better-quality RP of 13.5lb/gal HPHT was achieved in WBDF without affecting PV and YP.	[122]
Graphitized nanotubes	Homogenization	Decrease of PV with an increase in temperature from 25–85 °C. Value-added RP with an increase of temperature were presented.	[123]
Nano-graphite nanoparticles	Water-in-oil (w/o) micro emulsions	Decrease in FL and RP were enhanced for WBDF.	[124]
Graphene/CNT	Chemical method	Reduced mud filtrate volume up to 18%. Addition of CNT reduced FL, enhanced shale formation. Addition of graphene was decreased friction coefficient from 38–59%. Better lubricity was produced by CNT as compared to graphene at elevated temperature.	[125]
Graphene–SiO$_2$	-	Concentration of nanoparticles (0.75 wt.%) yielded better performance in both LPLT and HPHT filtration tests with a reduction of 20.93% and 27.21%, respectively, as compared to the base fluid.	[126]
GO-phosphorylated from welding waste	Chemical synthesis	Addition of GO was tested for improved RP such as PV was reduced from 10 to 7 cp, YP was increased from 11–15 lbs/100 ft^2, decreased filtrate volume (6 to 3.6 mL) and reduced MCT (1.06 to 0.33 mm), with enhanced lubricity were presented.	[127]

Table 2 includes the following designations; PV: plastic viscosity, SR: shear rate, RP: rheological properties, YP: yield point, ST: shear test, GS: gel strength, FL: fluid loss, MCT: mud cake thickness, CNT: carbon nanotubes, GNP: graphene nanoplates.

4.2. Graphene Oxide on Rheological Properties

Nanotechnology through the production of nanomaterials has appreciably uplifted a positive impact on the advancement of technology among innumerable industries. Nanomaterials have been found roughly in the size of 1 to 100 nm, provided many advantages upon their applications. In recent years, the effect of GO at HPHT and LPLT conditions have reported high-performance fluid loss control [128]. Upon the discovery of graphene back in 2004 [88], it was proved to be a promising material for many applications. However, due to the difficult top-down synthesis, poor solubility, and agglomeration problem in solution, GO was synthesized from graphite using Hummers method through oxidation [129]. It has retained worthy attributes such as electrical and thermal conductivity, mechanical stiffness and biocompatible properties [130]. Additionally, remarkable mechanical properties were obtained from interfacial interaction of GO with polymer matrix by increasing the functional group of GO sheets to fabricate GO-derived nanocomposites as presented in Figure 10. In this regard, Murphy et al. [131] evaluated important factors, such as plastic viscosity and yield point, through GO as an additive in drilling fluids. They examined rheological behavior as the main factor in adjusting the printability and structure of alginate hydrogels. Consequently, novel properties of GO and their application for modified rheological properties have solved the issues of poor mechanical strength and inadequate structural reliability [132].

Recently, the role of GO in featured rheology was analyzed such as an examination of GO-SiO$_2$ nanoparticles for unconventional reservoir shales to reduce cutting dispersion [128], suspension of cutting fluids using GO [133], improved thermal stability and inhibition capabilities of WBDF in Woodford shale [134], the effect of GO functionalization to improve heavy oil recovery [135], clean swelling inhibitor in WBDF [97], and GO as an additive to improve filtration, thermal conductivity, and rheological properties of drilling fluids [116,136].

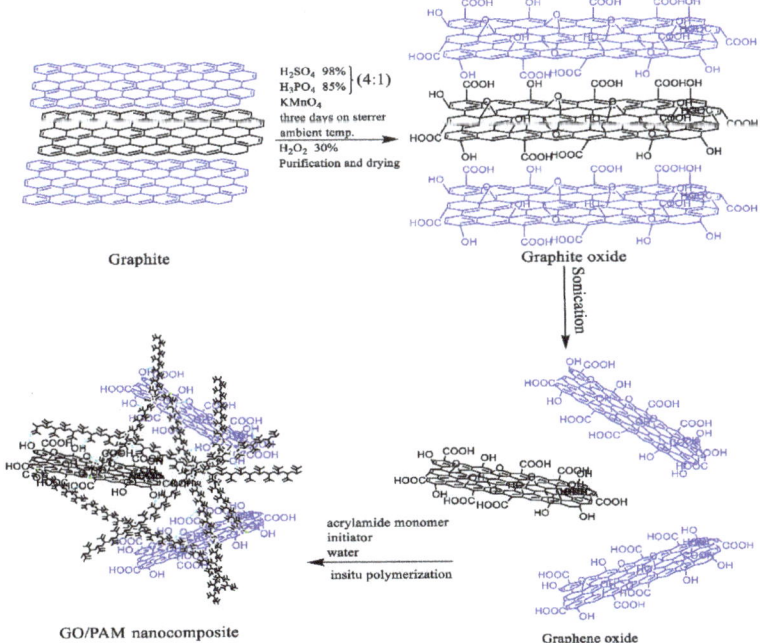

Figure 10. Structural representation of GO/PAM nanocomposite fabrication. Reprinted with permission from [116]. Copyright Elsevier, 2020.

4.3. Graphene Oxide in WBDF

Although, several nanomaterials are considered to maintain the stable rheological and filtration properties of WBDF [137], graphene contains a distinguished thickness of one atom and is known for its evident mechanical, thermal, electrical, and physical properties. However, deprived dispersion of graphene in water has prohibited its usage in WBDF [119].

Newly conducted studies by Kusrini et al. [127] have utilized GO as a superlative alternate and additive in drilling fluids. Production of GO from industrial waste using a modified Hummers method in which graphite goes through chemical oxidation and is followed by its reduction. Due to the highly hydrophilic properties of GO, it is dispersed well in aqueous solution with a wide range of concentrations [92]. From these studies, a comparative analysis of GO was observed to uplift fluid loss performance as compared to the base fluid. The filtrate loss volume was reduced from 6 mL to 3.6 mL and the mud cake thickness was decreased up to 70%. Also, GO produced positive results for the rheological properties of drilling fluid such as plastic viscosity, yield point, and gel strength. Addition of GO minimized the viscosity values. It was analyzed due to the reduction in GO size which was observed as nano-sized particles that triggered less friction and minimized the resistance of flow in drilling fluids. Low plastic viscosity values are desired to prevent high-pressure drops which result in the low circulation of fluid [122]. In contrast, yield point and gel strength are increased with the application of GO in drilling fluids. The increase of yield point was examined due to the twigging of particles together and, hence, to overcome the surface energy resulting from increased GO surface area. Furthermore, it improved the ability of the drilling fluid to carry the drill cuttings to the surface. The measured values of gel strength increased due to the attraction forces among well-dispersed GO with other particles presented in the drilling fluid system. An increase in gel strength is desirable, as it helps the drilling fluid system create a gel structure quicker than the base fluid [112]. Lastly, Alkinani et al. [138] directed an equivalent circulation density (ECD) simulation for the application of GO in drilling fluids. The ECD values of GO did not report a dominant difference with the base fluid. The ECD values indicated GO as more suitable for low-pressure wells to reduce the potential risk of circulation loss.

5. Limitations and Challenges

This review endorsed us to report several advantages of nanocomposites as additives for augmenting rheological performance and stable fluid loss control in WBDF. However, a few challenges should be addressed before they can be employed in drilling operations. Several rheological behaviors are uncovered through nanocomposites, for instance, wellbore strengthening, improved shale stability, and drill bit issues as displayed in Figure 11.

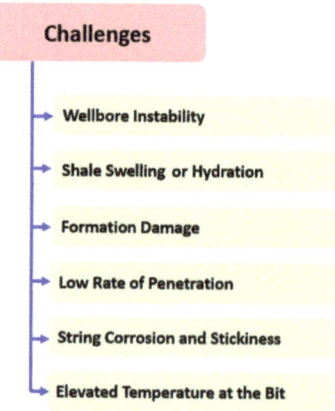

Figure 11. Challenges of drilling operations. Reprinted with permission from [139]. Copyright Elsevier, 2019.

The stability of nanocomposite dispersions reports a methodical challenge to maintain drilling fluids. A coherent method for an effective dispersion of nanomaterials in a liquid or solid medium is a crucial phase. In earlier studies, high-speed mixers, magnetic stirrers, ultrasonic baths, ball milling, and other homogenizers are presented [140]. The nano-size of composite mixtures tend to re-aggregate due to the presence of van der Waals forces and confine their high surface area control [141]. Therefore, additives, such as surfactants, stabilizers or ionic liquids, are required to increase the steric hindrance between nanocomposites fabrication and stable dispersions. A brief overview of the advantages and limitations of nanocomposites based on their processing and nanofiller content is presented in Table 3 [142].

Table 3. Advantages and disadvantages of nanocomposites synthesis routes [143].

Synthesis Routes	Nanofiller Content	Advantages	Limitations
In situ polymerization	5–70 wt.%	Fabrication and polymerization occur at the same time to produce an efficient interface between filler and polymer	Suitable for limited types of polymers
Shear press	60–70 wt.%	Fine alignment	Restricted to small-scale production
Vacuum-assisted polymer infiltration	5–70 wt.%	Competent at producing large and complex composites	Filler fractions and thickness are challenging to control
Spray winding	50–80 wt.%	Satisfactory alignment and large-scale production	Comparatively complex apparatus
Capillary rise infiltration	40–60 wt.%	User-friendly apparatus	Limited to thermoplastic polymers

Moreover, primary factors, including particle size and morphology, and other structural properties for nanocomposite dispersions are equally important. Researchers have identified challenges during quantitative studies of nanocomposites by addition of suitable additives for the chemical stability of fabricated nanocomposites in WBDF [144]. A comparative analysis of utilizing 0.3 ppb of nanosilica and graphene nanoplates established the ideal effects of graphene nanoplates on filtrate loss in WBDF [91]. It is equally challenging to find an optimized ratio of these nanocomposites for eco-friendly usage. For example, nanocomposites ratios with low concentrations of 0.5 wt.% have influenced the rheological behavior of drilling fluids [145].

Although many types of nanocomposites are commercially available, the cost of their synthesis is still an obstacle for targeted operations in oil and gas industry [146]. The American Petroleum Institute (API) has given particular specifications and procedures which provide difficulty for newly made-up nanocomposites and base fluid formulations due to the fact of their different properties, process parameters, and other requirements [147]. It is also recognized that preparation protocols and key measurements of these nanocomposites find difficulty in the combination with graphene-based materials [148]. In this regard, more research emphasis on graphene-derived materials could be a promising substitute. Significant factors such as mixing time, functionalization of additives, and order of materials doping are crucial and can knowingly affect the drilling fluids behavior. Vital procedures, consistent results, and the role of these nanocomposites in unconventional rheological studies are alarming for researchers and oil companies.

6. Future Prospects

An improved description of drilling fluids performance has advanced fundamental aspects in applied rheology. Therefore, potential analysis and extensive methodologies are essential for future research. More studies should endeavor for combinations of novel nanocomposites with graphene-based materials such as graphene-doped nano-additives and GO synthesized through novel analytical techniques. Moreover, a comprehensive quantitative analysis of these nanocomposites should be performed for drilling fluid operations. Further research should be focused on key mechanisms of interaction between nanocomposites and other additives available in drilling fluids. Apart from WBDF, optimizations of fluid formations can possibly be done by adding nanocomposites and their comparison with SBDF and OBDF. Deep analysis of these nanocomposites, for instance, bentonite or barite particles, should be compared to conventional base fluids at elevated pressure and temperature conditions. To reduce formation damages of mud cakes, advanced procedures should be utilized for the characterization and quantification of mud cakes. Several studies have entailed least exposed rheology through nanocomposites, i.e., plastic viscosity, shear rate, yield point, and filtrate loss. It is motivational for upcoming practices to take an account of complete rheograms for filtration tests. Incorporation of such nanocomposites shows a need to combat high pressure and temperature conditions encountering low shear rates and complicated information regarding cuttings, wellbore strengthening, and advancement of these techniques in several types of drilling fluids. Hereafter, the role of graphene-derived nanocomposites will have a major role in the preparation of novel additives for WBDF. It will advance a breakthrough in augmenting the efficiency of drilling operations and expand the overall competitiveness of industrial applications.

7. Conclusions

In summary, we demonstrate the usage of nanocomposites in drilling operations with their considerably amplified performance and functionality. It was shown that applications of nanocomposites by a combination of two or more nanomaterials were embedded in a matrix phase. However, challenges are required to resolve for advance production of nanocomposites. By comparing several types of nanocomposites behavior, it was revealed that graphene-derived nanocomposites, particularly, GO-nanocomposites, as additives enhanced the rheological properties of WBDF. This reflects the examination of key factors in producing nanocomposites, such as nanoparticles or other nanomaterials, combined with graphene, and examination of other rheological properties under extreme conditions. However, they have been observed to be expensive and found to be produced in small amounts. Promising attempts were displayed for the modification of plastic viscosity, yield point, gel strength, and filtrate loss at LPLH and HPHT by using graphene-derived nanocomposites. This leads the drilling industry to focus on the commercial production of nanocomposites, either through green synthesis or beyond the laboratories and consolidation of these materials into the end product preserving their nanostructures. Handling of these nanomaterials also paves the way to study a major role of these nanocomposites for OBDF as driving factors. Therefore, novel methods are needed to produce grapheneaceous nanocomposites on a large scale and at an affordable cost, prior to the applications of nanocomposites in WBDF. To conclude, this review will be helpful for researchers to discover novel routes of nanocomposites synthesis, their fabrication with graphene-grafted innovative nanomaterials, and their utilization in several unindicated rheological properties of drilling fluids. The addition of these methods could be equally helpful for future perspectives of modest comparison for a critical variety of drilling fluids.

Author Contributions: Conceptualization, R.I. and B.M.J.; Analysis, J.V. and B.M.J.; Investigation, B.M.J., J.V. and Z.Z.C.; Original draft preparation, R.I.; Writing—review and editing, R.I., B.M.J. and M.I.C.; Funding acquisition, J.V., B.M.J. and Z.Z.C. All authors have read and agreed to the published version of the manuscript.

Funding: This work was conducted with the support of Malaysia–Thailand Joint Authority under grant number IF062-2019, Fundamental Research Grant Scheme FP050-2019A, and IF073-2019 under University of Malaya, Malaysia and King Khalid University, Saudi Arabia, and Ministry of Education, Youth and Sports of the Czech Republic under Operational Programme Research, Development and Education, project Carbon Allotropes with Rationalized Nanointerfaces and Nanolinks for Environmental and Biomedical Applications (CARAT), number CZ.02.1.01/0.0/0.0/16_026/0008382.

Acknowledgments: Sincere gratitude to the University of Malaya for providing funds during the course of this study. The authors are thankful to anonymous referees for their valuable and considerate recommendations. R.I. would like to offer sincere appreciation to her father, deceased during this pandemic, who is a source of intrinsic motivation throughout her academic achievements.

Conflicts of Interest: The authors declare no conflict of interest.

References

1. Hajiabadi, S.H.; Aghaei, H.; Kalateh-Aghamohammadi, M.; Shorgasthi, M. An overview on the significance of carbon-based nanomaterials in upstream oil and gas industry. *J. Pet. Sci. Eng.* **2020**, *186*, 106783. [CrossRef]
2. Pastoriza-Santos, I.; Kinnear, C.; Pérez-Juste, J.; Mulvaney, P.; Liz-Marzán, L.M. Plasmonic polymer nanocomposites. *Nat. Rev. Mater.* **2018**, *3*, 375–391. [CrossRef]
3. Cha, G.D.; Lee, W.H.; Lim, C.; Choi, M.K.; Kim, D.H. Materials engineering, processing, and device application of hydrogel nanocomposites. *Nanoscale* **2020**, *12*, 10456–10473. [CrossRef] [PubMed]
4. Papageorgiou, D.G.; Kinloch, I.A.; Young, R.J. Mechanical properties of graphene and graphene-based nanocomposites. *Prog. Mater. Sci.* **2017**, *90*, 75–127. [CrossRef]
5. Kim, H.; Miura, Y.; Macosko, C.W. Graphene/polyurethane nanocomposites for improved gas barrier and electrical conductivity. *Chem. Mater.* **2010**, *22*, 3441–3450. [CrossRef]
6. Cai, D.; Song, M. Recent advance in functionalized graphene/polymer nanocomposites. *J. Mater. Chem.* **2010**, *20*, 7906–7915. [CrossRef]
7. Abdo, J.; Haneef, M. Clay nanoparticles modified drilling fluids for drilling of deep hydrocarbon wells. *Appl. Clay Sci.* **2013**, *86*, 76–82. [CrossRef]
8. Mao, H.; Yang, Y.; Zhang, H.; Zhang, J.; Huang, Y. A critical review of the possible effects of physical and chemical properties of subcritical water on the performance of water-based drilling fluids designed for ultra-high temperature and ultra-high pressure drilling applications. *J. Pet. Sci. Eng.* **2020**, *187*, 106795. [CrossRef]
9. Bageri, B.S.; Adebayo, A.R.; Al Jaberi, J.; Patil, S. Effect of perlite particles on the filtration properties of high-density barite weighted water-based drilling fluid. *Powder Technol.* **2020**, *360*, 1157–1166. [CrossRef]
10. Elgibaly, A.; Farahat, M.; Abd El Nabbi, M. The Optimum Types and Characteristics of Drilling Fluids Used During Drilling in The Egyption Westren Desert. *J. Pet. Min. Eng.* **2018**, *20*, 89–100. [CrossRef]
11. Neuberger, N.; Adidharma, H.; Fan, M. Graphene: A review of applications in the petroleum industry. *J. Pet. Sci. Eng.* **2018**, *167*, 152–159. [CrossRef]
12. Negin, C.; Ali, S.; Xie, Q. Application of nanotechnology for enhancing oil recovery–A review. *Petroleum* **2016**, *2*, 324–333. [CrossRef]
13. Werner, B.; Myrseth, V.; Saasen, A. Viscoelastic properties of drilling fluids and their influence on cuttings transport. *J. Pet. Sci. Eng.* **2017**, *156*, 845–851. [CrossRef]
14. Ahmad, H.M.; Kamal, M.S.; Al-Harthi, M.A. High molecular weight copolymers as rheology modifier and fluid loss additive for water-based drilling fluids. *J. Mol. Liq.* **2018**, *252*, 133–143. [CrossRef]
15. William, J.K.M.; Ponmani, S.; Samuel, R.; Nagarajan, R.; Sangwai, J.S. Effect of CuO and ZnO nanofluids in xanthan gum on thermal, electrical and high pressure rheology of water-based drilling fluids. *J. Pet. Sci. Eng.* **2014**, *117*, 15–27. [CrossRef]
16. Kosynkin, D.V.; Ceriotti, G.; Wilson, K.C.; Lomeda, J.R.; Scorsone, J.T.; Patel, A.D. Graphene Oxide as a High-Performance Fluid-Loss-Control Additive in Water-Based Drilling Fluids. *ACS Appl. Mater. Interfaces* **2012**, *4*, 222–227. [CrossRef] [PubMed]
17. Mao, H.; Qiu, Z.; Shen, Z.; Huang, W. Hydrophobic associated polymer based silica nanoparticles composite with core-shell structure as a filtrate reducer for drilling fluid at utra-high temperature. *J. Pet. Sci. Eng.* **2015**, *129*, 1–14. [CrossRef]
18. Echeverría, C.; Mijangos, C. A Way to Predict Gold Nanoparticles/Polymer Hybrid Microgel Agglomeration Based on Rheological Studies. *Nanomaterials* **2019**, *9*, 1499. [CrossRef]
19. Goyal, R.K. *Nanomaterials and Nanocomposites: Synthesis, Properties, Characterization Techniques, and Applications*; CRC Press: Boca Raton, FL, USA, 2017.
20. Camargo, P.H.C.; Satyanarayana, K.G.; Wypych, F. Nanocomposites: Synthesis, structure, properties and new application opportunities. *Mater. Res.* **2009**, *12*, 1–39. [CrossRef]
21. Oku, T. Carbon/Carbon Composites and Their Properties. In *Carbon Alloys*; Elsevier: Amsterdam, The Netherlands, 2003; pp. 523–544.
22. Lepak-Kuc, S.; Milowska, K.Z.; Boncel, S.; Szybowicz, M.; Dychalska, A.; Jozwik, I. Highly Conductive Doped Hybrid Carbon Nanotube–Graphene Wires. *ACS Appl. Mater. Interfaces* **2019**, *11*, 33207–33220. [CrossRef]
23. Jain, R.; Mahto, V. Evaluation of polyacrylamide/clay composite as a potential drilling fluid additive in inhibitive water based drilling fluid system. *J. Pet. Sci. Eng.* **2015**, *133*, 612–621. [CrossRef]

24. Saboori, R.; Sabbaghi, S.; Kalantariasl, A.; Mowla, D. Improvement in filtration properties of water-based drilling fluid by nanocarboxymethyl cellulose/polystyrene core-shell nanocomposite. *J. Pet. Explor. Prod. Technol.* **2018**, *8*, 445–454. [CrossRef]
25. Ahmad, H.M.; Kamal, M.S.; Hussain, S.M.S.; Al-Harthi, M. Synthesis of novel polymer nanocomposite for water-based drilling fluids. In *Proceedings of the 35th International Conference of the Polymer Processing Society*; AIP Publishing LLC: Melville, NY, USA, 2020.
26. Sadeghvaad, M.; Sabbaghi, S. The effect of the TiO_2/polyacrylamide nanocomposite on water-based drilling fluid properties. *Powder Technol.* **2015**, *272*, 113–119. [CrossRef]
27. Abdo, J.; Zaier, R.; Hassan, E.; Al-Sharji, H.; Al-Shabibi, A. ZnO-clay nanocomposites for enhance drilling at HTHP conditions. *Surf. Interface Anal.* **2014**, *46*, 970–974. [CrossRef]
28. Aftab, A.; Ismail, A.R.; Khokhar, S.; Ibupoto, Z.H. Novel zinc oxide nanoparticles deposited acrylamide composite used for enhancing the performance of water-based drilling fluids at elevated temperature conditions. *J. Pet. Sci. Eng.* **2016**, *146*, 1142–1157. [CrossRef]
29. Cheraghian, G.; Wu, Q.; Mostofi, M.; Li, M.C.; Afrand, M.; Sangwai, J.S. Effect of a novel clay/silica nanocomposite on water-based drilling fluids: Improvements in rheological and filtration properties. *Colloids Surf. A* **2018**, *555*, 339–350. [CrossRef]
30. Abdo, J.; AL-Sharji, H.; Hassan, E. Effects of nano-sepiolite on rheological properties and filtration loss of water-based drilling fluids. *Surf. Interface Anal.* **2016**, *48*, 522–526. [CrossRef]
31. Saboori, R.; Sabbaghi, S.; Kalantariasl, A. Improvement of rheological, filtration and thermal conductivity of bentonite drilling fluid using copper oxide/polyacrylamide nanocomposite. *Powder Technol.* **2019**, *353*, 257–266. [CrossRef]
32. Liu, M.; Lin, M.C.; Wang, C. Enhancements of thermal conductivities with Cu, CuO, and carbon nanotube nanofluids and application of MWNT/water nanofluid on a water chiller system. *Nanoscale Res. Lett.* **2011**, *6*, 297. [CrossRef]
33. Sahooli, M.; Sabbaghi, S. Investigation of thermal properties of CuO nanoparticles on the ethylene glycol–water mixture. *Mater. Lett.* **2013**, *93*, 254–257. [CrossRef]
34. Das, P.K.; Mallik, A.K.; Ganguly, R.; Santra, A.K. Synthesis and characterization of TiO_2–water nanofluids with different surfactants. *Int. Commun. Heat Mass Transfer.* **2016**, *75*, 341–348. [CrossRef]
35. Rafati, R.; Smith, S.R.; Haddad, A.S.; Novara, R.; Hamidi, H. Effect of nanoparticles on the modifications of drilling fluids properties: A review of recent advances. *J. Pet. Sci. Eng.* **2018**, *161*, 61–76. [CrossRef]
36. Krishnakumar, T.; Sheeba, A.; Mahesh, V.; Prakash, M.J. Heat transfer studies on ethylene glycol/water nanofluid containing TiO_2 nanoparticles. *Int. J. Refrig.* **2019**, *102*, 55–61. [CrossRef]
37. Ismail, A.R.; Alias, A.; Sulaiman, W.; Jaafar, M.; Ismail, I. Drilling fluid waste management in drilling for oil and gas wells. *Chem. Eng. Trans.* **2017**, *56*, 1351–1356.
38. Onwukwe, S.; Nwakaudu, M. Drilling wastes generation and management approach. *Int. J. Environ. Sci. Dev.* **2012**, *3*, 252. [CrossRef]
39. Al-Hameedi, A.T.T.; Alkinani, H.H.; Dunn-Norman, S.; Al-Alwani, M.A.; Alshammari, A.F.; Albazzaz, H.W. Insights into the application of new eco-friendly drilling fluid additive to improve the fluid properties in water-based drilling fluid systems. *J. Pet. Sci. Eng.* **2019**, *183*. [CrossRef]
40. Song, Y.; Li, Z.; Jiang, L.; Hong, F. The concept and the accumulation characteristics of unconventional hydrocarbon resources. *Pet. Sci.* **2015**, *12*, 563–572. [CrossRef]
41. Oseh, J.O.; Mohd Norddin, M.N.A.; Ismail, I.; Gbadamosi, A.O.; Agi, A.; Mohammed, H.N. A novel approach to enhance rheological and filtration properties of water–based mud using polypropylene–silica nanocomposite. *J. Pet. Sci. Eng.* **2019**, *181*. [CrossRef]
42. Borthakur, A.; Choudhury, S.; Sengupta, P.; Rao, K.; Nihalani, M. Synthesis and evaluation of partially hydrolysed polyacrylamide (PHPA) as viscosifier in water based drilling fluids. *Ind. J. Chem. Technol.* **1997**, *4*, 83–88.
43. Hale, A.; Mody, F. Partially hydrolyzed polyacrylamide (PHPA) mud systems for Gulf of Mexico deepwater prospects. SPE International Symposium on oilfield chemistry. *Soc. Pet. Eng.* **1993**.
44. Nizamani, A.; Ismail, A.R.; Junin, R.; Dayo, A.; Tunio, A.; Ibupoto, Z. Synthesis of titaniabentonite nano composite and its applications in water-based drilling fluids. *Chem. Eng. Trans.* **2017**, *56*, 949–954.

45. Bhasney, S.M.; Kumar, A.; Katiyar, V. Microcrystalline cellulose, polylactic acid and polypropylene biocomposites and its morphological, mechanical, thermal and rheological properties. *Compos. Part B* **2020**, *184*, 107717. [CrossRef]
46. Jahed, M.; Naderi, G.; Hamid Reza Ghoreishy, M. Microstructure, mechanical, and rheological properties of natural rubber/ethylene propylene diene monomer nanocomposites reinforced by multi-wall carbon nanotubes. *Polym. Compos.* **2018**, *39*, E745–E753. [CrossRef]
47. Zhao, Q.; Finlayson, C.E.; Snoswell, D.R.; Haines, A.; Schäfer, C.; Spahn, P. Large-scale ordering of nanoparticles using viscoelastic shear processing. *Nat. Commun.* **2016**, *7*, 1–10. [CrossRef] [PubMed]
48. Srivatsa, J.T.; Ziaja, M.B. An experimental investigation on use of nanoparticles as fluid loss additives in a surfactant–polymer based drilling fluid. In *Conference Proceedings, IPTC 2012: International Petroleum Technology Conference*; European Association of Geoscientists & Engineers: Amsterdam, The Netherlands, 2012; p. 280.
49. Needaa, A.M.; Pourafshary, P.; Hamoud, A.-H.; Jamil, A. Controlling bentonite-based drilling mud properties using sepiolite nanoparticles. *Pet. Explor. Dev.* **2016**, *43*, 717–723.
50. Jain, R.; Mahto, T.K.; Mahto, V. Rheological investigations of water based drilling fluid system developed using synthesized nanocomposite. *Korea-Aust. Rheol. J.* **2016**, *28*, 55–65. [CrossRef]
51. Xu, J.G.; Qiu, Z.; Zhao, X.; Huang, W. Hydrophobic modified polymer based silica nanocomposite for improving shale stability in water-based drilling fluids. *J. Pet. Sci. Eng.* **2017**, *153*, 325–330. [CrossRef]
52. Xu, J.G.; Qiu, Z.S.; Zhao, X.; Zhong, H.Y.; Li, G.R.; Huang, W.A. Synthesis and characterization of shale stabilizer based on polyethylene glycol grafted nano-silica composite in water-based drilling fluids. *J. Pet. Sci. Eng.* **2018**, *163*, 371–377. [CrossRef]
53. Qiu, Z.; Xu, J.; Yang, P.; Zhao, X.; Mou, T.; Zhong, H. Effect of Amphiphilic Polymer/Nano-Silica Composite on Shale Stability for Water-Based Muds. *Appl. Sci.* **2018**, *8*, 1839. [CrossRef]
54. Khan, M.A.; Al-Saliml, H.; Arsanjani, L.N. Development of high temperature high pressure (HTHP) water based drilling mud using synthetic polymers, and nanoparticles. *J. Adv. Res. Fluid Mech. Thermal Sci.* **2018**, *45*, 99–108.
55. Abdollahi, M.; Pourmahdi, M.; Nasiri, A.R. Synthesis and characterization of lignosulfonate/acrylamide graft copolymers and their application in environmentally friendly water-based drilling fluid. *J. Pet. Sci. Eng.* **2018**, *171*, 484–494. [CrossRef]
56. Mohamadian, N.; Ghorbani, H.; Wood, D.A.; Khoshmardan, M.A. A hybrid nanocomposite of poly (styrene-methyl methacrylate-acrylic acid)/clay as a novel rheology-improvement additive for drilling fluids. *J. Polym. Res.* **2019**, *26*, 33. [CrossRef]
57. Davoodi, S.; SA, A.R.; Soleimanian, A.; Jahromi, A.F. Application of a novel acrylamide copolymer containing highly hydrophobic comonomer as filtration control and rheology modifier additive in water-based drilling mud. *J. Pet. Sci. Eng.* **2019**, *180*, 747–755. [CrossRef]
58. Kök, M.V.; Bal, B. Effects of silica nanoparticles on the performance of water-based drilling fluids. *J. Pet. Sci. Eng.* **2019**, *180*, 605–614. [CrossRef]
59. Kang, Y.; She, J.; Zhang, H.; You, L.; Song, M. Strengthening shale wellbore with silica nanoparticles drilling fluid. *Petroleum* **2016**, *2*, 189–195. [CrossRef]
60. Quintero, L.; Cardenas, A.E.; Clark, D.E. Nanofluids and Methods of Use for Drilling and Completion Fluids. U.S. Patent No. 8,822,386, 2 September 2014.
61. Putra, P.H.M.; Jan, B.M.; Patah, M.F.A.; Junaidi, M.U.M. Effect of temperature and concentration of industrial waste graphene on rheological properties of water based mud. *IOP Conf. Ser. Mater. Sci. Eng.* **2020**, *778*, 012120.
62. Razi, M.M.; Mazidi, M.; Razi, F.M.; Aligolzadeh, H.; Niazi, S. Artificial neural network modeling of plastic viscosity, yield point, and apparent viscosity for water-based drilling fluids. *J. Dispers. Sci. Technol.* **2013**, *34*, 822–827. [CrossRef]
63. Sajjadian, M.; Sajjadian, V.A.; Rashidi, A. Experimental evaluation of nanomaterials to improve drilling fluid properties of water-based muds HP/HT applications. *J. Pet. Sci. Eng.* **2020**, *190*. [CrossRef]
64. Amani, M.; Al-Jubouri, M. The effect of high pressures and high temperatures on the properties of water based drilling fluids. *Energy Sci. Technol.* **2012**, *4*, 27–33.

65. Smith, S.R.; Rafati, R.; Haddad, A.S.; Cooper, A.; Hamidi, H. Application of aluminium oxide nanoparticles to enhance rheological and filtration properties of water based muds at HPHT conditions. *Colloids Surf. A* **2018**, *537*, 361–371. [CrossRef]
66. Ghayedi, A.; Khosravi, A. Laboratory investigation of the effect of GO-ZnO nanocomposite on drilling fluid properties and its potential on H2S removal in oil reservoirs. *J. Pet. Sci. Eng.* **2020**, *184*, 106684. [CrossRef]
67. Barry, M.M.; Jung, Y.; Lee, J.K.; Phuoc, T.X.; Chyu, M.K. Fluid filtration and rheological properties of nanoparticle additive and intercalated clay hybrid bentonite drilling fluids. *J. Pet. Sci. Eng.* **2015**, *127*, 338–346. [CrossRef]
68. Abduo, M.; Dahab, A.; Abuseda, H.; AbdulAziz, A.M.; Elhossieny, M. Comparative study of using water-based mud containing multiwall carbon nanotubes versus oil-based mud in HPHT fields. *Egypt. J. Pet.* **2016**, *25*, 459–464. [CrossRef]
69. Vallejo, J.P.; Żyła, G.; Fernández-Seara, J.; Lugo, L. Influence of six carbon-based nanomaterials on the rheological properties of nanofluids. *Nanomaterials* **2019**, *9*, 146. [CrossRef] [PubMed]
70. Monjezi, S.; Jones, J.D.; Nelson, A.K.; Park, J. The effect of weak confinement on the orientation of nanorods under shear flows. *Nanomaterials* **2018**, *8*, 130. [CrossRef]
71. Vallejo, J.P.; Żyła, G.; Fernández-Seara, J.; Lugo, L. Rheological behaviour of functionalized graphene nanoplatelet nanofluids based on water and propylene glycol: Water mixtures. *Int. Commun. Heat Mass Transfer.* **2018**, *99*, 43–53. [CrossRef]
72. Sayindla, S.; Lund, B.; Ytrehus, J.D.; Saasen, A. Hole-cleaning performance comparison of oil-based and water-based drilling fluids. *J. Pet. Sci. Eng.* **2017**, *159*, 49–57. [CrossRef]
73. Mehrali, M.; Sadeghinezhad, E.; Latibari, S.T.; Kazi, S.N.; Mehrali, M.; Zubir, M. Investigation of thermal conductivity and rheological properties of nanofluids containing graphene nanoplatelets. *Nanoscale Res. Lett.* **2014**, *9*, 15. [CrossRef]
74. Liang, Y.; Wu, D.; Feng, X.; Müllen, K. Dispersion of graphene sheets in organic solvent supported by ionic interactions. *Adv. Mater.* **2009**, *21*, 1679–1683. [CrossRef]
75. Hummers, W.S., Jr.; Offeman, R.E. Preparation of graphitic oxide. *J. Am. Chem. Soc.* **1958**, *80*, 1339. [CrossRef]
76. Park, S.; An, J.; Piner, R.D.; Jung, I.; Yang, D.; Velamakanni, A. Aqueous suspension and characterization of chemically modified graphene sheets. *Chem. Mater.* **2008**, *20*, 6592–6594. [CrossRef]
77. Li, D.; Müller, M.B.; Gilje, S.; Kaner, R.B.; Wallace, G.G. Processable aqueous dispersions of graphene nanosheets. *Nat. Nanotechnol.* **2008**, *3*, 101. [CrossRef]
78. Lawal, A.T. Graphene-based nano composites and their applications. A review. *Biosens. Bioelectron.* **2019**, *141*, 111384. [CrossRef] [PubMed]
79. Guardia, L.; Fernández-Merino, M.; Paredes, J.; Solis-Fernandez, P.; Villar-Rodil, S.; Martinez-Alonso, A. High-throughput production of pristine graphene in an aqueous dispersion assisted by non-ionic surfactants. *Carbon* **2011**, *49*, 1653–1662. [CrossRef]
80. Su, Y.; Kravets, V.; Wong, S.; Waters, J.; Geim, A.; Nair, R. Impermeable barrier films and protective coatings based on reduced graphene oxide. *Nat. Commun.* **2014**, *5*, 1–5. [CrossRef] [PubMed]
81. Williams, G.; Seger, B.; Kamat, P.V. TiO_2-graphene nanocomposites. UV-assisted photocatalytic reduction of graphene oxide. *ACS Nano.* **2008**, *2*, 1487–1491. [CrossRef]
82. Moon, I.K.; Lee, J.; Ruoff, R.S.; Lee, H. Reduced graphene oxide by chemical graphitization. *Nat. Commun.* **2010**, *1*, 1–6. [CrossRef]
83. George, M.; Chae, M.; Bressler, D.C. Composite materials with bast fibres: Structural, technical, and environmental properties. *Prog. Mater. Sci.* **2016**, *83*, 1–23. [CrossRef]
84. Tesh, S.J.; Scott, T.B. Nano-composites for water remediation: A review. *Adv. Mater.* **2014**, *26*, 6056–6068. [CrossRef]
85. Ramakrishnan, S.; Pradeep, K.; Raghul, A.; Senthilkumar, R.; Rangarajan, M.; Kothurkar, N.K. One-step synthesis of Pt-decorated graphene–carbon nanotubes for the electrochemical sensing of dopamine, uric acid and ascorbic acid. *Anal. Methods.* **2015**, *7*, 779–786. [CrossRef]
86. Jiang, L.; Wen, Y.; Zhu, Z.; Su, C.; Ye, S.; Wang, J. Construction of an efficient nonleaching graphene nanocomposites with enhanced contact antibacterial performance. *Chem. Eng. J.* **2020**, *382*, 122906. [CrossRef]
87. Alsaba, M.; Al Marshad, A.; Abbas, A.; Abdulkareem, T.; Al-Shammary, A.; Al-Ajmi, M. Laboratory evaluation to assess the effectiveness of inhibitive nano-water-based drilling fluids for Zubair shale formation. *J. Pet. Explor. Prod. Technol.* **2020**, *10*, 419–428. [CrossRef]

88. Novoselov, K.S.; Geim, A.K.; Morozov, S.V.; Jiang, D.; Zhang, Y.; Dubonos, S.V. Electric Field Effect in Atomically Thin Carbon Films. *Science* **2004**, *306*, 666. [CrossRef]
89. Qalandari, R.; Qalandari, E. A review on the potential application of nano graphene as drilling fluid modifier in petroleum industry. *Int. Refereed J. Eng. Sci.* **2018**, *7*, 1–7.
90. Dreyer, D.R.; Park, S.; Bielawski, C.W.; Ruoff, R.S. The chemistry of graphene oxide. *Chem. Soc. Rev.* **2010**, *39*, 228–240. [CrossRef]
91. Ridha, S.; Ibrahim, A.; Shahari, R.; Fonna, S. Graphene nanoplatelets as high-performance filtration control material in water-based drilling fluids. *IOP Conf. Ser. Mater. Sci. Eng.* **2018**, *352*, 012025. [CrossRef]
92. Dideikin, A.T.; Vul, A.Y. Graphene Oxide and Derivatives: The Place in Graphene Family. *Front. Phys.* **2019**, *6*, 149. [CrossRef]
93. Friedheim, J.E.; Young, S.; De Stefano, G.; Lee, J.; Guo, Q. Nanotechnology for oilfield applications-hype or reality? In Proceedings of the SPE International Oilfield Nanotechnology Conference and Exhibition, Noordwijk, The Netherlands, 12–14 June 2012.
94. Liu, B.; Zhou, K. Recent progress on graphene-analogous 2D nanomaterials: Properties, modeling and applications. *Prog. Polym. Sci.* **2019**, *100*, 99–169. [CrossRef]
95. Hu, K.; Kulkarni, D.D.; Choi, I.; Tsukruk, V.V. Graphene-polymer nanocomposites for structural and functional applications. *Prog. Polym. Sci.* **2014**, *39*, 1934–1972. [CrossRef]
96. Sharma, B.; Malik, P.; Jain, P. Biopolymer reinforced nanocomposites: A comprehensive review. *Mater. Today Commun.* **2018**, *16*, 353–363. [CrossRef]
97. Rana, A.; Arfaj, M.K.; Yami, A.S.; Saleh, T.A. Cetyltrimethylammonium modified graphene as a clean swelling inhibitor in water-based oil-well drilling mud. *J. Environ. Chem. Eng.* **2020**, *8*, 103802. [CrossRef]
98. Li, R.; Liu, Y.; Cheng, L.; Yang, C.; Zhang, J. Photoelectrochemical aptasensing of kanamycin using visible light-activated carbon nitride and graphene oxide nanocomposites. *Anal. Chem.* **2014**, *86*, 9372–9375. [CrossRef]
99. Liu, C.; Wang, K.; Luo, S.; Tang, Y.; Chen, L. Direct electrodeposition of graphene enabling the one-step synthesis of graphene-metal nanocomposite films. *Small* **2011**, *7*, 1203–1206. [CrossRef] [PubMed]
100. Ali, J.A.; Kalhury, A.M.; Sabir, A.N.; Ahmed, R.N.; Ali, N.H.; Abdullah, A.D. A state-of-the-art review of the application of nanotechnology in the oil and gas industry with a focus on drilling engineering. *J. Pet. Sci. Eng.* **2020**, *191*, 107118. [CrossRef]
101. Li, M.C.; Wu, Q.; Song, K.; De Hoop, C.F.; Lee, S.; Qing, Y. Cellulose nanocrystals and polyanionic cellulose as additives in bentonite water-based drilling fluids: Rheological modeling and filtration mechanisms. *Ind. Eng. Chem. Res.* **2016**, *55*, 133–143. [CrossRef]
102. Potts, J.R.; Dreyer, D.R.; Bielawski, C.W.; Ruoff, R.S. Graphene-based polymer nanocomposites. *Polymer* **2011**, *52*, 5–25. [CrossRef]
103. Das, S.; Irin, F.; Ma, L.; Bhattacharia, S.K.; Hedden, R.C.; Green, M.J. Rheology and morphology of pristine graphene/polyacrylamide gels. *ACS Appl. Mater. Interfaces* **2013**, *5*, 8633–8640. [CrossRef]
104. Nelson, E.; Guillot, D. *Well Cementing 773*; Sugar Land: Schlumberger, TX, USA, 2006.
105. Cheng, X.; Yokozeki, T.; Wu, L.; Wang, H.; Zhang, J.; Koyanagi, J. Electrical conductivity and interlaminar shear strength enhancement of carbon fiber reinforced polymers through synergetic effect between graphene oxide and polyaniline. *Compos. Part A Appl. Sci. Manuf.* **2016**, *90*, 243–249. [CrossRef]
106. Yao, L.; Lu, Y.; Wang, Y.; Hu, L. Effect of graphene oxide on the solution rheology and the film structure and properties of cellulose carbamate. *Carbon* **2014**, *69*, 552–562. [CrossRef]
107. Chen, Y.; Yin, Q.; Zhang, X.; Xue, X.; Jia, H. The crystallization behaviors and rheological properties of polypropylene/graphene nanocomposites: The role of surface structure of reduced graphene oxide. *Thermochim. Acta* **2018**, *661*, 124–136. [CrossRef]
108. Gao, C.; Zhang, S.; Wang, F.; Wen, B.; Han, C.; Ding, Y. Graphene networks with low percolation threshold in ABS nanocomposites: Selective localization and electrical and rheological properties. *ACS Appl. Mater. Interfaces* **2014**, *6*, 12252–12260. [CrossRef] [PubMed]
109. Kim, H.; Macosko, C.W. Morphology and properties of polyester/exfoliated graphite nanocomposites. *Macromolecules* **2008**, *41*, 3317–3327. [CrossRef]
110. Sadasivuni, K.K.; Ponnamma, D.; Kumar, B.; Strankowski, M.; Cardinaels, R.; Moldenaers, P. Dielectric properties of modified graphene oxide filled polyurethane nanocomposites and its correlation with rheology. *Compos. Sci. Technol.* **2014**, *104*, 18–25. [CrossRef]

111. Gaska, K.; Kádár, R.; Rybak, A.; Siwek, A.; Gubanski, S. Gas barrier, thermal, mechanical and rheological properties of highly aligned graphene-LDPE nanocomposites. *Polymers* **2017**, *9*, 294. [CrossRef]
112. Mohideen, A.A.M.; Saheed, M.S.M.; Mohamed, N.M. Multiwalled carbon nanotubes and graphene oxide as nano-additives in water-based drilling fluid for enhanced fluid-loss-control & gel strength. *AIP Conf. Proc.* **2019**, *2151*, 020001.
113. Almoshin, A.M.; Alsharaeh, E.; Fathima, A.; Bataweel, M. A novel polymer nanocomposite graphene based gel for high temperature water shutoff applications. SPE Kingdom of Saudi Arabia annual technical symposium and exhibition. *Soc. Pet. Eng.* **2018**. [CrossRef]
114. Chawhan, S.S.; Barai, D.P.; Bhanvase, B.A. Sonochemical preparation of rGO-SnO2 nanocomposite and its nanofluids: Characterization, thermal conductivity, rheological and convective heat transfer investigation. *Mater. Today Commun.* **2020**, 101148. [CrossRef]
115. Manivel, P.; Kanagaraj, S.; Balamurugan, A.; Ponpandian, N.; Mangalaraj, D.; Viswanathan, C. Rheological behavior and electrical properties of polypyrrole/thermally reduced graphene oxide nanocomposite. *Colloids Surf. A* **2014**, *441*, 614–622. [CrossRef]
116. Gudarzifar, H.; Sabbaghi, S.; Rezvani, A.; Saboori, R. Experimental investigation of rheological & filtration properties and thermal conductivity of water-based drilling fluid enhanced. *Powder Technol.* **2020**. [CrossRef]
117. Haruna, M.A.; Pervaiz, S.; Hu, Z.; Nourafkan, E.; Wen, D. Improved rheology and high-temperature stability of hydrolyzed polyacrylamide using graphene oxide nanosheet. *J. Appl. Polym. Sci.* **2019**, *136*, 47582. [CrossRef]
118. Jamrozik, A.; Wiśniowski, R.; Czekaj, L.; Pintal, K. Analyses of graphene oxide applicability in drilling mud technology. *Int. Multidiscip. Sci. GeoConf. SGEM* **2017**, *17*, 515–522.
119. Xuan, Y.; Jiang, G.; Li, Y. Nanographite oxide as ultrastrong fluid-loss-control additive in water-based drilling fluids. *J. Dispers. Sci. Technol.* **2014**, *35*, 1386–1392. [CrossRef]
120. Mahmood, D.; Al-Zubaidi, N.; Alwasiti, A. Improving Drilling Fluid Properties by Using Nano-Additives. *Eng. Technol. J.* **2017**, *35*, 1034–1041.
121. Al-Yasiri, M.; Wen, D. Gr-Al2O3 Nanoparticles-Based Multifunctional Drilling Fluid. *Ind. Eng. Chem. Res.* **2019**, *58*, 10084–10091. [CrossRef]
122. Taha, N.M.; Lee, S. Nano graphene application improving drilling fluids performance. *Int. Pet. Technol. Conf.* **2015**. [CrossRef]
123. Ahmad, H.M.; Kamal, M.S.; Murtaza, M.; Al-Harthi, M.A. Improving the drilling fluid properties using nanoparticles and water-soluble polymers. SPE Kingdom of Saudi Arabia annual technical symposium and exhibition. *Soc. Pet. Eng.* **2017**.
124. Nasser, J.; Jesil, A.; Mohiuddin, T.; Al Ruqeshi, M.; Devi, G.; Mohataram, S. Experimental investigation of drilling fluid performance as nanoparticles. *World J. Nano Sci.* **2013**, *3*, 57. [CrossRef]
125. Ismail, A.; Rashid, M.; Thameem, B. Application of nanomaterials to enhanced the lubricity and rheological properties of water based drilling fluid. *IOP Conf. Ser. Mat. Sci. Eng.* **2018**, *380*, 012021. [CrossRef]
126. Aramendiz, J.; Imqam, A. Water-based drilling fluid formulation using silica and graphene nanoparticles for unconventional shale applications. *J. Pet. Sci. Eng.* **2019**, *179*, 742–749. [CrossRef]
127. Kusrini, E.; Oktavianto, F.; Usman, A.; Mawarni, D.P.; Alhamid, M.I. Synthesis, characterization, and performance of graphene oxide and phosphorylated graphene oxide as additive in water-based drilling fluids. *Appl. Surf. Sci.* **2020**, *506*. [CrossRef]
128. Aramendiz, J.; Imqam, A.; Fakher, S.M. Design and Evaluation of a Water-Based Drilling Fluid Formulation Using SiO and Graphene Oxide Nanoparticles for Unconventional Shales. *Int. Pet. Technol. Conf.* **2019**.
129. Smith, A.T.; LaChance, A.M.; Zeng, S.; Liu, B.; Sun, L. Synthesis, properties, and applications of graphene oxide/reduced graphene oxide and their nanocomposites. *Nano Mater. Sci.* **2019**, *1*, 31–47. [CrossRef]
130. Singh, P.; Chamoli, P.; Sachdev, S.; Raina, K.; Shukla, R.K. Structural, optical and rheological behavior investigations of graphene oxide/glycerol based lyotropic liquid crystalline phases. *Appl. Surf. Sci.* **2020**, *509*, 144710. [CrossRef]
131. Murphy, S.V.; Atala, A. 3D bioprinting of tissues and organs. *Nat. Biotechnol.* **2014**, *32*, 773. [CrossRef]
132. Li, H.; Liu, S.; Lin, L. Rheological study on 3D printability of alginate hydrogel and effect of graphene oxide. *Int. J. Bioprint.* **2016**, *2*, 54–66. [CrossRef]
133. Yi, S.; Li, G.; Ding, S.; Mo, J. Performance and mechanisms of graphene oxide suspended cutting fluid in the drilling of titanium alloy Ti-6Al-4V. *J. Manuf. Process.* **2017**, *29*, 182–193. [CrossRef]

134. Aramendiz, J.; Imqam, A. Silica and Graphene Oxide Nanoparticle Formulation to Improve Thermal Stability and Inhibition Capabilities of Water-Based Drilling Fluid Applied to Woodford Shale. *SPE Drill. Complet.* **2019**. [CrossRef]
135. Aliabadian, E.; Sadeghi, S.; Moghaddam, A.R.; Maini, B.; Chen, Z.; Sundararaj, U. Application of graphene oxide nanosheets and HPAM aqueous dispersion for improving heavy oil recovery: Effect of localized functionalization. *Fuel* **2020**, *265*, 116918. [CrossRef]
136. Le Ba, T.; Mahian, O.; Wongwises, S.; Szilágyi, I.M. Review on the recent progress in the preparation and stability of graphene-based nanofluids. *J. Therm. Anal. Calorim.* **2020**, 1–28. [CrossRef]
137. Hoelscher, K.P.; Young, S.; Friedheim, J.; De Stefano, G. Nanotechnology application in drilling fluids. In Proceedings of the Offshore Mediterranean Conference and Exhibition, Ravenna, Italy, 20–22 March 2013.
138. Alkinani, H.H.; Al-Hameedi, A.T.T.; Dunn-Norman, S.; Lian, D. Application of artificial neural networks in the drilling processes: Can equivalent circulation density be estimated prior to drilling? *Egypt. J. Pet.* **2019**. [CrossRef]
139. Saleh, T.A.; Ibrahim, M.A. Advances in functionalized Nanoparticles based drilling inhibitors for oil production. *Energy Rep.* **2019**, *5*, 1293–1304. [CrossRef]
140. Rishi, K.; Narayanan, V.; Beaucage, G.; McGlasson, A.; Kuppa, V.; Ilavsky, J. A thermal model to describe kinetic dispersion in rubber nanocomposites: The effect of mixing time on dispersion. *Polymer* **2019**, *175*, 272–282. [CrossRef]
141. Zhang, J.; Xu, Q.; Gao, L.; Ma, T.; Qiu, M.; Hu, Y. A molecular dynamics study of lubricating mechanism of graphene nanoflakes embedded in Cu-based nanocomposite. *Appl. Surf. Sci.* **2020**, *511*, 145620. [CrossRef]
142. Viswanathan, V.; Laha, T.; Balani, K.; Agarwal, A.; Seal, S. Challenges and advances in nanocomposite processing techniques. *Mater. Sci. Eng. R Rep.* **2006**, *54*, 121–285. [CrossRef]
143. Harito, C.; Bavykin, D.V.; Yuliarto, B.; Dipojono, H.K.; Walsh, F.C. Polymer nanocomposites having a high filler content: Synthesis, structures, properties, and applications. *Nanoscale* **2019**, *11*, 4653–4682. [CrossRef]
144. Morgani, M.S.; Saboori, R.; Sabbaghi, S. Hydrogen sulfide removal in water-based drilling fluid by metal oxide nanoparticle and ZnO/TiO$_2$ nanocomposite. *Mater. Res. Express* **2017**, *4*, 075501. [CrossRef]
145. Amr, I.T.; Al-Amer, A.; Al-Harthi, M.; Girei, S.A.; Sougrat, R.; Atieh, M.A. Effect of acid treated carbon nanotubes on mechanical, rheological and thermal properties of polystyrene nanocomposites. *Compos. Part B* **2011**, *42*, 1554–1561. [CrossRef]
146. Tang, W.; Zou, C.; Da, C.; Cao, Y.; Peng, H. A review on the recent development of cyclodextrin-based materials used in oilfield applications. *Carbohydr. Polym.* **2020**, 116321. [CrossRef]
147. Perween, S.; Beg, M.; Shankar, R.; Sharma, S.; Ranjan, A. Effect of zinc titanate nanoparticles on rheological and filtration properties of water based drilling fluids. *J. Pet. Sci. Eng.* **2018**, *170*, 844–857. [CrossRef]
148. Husin, H.; Elraies, K.A.; Choi, H.J.; Aman, Z. Influence of Graphene Nanoplatelet and Silver Nanoparticle on the Rheological Properties of Water-Based Mud. *Appl. Sci.* **2018**, *8*, 1386. [CrossRef]

© 2020 by the authors. Licensee MDPI, Basel, Switzerland. This article is an open access article distributed under the terms and conditions of the Creative Commons Attribution (CC BY) license (http://creativecommons.org/licenses/by/4.0/).

Review

A Review of Microscale, Rheological, Mechanical, Thermoelectrical and Piezoresistive Properties of Graphene Based Cement Composite

Sardar Kashif Ur Rehman [1,*], Sabina Kumarova [2], Shazim Ali Memon [2,*], Muhammad Faisal Javed [1] and Mohammed Jameel [3]

1. Department of Civil Engineering, Abbottabad Campus, COMSATS University Islamabad, Abbottabad 22060, Pakistan; arbabfaisal@cuiatd.edu.pk
2. Department of Civil and Environmental Engineering, School of Engineering and Digital Sciences, Nazarbayev University, Nur-Sultan 010000, Kazakhstan; Sabina.kumarova@nu.edu.kz
3. Department of Civil Engineering, King Khalid University, Abha 61421, Saudi Arabia; Jamoali@kku.edu.sa
* Correspondence: skashif@cuiatd.edu.pk (S.K.U.R.); shazim.memon@nu.edu.kz (S.A.M.)

Received: 26 August 2020; Accepted: 14 October 2020; Published: 21 October 2020

Abstract: Extensive research on functionalized graphene, graphene oxide, and carbon nanotube based cement composites has been carried out to strengthen and overcome the shortcomings of construction materials. However, less literature is available on the pure graphene based cement composite. In this review paper, an in-depth study on a graphene-based cement composite was performed. Various structural forms of graphene and classifications of graphene-based nanomaterial have been presented. The dispersion mechanism and techniques, which are important for effective utilization in the construction industry, are reviewed critically. Micro-scale characterization of carbon-based cement composite using thermogravimetric analysis (TGA), infrared (IR) spectroscopic analysis, x-ray diffractometric (XRD) analysis, and morphological analysis has also been reviewed. As per the authors' knowledge, for the first time, a review of flow, energy harvesting, thermoelectrical, and self-sensing properties of graphene and its derivatives as the bases of cement composite are presented. The self-sensing properties of the composite material are reported by exploring physical applications by reinforcing graphene nanoplatelets (GNPs) into concrete beams.

Keywords: graphene; cement composite; characterization; rheological; application; energy harvesting

1. Introduction

The construction industry makes a significant contribution to economic growth in every part of the world. Every year, 20–35 billion tons of concrete is used globally, making it the most widely used construction material [1], and its advantages, including high strength, durability, fire resistance etc., increase the consumption of concrete [2,3]. However, the main drawbacks of concrete are known to be its brittleness and low tensile strength [4]. In order to overcome the shortcomings of the concrete, researchers used different materials and techniques [5–10]. Chemical admixtures [5–8], supplementary cementitious materials [11–14], and fibres [15–19] were used to halt the propagation of micro-cracks and improve tensile strength. The size of these fillers ranged from the macro-scale to micro and nano scales. Currently, advancements in nanotechnology made it possible to control nano size cracks (pores with diameter <20 nm) before micro size cracks are developed [20]. Hence, nanomaterials like graphene, carbon nanotubes, and graphene oxide were studied by many researchers [21–23].

The literature on the review of nanotechnology in concrete [24] has highlighted several key findings. As for nano-reinforcements, the addition of carbon nanotubes/nanofibers (CNT/CNF) is widely recognised as an appropriate way to enhance the mechanical properties of the cement composite and to resist crack propagation. Nanotechnology is considered to be effective in terms of compatibility, cost and safety. Fraga et al. [25] observed that the incorporation of finely distributed CNTs in cement matrix reduced crack development. In addition, a small amount of multiple walls concentrically arranged in CNTs are able to enhance mechanical properties significantly in terms of tensile strength, brittleness and strain capacity. Han et al. [26] reported the enhancement mechanism of CNTs/CNFs and graphene nanoplatelets (GNPs). CNTs/CNFs have remarkable effect on the mechanical characteristics of cement-based composite, such as pore filling between hydration products (ettringite and CSH gel) which as a result enhances bond strength between the matrix and CNTs/CNFs while the GNPs showed an augment of the compactness and the homogeneity of hardened cementitious composite. Also, the addition of CNTs/CNFs and GNPs led to the improvement of piezoelectric and dielectric properties.

Qureshi and Panesar [27] in their literature review emphasized a nanofibrous material, graphene oxide (GO), which potentially improves the behavior of cement-based materials. According to the authors [27], addition of 0.05% of GO by weight of cement in portland cement paste increased its compressive strength and flexural strength by 15–33% and 41–59%, respectively. Moreover, a 70.5% increase in flexural strength was found when GO was dispersed with a superplasticizer. The authors also underlined the need to determine the effect of GO on cement hydration process, life-cycle cost and carbon release. Yang et al. [28] reviewed the size of GO particles and observed that GO of the smaller size showed a better performance when compared to the larger size. They also reported the effect of hybrid GO cement-based composite materials and stated that cement composites with both GO and CNTs perform better than a cement composite with only GO or CNTs. This was due to the reason that CNTs in GO solution are dispersed better because of the huge electrostatic repulsion in the CNTs/GO mix. The authors also considered it obligatory to concentrate on the synergic effects of nanomaterials and GO, since dispersion issue restricts the application of nanomaterials in civil engineering. The application of graphene and GO in geopolymer cement composites has been reviewed by [29]. The authors concluded that the experimental results of addition of graphene in geopolymers cement composite reported by different researchers vary and are inconsistent. This might be related to different ways of treating graphene nanomaterials which will results in more interference factors in addition to varying content of aluminium-silicon of geopolymers. Recently, Zhao et al. [30] reviewed the impact of GO on cement composite. The authors observed inconsistency amongst the various published literature due to complex nature of hydrated cement matrix and varying characteristics of graphene oxide.

In this review article, a detailed study on the graphene-based cement composite is performed. A brief explanation of graphene nanostructures is presented in Section 2. In Section 3, various structural forms of graphene are discussed, and classification of graphene-based nanomaterial is presented. As the dispersion of graphene-based nanomaterials is one of the substantial challenges for their employment in the construction industry, dispersion techniques along with mechanisms are reviewed in Section 4. Micro-scale characteristics of the graphene and its derivatives-based cement composite is essential as it helps in scientific understanding of the material. Hence, in Section 5, various characterization techniques, including thermogravimetric analysis (TGA), infrared (IR) spectroscopic analysis, X-ray diffractometric (XRD) analysis, and morphological analysis, are reviewed. In Sections 6, 8 and 9, for the first time per the authors' best knowledge, a review of flow properties, energy harvesting, thermoelectrical, and self-sensing properties of graphene based cement composites are presented. Finally, the research gaps are highlighted in Section 10 while conclusions are presented in Section 11.

2. Brief Description of Graphene Nanostructures

Graphene is a single layer carbon sheet and one of the most promising nanofiller used to enhance cementitious materials [31,32]. Practically, multi-layered GNPs are very commonly used since they can be easily manufactured from graphite oxide or graphite. GNPs consist of layers of graphene having a thickness from 3 to 100 nm. Thus, their morphological structure makes it a remarkable reinforcing material [33]. GO is a layered material, which is oxidized from graphite, and oxygen particles are interspersed on the edges and its basal surfaces. In fact, graphene sheets are naturally available and it is only required to exfoliate them from the graphite [34]. The exfoliation of graphite into GNPs can be made by using chemical and mechanical techniques [35]. The mass scale production of graphene is possible by means of chemical oxidation and reduction of graphite [36]. This method is considered to be faster, easier, more scalable, economic, facile, and dynamic as compared with other methods [36]. Figure 1 presents the chemical procedure for production of graphene from naturally available graphite.

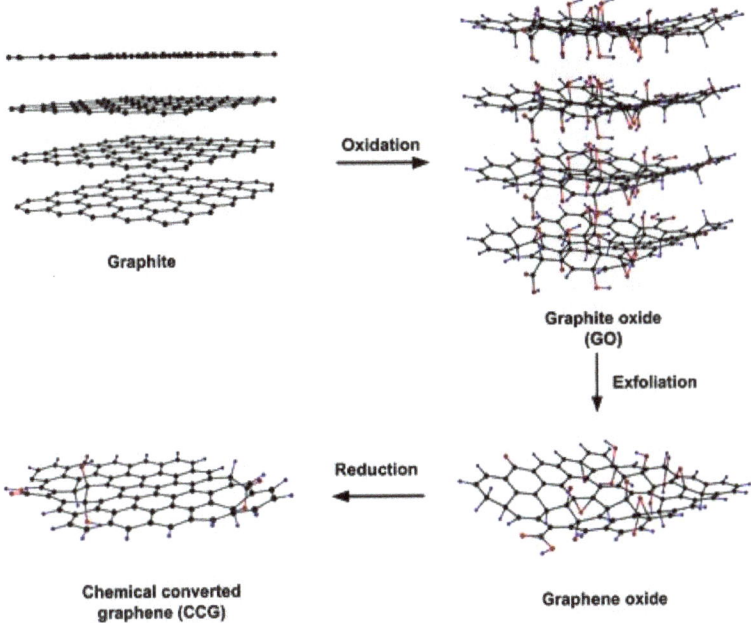

Figure 1. Preparation of graphene nanosheets from graphite. Reproduced with permission from [37], John Wiley and Sons, 2011.

This section may be divided by subheadings. It should provide a concise and precise description of the experimental results, their interpretation as well as the experimental conclusions that can be drawn.

3. Classification of Graphene Based on Nanostructure

These newly developed engineered nanomaterials are characterised by their morphology: zero-dimensional (0D) nanoparticles (spherical shape and low aspect ratio), i.e., carbon black; one-dimensional (1D) fibers (straight and high aspect ratio), i.e., carbon nanotubes; and two-dimensional (2D) sheets, i.e., graphene and GO [21]. Figure 2 shows the schematic of these nanomaterials. These engineered materials are used in the construction industry to overcome weaknesses of building materials, i.e., cement paste, mortar and concrete. Besides enhancing the mechanical abilities of cement composites, the addition of the nanomaterials also improves their electrical, thermal, and electromagnetic properties [38].

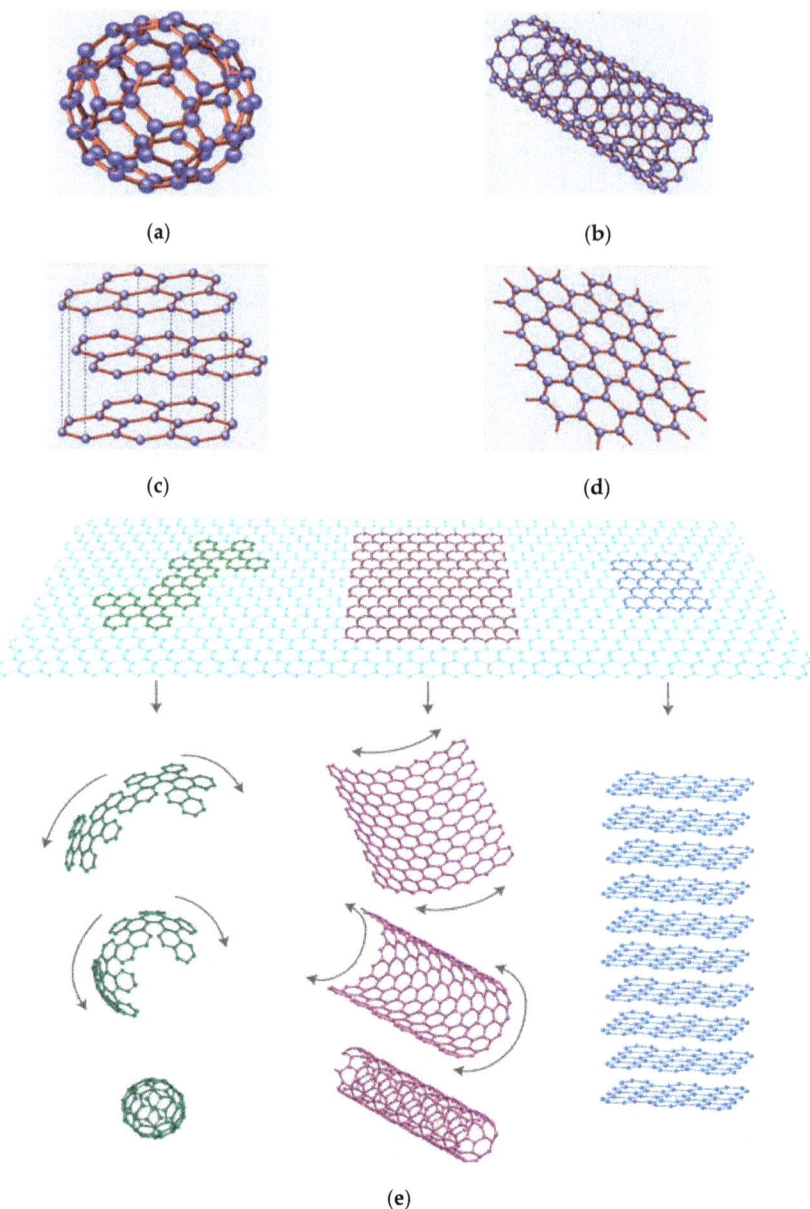

Figure 2. Various structural form of graphene; (**a**) wrapped honeycomb structure of zero-dimensional graphene nanoparticle; (**b**) Rolled honeycomb structure forms of one-dimensional graphene; (**c**) single planar two-dimensional graphene sheet; (**d**) stacked honeycomb structure of three-dimensional graphite, and (**e**) extraction of various structural forms from 2D graphene sheet. Modified from [39,40], Elsevier, 2010.

3.1. Zero Dimensional Graphene Nanoparticles

Molecules, which consist of wrapped graphene by means of the introduction of pentagons on the hexagonal lattice are called zero dimensional nanoparticles. These allotropes of carbon were discovered by Kroto et al. [41]. They are like Buckyball's; common examples are Fullerenes (C60) and Carbon Black (CB). Several researchers have explored the mechanical and electrical properties of carbon black cement composite as well as its applications for structural health monitoring. According to Xi et al. [42], CB particles having about 33 nm diameter provides cheaper solution for piezo-resistive effects as compared with carbon fibers mixed concrete. They found that CB filled cement matrix is promising candidate for strain sensing. Gong et al. [43] observed that piezoelectric sensitivity of the cement composite enhanced dramatically by addition of 1% volume of CB. Wang et al. [44] dispersed the CB particles in high density polyethylene matrix and silicone rubber, respectively. They also presented a mathematical piezoresistivity model of the CB filled composite material based on the general effective medium theory. The authors found that the results predicted by the mathematical model were in alignment with the experimental results. CB particles are also found to interact with air-entraining admixtures and smaller particle sizes with more surface area have shown optimum interaction results [45]. Figure 2a presents the wrapped honeycomb structure and schematic of zero-dimensional graphene nanoparticle. Zero-dimensional nanoparticles lack the ability to arrest micro-cracks due to non-uniform mixing, a low aspect ratio, and the formation of weak zones in concrete, especially when used in a large amount.

3.2. One Dimensional Graphene Nanotubes

Compared to zero-dimensional nanoparticles, spherical shape one-dimensional nanofibers have a high aspect ratio i.e., carbon nanotubes. Exfoliated GNP (xGnP) and carbon nanotubes (CNT) share the same chemical structure [32]. CNTs are carbon allotropes of cylindrical shape, made of rolled graphene layers. Based on the number of walls, CNTs are classified as single wall CNTs (SWCNTs) and multi walls CNTs (MWCNTs), i.e., 10–100 walls. The diameter of SWCNTs varies from 1 nm to 3 nm while the diameter of MWCNTs varies from 5 nm to 50 nm. [46]. MWCNTs have a surface area of around 400 m^2/g and aspect ratio of more than 1000, due to varying length of carbon nanotubes [26]. CNTs have high elastic modulus of 1TPa, strength of 10–60 GPa for SWCNTs and 50–500 GPa for MWCNTs, and electrical resistance of 5–50 $\mu\Omega$cm. Such impressive properties of CNTs enhanced the properties of cementitious materials, when mixed with cement [47]. Konsta et al. [47] noted 25% rise in the flexural strength of CNTs-cement composite. According to Li et al. [48], when 0.5% functionalized CNTs was added to plain cement concrete, the compressive strength and flexural strength increased by 19% and 25%, respectively as compared with control specimens, while porosity decreased by 64%. Moreover, pores with a size of more than 50 nm in diameter were 82% less as compared with plain cement concrete. Nevertheless, the problem with CNTs is non-uniform dispersion and weak connection between CNTs and the cement matrix. The arrangement of CNTs is complicated because strong Van der Waals forces exist between individual CNTs, which may cause the formation of agglomeration and bundles in the composite. As a result, these agglomerates form defects and limit the influence of CNTs on cement composite [49]. That is the reason why, even after decades of research on CNTs, their full potential as reinforcement has been severely limited [36]. Several researchers [50,51] noted the decline in mechanical properties of the CNT based cement composite because of a non-uniform dispersion, worst workability, higher inhomogeneity, and porosity. Research performed by Cwirzen et al. [50] states that no major effect was recorded on the mechanical properties of the CNT based cement composite mix in contrast with pure cement mix even by using surfactants and achieving uniform dispersion in the mix. This was most probably because of the very low bond strength between CNTs and cement matrix, due to which CNTs were easily pulled out in fractured cement paste specimens.

3.3. Two Dimensional Graphene Sheets

In contrast to CNTs, graphene and GO are the two-dimension sheet-like structures and have a considerable surface area. GO has a thickness of a single atomic layer while having lateral thickness reaching to tens of micrometres, which provides large surface area and immense aspect ratio [52]. It has been observed by researchers that by incorporating graphene sheets in cement composite, electrical, mechanical and thermal properties remarkably enhanced [22,53,54]. Pan et al. [55] reported an increase in tensile strength by 78.6%, flexure strength by 60.7% and compressive strength by 38.9% when 0.03% dosage of GO by weight of cement was incorporated in cement mortar. At the microscopic level, they observed the flower-like crystals, which enhanced toughness. Moreover, Pan et al. [55] found that 41.7% decline in slump size when 0.05% GO by weight of cement was used in the cement paste mix. The possible reason was considered to be the huge surface area of GO, which reduces the accessible moisture content in mix design from wetting the GO sheets. GO is the carbon antecedent combined with carboxyl, epoxy and/or hydroxyl groups [56]. At nanoscale, the spacing between atoms in GO is almost identical to graphene [56]. Extensive research has been conducted on GO cement composite. However, much less focus has been given to graphene cement composite.

In 2004, Novoselov et al. [31] derived single atom thick crystallites of graphene from bulk graphite. They obtained the graphene layer in a repeated pealing Scotch Tape technique process [31]. According to Zhang et al. [57], by using this method, thickness of graphene up to 300 nm can be achieved. Graphene is known to be the thinnest material [58,59]. Boehm et al. [60] concluded that graphene is one of the carbon allotropes with 2D properties. Figure 2c shows that graphene is arranged in a single planar sheet while Figure 2d shows the stacked honeycomb structure forms of three-dimensional graphite. The structural relationship between graphene and various other forms are shown in Figure 2e.

Recently, the 2D flat graphene sheet has gained an enormous attention in science for its promising and outstanding properties: high intrinsic strength (130 GPa) [59], large specific surface area (2630 m^2 g^{-1}) [61], high thermal conductivity (~5000 Wm^{-1}K^{-1}) [62,63] and firm Young's module (~1.0 TPa) [64–66]. This unique and tremendous behaviour of graphene opened a new window for a wide range of applications. A large exposed surface area of graphene sheets has a strong capability to make a great physical and chemical bond with the cement matrix. Rafiee et al. [67] mentioned that unzipping the MWCNTs into graphene nanoribbons results in a significant improvement. This was due to an extraordinary increase in interfacial area and geometry of graphene sheets as compared with multi-walled carbon nanotubes. Rafiee et al. [67] also found a 30% increment in Young's modulus and 22% rise in the ultimate tensile strength of graphene composite against the same amount of multi-walled carbon nanotubes composite. Conversely, graphene has a very high cost of manufacturing. Therefore, the application of graphene is restricted and limited in the construction industry due to production in very small quantities [32].

4. Dispersion of Graphene Based Nanomaterials

Dispersion is a primary problem related to fabricating cementitious nanocomposite because Van der Waals force forms an agglomeration of the nanoparticles [68]. Dispersion of nanomaterials in aqueous solution is an important step and significantly alters the final results [69,70]. In aqueous solution, nanomaterials tend to precipitate or float on the surface. Naturally, graphene is hydrophobic, and it forms the agglomerates in aqueous solution, causing non-uniform dispersion [71]. According to Hamann and Clemens [72], a weak dispersion will result in the formation of a defect in the composite matrix and restrain the effect of the nanoparticles. Most of the studies about the dispersion of graphene and graphite nanoparticles were directed to surface adjustment, such as oxidation, or the inclusion of other nanoparticles in the suspension [73], oxidation of GNPs to form GO and then reduction to GO particles and finally, dispersed them in water. In addition, oxygen-containing functional groups on the GO could result in stable dispersions due to electrostatic repulsion between the oxidized GO particles [74]. Peyvandi et al. [75] noted that covalent surface modifications cause destructive effects on the atomic structure of the graphite nanoplatelets and are able to reduce the strength of these nanoparticles. In order to preserve the structure of the graphene flakes, other methods of dispersion are required. Still, this area is relatively new and least research on the dispersion of non-covalently modified GNPs in water has been conducted. Hence, to improve the dispersion of graphene flakes keeping the atomic structure safe in the cement composite, numerous techniques and the results of various solvent types have been reviewed in this section.

4.1. Dispersion Using Dispersant

The following section is regarding the application of dispersant to the Graphene and its derivatives. As graphene flakes are hydrophobic and tend to form coagulation in the aqueous solution, therefore, numerous researchers used different dispersants to obtain a stable suspension. Stankovich et al. [76] used reduced graphite oxide (rGO) flakes with polysodium styrenesulfonate. These graphite nanoparticles, which were coated with the polysodium styrenesulfonate, remained in suspension. Jue et al. [77] noted that using of polyelectrolytes with GNPs in water retains the structure of the GNPs and provides a complete utilization of the features of these nanoparticles. Wotring [78] evaluated the performance of GNPs dispersion in water with high-range water-reducing admixture (WRA). Water-cement (w/c) ratio remained as 0.5, graphene dosage was 0.1% by weight of cement paste and AdvaCast 575 polycarboxylates based high range WRA in the range of 0–10 times the weight of graphene flakes were used. According to Wotring [78], the stability of graphene in the solvent was preserved by the WRA. When the dosage of water reducing agents increased, the stability also increased as shown in Figure 3. By visual observations, it was found that the WRA to GNPs ratio of 3% was stable until seven days, as shown in Figure 3 [78]. Meanwhile, good stability was observed after 24 h for other ratios of WRA to GNPs.

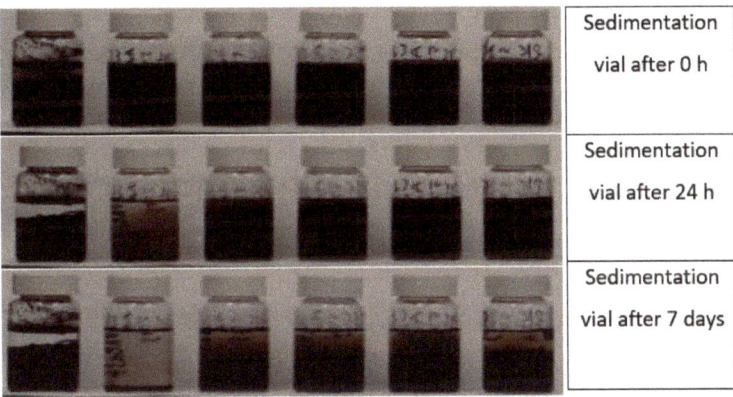

Figure 3. Effect of the presence of WRA to the dispersion of graphene sedimentation at different times; graphene sedimentation jars with WRA-to-GNPs ratios of 0, 1, 1.5, 2, 2.5 and 3 from left to right. Reproduced with permission from [78], Ph.D. Thesis, University of Illinois, 2014.

Sixuan [79] used acetone, particularly Gum Arabic and Darex Super 20 solvent, for the dispersion of graphite nanoplatelets in water solution. The stability and homogeneity of graphite flakes suspensions with respect to time using various dispersing agents were tested by visual inspection. Figure 4 presents the dispersion of graphite flakes in different dispersants. The stability of graphite flakes was assessed by observing the colour of the solution. Usually, the dispersant and distilled water are colourless and after incorporating the graphite, the colour changed to black. Based on Figure 4, the graphite nanoplatelets in Darex Super 20 and Gum Arabic presented good stability and the colour of the solution remained unchanged after 30 min of mixing. However, in acetone and water solution the graphite accumulated at the bottom of the test tube. Gum Arabic is a natural polysaccharide available as white powder and is remarkably soluble in the water. The weak acidity of Gum Arabic makes it reactive in the alkaline cementitious environment, thus leading to a production of surplus water while mixing. Finally, a watery mixture containing light graphite flakes on top was formed. Likewise, Darex Super 20, which is the naphthalene sulfonate-based high water-reducing superplasticizer, exhibited a good stability as dispersant with minimal alteration on the fresh mixture. Therefore, based on test results, Sixuan [79] concluded that Darex Super 20 was found to be the best dispersant among acetone, tap water, and Gum Arabic for the graphite nanoplatelets in the cement matrix.

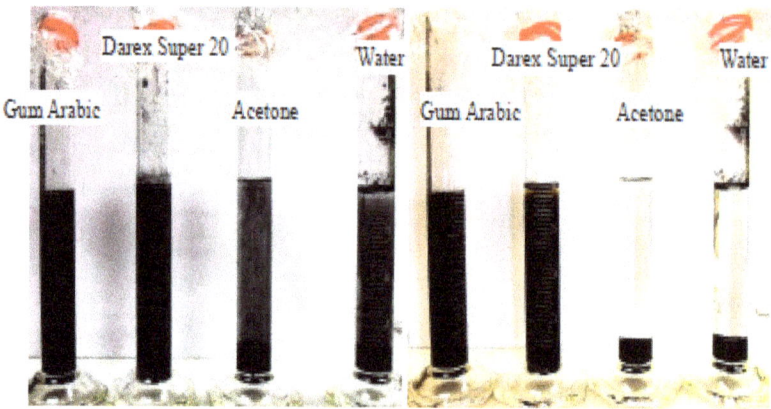

Figure 4. Stability test for graphite suspension in various dispersant of liquid at 5 min (left) and 30 min (right). Reproduced with permission from [79], ME. Thesis, National University of Singapore, 2012.

Sharif et al. [80] examined the dispersion of CB, i.e., zero dimensional graphene in the presence of five dispersants, including Dispex G40, sodium dodecylbenzene sulfonate (NaDDBS, MW = 348.47), Tween 80, *t*-octylphenol decaethylene glycol ether (Triton X-100, MW = 647), and Dispex N40. They concluded that most favourable dispersant was Triton X-100, a surfactant molecule, as it showed the highest absorption peak in the UV-vis absorption spectra, as presented in Figure 5. The affixing of surface-active agents to the surface of carbon nanomaterial is mainly attributed to hydrophobic synergy.

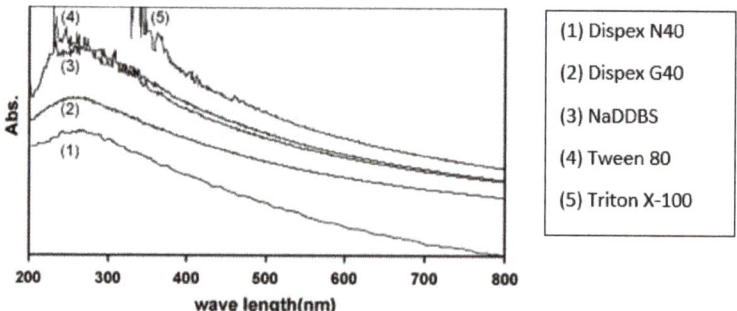

Figure 5. UV–vis absorption spectra of carbon black in presence of various dispersant. Reproduced with permission from [80], Elsevier, 2009.

Recently, Silva et al. [81] used isopropanol alcohol with the expanded graphene structures in a ratio of 1:1. The authors found that the combination of multilayer graphene sheets and isopropanol produced excellent dispersion. Han et al. [82] used the polycarboxylate superplasticizer (Sike ViscoCrete 3301E) to disperse multi-layered graphene in aqueous solution and found that the graphene flakes did not form agglomerations.

The dispersion mechanism for dispersant and graphite nanoplatelets was explained by Sixuan [79]. The authors stated that the organic molecules in the dispersant are negatively-charged and absorbed mainly at the interface of water and graphite. The graphite surface initially possessed the residual charges on their surfaces. When these graphite nanoparticles were mixed in liquid solution, they formed the flocculated structures. The flocculation of the graphite particles occurred due to the electrostatics interactions exerted by the adjacent graphite particles of the opposite charges, as seen in Figure 6a. After that, dispersant was used to neutralize these residual charges and made the entire surface to carry the same charges. Lastly, the particles of graphite nanoplatelets remained fully dispersed in the suspension of the liquid because of the repulsion of the graphite nanoparticles (Figure 6b) [79].

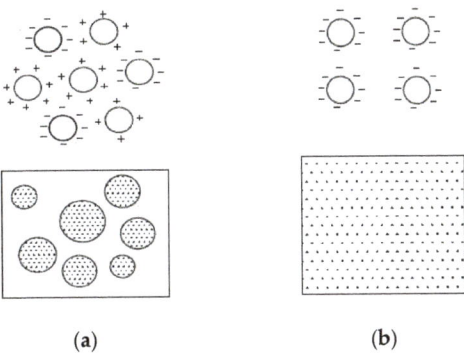

Figure 6. Action of dispersion for graphite nanoparticles; (**a**) schematic diagram of flocculated graphite flakes in aqueous solution; and (**b**) and schematic diagram of uniformly dispersed graphite nanoparticles. Reproduced with permission from [79], ME. Thesis, National University of Singapore, 2012.

4.2. Dispersion Using Ultrasonication

Another way to deal with dispersion problem with graphene and its derivatives to apply sonication. The sonication is the act of agitation of particles by means of applying energy. The ultrasonication is referred to as the waves having a frequency of more than 20 KHz. The ultrasonic electric generator takes the biggest part of sonication device and generates a signal, normally about 20 KHz, which charges a transducer and it transforms the electric signal to mechanical vibrations. These vibrations are further augmented by the sonicator and transmitted to the probe, which transfers the vibrations to the solution. The quick movement of the probe produces a cavitation event. It takes place when the vibrations generate multiple microscopic bubbles in the solution, some wedged intermolecular space is developed and breaks down continuously under the influence of the weight of the solution. Constant generation and breakdown of thousands of these bubbles develop the robust waves of vibration, which pass through the solution and crush the particles [83]. The energy, which had been transmitted to the GNPs resulted in the collapse of the interlayer π-bond. Hence, exfoliated GNPs can be attained with higher aspect ratio, decreased thickness and improved mobility of particles as shown in Figure 7. The maximum size of the bubble being produced in the liquid is dependent on the frequency of ultrasonication. A low-frequency ultrasonication will generate large size bubbles and vice versa. Higher energy forces are being produced upon the collapse of the large-sized bubbles in the solution [79].

Mehrali et al. [84] applied sonication to GNPs in distilled water with a high-powered probe sonicator. The stability of the GNPs was reported to remain in suspension for 600 h. Han et al. [82] used the ultrasonication for 1 h to achieve the uniform dispersion of cementitious materials with multi-layered graphene (MLGs). Silva et al. [81] employed the isopropanol alcohol blended with the expanded graphite structures in 1:1. The solution was then ultra-sonicated for the next 2 h and achieved excellent dispersion.

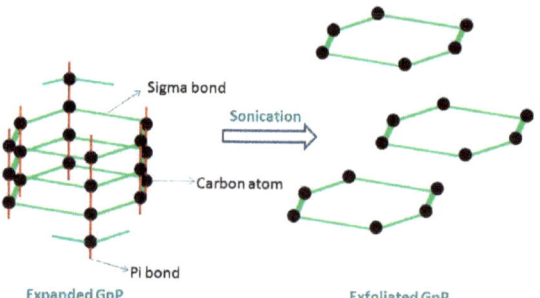

Figure 7. Mechanism of graphene dispersant after ultrasonication from expanded GNP to exfoliated GNP. Reproduced with permission from [79], ME. Thesis, National University of Singapore, 2012.

4.3. Assessment of Dispersion Efficiency Using UV-Vis Spectrometry

Ultraviolet-visible spectroscopy (UV-vis) is defined as an absorption or reflectance spectroscopy in the UV spectral region. It is also known as a competent method to examine the dispersion of graphene structure, particularly, the GNPs and CNTs. Jiang et al. [85] used the UV-vis measurements for the quantitative evaluation of the colloidal stability of CNTs dispersion. In UV-vis spectral range, carbon nanomaterials showed the absorption characteristics and it is attributed to the electronic changes between the bonding and antibonding π orbital [86]. Jager et al. [87] stated that the σ–σ* transitions are anticipated in the 60–100 nm ultraviolet range, meanwhile the π–π* transitions are observed in 180–280 nm spectrum. Due to this reason, this method has been utilized by the many researchers to evaluate the dispersion of rGO as shown in Table 1. Wang et al. [33] used graphene flakes in cement composite with w/c of 0.35 and 0.05% GNPs by weight of cement with different concentrations of dispersant Methylcellulose (MC) within the range of 0.2–1.0 g/L. It was noted (Figure 8) that, for different mixes, the highest peak of GNPs-suspension was found at a wavelength of 260 nm in UV-vis spectra. Similarly, a peak of the absorption at 270 nm was observed in the UV-vis absorption spectrum of graphene, which is generally regarded as the agitation of the π-plasmon of the graphitic structure [88]. Moreover, Aunkor et al. [88] research state that the absorption peak value is a function of the concentration of dispersed graphene sheets. Sharif et al. [80] investigated the dispersion of CB applying UV-vis spectroscopy and determined the rise in UV-vis absorption related to the surface area of nanomaterials (Figure 9). They used four types of carbon black with varying surface areas and determined the highest dispersion values for CB having the highest surface area.

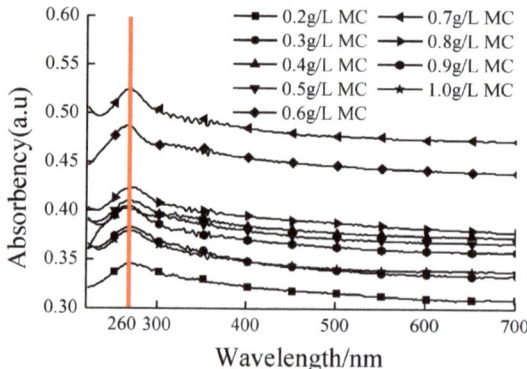

Figure 8. UV-Vis spectroscopy absorbance range of GNPs suspension with different MC amounts. Reproduced with permission from [33], MDPI, 2016.

Table 1. Peak (C=C bonds) of UV-vis absorbance of rGO from various reduction approaches.

π-π* Transition of rGO	References
275 nm	[89]
273 nm	[90]
273 nm	[89]
272 nm	[91]
271 nm	[92]
271 nm	[93]
270.9 nm	[94]
270 nm	[95]
269 nm	[96]
269 nm	[97]
269 nm	[98]
267 nm	[99]
266 nm	[100]
265 nm	[101]
264 nm	[102]
263 nm	[103]
261 nm	[104]
260 nm	[105]

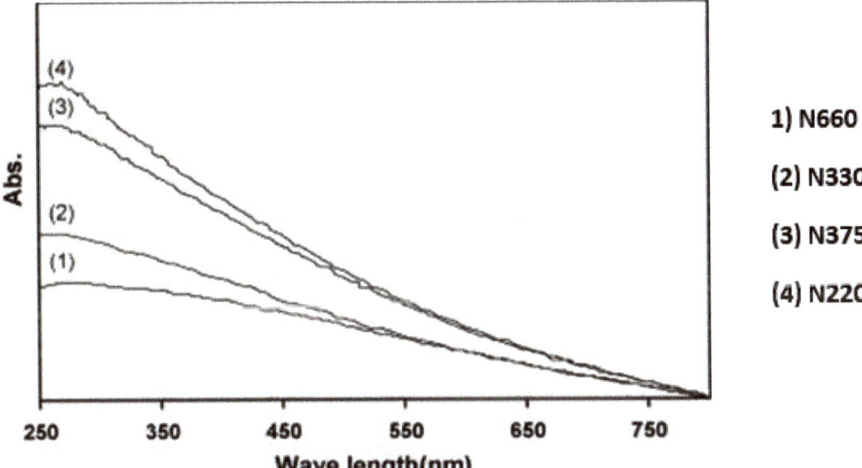

Figure 9. UV–vis absorption spectra for dispersion efficiency of various types of CB in aqueous solution. Reproduced with permission from [80], Elsevier, 2009.

Therefore, it can be concluded from the literature of the dispersion section that the ultra-sonication technique provides uniform dispersion to graphene flakes. The use of dispersant, i.e., polycarboxylate based high range water reducing admixture will help in exfoliation of graphene flakes. Additionally, the UV-vis spectroscopy method is extensively applied to monitor and assess the dispersion of nanomaterials in aqueous solution.

5. Characterization of Graphene Cement Composite

Characterization refers to the procedures by which the material's properties and structure are explored and measured. The graphene-based nanomaterials interact with hydrated cement products and influence the hardened properties of the composite material. Thus, it is important to study the micro-scale characterization of graphene-based cement composite. In this section, the characteristics of graphene-based cement composite are explored by (a) thermogravimetric analysis (TGA), (b) infrared spectroscopic analysis, (c) X-ray diffractometric (XRD) analysis, and (d) morphological analysis.

5.1. Thermogravimetric Analysis (TGA)

Thermal analysis is a method which estimates the change in the materials properties depending on the temperature [106]. The effects of pristine graphene oxide (PGO) and graphene oxide nanoplatelets (GONPs) (produced from ball-milling) were investigated by Sharma and Kothiyal [107]. They used 0.10% and 0.125% of PGO and GONPs by weight of cement in mix design with a w/c ratio of 0.45. Figure 10 presents the TGA curves for the control sample, 0.125% of PGO-cement mortar nanocomposites (PGO-CNCs) and GONPs-CNCs obtained after 90 days of curing. The weight loss corresponding to CH in the control mix, pristine graphene oxide (0.125PGO-CNC) and GO nanoplatelets (0.125GONP-CNC), appeared to be 12.7%, 10.8%, and 5.3% respectively. The final weight loss in the TGA curve of GO-based cement mortar was slightly greater as compared with the control mix due to the inherent thermal conductive properties of GO as shown in Figure 10 [107].

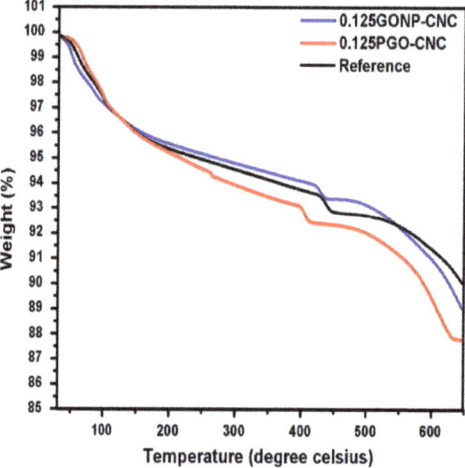

Figure 10. TGA curves of plain mix, pristine graphene oxide mortar composite and graphene oxide mortar composite after 90 days of curing. Reproduced with permission from [107], Elsevier, 2016.

Wang et al. [33] observed the variation in the TGA curve after the addition of graphene nanocomposites to the cement paste. Figure 11 showed the TGA curve for cement composites with and without GNPs at 7-day and 28-day respectively. Both samples showed similar trends at seven and 28 days, however, the graphene flakes accelerated the hydration process of cement. On the 7th day, the amount of amorphous phases, i.e., CSH and calcium hydroxide in GNP-cement composite was greater than the control cement phase. However, the content of CSH gel and calcium hydroxide in both samples was nearly equal after 28 days of curing. It was concluded that GNPs enhanced the hydrated cement products at an early age.

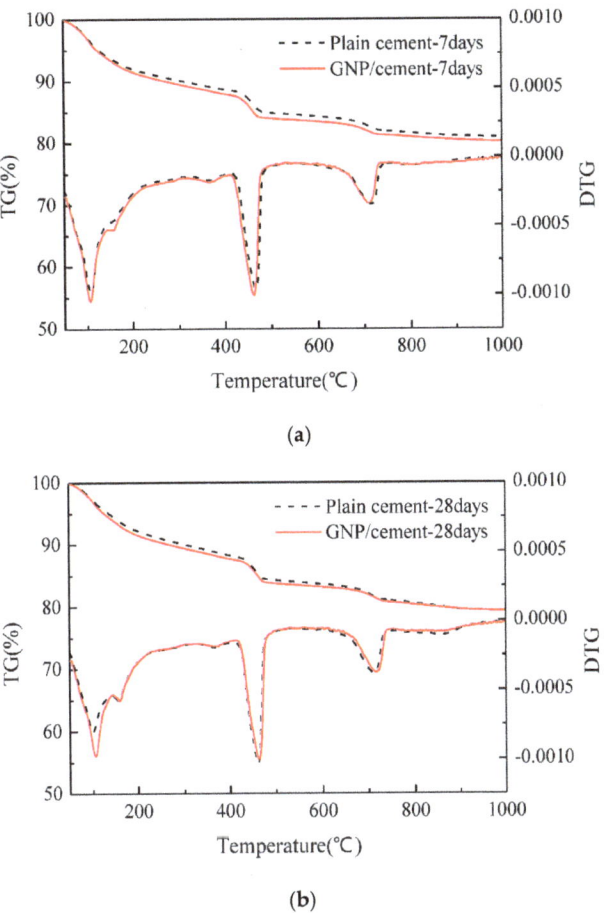

Figure 11. TGA curves of cement composites with and without GNPs (**a**) 7 days, and (**b**) 28 days. Reproduced with permission from [33], MDPI, 2016.

5.2. Infrared Spectroscopic Analysis

Vibrational spectroscopy is an approach for the assessment of the molecular structure. It provides useful information about possible chemical and physical interaction [108]. FTIR spectra of 28-day cement paste and graphene are given in Figure 12. The spectrum of control sample is shown in Figure 12a while the observed maximum peak values are listed in Table 2 [108–111]. In graphene spectra (Figure 12b), besides these peaks, two additional peaks are observed at 1570 cm^{-1} (sp2 hybridized C=C), 2918 cm^{-1} and 2850 cm^{-1} (symmetric and asymmetric stretching vibration of –CH$_2$), pointing to the sp2 network [112]. According to Mollah et al. [108], FTIR spectra can be divided into three regions; (a) water region (>1600 cm^{-1}); (b) the sulphate region (1100–1150 cm^{-1}); and (c) the material region (<1000 cm^{-1}). Spectral data vary by the incorporation of different nanomaterials and assessments of these transitions will yield useful information. Shifting of these bands implies the stronger bonding, formation of hydrated products and polymerization in hydrated products [108,113] while the variation in the absorbance intensities provides information about the quantity of material [113].

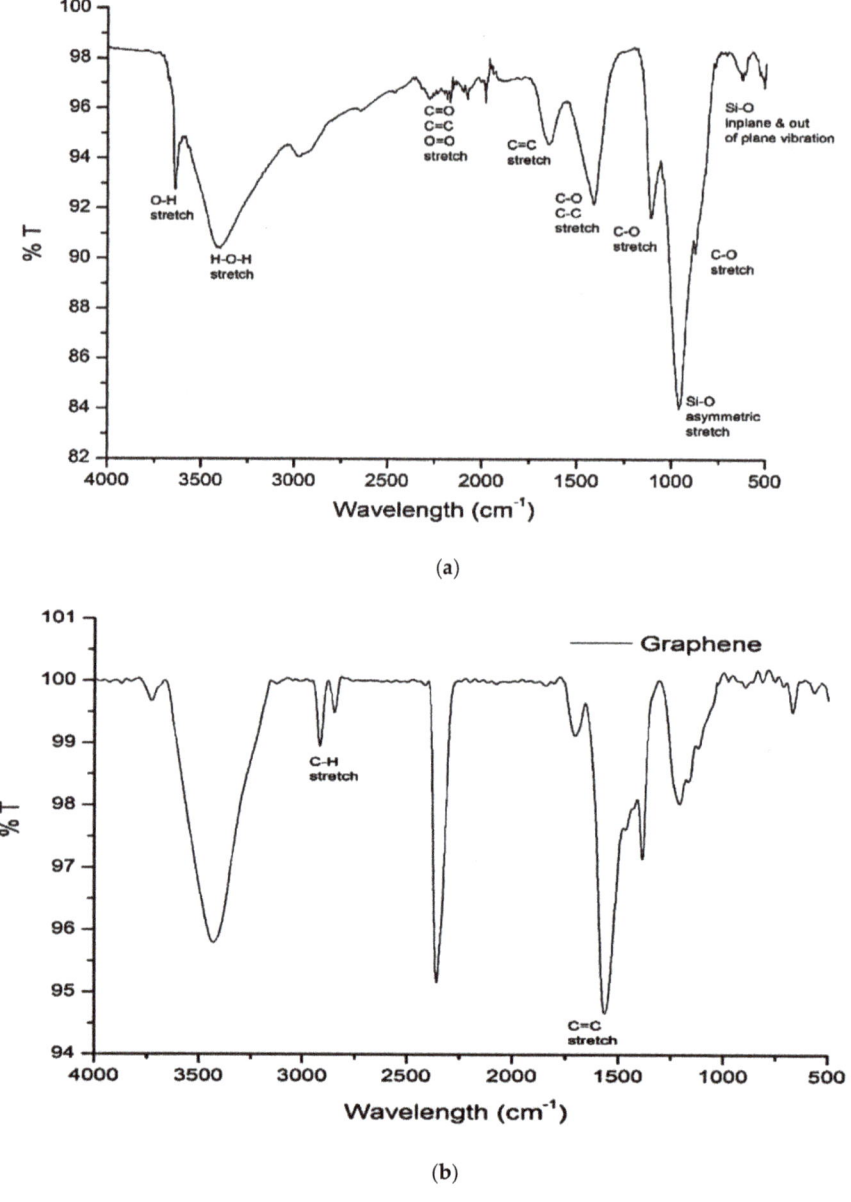

Figure 12. Analysis of FTIR spectra of (**a**) control sample adapted from [48] Elsevier, 2005 and (**b**) graphene adapted from [88], RSC Advances, 2015.

Geng et al. [48] employed FTIR analysis on three different blends of cement paste. The w/c was maintained at 0.45 and CNTs and carbon fibers were added 2% by weight of cement. Figure 13 shows four FTIR spectra, and three spectra are associated with cement pastes prepared with and without CNTs and carbon fibers while the remaining one is related to treated carbon nanotubes. The authors noted that untreated carbon fibers cement paste spectra were similar to plain cement paste as presented in Figure 13c. Hence, no chemical interaction and new phase formation were recorded. The spectra in Figure 13a displayed the peaks at 1733 cm^{-1} and 1118 cm^{-1} corresponds to C=O stretching of carboxylic acid and C-OH stretch of hydroxyl. These peaks confirmed that various oxygen-containing groups are attached to the surface of CNTs. When these surface treated CNTs were added to cement paste, a chemical interaction took place, as shown in Figure 13b. A positive shift in a spectral peak at 1756 cm^{-1} by 22 cm^{-1} indicated the probable existence of carboxylate while the disappearance of peak at 3643 cm^{-1} specified the chemical synergy between hydrated cement products with oxygen groups of carboxylic acid attached with the CNTS. In addition, the variation in spectral shape in the CSH region indicates the variation in CSH phases due to the functionalization of carbon nanotubes.

Table 2. Observed peak values in FTIR spectra of control sample. Adapted from [48].

Bond Type	Wavelength (cm^{-1})
H-O-H stretching of CSH	3375
Si-O asymmetric stretching vibrations of CSH	1014
Si-O in-plane vibration of CSH	460
Si-O out of plane vibration of CSH	690
C-O Stretching of CO$_3^{-2}$	1410
C-O Stretching of CO$_3^{-2}$	874
C-O Stretching of CO$_3^{-2}$	712
H-O-H stretching of ettringite	1630
H-O-H stretching of ettringite	3430
S-O bending vibration of SO$_4^{-2}$	695
C=O, C=C, O=O	2299
C=O, C=C, O=O	2075

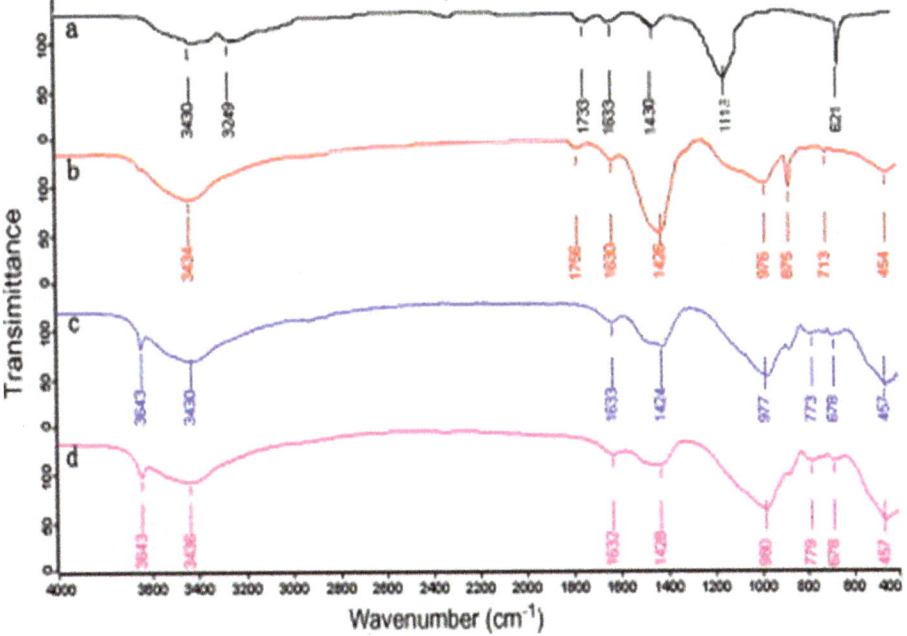

Figure 13. FT-IR spectra of various cement composites at age of 90 days. (**a**) carboxylic carbon nanotubes; (**b**) carboxylic carbon nanotubes cement paste; (**c**) carbon fibres cement paste and; (**d**) plain cement paste. Reproduced with permission from [48], Elsevier, 2005.

The effect of rGO, n-Al$_2$O$_3$, and n-SiO$_2$ to cement composite was investigated by Murugan et al. [114]. Figure 14 presents the FTIR spectra of these cement pastes. The authors did not notice any major difference in FTIR spectra. In these spectra of cement pastes observed after 28 days of curing, free water peak found at ~1645 cm^{-1} was attributed to O–H bend, ettringite peak at ~1118 cm^{-1} was due to S–O stretch, silicates peaks at ~950 cm^{-1} was due to Si–O asymmetric stretch. Calcite peaks, which are indicative of carbonation, were recorded in all mixes at ~1414 cm^{-1} and ~874 cm^{-1}, attributed to C–O stretching and C–O bend vibration respectively.

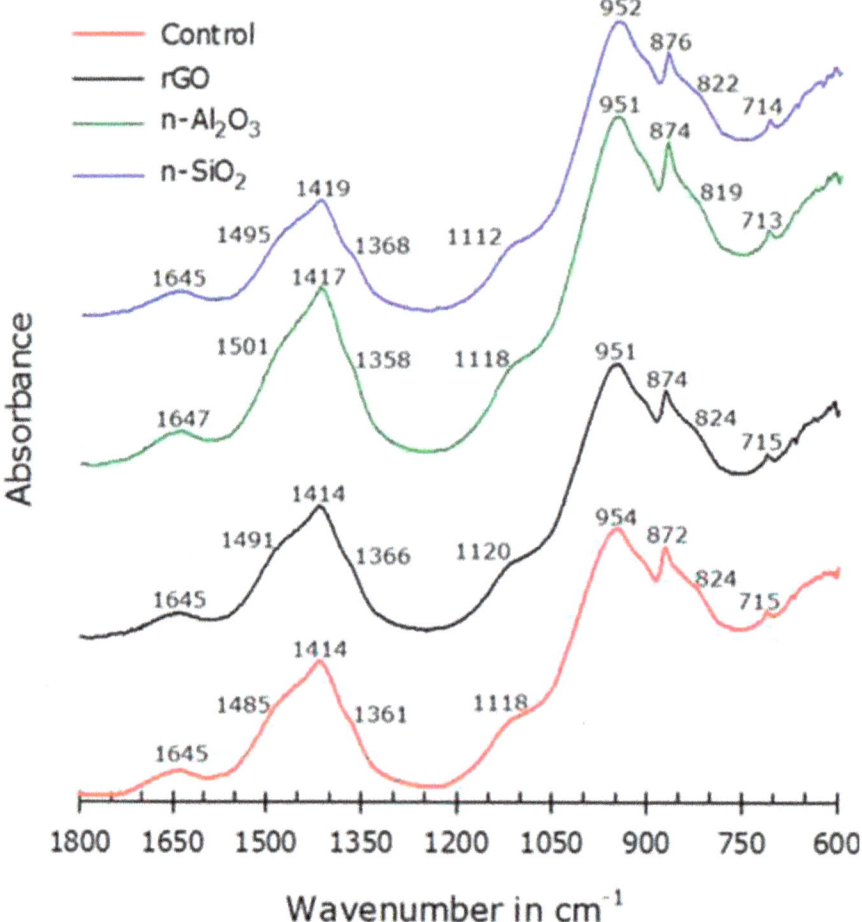

Figure 14. FTIR curve of the various pastes cured for 28 days. Reproduced with permission from [114], Elsevier, 2016.

5.3. XRD Analysis

X-ray diffractometer was broadly adapted for determining the crystalline structure of materials [106]. Wang et al. [33] evaluated the influence of graphene flakes in cement composite keeping w/c constant as 0.35 and GNPs/cement of 0.05%. The XRD patterns of control sample and cement paste with graphene flakes at 7 and 28 days are presented in Figure 15. The authors observed that no new phases were found after the incorporation of graphene in the cement mix. Hence, the type and structure of the final hydration products remained unchanged. However, the XRD spectra characterize the degree of the hydration process. Peak intensities of the hydrated cement products, i.e., calcium hydroxide and ettringite (AFt) were higher in graphene cement composite as compared with plain cement. The intensity of unhydrated cement content, i.e., alite (C_3S) was also found to be lower in graphene cement composite. In addition to this, Murugan et al. [114] found that mineralogical composition also remained the same.

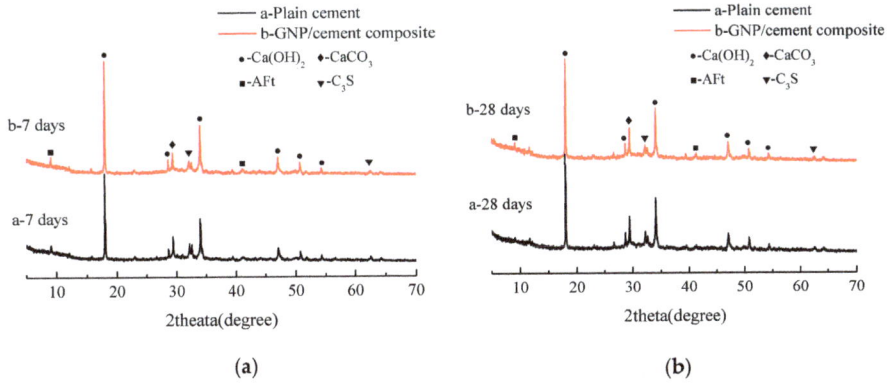

Figure 15. XRD patterns of GNP cement composite and plain cement at (**a**) 7-day and (**b**) 28 days. Reproduced with permission from [33], MDPI, 2016.

The XRD patterns for control mix, 0.125 PGO-CNC, and 0.125 GONO-CNC after 90 days of curing were obtained by Sharma and Kothiyal [107] and presented in Figure 16. The quantity of portlandite was found to be greater with the incorporation of GO in cement-based composite. The authors found that peaks of C_3S and C_2S were reduced significantly by incorporating the graphene in the cement mix. Brittleness of composite mix was found less as compared with the plain sample because of the absence of Aft peak. It was suggested that the relative decrement of the C_4AF (Tetra calcium aluminate ferrite) peak in the composite mix indicates the higher hydration rates due to nanocomposites.

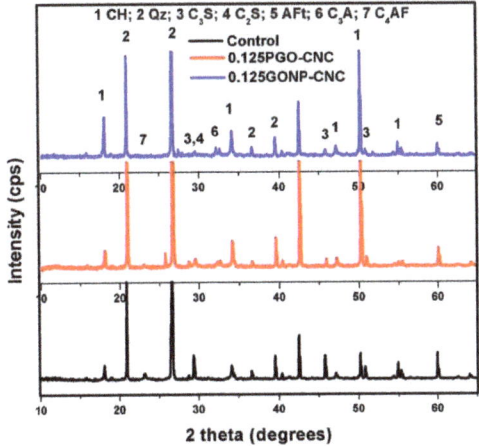

Figure 16. X-ray diffractograms of pristine graphene oxide mortar cement nanocomposite, graphene oxide nanoplatelets mortar composite and control samples. Reproducde with permission from [107], Elsevier, 2016.

5.4. Morphological Analysis

Various researchers have used field emission scanning electron microscopy (FESEM) to evaluate the morphology of cement-based composite prepared with graphene and its derivatives. The microstructure and pattern of hydrated crystals of graphene oxide based cement composite were investigated by Shenghua et al. [115]. The SEM of fractured surfaces are presented in Figure 17. The authors reported that disorderly stacked hydration products were formed in plain cement paste (Figure 17a). In contrast, with an increasing percentage of GO in cement composite, flower-like crystals as shown in Figure 17b,c) were formed. These flower-like crystals became denser with the rise in GO percentage up to 0.03% (Figure 17d). For a 0.04% dosage of GO, an irregular polyhedral resembling shape appeared in the hydrated products (Figure 17e), and for 0.05%, they became regular and complete polyhedral shapes (Figure 17f). According to authors, these flower-like crystals improved the toughness while polyhedron like crystals contributed to compressive strength. Shenghua et al. [115] concluded that GO regulated the cement hydration process and formed the regular flower-like and polyhedral like crystals. Furthermore, Shenghua et al. [115] evaluated the effect of hydration time for 0.03% of GO cement composite. According to the authors, GO contributes to the formation of the flower-like structure after 1 day of casting and on 28-day, these crystals became perfect and large flower-like shape. It was confirmed that graphene oxide regulated the hydration crystals in flower-like shape and these shapes tended to form a massive and compact structure through the cross-linking of flower-like crystals.

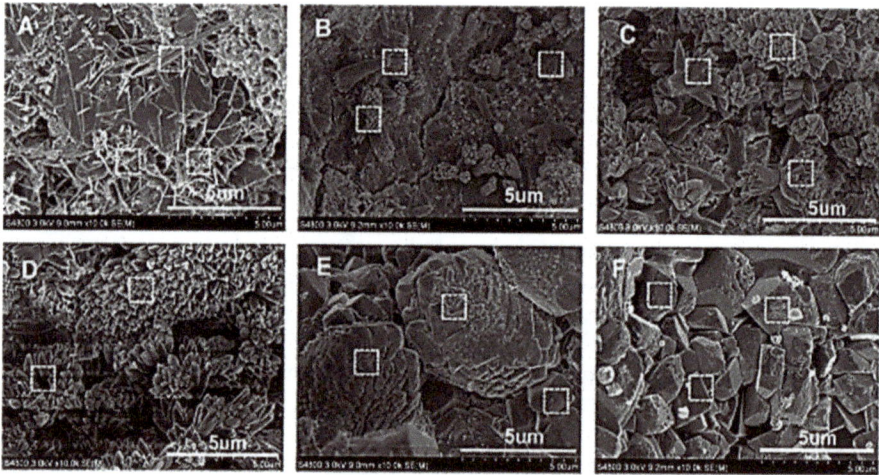

Figure 17. SEM images of cement composite at 28 days mixed with graphene oxide at various percentages: (**A**) no GO; (**B**) GO 0.01%; (**C**) 0.02%; (**D**) 0.03%; (**E**) 0.04%; and (**F**) 0.05%. Reproduced with permission from [115], Elsevier, 2013.

Cui et al. [116] studied the chemical composition of flower-shaped crystal and polyhedral like hydrated crystal as reported in research of Shenghua et al. [115]. They proposed that these are calcium carbonates, which are due to the carbonation of cementitious hydrates and are not the product formed by cement hydration. Cui et al. [116] anticipated a potential pitfall in sample preparation for the scanning electron microscopy by Shenghua et al. [115], which resulted in the production of flower-like crystals. In order to prove this statement, Cui et al. [116] performed experimental work using carbon nanotubes with –COOH functional group. They collected two samples for SEM analysis from the same mother cube using two methods. For the method I, the obtained sample was oven dried for 24-h, after which, the sample was cooled down to room temperature for SEM analysis. For method II, the obtained sample was placed in a natural environment for seven days. After that, sample was oven dried and then cooled for SEM analysis. Figure 18 showed the SEM image and XRD pattern of the sample obtained from method II. The calcite peaks were distinguished and prominent in Figure 18b. Thus, Cui et al. [116] recommended further research on the regulation mechanism of GO on cement hydration as suggested by Shenghua et al. [115].

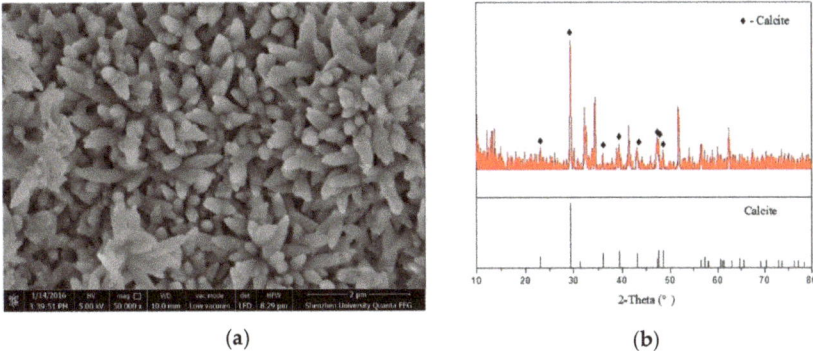

(a)   (b)

Figure 18. Description of the flow-like crystal (**a**) SEM image; and (**b**) XRD results of the surface of the sample obtained by method II. Reproduced with permission from [116], Elsevier, 2017.

Cao et al. [117] stated that the structure of the hydrated cement products showed disorderly production of needle-shaped ettringite and hexagonal calcium hydroxide as shown in Figure 19a. After the addition of functionalized graphene nanosheets (FGN) in mix design, the structure of hydrated cement products became more compact and less needle-shaped crystals (Figure 19b). When FGN percentage increased to 0.02%, it formed the polyhedron shape (Figure 19c) which represents the compact structure. However, the high content of functionalized graphene nanosheets (0.03% to 0.05%) led to the decrease in degree of hydration product because of the attachment of hydrophilic groups on the surface of FGN, which absorbed some part of the available water and prevented the full hydration process of cement pastes (Figure 19d–f) [117].

Figure 19. The effect of functionalized FGN to the cement composite after 28 days with varying FGN content (**a**) No FGN; (**b**) 0.01% FGN; (**c**) 0.02% FGN; (**d**) 0.03% FGN; (**e**) 0.04 % FGN and (**f**) 0.05% FGN. Same scale bar was used in (**a–f**). Reproduced with permission from [117], Springer, 2016.

Therefore, it can be concluded form a literature of microscale characterization that no chemical interaction and new phase formation was observed in graphene cement composite. However, graphene-based nanomaterials act as accelerator and dense formation of hydrated cement product was found near the nanomaterials. A detailed and in-depth study is still required to completely understand and explore the influence of graphene-based nanomaterials on hydrated cement products at the microscale level.

6. Rheological Properties of Graphene Cement Paste

Concrete possesses great advantages which makes it the most extensively used building material [118]. During mixing and placing of concrete, key properties are determined to be fluidity, homogeneity, consistency, and workability [118]. Any deficiencies in these properties are prone to contribute to laitance, segregation, bleeding and cracking of the concrete [119]. It is known that the mechanical and durability properties of cementitious construction components depend on viscosity and fresh properties of cement composite. Any variation in viscosity may lead to defects in concrete structures [120]. Hence, the rheological properties of concrete are extremely significant to attain homogeneity and obtain improved workability. The rheology of cement paste firmly impacts the overall fresh properties of concrete [121]. Typically, shear stress and shear rate are the indicators of the flow properties of the cement paste. Next, using flow curves and mathematical models, viscosity and other flow parameters are determined. With the improvement in nanotechnology, researchers are now focusing more on evaluating the influence of nanomaterials on cement composite [122].

Many researchers examined the influence of different nanomaterials on cement paste flow properties [123–128]. Ormbsy et al. [123] used the parallel plate geometry for rheological investigation and found that MWCNTs meaningfully influenced the rheological behaviour of polymerizing cement. Konsta et al. [47] used four different length of MWCNTs while the surfactant to MWCNTs weight ratio was kept as 1.5, 4.0, 5.0, and 6.25. In the aqueous solution, the MWCNTs content was kept constant at an amount of 0.16% by weight of water. The MWCNTs were treated using with and without sonicated energy. After that, MWCNTs aqueous-surfactant suspensions were mixed with cement and rheological properties were investigated. In preparation of MWCNTs cement composite, MWCNTs were used with the content of 0.08% of cement (by weight) and w/c = 0.5. Preliminary rheological results indicated that long MWCNTs are difficult to disperse. They observed the shear thinning response of cement paste and at high shear stress (above 70 Pa), approximately constant viscosity independent of sonication energy was obtained. Shang et al. [125] investigated the rheological properties of GO and GO encapsulated silica fume-based cement pastes using Bingham model. The authors found that GO reduced the fluidity of the cement paste by 36.2% when compared to control cement paste. A rise in the value of yield stress and plastic viscosity was obtained when GO was added to the cement paste. Wang et al. [126] mentioned that the addition of GO to cement paste forms flocculation particles, which had a dependence on the GO concentration and it, consequently, improved the yield stress, plastic viscosity, and area of the hysteresis loop of the flow curve. Also, they investigated the effect of fly ash on the GO cement paste and found that for 0.01 wt.% of GO and 20 wt.% of fly ash, the yield stress and plastic viscosity of the cement paste dropped when comparing with control cement paste by 85.81% and 29.53%, respectively. For cement specimen with 0.03 wt.% of GO and same amount of fly ash, the yield stress and plastic viscosity of the cement paste was lowered by 50.33% and 5.58%, respectively (Figure 20).

Yahia et al. [129] found that estimated viscosity and yield stress values varies according to the mathematical model used. Various researchers used different mathematical models to determine the yield stress and plastic viscosity values and predicted the specific tendency of the flow. Due to statistical errors [130], it is impossible for one model to predict precisely the trend of the flow behaviour of cement paste [129].

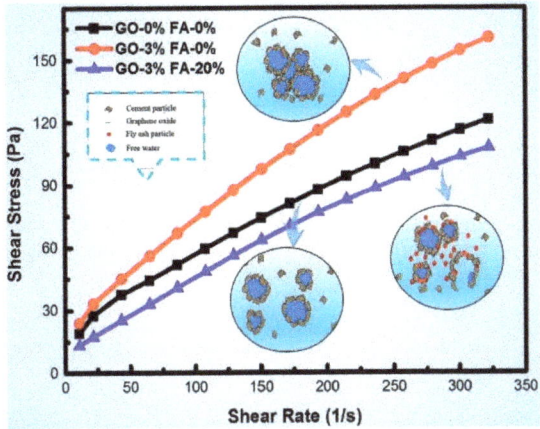

Figure 20. Impact of fly ash on flow curves of GO cement paste. Reproduced with permission from [127], Elsevier, 2017.

The rheological properties of fresh cement paste with different surface area of GNPs, resting time and shear rate cycles were examined by Rehman et al. [118,131] using various rheological mathematical models. The authors observed the increase in plastic viscosity and yield stress with the increase in intensity of graphene in the composite and resting time but a decrease was noted for higher shear rate cycle. Furthermore, it was found that the highest values of yield stress were obtained when measured by the modified Bingham model (BM) and the lowest values were estimated by the Casson model. Figure 21 shows the plastic viscosity values for graphene cement composite and the influence of different parameters on it. It was concluded that the modified BM fits the experimental flow curves best and the Casson model demonstrated greater standard error values.

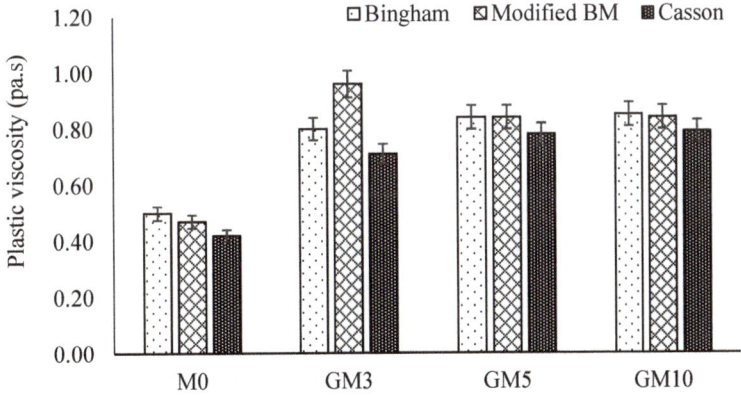

(a)

Figure 21. *Cont.*

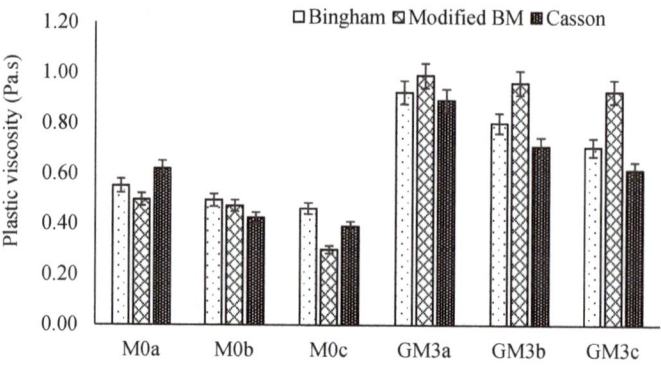

(b)

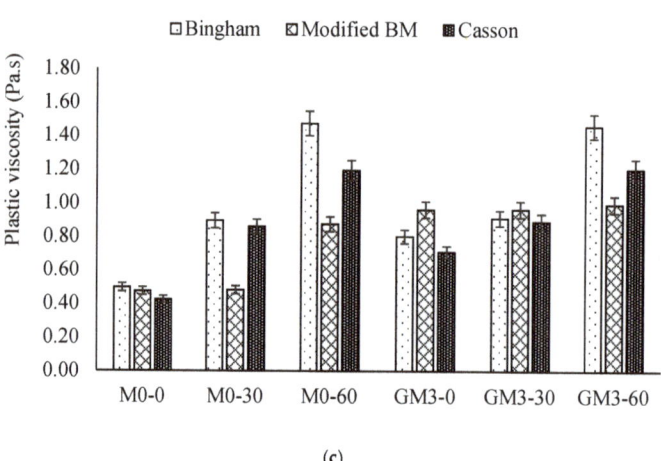

(c)

Figure 21. Plastic viscosity values (Pa·s) calculated by various mathematical models with varying (**a**) GNP amount, (**b**) shear rate and (**c**) resting time. Reproduced with permission from [118], MDPI, 2017.

It can be concluded from the existing available literature that new mathematical models need to be developed which will predict the rheological properties of nanomaterial-based cement composite. Moreover, the influence of graphene-based nanomaterials requires further exploration so that it can be successfully used in 3D printing applications of the construction industry.

7. Mechanical Properties of Graphene Cement Composite

Consumption of cementitious building materials i.e., concrete is increasing due to its various advantageous characteristics such as high strength, durability, and resistance to fire [2]. Sixuan [79] conducted a study on the mechanical properties of both cement paste and mortar by adding a different concentration of GNPs. Their study revealed that there was no increment in compressive strength for the cement paste incorporated with 0.05% and 0.25% of GNPs. However, an increment of 20% in compressive strength was observed for 0.50% of GNP added to cement mortar. As for the flexural strength, the maximum increment up to 82% was observed for the cement paste with 0.05% of GNP [79].

Table 3 provides the effect of graphene nanomaterials and its derivatives on the mechanical properties of cement composite. The incorporation of nanomaterials in cement composite significantly increased the mechanical properties of cement composite both at early and later age. Wengui et al. [132] proposed that due to the nucleation and filling effect of GO, it can speed-up the hydration process at an early stage. Yet, the complete mechanisms have not been described in the available literature. For example, the compressive and flexural strength reported by Shenghua [133] and Mokhtar et al. [134] are considerably different. Both researchers used the same water-to-binder ratio (w/b), the GO dosage and almost similar curing conditions. Moreover, Wang et al. [127] and Wang et al. [135] used the same w/b and GO dosage yet the difference in the rise of compressive strength was almost double. Under similar conditions, the increase in compressive strength reported by Kothiyal et al. [136] and Shang et al. [125] is twice as that reported by Sharma and Kothiyal [107] and Wengui et al. [132] respectively. It is commonly known that the factors influencing the mechanical strength in the composite matrix are w/b, type of nanomaterial, its dosage and curing duration of the specimen [137]. Moreover, the results of dispersion, agglomeration, size, and functional groups attached with nanomaterials have considerable influence on mechanical properties [28]. It can be observed from Table 3 that the nanomaterials have more influence on the flexural strength of cement composite as compared with compressive strength. According to Sharma and Kothiyal [138], the bridging and bonding effect of GO with cement matrix and dense microstructure of cement matrix are the factors contributing to flexural strength. Figure 22 shows the schematic of growth of cement hydrates on the templates of GO. Furthermore, the use of dispersant, surface modification, reduction in size and thickness of graphene has been employed to improve the performance of GNDs on mechanical properties of cement-based materials [138–140]. It is understood that the key factors controlling the porosity and the mechanical properties of cement-based materials are the availability of more hydrated cement products, filling of pores and bonding between hydrated cement products and nanomaterial [138,141,142]. Nevertheless, the tremendous increment in the mechanical properties has been recorded, yet the role of graphene nanomaterial is still unclear in the literature.

Table 3. The impact of graphene based nanomaterials on the mechanical properties of cement composite modified from [28].

Matrix	Nanomaterial Type/Dosage (wt.%)	w/b	Percentage Rise in Compressive Strength/Age	Percentage Rise in Flexural Strength/Age	Reference
Cement Paste specimens	GO/0.02	0.3	13.0/28 d	41.0/28 d	[134]
			60.1/28 d	84.5/28 d	[133]
	rGO/0.02		22.0/28 d	70.0/7 d	[114]
	GO/0.03		18.8/28 d	56.6/28 d	[135]
			42.5/28 d	55.0/28 d	[127]
	FGON/0.03		51.3/28 d	65.5/28 d	[140]]
	GO/0.04		28.6/28 d	43.2/28 d	[144]
	GO/0.05		66.4/7 d	69.4/7 d	[145]
			52.4/3 d	90.5/28 d	[146]
	GNPs/0.15		49.4/28 d	27.5/28 d	[147]
	GO/0.022		27.6/3 d	26.7/3 d	[148]
	GNPs/0.03		1.3/28 d	16.8/28 d	[33]
			30/28d d	-	[118]
	GO/0.04	0.4	37.0/28 d	14.2/28 d	[132]
			15.1/28 d	–	[125]
	GO/0.05		11.0/15 d	16.2/15 d	[149]
	GO-CNT/0.05		21.1/15 d	24.1/15 d	[149]
	GO-CNFs/0.05		2.89/28 d	25.0/28 d	[150]
	GO-CNT/0.06		23.9/28 d	16.7/28 d	[151]
	GO/0.03	0.5	40.0/28 d	–	[152]

Table 3. Cont.

Matrix	Nanomaterial Type/Dosage (wt.%)	w/b	Percentage Rise in Compressive Strength/Age	Percentage Rise in Flexural Strength/Age	Reference
Cement Mortar Specimens	GO/0.022	0.4	–	34.1/7 d	[153]
	GO/0.03		–	18.7/7 d	[154]
			45.1/3 d	70.7/3 d	[155]
	GO/0.05		43.2/3 d	106.4/14 d	[146]
	GO/0.02		–	36.7/3 d	[156]
	GO/0.03	0.5	30.0/28 d	–	[157]
	FGON/0.03		20.3/28 d	32.0/28 d	[158]
	FGON/0.1		39.0/15 d	70.8/15 d	[139]
	GNPs/0.1		19.9/28 d	–	[159]
	GO/0.125		110.7/3 d	–	[136]
			53.0/3 d	–	[107]
	GNPs/0.8		87.5/28 d	–	[160]
	GO/1.0		114.1/14 d	–	[138]
	GNPs/0.08	0.6	55.3/7 d	–	[161]
Concrete	GO/0.1	0.5	14.2/7 d	4.0/3 d	[162]

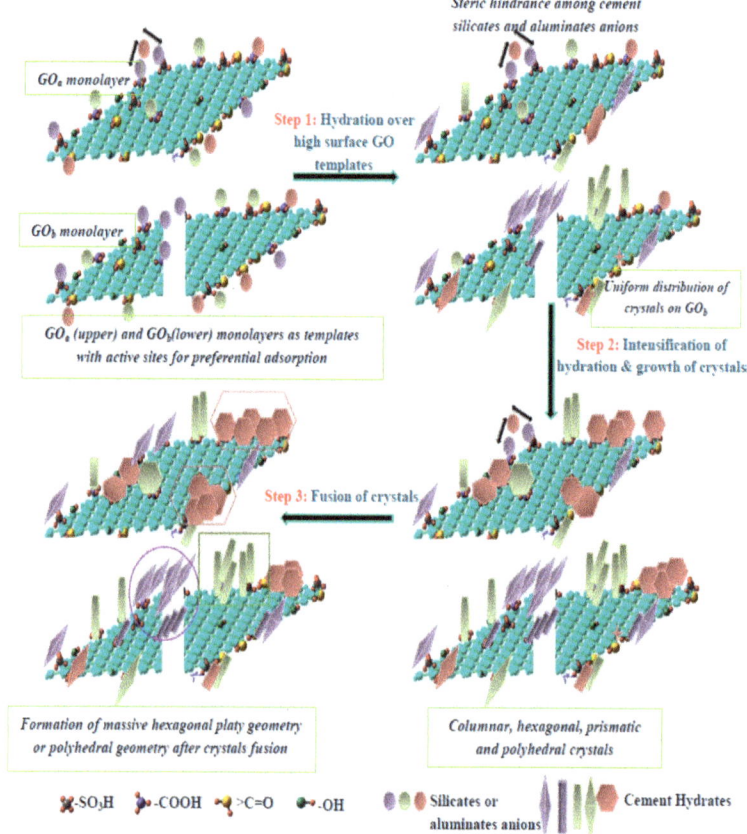

Figure 22. Schematic diagram of growing cement hydrates on templates of GO monolayers. Reproduced with permission from [138], RSC Advances, 2015.

Enhanced mechanical properties of graphene cement composite are attributed mainly to the bonding of graphene and its functional groups with CSH gel [138]. Hou et al. [143] performed simulation work using ReaxFF forced field to explore the molecular-scale structure and chemical interaction between the functional groups of graphene and CSH. In order to strength the simulation results, Hou et al. [143] also performed the experimental investigation. The experimental results showed that by incorporation of 0.16% GO by weight of cement in GO cement composite, the compressive and flexural strength enhanced by 3.21% an 11.62% respectively. For further detailed analysis, the mechanical behaviour of G_CSH, GO_CSH and GO_CASH models under tension loading was studied using a stress–strain curve. They found that graphene without chemical bonding with CSH in G_CSH model showed a small increase in compressive stress. The authors concluded that functionalization significantly enhanced the mechanical properties due to interfacial strength between functionalized GO and CSH gel. In addition, weak bonding and instability of atoms in the interface region resulted in the weakest mechanical behaviour.

8. Energy Harvesting and Thermoelectrical Properties of Graphene Cement Composite

Energy harvesting is a concept through which existing ambient energy in the environment e.g., solar, wind, hydro energy has been converted into useful forms i.e., electrical energy [163]. According to Francisco and Adelino [163], micro energy harvesting, which is associated with small scale energy harvesting is gaining interest in the scientific community. In micro energy harvesting, the main energy sources are heat, thermal variations, acoustic emissions, electromagnetic variations and mechanical vibrations. In this section, the application of graphene cement composite for energy harvesting using its thermoelectrical properties will be discussed. A schematic of energy harvesting in buildings by converting the abundant solar energy into electrical energy is demonstrated in Figure 23. Graphene cement composite have been used by researches for energy harvesting in buildings by enhancing the thermoelectrical properties of cement composite [164]. Thermoelectrical properties are measured in terms of dimensionless figure of merit, ZT, which is equal to ($S^2\sigma T/\kappa$), where S is Seebeck coefficient in μVK^{-1}, σ is electrical conductivity in Scm^{-1}, T temperature in K and thermal conductivity κ in $Wm^{-1}K^{-1}$ respectively. In addition, ZT ≥ 1 is recommended for energy harvesting application purposes in buildings. Moreover, graphene has the capability to transform the non conductive material into conductive material. According to Balandin [165], graphene enhanced the electrical conductive properties of cement composite and reduced the thermal conductive properties.

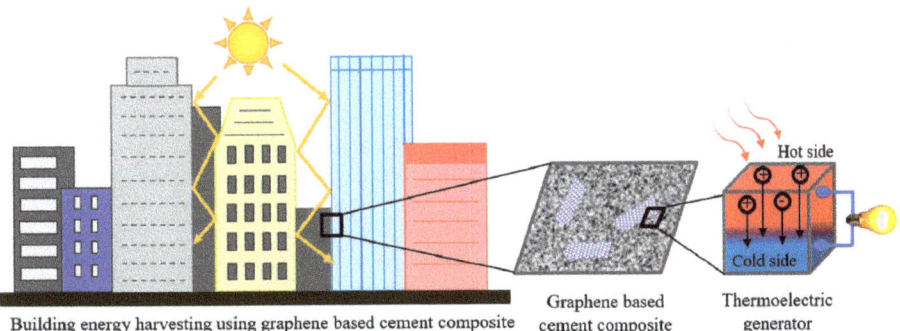

Figure 23. Building energy harvesting using thermoelectric generator of graphene cement composite. Reproduced with permission from [164], Elsevier, 2019.

It was Wei et al. [166], who for the first time measured the thermoelectrical properties of expanded graphite cement composite for large scale energy harvesting and climate adaptation. Expanded graphite was added as 5, 10 and 15 by mass of cement for the preparing the graphite cement composite.

Figure 24 shows experimental setup for determining the thermoelectrical properties of graphite cement composite. The authors found that ZT values depend on the temperature. The values of ZT increased from 30–75 °C while it decreased from 75–100 °C. The maximum ZT value of 6.82×10^{-4} was noted for 15% addition of graphite in cement composite at 75 °C. Overall, ZT values were found to be smaller due to lower electrical conductivity and seebeck coefficient of graphite cement composite. However, due to large areas of graphite cement composite in pavements and roofs being exposed to solar radiation, the solar energy can be converted into electrical energy during the summer season around the globe. Table 4 presents the thermoelectrical properties of graphene based cement composite determined by various researchers.

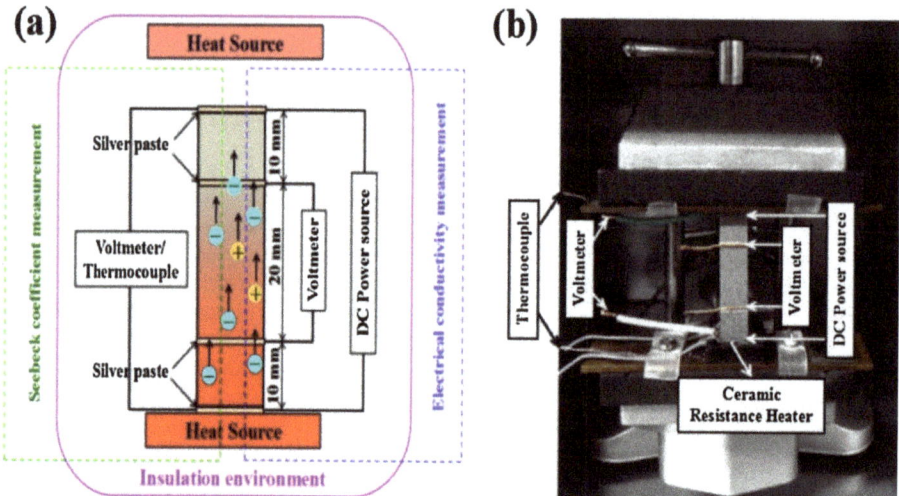

Figure 24. Laboratory setup for evaluating the seebeck coefficient and electrical conductivity (**a**) The test principle diagram and, (**b**) the experimental setup. Reproduced with permission from [166], Elsevier, 2018.

Ghosh et al. [164] performed experimental study using graphene nanoplatelets in cement composite to determine the thermoelectrical properties and energy harvesting capability of composite material. They employed GNP as 5%, 10%, 15%, and 20% by mass of cement to prepare graphene cement composite. Four-probe electrical conductivity and seebeck coefficient measurement system was used to determine the Seebeck coefficient and electrical conductivity simultaneously. For testing, rectangular specimens were used and experiment was performed with varying temperature from 25 °C to 75 °C with heating rate of 0.01 °C/s. Furthermore, differential scanning calorimeter (DSC) was used by the authors to monitor the specific heat capacities within the temperature range of 25 °C to 75 °C. The authors observed largest seebeck coefficient of 34 $\mu V K^{-1}$ for 15% GNP inclusion at 70 °C, while electrical conductivity of 16.2 Scm^{-1} and power factor of 1.6 $\mu W m^{-1} K^{-2}$ for 20% GNP inclusion as shown in Figure 25. The linear relationship between temperature and power factor was also observed. Highest specific heat capacity value 0.88 $Jg^{-1} K^{-1}$ was found for 10% GNP cement composite while 20% graphene cement composite showed highest thermal diffusivity as compared to other percentages of graphene cement composites. Moreover, the highest ZT of 0.44×10^{-3} was found for the 15% GNP cement composite at 70 °C. According to authors, the higher amount of GNP in the cement composite was beneficial for establishing the conductive network. It was concluded that graphene based cement composite significantly contribute to energy harvesting application in addition to improving the quality of indoor environment of buildings. With this application, graphene cement composite can harvest the energy, reduce electric consumption, and provide a substantial financial benefit [164].

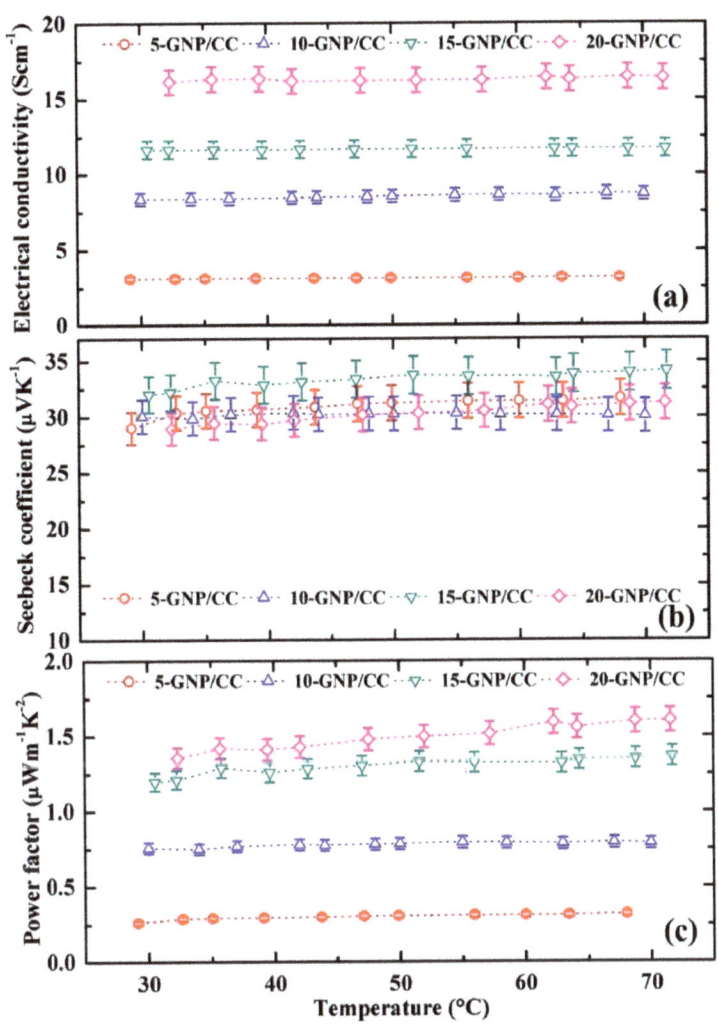

Figure 25. Estimation of (**a**) electrical conductivity, (**b**) seebeck coefficient and (**c**) power factor of graphene cement composite. Reproduced with permission from [164], Elsevier, 2019.

Table 4. Thermoelectrical properties of graphene based cement composite.

Materials	Concentration (wt.%)	S (μV/°C)	σ (Scm^{-1})	K (Wm^{-1}K^{-1})	Power Factor (μWm^{-1}K^{-2})	Reference
Control sample without nanomaterial	0	10^{-5}–10^{-4}	10^{-7}	0.53		[166]
Expanded Graphite	5	−54.5	0.2	1.619	0.1	[166]
	10	−51.5	7.4	2.594	1.9	[166]
	15	−50.1	24.8	3.213	6.38	[166]

Table 4. Cont.

Materials	Concentration (wt.%)	S (μV/°C)	σ (Scm^{-1})	K (Wm^{-1}K^{-1})	Power Factor (μWm^{-1}K^{-2})	Reference
Carbon nanotubes	15	57.98	0.818	0.947	-	[167]
Graphene nanoplatelets	5	32	3.13	0.743	0.4	[164]
	10	30	8.5	0.947	0.7	[164]
	15	34	11.68	1.067	1.25	[164]
	20	31	16.2	1.327	1.6	[164]
n-doped CNT (before drying)	1	−500	0.0173	-	0.435	[168]
n-doped CNT (after drying)	1	−58	0.0219	-	0.007	[168]
p-doped CNT (before drying)	1	−112	0.0054	-	0.009	[168]
n-doped CNT (after drying)	1	20	0.0069	-	0.0002	[168]

9. Piezoresistive Properties of Graphene Cement Composite

Nanomaterials are gifted with the amazing characteristics and their incorporation in cement composite improved the properties of nanomaterial based cement composite. The reason for the presence of electrical properties of graphene is the half-filled band that permits free-movement of electrons. The π-bonds hybridize together to form the π-band and π*-band, which are responsible for the electrical properties of graphene [169]. Self-sensing and damage memorizing capability of graphene-based cement composite are one of the characteristics, which graphene brought in cement composite.

The electrical properties of the cement composites are also of great importance and may be used to control damage in a concrete structure. The cement-based composite reinforced with conducting fillers can recognise its own strain by indicating the variations in the electrical resistivity values. According to Rehman et al. [118], the piezo-resistive properties are the result of a change in electrical resistance in the specimen subjected to mechanical strain [170]. Hence, electrical resistance of the cement composite is measured. By definition, the electrical resistance is the strength of a material in opposing the electrical current flowing through it. Researchers have mostly used the four-probe method to investigate the electoral properties of cement composite. In four-probe method, voltage is measured using the inner two electrical contacts while the current is measured using the outer two electrical contacts [171]. In comparison with the two-probe method, the four-probe technique is better since the calculated resistance does not include contact resistance [172].

Han et al. [172] reported that the distance between the current and voltage poles has a great importance but its impact is insignificant if the space is larger than 7.5 mm. Numerous researchers performed experiments with various spacing values. For instance, Geng et al. [173] used 10 mm while Liang et al. [174] used 40 mm distance between voltage measuring probes and current. Geng et al. [173] used the 40-mm gap between two measuring probes while Liang et al. [174] used an 80-mm gap. For unequal spacing, the electrical resistivity values are calculated by using Equation (1) [175].

$$\rho = \frac{V}{I} \times 2\pi \times \frac{1}{\left(\frac{1}{S1} + \frac{1}{S3} - \frac{1}{S1+S2} - \frac{1}{S2+S3}\right)} \quad (1)$$

where, S_1, S_2, and S_3 are the spacing in cm and calculated from current carrying probe to the voltage measuring probe. Various researchers used the piezoresistive characteristics of nanomaterials based cement composite for self-sensing prose [173,176,177]. Hence, the piezoresistive properties of nanomaterial-based cement composite are critically reviewed in the following paragraphs.

According to Hui et al. [178], the self-sensing property of the nanomaterials with cement composite may be used for structural health monitoring (SHM) purposes with no need of any additionally attached or embedded sensors. It will further open the application of nanomaterials based cement composites for fabricating smart sensors [178]. Geng et al. [173] conducted research on the functionalization of CNTs using carboxyl. The effect of functionalization of CNTs on the piezoresistive and electrical properties of the cement composite was observed. It was noted that the electrical resistance of cement composite with 0.5% of carboxyl functionalized CNTs (SPCNT) was 149 Ω.cm and for plain CNTs (PCNT) was 130 Ω.cm. Thereafter, the fractional change in electric resistivity for both types of cement composited was recorded against the cyclic compressive loading. The results showed that both cement composites were capable of monitoring the applied compressive cyclic loading, however, SPCNT specimen showed better response as compared with PCNT.

Xun and Kwon [179] studied the piezoresistive behavior of the CNT based cement composite and investigated its potential application as an embedded sensor in civil infrastructure. It was shown that the electrical resistance of CNT-based cement composite was following the same trend as applied compressive loading (Figure 26). The authors proposed that the fabrication method needs to be further optimized and the response of these composite materials must be investigated in concrete. Furthermore, they proposed that if these composite stress sensors were embedded in civil infrastructure, e.g., pavements or bridges, then due to their compatibility with concrete, they will have a long service life with the least maintenance.

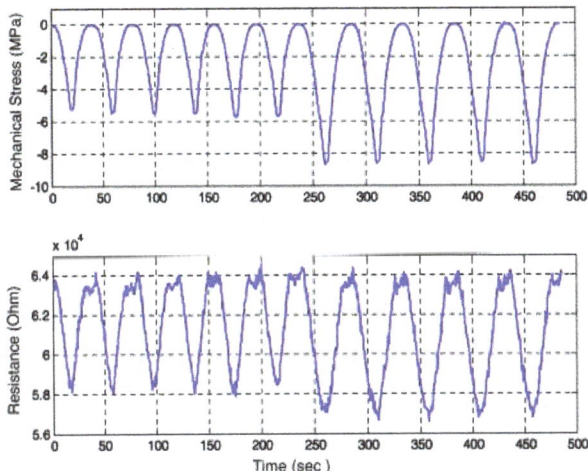

Figure 26. The piezo resistivity behaviour of cement paste enhanced with 0.1% of carboxyl functionalized MWCNTs. Reproduced with permission from [179], IOP Sciences, 2009.

The influence of different amounts of GNPs on the sensing behaviour of the cementitious composite was investigated by Hingiian and Pang [180]. The authors noted a decrease in the electrical resistivity of cement composite when GNPs was increased from 2.4% to 3.6%. Liang et al. [174] investigated the electrical resistivity values of cement mortar with increasing content of graphene flakes. For this purpose, the specimen was subjected to two different environmental conditions, i.e., one specimen was air-dried for one year, and another specimen was oven-dried at 1-day after casting. The authors proposed a link between the electrical resistivity values of the specimen with increasing content of GNP as presented in Figure 27. The electrical resistivity values decreased as GNPs content increased. 2.4% of GNP was estimated as the percolation threshold value for GNP-composite mortar [174]. The decrement of more than one order occurred when GNP was increased from 2.4% to 3.6% in cement mortar, as shown in Figure 27b.

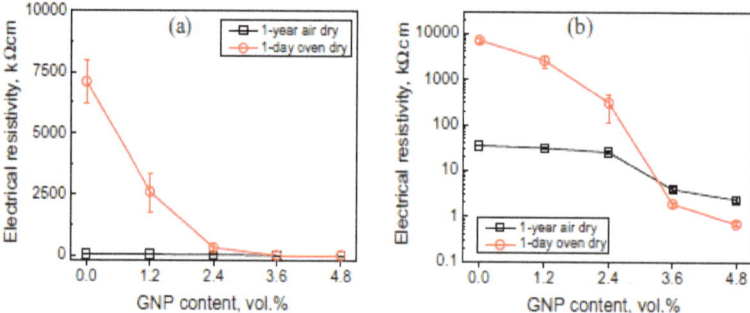

Figure 27. The relation between the electrical resistivity of the GNPs cement mortar composite and its respective GNPs content in (**a**) linear scale and (**b**) log scale. Reproduced with permission from [174], Elsevier, 2014.

Rehman et al. [118] investigated the application of graphene cement smart sensor in a GNPs reinforced concrete beam. This beam was subjected to flexural loading and the values of fractional change in resistance (FCR) of graphene cement composite specimen were recorded as shown in Figure 28. The authors observed that FCR values varied as the applied loading on the beam increased and a sudden response was indicated when the beam failed. The conclusion of the experiment was that the cement composite containing graphene favourably responded against crack propagation. The authors concluded that the graphene cement composite specimen is an inexpensive and effective way to control the structural health of the members during its shelf life.

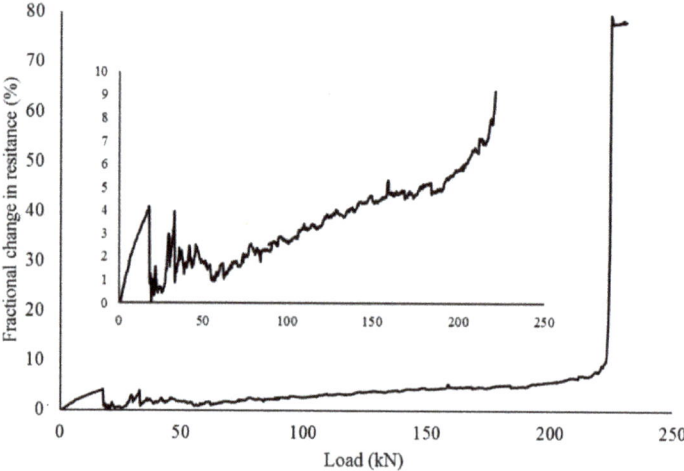

Figure 28. Fractional change in resistance of graphene cement smart sensor specimen subjected to compressive loading on the reinforced concrete beam. Reproduced with permission from [118], MDPI, 2017.

As existing well-known methods of health monitoring have some limitations in the application [9] and most importantly, these methods require additional cost for health monitoring purposes. Therefore, it is required that smart sensors with self-sensing characteristics should be developed. These smart sensors are required to reduce health monitoring costs and make construction projects financially more economical. Furthermore, they should be compatible with the cement-based building materials and should have the capability to be used for health monitoring purpose.

10. Discussion and Research Gaps

In the literature, various nomenclature has been used which could create confusion for the readers. Therefore, a detailed description of graphene and its derivation including its precursors is provided in Table 5.

Table 5. Description of Graphene and its derivatives.

Sr. No.	Nomenclature	Description	Reference
		Nanomaterials	
1.	Graphene	Graphene is single layer of densely packed carbon atoms in benzene- ring structure in 2-dimensions (2D). sp2 interacted carbon atoms are conned tightly and forms honeycomb lattice.	[88]
2.	Rolled sheets of graphene (CNT)	Carbon nanotubes (CNT) consist of sp2 carbon atoms arranged in honeycomb lattice in 1-dimension (1D) and capped with fullerene-like hemisphere at each end. They are also conceptualized as 1D rolled sheets of graphene.	[181,182]
3.	Graphene Oxide (GO);	Graphene Oxide (GO) is derived from graphene sheets and known as oxidized form of graphene. A mixture of carboxyl, hydroxyl and epoxide groups are attached chemically (covalent linkage) with graphene sheets.	[21,55,183]
4.	Graphite oxide,	Graphite oxide is identical chemically to Graphene oxide, However, inter-planar spacing between them vary due to oxidation process.	[184,185]
5.	Functionalized Graphene	Attaching various functional groups like epoxide, hydroxyl and carboxyl group with Graphene through covalent linkage is known as Functionalized graphene.	[21]
6.	Graphite/Three- dimensional graphite/Graphite structure/ Expanded graphite structure	3-D carbon allotrope containing minute crystallite of graphite. It is made of stacked graphene sheets with 0.345 nm spacing.	[186–188]
7.	2D flat sheet of carbon nanomaterial	Single layer graphene sheet is also known as 2D flat sheet of carbon nanomaterial.	
8.	Pristine graphene oxide (PGO)	Graphene oxide in its original condition (i.e., Ideal) and does not have a single defect.	
9.	Reduced graphene oxide (RGO)	Reduced graphene oxide (RGO) are prepared from oxidation, exfoliation and chemical reduction of graphite oxide.	[189]
10.	Reduced graphite oxide (rGO)	Main difference between reduced graphene oxide and Reduced graphite oxide is inter-planer spacing between atomic layers of the compound.	[184,185]
11.	Graphene and its derivatives (GNDs)	Various structural form of graphitic structure in 0D (carbon materials), 1D (CNTs), 2D (graphene sheets) and 3D (Naturally available graphite).	

Table 5. Cont.

Sr. No.	Nomenclature	Description	Reference
		Flakes	
12.	Graphene flakes (GNFs)	Graphene flakes are small tiny particles, which are much easier to form and handle in solution and powder form as compared with large graphene sheets. These GNFs possess some of the properties of large graphene sheets however; they vary with shape and size of tiny particles. Furthermore, maintaining uniform consistency in shape and sizes is not easy during synthesis process. Graphene flakes (GNFs) are also labelled as graphene nanoplatelets (GNPs).	[190]
13.	Graphene oxide flakes	Formation and production of graphene oxides in small tiny particles in powder or solution form is known as Graphene Oxide flakes.	[190]
14.	Exfoliated graphene flakes	Exfoliated large size graphene flakes/particles/sheets	
		Nanoplatelets	
15.	Graphene Nanoplatelets (GNPs)	Graphene Nanoplatelets (GNPs) are small rounded disk-shaped tiny particles, However, it is difficult to uniformly produce them even if synthesized artificially. Theoretically, all GNPs are not disk-shaped particles, therefore should be considered as Graphene flakes (GNFs).	[191]
16.	Graphite Nanoplatelets (GnPs or GPs)	Graphite Nanoplatelets are composed of mixture of graphene layers with thicker graphite particles. Their chemical structure is in-between graphene and graphite. Some researchers also consider them as graphene nanoplatelets, which is not true as per authors' point of view. Hence, their common nomenclature consist of Graphite Nanoplatelets (GnPs) and Graphite platelets (GPs).	[191]
17.	Graphene Oxide nanoplatelets (GONPs)	Graphene Oxide nanoplatelets (GONPs) consist of small rounded disk-shaped particles of graphene oxide in powder or solution form.	[190]
18.	Functionalized Graphene Sheets/flakes and nanoplatelets	Attaching various functional groups like epoxide, hydroxyl and carboxyl group with Graphene sheets/flakes/nanoplatelets through covalent linkage is known as Functionalized graphene sheets/flakes/nanoplatelets.	[21]
19.	Graphite nanoparticles, Graphite flakes, Graphite particles, Nano graphite platelets	Graphite flakes have the large size while graphite nanoparticles possess small size. Graphite nanoplatelets (GNPs) with a thickness in nanometer scale, can be obtained by exfoliation of natural graphite flakes.	[73]

It has been noted that GNPs enhance the hydrated cement products at an early age of 7-days in contrast to 28-days. Furthermore, no chemical interaction and new phase formation in hydrated cement product took place in the production of GNPs and hydrated cement products. However, from infrared spectral analysis Li et al. [48] found that a chemical interaction exist between hydrated cement product and functional groups (hydroxyl and carboxyl) of carbon nanotubes. The variation in CSH phases was also monitored. Some researchers like Murugan et al. [114] did not find any major difference in IR spectra. Several researchers [33,158] also noted that the type and structure of hydrated cement products remained undisturbed. Yet, they noted that GNPs acted as accelerator in the hydration process and enhanced early age strength. Some researchers like Alkhateb et al. [32] found high density CSH near GNPs using SEM images and concluded that high interfacial strength is available between graphene and CSH gel. Shenghua et al. [115] observed the generation of flower-shaped and polyhedral-like crystals on GO sheets and concluded that GO regulates these hydration crystals. However, Cui et al. [116] investigated the chemical composition of these flower like crystals and noted that these crystal formed due to carbonation of hydrated cementitious products not by the graphene oxide. Various researchers have also stated that due to nucleation and filling effect of graphene, early age hydration process accelerated. Yet, the complete mechanism has not been explained. Graphene and its derivatives also influenced the rheological behaviour of cement composite. According to Wang et al. [126] and Rehman et al. [118] these nanoparticles form the flocculated structures and entrapped the water molecules. However, the trend of flow behaviour of a composite material significantly depends on the mathematical model and its assumptions. Mechanical properties of cement-based composite are also important. Due to nucleation and filling effect of graphene, it has been found that the mechanical properties enhance significantly. Bridging and blockage in crack propagation at nanoscale level also contribute to and influence the mechanical properties of cement-based composite. The molecular modelling approach suggest that good bonding and stability of atoms in interface region play an important role for enhancing the mechanical properties of graphene cement composite. Furthermore, energy harvesting and the thermoelectrical properties of graphene cement composite will benefit the socioeconomic system and reduce the electricity consumption in buildings. Some incredible characteristics of graphene like self-sensing, crack monitoring and damage memorization properties outshine it. Four probe electrical resistivity method has been utilized by various researchers for self-sensing characteristics. Spacing between current pole and voltage pole is important for development of sensors to be used for health monitoring of concrete structures. Various researchers found fractional changes in resistance against the cyclic compressive load and this has opened the gateway for the use of these sensors in the highway and transportation industry.

It was found that the graphene based nanomaterials have several issues which, needs to be resolved before its usage in the construction industry. The following research gaps have been identified.

(1) Very limited research has been found regarding the manufacture of nano-size cement particles and nano-binders however, it has great potential for the development of novel admixtures, nanoparticles, and nano-reinforcements.

(2) Most of the researches on the dispersion of graphene and its derivatives were focused on surface modification, functionalization, and oxidation process. However, these processes damaged the atomic structure of graphene. Therefore, other methods are indispensably required for the dispersion of GNDs and preserving the atomic structure of graphene.

(3) The enhancement mechanism in properties is not completely described as yet. Further research is required to study the regulating mechanism of graphene and its derivatives on hydrated cement crystals.

(4) Flow properties of graphene cement composite and its dependence on various factors are missing in the existing literature. The variation of geometric flow with time, dispersing agent, shear rate and various types of graphene sheets is required to explore the flow of cement paste in the plastic state. Moreover, the best optimized rheological mathematical model needs to be sorted out, as a single rheological model cannot predict the flow behaviour of cement paste accurately. Thus, graphene cement composite demands further exploration to achieve maximum benefit from graphene.

(5) The complete mechanism for accelerating the hydration reactions and enhancing overall mechanical properties has not been explained. Prominent differences in mechanical properties were found in experimental work conducted by various researchers under similar experimental conditions.

(6) The distance between the current and voltage poles is required to be optimized for four-probe method. The potential application of graphene-based cement composite as an embedded smart sensor in the real concrete structure was seldom found in the literature. Moreover, the suitability, industrial demand, and compatibility of graphene cement smart sensors with the existing non-destructive health monitoring methods need to be determined.

11. Conclusions

In this review paper, we have critically reviewed recent research on graphene based cement composites. It was demonstrated that dispersion of graphene is critical for its effective utilization in cement based composites. Several characterization techniques, such as TGA, IR, XRD, and SEM, were discussed in detail to evaluate the impact of graphene on the performance of cement-based composites. Conclusively, graphene was found to enhance the rheological, mechanical, thermoelectrical, piezoresistive, self-sensing, and damage memorizing capabilities of cement-based composites.

Author Contributions: S.K.U.R. and S.A.M. conceived and designed the idea; S.K.U.R., S.K., and S.A.M. analyzed the data; M.F.J. and M.J. contributed reagents/materials/analysis tools; S.K.U.R., S.K., and S.A.M. wrote the original draft of paper, S.K.U.R., and S.A.M. revised the draft of paper. All authors have read and agreed to the published version of the manuscript.

Funding: This research was supported by Nazarbayev University Faculty development competitive research grants numbers (090118FD5316 and SEDS2021006).

Acknowledgments: The authors are thankful to the editor and reviewers who provide valuable suggestion to improve this manuscript.

Conflicts of Interest: The authors declare no conflict of interest.

References

1. Barcelo, L.; Kline, J.; Walenta, G.; Gartner, E. Cement and carbon emissions. *Mater. Struct.* **2014**, *47*, 1055–1065. [CrossRef]
2. Kodur, V.; Sultan, M. Effect of temperature on thermal properties of high-strength concrete. *J. Mater. Civ. Eng.* **2003**, *15*, 101–107. [CrossRef]
3. Javed, M.F.; Hafizah, N.; Memon, S.A.; Jameel, M.; Aslam, M. Recent research on cold-formed steel beams and columns subjected to elevated temperature: A review. *Constr. Build. Mater.* **2017**, *144*, 686–701. [CrossRef]
4. Birchall, J.; Howard, A.; Kendall, K. Flexural strength and porosity of cements. *Nature* **1981**, *289*, 388–390. [CrossRef]
5. Hanehara, S.; Yamada, K. Interaction between cement and chemical admixture from the point of cement hydration, absorption behaviour of admixture, and paste rheology. *Cem. Concr. Res.* **1999**, *29*, 1159–1165. [CrossRef]
6. Zhang, M.-H.; Sisomphon, K.; Ng, T.S.; Sun, D.J. Effect of superplasticizers on workability retention and initial setting time of cement pastes. *Constr. Build. Mater.* **2010**, *24*, 1700–1707. [CrossRef]
7. Bessaies-Bey, H.; Baumann, R.; Schmitz, M.; Radler, M.; Roussel, N. Organic admixtures and cement particles: Competitive adsorption and its macroscopic rheological consequences. *Cem. Concr. Res.* **2016**, *80*, 1–9. [CrossRef]
8. Flatt, R.J.; Houst, Y.F. A simplified view on chemical effects perturbing the action of superplasticizers. *Cem. Concr. Res.* **2001**, *31*, 1169–1176. [CrossRef]
9. Rehman, S.K.U.; Ibrahim, Z.; Memon, S.A.; Jameel, M. Nondestructive test methods for concrete bridges: A review. *Constr. Build. Mater.* **2016**, *107*, 58–86. [CrossRef]
10. Javed, M.F.; Sulong, N.H.R.; Memon, S.A.; Rehman, S.K.U.; Khan, N.B. FE modelling of the flexural behaviour of square and rectangular steel tubes filled with normal and high strength concrete. *Thin-Walled Struct.* **2017**, *119*, 470–481. [CrossRef]

11. Lothenbach, B.; Scrivener, K.; Hooton, R.D. Supplementary cementitious materials. *Cem. Concr. Res.* **2011**, *41*, 1244–1256. [CrossRef]
12. Richardson, I.; Groves, G. The structure of the calcium silicate hydrate phases present in hardened pastes of white Portland cement/blast-furnace slag blends. *J. Mater. Sci.* **1997**, *32*, 4793–4802. [CrossRef]
13. Shi, C.; Qian, J. High performance cementing materials from industrial slags—A review. *Resour. Conserv. Recycl.* **2000**, *29*, 195–207. [CrossRef]
14. Yoo, D.-Y.; Banthia, N.; Fujikake, K.; Borges, P.H.; Gupta, R. Advanced Cementitious Materials: Mechanical Behavior, Durability, and Volume Stability. *Adv. Mater. Sci. Eng.* **2017**, *2017*, 1–2. [CrossRef] [PubMed]
15. Dawood, E.T.; Ramli, M. High strength characteristics of cement mortar reinforced with hybrid fibres. *Constr. Build. Mater.* **2011**, *25*, 2240–2247. [CrossRef]
16. Chung, D.D.L. Comparison of submicron-diameter carbon filaments and conventional carbon fibers as fillers in composite materials. *Carbon* **2001**, *39*, 1119–1125. [CrossRef]
17. Juárez, C.; Valdez, P.; Durán, A.; Sobolev, K. The diagonal tension behavior of fiber reinforced concrete beams. *Cem. Concr. Compos.* **2007**, *29*, 402–408. [CrossRef]
18. Topçu, İ.B.; Canbaz, M. Effect of different fibers on the mechanical properties of concrete containing fly ash. *Constr. Build. Mater.* **2007**, *21*, 1486–1491. [CrossRef]
19. Yoo, D.-Y.; Banthia, N.; Fujikake, K.; Kim, Y.H.; Gupta, R. Fiber-Reinforced Cement Composites: Mechanical Properties and Structural Implications 2019. *Adv. Mater. Sci. Eng.* **2019**, *2019*, 1–2. [CrossRef]
20. Konsta, G.M.S.; Metaxa, Z.S.; Shah, S.P. Multi-scale mechanical and fracture characteristics and early-age strain capacity of high performance carbon nanotube/cement nanocomposites. *Cem. Concr. Compos.* **2010**, *32*, 110–115. [CrossRef]
21. Chuah, S.; Pan, Z.; Sanjayan, J.G.; Wang, C.M.; Duan, W.H. Nano reinforced cement and concrete composites and new perspective from graphene oxide. *Constr. Build. Mater.* **2014**, *73*, 113–124. [CrossRef]
22. Xu, Y.; Fan, Y. Effects of Graphene Oxide Dispersion on Salt-Freezing Resistance of Concrete. *Adv. Mater. Sci. Eng.* **2020**, *2020*, 1–9. [CrossRef]
23. Liu, C.; Liu, G.; Ge, Z.; Guan, Y.; Cui, Z.; Zhou, J. Mechanical and Self-Sensing Properties of Multiwalled Carbon Nanotube-Reinforced ECCs. *Adv. Mater. Sci. Eng.* **2019**, *2019*, 1–9. [CrossRef]
24. Sanchez, F.; Sobolev, K. Nanotechnology in concrete—A review. *Constr. Build. Mater.* **2010**, *24*, 2060–2071. [CrossRef]
25. Fraga, J.L.; del Campo, J.M.; García, J.Á. Carbon nanotube-cement composites in the construction industry: 1952–2014. A state of the art review. In Proceedings of the 2nd International Conference on Emerging Trends in Engineering and Technology (ICETET'2014), London, UK, 30–31 May 2014.
26. Han, B.; Sun, S.; Ding, S.; Zhang, L.; Yu, X.; Ou, J. Review of nanocarbon-engineered multifunctional cementitious composites. *Compos. Part A Appl. Sci. Manuf.* **2015**, *70*, 69–81. [CrossRef]
27. Qureshi, T.S.; Panesar, D.K. A review: The effect of graphene oxide on the properties of cement-based composites. In Proceedings of the CSCE Annual Conference, Vancouver, BC, Canada, 31 May–3 June 2017; p. 642-1.
28. Yang, H.; Cui, H.; Tang, W.; Li, Z.; Han, N.; Xing, F. A critical review on research progress of graphene/cement based composites. *Compos. Part A Appl. Sci. Manuf.* **2017**, *102*, 273–296. [CrossRef]
29. Liu, C.; Huang, X.; Wu, Y.-Y.; Deng, X.; Liu, J.; Zheng, Z.; Hui, D. Review on the research progress of cement-based and geopolymer materials modified by graphene and graphene oxide. *Nanotechnol. Rev.* **2020**, *9*, 155. [CrossRef]
30. Zhao, L.; Guo, X.; Song, L.; Song, Y.; Dai, G.; Liu, J. An intensive review on the role of graphene oxide in cement-based materials. *Constr. Build. Mater.* **2020**, *241*, 117939. [CrossRef]
31. Novoselov, K.; Geim, A.K.; Morozov, S.V.; Jiang, D.; Zhang, Y.; Dubonos, S.V.; Grigorieva, I.V.; Firsov, A.A. Electric field effect in atomically thin carbon films. *Science* **2004**, *306*, 666–669. [CrossRef]
32. Alkhateb, H.; Al-Ostaz, A.; Cheng, A.H.-D.; Li, X. Materials genome for graphene-cement nanocomposites. *J. Nanomech. Micromech.* **2013**, *3*, 67–77. [CrossRef]
33. Wang, B.; Jiang, R.; Wu, Z. Investigation of the mechanical properties and microstructure of graphene nanoplatelet-cement composite. *Nanomaterials* **2016**, *6*, 200. [CrossRef]
34. Segal, M. Selling graphene by the ton. *Nat Nano* **2009**, *4*, 612–614. [CrossRef]
35. Novoselov, K.; Fal, V.; Colombo, L.; Gellert, P.; Schwab, M.; Kim, K. A roadmap for graphene. *Nature* **2012**, *490*, 192–200. [CrossRef]

36. Rehman, S.K.U.; Imtiaz, L.; Aslam, F.; Khan, M.K.; Haseeb, M.; Javed, M.F.; Alyousef, R.; Alabduljabbar, H. Experimental Investigation of NaOH and KOH Mixture in SCBA-Based Geopolymer Cement Composite. *Materials* **2020**, *13*, 3437. [CrossRef]
37. Bai, H.; Li, C.; Shi, G. Functional composite materials based on chemically converted graphene. *Adv. Mater.* **2011**, *23*, 1089–1115. [CrossRef] [PubMed]
38. Sun, S.; Yu, X.; Han, B.; Ou, J. In situ growth of carbon nanotubes/carbon nanofibers on cement/mineral admixture particles: A review. *Constr. Build. Mater.* **2013**, *49*, 835–840. [CrossRef]
39. Geim, A.K.; Novoselov, K.S. The rise of graphene. In *Nanoscience and Technology: A Collection of Reviews from Nature Journals*; World Scientific: London, UK, 2010; pp. 11–19.
40. Kuilla, T.; Bhadra, S.; Yao, D.; Kim, N.H.; Bose, S.; Lee, J.H. Recent advances in graphene based polymer composites. *Prog. Polym. Sci.* **2010**, *35*, 1350–1375. [CrossRef]
41. Kroto, H.W.; Heath, J.R.; O'Brien, S.C.; Curl, R.F.; Smalley, R.E. C60: Buckminsterfullerene. *Nature* **1985**, *318*, 162–163. [CrossRef]
42. Long, X.; Sun, M.-Q.; Li, Z.-Q.; Song, X.-H. Piezo-resistive Effects in Carbon Black-filled Cement-matrix Nanocomposites. *J. Wuhan Univ. Technol.* **2008**, *3*, 57–59.
43. Gong, H.; Li, Z.; Zhang, Y.; Fan, R. Piezoelectric and dielectric behavior of 0-3 cement-based composites mixed with carbon black. *J. Eur. Ceram. Soc.* **2009**, *29*, 2013–2019. [CrossRef]
44. Wang, P.; Ding, T.; Xu, F.; Qin, Y. Piezoresistivity of conductive composites filled by carbon black particles. *Fuhe Cailiao Xuebao (Acta Mater. Compos. Sin.)* **2004**, *21*, 34–38.
45. Gao, Y.-M.; Shim, H.-S.; Hurt, R.H.; Suuberg, E.M.; Yang, N.Y. Effects of carbon on air entrainment in fly ash concrete: The role of soot and carbon black. *Energy Fuels* **1997**, *11*, 457–462. [CrossRef]
46. Agrawal, S.; Raghuveer, M.S.; Ramprasad, R.; Ramanath, G. Multishell Carrier Transport in Multiwalled Carbon Nanotubes. *Nanotechnol. IEEE Trans.* **2007**, *6*, 722–726. [CrossRef]
47. Konsta, G.M.S.; Metaxa, Z.S.; Shah, S.P. Highly dispersed carbon nanotube reinforced cement based materials. *Cem. Concr. Res.* **2010**, *40*, 1052–1059. [CrossRef]
48. Li, G.Y.; Wang, P.M.; Zhao, X. Mechanical behavior and microstructure of cement composites incorporating surface-treated multi-walled carbon nanotubes. *Carbon* **2005**, *43*, 1239–1245. [CrossRef]
49. Ma, P.-C.; Siddiqui, N.A.; Marom, G.; Kim, J.-K. Dispersion and functionalization of carbon nanotubes for polymer-based nanocomposites: A review. *Compos. Part A Appl. Sci. Manuf.* **2010**, *41*, 1345–1367. [CrossRef]
50. Cwirzen, A.; Habermehl-Cwirzen, K.; Nasibulin, A.G.; Kaupinen, E.I.; Mudimela, P.R.; Penttala, V. SEM/AFM studies of cementitious binder modified by MWCNT and nano-sized Fe needles. *Mater. Charact.* **2009**, *60*, 735–740. [CrossRef]
51. Cwirzen, A. Controlling physical properties of cementitious matrixes by nanomaterials. *Adv. Mater. Res.* **2010**, *123*, 639–642. [CrossRef]
52. Kim, J.; Cote, L.J.; Kim, F.; Yuan, W.; Shull, K.R.; Huang, J. Graphene Oxide Sheets at Interfaces. *J. Am. Chem. Soc.* **2010**, *132*, 8180–8186. [CrossRef]
53. Ramanathan, T.; Abdala, A.A.; Stankovich, S.; Dikin, D.A.; Herrera Alonso, M.; Piner, R.D.; Adamson, D.H.; Schniepp, H.C.; Chen, X.; Ruoff, R.S.; et al. Functionalized graphene sheets for polymer nanocomposites. *Nat. Nano* **2008**, *3*, 327–331. [CrossRef]
54. Kim, H.; Miura, Y.; Macosko, C.W. Graphene/Polyurethane Nanocomposites for Improved Gas Barrier and Electrical Conductivity. *Chem. Mater.* **2010**, *22*, 3441–3450. [CrossRef]
55. Pan, Z.; He, L.; Qiu, L.; Korayem, A.H.; Li, G.; Zhu, J.W.; Collins, F.; Li, D.; Duan, W.H.; Wang, M.C. Mechanical properties and microstructure of a graphene oxide–cement composite. *Cem. Concr. Compos.* **2015**, *58*, 140–147. [CrossRef]
56. Wilson, N.R.; Pandey, P.A.; Beanland, R.; Young, R.J.; Kinloch, I.A.; Gong, L.; Liu, Z.; Suenaga, K.; Rourke, J.P.; York, S.J. Graphene oxide: Structural analysis and application as a highly transparent support for electron microscopy. *ACS Nano* **2009**, *3*, 2547–2556. [CrossRef] [PubMed]
57. Zhang, Y.; Tan, Y.-W.; Stormer, H.L.; Kim, P. Experimental observation of the quantum Hall effect and Berry's phase in graphene. *Nature* **2005**, *438*, 201–204. [CrossRef]
58. Ferrari, A.C.; Meyer, J.C.; Scardaci, V.; Casiraghi, C.; Lazzeri, M.; Mauri, F.; Piscanec, S.; Jiang, D.; Novoselov, K.S.; Roth, S.; et al. Raman Spectrum of Graphene and Graphene Layers. *Phys. Rev. Lett.* **2006**, *97*, 187401. [CrossRef]

59. Lee, C.; Wei, X.; Kysar, J.W.; Hone, J. Measurement of the elastic properties and intrinsic strength of monolayer graphene. *Science* **2008**, *321*, 385–388. [CrossRef]
60. Boehm, H.; Clauss, A.; Fischer, G.; Hofmann, U. Surface properties of extremely thin graphite lamellae. In *Proceedings of Fifth Conference on Carbon*; Pergamon Press: New York, NY, USA, 1962; pp. 73–80.
61. Zhu, Y.; Murali, S.; Cai, W.; Li, X.; Suk, J.W.; Potts, J.R.; Ruoff, R.S. Graphene and graphene oxide: Synthesis, properties, and applications. *Adv. Mater.* **2010**, *22*, 3906–3924. [CrossRef]
62. Balandin, A.A.; Ghosh, S.; Bao, W.; Calizo, I.; Teweldebrhan, D.; Miao, F.; Lau, C.N. Superior thermal conductivity of single-layer graphene. *Nano Lett.* **2008**, *8*, 902–907. [CrossRef]
63. Lin, J.; Teweldebrhan, D.; Ashraf, K.; Liu, G.; Jing, X.; Yan, Z.; Li, R.; Ozkan, M.; Lake, R.K.; Balandin, A.A. Gating of Single-Layer Graphene with Single-Stranded Deoxyribonucleic Acids. *Small* **2010**, *6*, 1150–1155. [CrossRef]
64. Geim, A.K.; Novoselov, K.S. The rise of graphene. *Nat. Mater.* **2007**, *6*, 183–191. [CrossRef]
65. Lee, C.; Wei, X.; Li, Q.; Carpick, R.; Kysar, J.W.; Hone, J. Elastic and frictional properties of graphene. *Phys. Status Solidi (B)* **2009**, *246*, 2562–2567. [CrossRef]
66. Mak, K.F.; Lee, C.; Hone, J.; Shan, J.; Heinz, T.F. Atomically thin MoS 2: A new direct-gap semiconductor. *Phys. Rev. Lett.* **2010**, *105*, 136805. [CrossRef] [PubMed]
67. Rafiee, M.A.; Lu, W.; Thomas, A.V.; Zandiatashbar, A.; Rafiee, J.; Tour, J.M.; Koratkar, N.A. Graphene nanoribbon composites. *ACS Nano* **2010**, *4*, 7415–7420. [CrossRef] [PubMed]
68. Grobert, N. Carbon nanotubes—Becomng clean. *Materialstoday* **2007**, *10*, 28–35.
69. Ubertini, F.; Laflamme, S.; Ceylan, H.; Materazzi, A.L.; Cerni, G.; Saleem, H.; D'Alessandro, A.; Corradini, A. Novel nanocomposite technologies for dynamic monitoring of structures: A comparison between cement-based embeddable and soft elastomeric surface sensors. *Smart Mater. Struct.* **2014**, *23*, 045023. [CrossRef]
70. Chen, J.; Gao, X.; Xu, D. Recent Advances in Characterization Techniques for the Interface in Carbon Nanotube-Reinforced Polymer Nanocomposites. *Adv. Mater. Sci. Eng.* **2019**, *2019*, 46. [CrossRef]
71. Munz, M.; Giusca, C.E.; Myers-Ward, R.L.; Gaskill, D.K.; Kazakova, O. Thickness-dependent hydrophobicity of epitaxial graphene. *ACS Nano* **2015**, *9*, 8401–8411. [CrossRef]
72. Hamann, H.; Clemens, D. Suspension Polymerization of Uniform Polymer Beads. U.S. Patent No. 3,728,318, 17 April 1973.
73. Geng, Y.; Wang, S.J.; Kim, J.-K. Preparation of graphite nanoplatelets and graphene sheets. *J. Colloid Interface Sci.* **2009**, *336*, 592–598. [CrossRef]
74. Li, P.; Yao, H.; Wong, M.; Sugiyama, H.; Zhang, X.; Sue, H.-J. Thermally stable and highly conductive free-standing hybrid films based on reduced graphene oxide. *J. Mater. Sci.* **2014**, *49*, 380–391. [CrossRef]
75. Peyvandi, A.; Soroushian, P.; Abdol, N.; Balachandra, A.M. Surface-modified graphite nanomaterials for improved reinforcement efficiency in cementitious paste. *Carbon* **2013**, *63*, 175–186. [CrossRef]
76. Stankovich, S.; Piner, R.D.; Chen, X.; Wu, N.; Nguyen, S.T.; Ruoff, R.S. Stable aqueous dispersions of graphitic nanoplatelets via the reduction of exfoliated graphite oxide in the presence of poly (sodium 4-styrenesulfonate). *J. Mater. Chem.* **2006**, *16*, 155–158. [CrossRef]
77. Lu, J.; Do, I.; Fukushima, H.; Lee, I.; Drzal, L.T. Stable aqueous suspension and self-assembly of graphite nanoplatelets coated with various polyelectrolytes. *J. Nanomater.* **2010**, *2010*, 2. [CrossRef]
78. Wotring, E. Dispersion of Graphene Nanoplatelets in Water with Surfactant and Reinforcement of Mortar with Graphene Nanoplatelets. Ph.D. Thesis, University of Illinois at Urbana-Champaign, Champaign, IL, USA, 2014.
79. Sixuan, H. Multifunctional Graphite Nanoplatelets (GNP) Reinforced Cementitious Composites. Master's Thesis, National University of Singapore, Singapore, 2012.
80. Sharif, M.S.; Fard, F.G.; Khatibi, E.; Sarpoolaky, H. Dispersion and stability of carbon black nanoparticles, studied by ultraviolet–visible spectroscopy. *J. Taiwan Inst. Chem. Eng.* **2009**, *40*, 524–527. [CrossRef]
81. Silva, R.A.; de Castro Guetti, P.; da Luz, M.S.; Rouxinol, F.; Gelamo, R.V. Enhanced properties of cement mortars with multilayer graphene nanoparticles. *Constr. Build. Mater.* **2017**, *149*, 378–385. [CrossRef]
82. Han, B.; Zheng, Q.; Sun, S.; Dong, S.; Zhang, L.; Yu, X.; Ou, J. Enhancing mechanisms of multi-layer graphenes to cementitious composites. *Compos. Part A Appl. Sci. Manuf.* **2017**, *101*, 143–150. [CrossRef]
83. Li, J.; Sham, M.L.; Kim, J.-K.; Marom, G. Morphology and properties of UV/ozone treated graphite nanoplatelet/epoxy nanocomposites. *Compos. Sci. Technol.* **2007**, *67*, 296–305. [CrossRef]

84. Mehrali, M.; Sadeghinezhad, E.; Latibari, S.T.; Kazi, S.N.; Mehrali, M.; Zubir, M.N.B.M.; Metselaar, H.S.C. Investigation of thermal conductivity and rheological properties of nanofluids containing graphene nanoplatelets. *Nanoscale Res. Lett.* **2014**, *9*, 15. [CrossRef]
85. Jiang, L.; Gao, L.; Sun, J. Production of aqueous colloidal dispersions of carbon nanotubes. *J. Colloid Interface Sci.* **2003**, *260*, 89–94. [CrossRef]
86. Liu, M.; Horrocks, A. Effect of carbon black on UV stability of LLDPE films under artificial weathering conditions. *Polym. Degrad. Stab.* **2002**, *75*, 485–499. [CrossRef]
87. Jäger, C.; Henning, T.; Schlögl, R.; Spillecke, O. Spectral properties of carbon black. *J. Non-Cryst. Solids* **1999**, *258*, 161–179.
88. Aunkor, M.; Mahbubul, I.; Saidur, R.; Metselaar, H. Deoxygenation of graphene oxide using household baking soda as a reducing agent: A green approach. *RSC Adv.* **2015**, *5*, 70461–70472. [CrossRef]
89. Lei, Y.; Tang, Z.; Liao, R.; Guo, B. Hydrolysable tannin as environmentally friendly reducer and stabilizer for graphene oxide. *Green Chem.* **2011**, *13*, 1655–1658. [CrossRef]
90. Li, J.; Xiao, G.; Chen, C.; Li, R.; Yan, D. Superior dispersions of reduced graphene oxide synthesized by using gallic acid as a reductant and stabilizer. *J. Mater. Chem. A* **2013**, *1*, 1481–1487. [CrossRef]
91. Mei, X.; Ouyang, J. Ultrasonication-assisted ultrafast reduction of graphene oxide by zinc powder at room temperature. *Carbon* **2011**, *49*, 5389–5397. [CrossRef]
92. Xu, L.Q.; Yang, W.J.; Neoh, K.-G.; Kang, E.-T.; Fu, G.D. Dopamine-induced reduction and functionalization of graphene oxide nanosheets. *Macromolecules* **2010**, *43*, 8336–8339. [CrossRef]
93. Wang, Y.; Shi, Z.; Yin, J. Facile synthesis of soluble graphene via a green reduction of graphene oxide in tea solution and its biocomposites. *ACS Appl. Mater. Interfaces* **2011**, *3*, 1127–1133. [CrossRef]
94. Thakur, S.; Karak, N. Green reduction of graphene oxide by aqueous phytoextracts. *Carbon* **2012**, *50*, 5331–5339. [CrossRef]
95. Chen, D.; Li, L.; Guo, L. An environment-friendly preparation of reduced graphene oxide nanosheets via amino acid. *Nanotechnology* **2011**, *22*, 325601. [CrossRef]
96. Khai, T.V.; Kwak, D.S.; Kwon, Y.J.; Cho, H.Y.; Huan, T.N.; Chung, H.; Ham, H.; Lee, C.; Dan, N.V.; Tung, N.T. Direct production of highly conductive graphene with a low oxygen content by a microwave-assisted solvothermal method. *Chem. Eng. J.* **2013**, *232*, 346–355. [CrossRef]
97. Peng, H.; Meng, L.; Niu, L.; Lu, Q. Simultaneous reduction and surface functionalization of graphene oxide by natural cellulose with the assistance of the ionic liquid. *J. Phys. Chem. C* **2012**, *116*, 16294–16299. [CrossRef]
98. Tran, D.N.; Kabiri, S.; Losic, D. A green approach for the reduction of graphene oxide nanosheets using non-aromatic amino acids. *Carbon* **2014**, *76*, 193–202. [CrossRef]
99. Bose, S.; Kuila, T.; Mishra, A.K.; Kim, N.H.; Lee, J.H. Dual role of glycine as a chemical functionalizer and a reducing agent in the preparation of graphene: An environmentally friendly method. *J. Mater. Chem.* **2012**, *22*, 9696–9703. [CrossRef]
100. Pham, V.H.; Pham, H.D.; Dang, T.T.; Hur, S.H.; Kim, E.J.; Kong, B.S.; Kim, S.; Chung, J.S. Chemical reduction of an aqueous suspension of graphene oxide by nascent hydrogen. *J. Mater. Chem.* **2012**, *22*, 10530–10536. [CrossRef]
101. Gurunathan, S.; Han, J.; Kim, J.H. Humanin: A novel functional molecule for the green synthesis of graphene. *Colloids Surf. B Biointerfaces* **2013**, *111*, 376–383. [CrossRef] [PubMed]
102. Zhang, J.; Yang, H.; Shen, G.; Cheng, P.; Zhang, J.; Guo, S. Reduction of graphene oxide via L-ascorbic acid. *Chem. Commun.* **2010**, *46*, 1112–1114. [CrossRef]
103. Khanra, P.; Kuila, T.; Kim, N.H.; Bae, S.H.; Yu, D.-S.; Lee, J.H. Simultaneous bio-functionalization and reduction of graphene oxide by baker's yeast. *Chem. Eng. J.* **2012**, *183*, 526–533. [CrossRef]
104. Zhu, C.; Guo, S.; Fang, Y.; Dong, S. Reducing sugar: New functional molecules for the green synthesis of graphene nanosheets. *ACS Nano* **2010**, *4*, 2429–2437. [CrossRef]
105. Liu, S.; Tian, J.; Wang, L.; Sun, X. A method for the production of reduced graphene oxide using benzylamine as a reducing and stabilizing agent and its subsequent decoration with Ag nanoparticles for enzymeless hydrogen peroxide detection. *Carbon* **2011**, *49*, 3158–3164. [CrossRef]
106. Leng, Y. *Materials Characterization: Introduction to Microscopic and Spectroscopic Methods*; John Wiley & Sons: Hoboken, NJ, USA, 2009.
107. Sharma, S.; Kothiyal, N. Comparative effects of pristine and ball-milled graphene oxide on physico-chemical characteristics of cement mortar nanocomposites. *Constr. Build. Mater.* **2016**, *115*, 256–268. [CrossRef]

108. Mollah, M.Y.A.; Yu, W.; Schennach, R.; Cocke, D.L. A Fourier transform infrared spectroscopic investigation of the early hydration of Portland cement and the influence of sodium lignosulfonate. *Cem. Concr. Res.* **2000**, *30*, 267–273. [CrossRef]
109. Ortego, J.D.; Jackson, S.; Yu, G.S.; McWhinney, H.; Cocke, D.L. Solidification of hazardous substances-a TGA and FTIR study of Portland cement containing metal nitrates. *J. Environ. Sci. Health Part A* **1989**, *24*, 589–602.
110. Horgnies, M.; Chen, J.; Bouillon, C. Overview about the use of Fourier Transform Infrared spectroscopy to study cementitious materials. In Proceedings of the 6th International Conference on Computational Methods and Experiments in Materials Characterization, Siena, Italy, 4–6 June 2013; WIT Transactions on Engineering Sciences: Southampton, UK, 2013; pp. 251–262.
111. Bensted, J.; Varma, S.P. Some applications of infrared and Raman spectroscopy in cement chemistry. Part 3-hydration of Portland cement and its constituents. *Cem. Technol.* **1974**, *5*, 440–445.
112. Naeimi, H.; Golestanzadeh, M. Microwave-assisted synthesis of 6,6'-(aryl (alkyl) methylene) bis (2,4-dialkylphenol) antioxidants catalyzed by multi-sulfonated reduced graphene oxide nanosheets in water. *New J. Chem.* **2015**, *39*, 2697–2710. [CrossRef]
113. Smith, B.C. *Fundamentals of Fourier Transform Infrared Spectroscopy*; CRC press: Boca Raton, FL, USA, 2011.
114. Murugan, M.; Santhanam, M.; Gupta, S.S.; Pradeep, T.; Shah, S.P. Influence of 2D rGO nanosheets on the properties of OPC paste. *Cem. Concr. Compos.* **2016**, *70*, 48–59. [CrossRef]
115. Lv, S.; Ma, Y.; Qiu, C.; Sun, T.; Liu, J.; Zhou, Q. Effect of graphene oxide nanosheets of microstructure and mechanical properties of cement composites. *Constr. Build. Mater.* **2013**, *49*, 121–127. [CrossRef]
116. Cui, H.; Yan, X.; Tang, L.; Xing, F. Possible pitfall in sample preparation for SEM analysis—A discussion of the paper "Fabrication of polycarboxylate/graphene oxide nanosheet composites by copolymerization for reinforcing and toughening cement composites" by Lv et al. *Cem. Concr. Compos.* **2017**, *77*, 81–85. [CrossRef]
117. Cao, M.-L.; Zhang, H.-X.; Zhang, C. Effect of graphene on mechanical properties of cement mortars. *J. Cent. South Univ.* **2016**, *23*, 919–925. [CrossRef]
118. Rehman, S.K.U.; Ibrahim, Z.; Memon, S.A.; Javed, M.F.; Khushnood, R.A. A sustainable graphene based cement composite. *Sustainability* **2017**, *9*, 1229. [CrossRef]
119. Zhang, Y.; Kong, X.; Gao, L.; Lu, Z.; Zhou, S.; Dong, B.; Xing, F. In-situ measurement of viscoelastic properties of fresh cement paste by a microrheology analyzer. *Cem. Concr. Res.* **2016**, *79*, 291–300. [CrossRef]
120. Nazar, S.; Yang, J.; Thomas, B.S.; Azim, I.; Ur Rehman, S.K. Rheological properties of cementitious composites with and without nano-materials: A comprehensive review. *J. Clean. Prod.* **2020**, *272*, 122701. [CrossRef]
121. Ferraris, C.F. Measurement of the rheological properties of cement paste: A new approach. In Proceedings of the International RILEM Conference the Role of Admixtures in High Performance Concrete, Monterrey, Mexico, 21–26 March 1999; pp. 333–342.
122. Kawashima, S.; Hou, P.; Corr, D.J.; Shah, S.P. Modification of cement-based materials with nanoparticles. *Cem. Concr. Compos.* **2013**, *36*, 8–15. [CrossRef]
123. Ormsby, R.; McNally, T.; Mitchell, C.; Halley, P.; Martin, D.; Nicholson, T.; Dunne, N. Effect of MWCNT addition on the thermal and rheological properties of polymethyl methacrylate bone cement. *Carbon* **2011**, *49*, 2893–2904. [CrossRef]
124. Bilal, H.; Yaqub, M.; Rehman, S.K.U.; Abid, M.; Alyousef, R.; Alabduljabbar, H.; Aslam, F. Performance of Foundry Sand Concrete under Ambient and Elevated Temperatures. *Materials* **2019**, *12*, 2645. [CrossRef] [PubMed]
125. Shang, Y.; Zhang, D.; Yang, C.; Liu, Y.; Liu, Y. Effect of graphene oxide on the rheological properties of cement pastes. *Constr. Build. Mater.* **2015**, *96*, 20–28. [CrossRef]
126. Wang, Q.; Wang, J.; Lv, C.-X.; Cui, X.-Y.; Li, S.-Y.; Wang, X. Rheological behavior of fresh cement pastes with a graphene oxide additive. *New Carbon Mater.* **2016**, *31*, 574–584. [CrossRef]
127. Wang, Q.; Cui, X.; Wang, J.; Li, S.; Lv, C.; Dong, Y. Effect of fly ash on rheological properties of graphene oxide cement paste. *Constr. Build. Mater.* **2017**, *138*, 35–44. [CrossRef]
128. Rehman, S.K.U.; Ibrahim, Z.; Memon, S.A.; Aunkor, M.; Hossain, T.; Javed, M.F.; Mehmood, K.; Shah, S.M.A. Influence of Graphene Nanosheets on Rheology, Microstructure, Strength Development and Self-Sensing Properties of Cement Based Composites. *Sustainability* **2018**, *10*, 822. [CrossRef]
129. Yahia, A.; Khayat, K.H. Analytical models for estimating yield stress of high-performance pseudoplastic grout. *Cem. Concr. Res.* **2001**, *31*, 731–738. [CrossRef]

130. Nehdi, M.; Rahman, M.A. Estimating rheological properties of cement pastes using various rheological models for different test geometry, gap and surface friction. *Cem. Concr. Res.* **2004**, *34*, 1993–2007. [CrossRef]
131. Rehman, S.K.U.; Ibrahim, Z.; Jameel, M.; Memon, S.A.; Javed, M.F.; Aslam, M.; Mehmood, K.; Nazar, S. Assessment of Rheological and Piezoresistive Properties of Graphene based Cement Composites. *Int. J. Concr. Struct. Mater.* **2018**, *12*, 64. [CrossRef]
132. Wengui, L.; Li, X.; Chen, S.J.; Liu, Y.M.; Duan, W.H.; Shah, S.P. Effects of graphene oxide on early-age hydration and electrical resistivity of Portland cement paste. *Constr. Build. Mater.* **2017**, *136*, 506–514.
133. Shenghua, L.; Ma, Y.; Qiu, C.; Zhou, Q. Regulation of GO on cement hydration crystals and its toughening effect. *Mag. Concr. Res.* **2013**, *65*, 1246–1254.
134. Mokhtar, M.; Abo-El-Enein, S.; Hassaan, M.; Morsy, M.; Khalil, M. Mechanical performance, pore structure and micro-structural characteristics of graphene oxide nano platelets reinforced cement. *Constr. Build. Mater.* **2017**, *138*, 333–339. [CrossRef]
135. Wang, M.; Wang, R.; Yao, H.; Farhan, S.; Zheng, S.; Du, C. Study on the three dimensional mechanism of graphene oxide nanosheets modified cement. *Constr. Build. Mater.* **2016**, *126*, 730–739. [CrossRef]
136. Kothiyal, N.; Sharma, S.; Mahajan, S.; Sethi, S. Characterization of reactive graphene oxide synthesized from ball-milled graphite: Its enhanced reinforcing effects on cement nanocomposites. *J. Adhes. Sci. Technol.* **2016**, *30*, 915–933. [CrossRef]
137. Shamsaei, E.; de Souza, F.B.; Yao, X.; Benhelal, E.; Akbari, A.; Duan, W. Graphene-based nanosheets for stronger and more durable concrete: A review. *Constr. Build. Mater.* **2018**, *183*, 642–660. [CrossRef]
138. Sharma, S.; Kothiyal, N. Influence of graphene oxide as dispersed phase in cement mortar matrix in defining the crystal patterns of cement hydrates and its effect on mechanical, microstructural and crystallization properties. *RSC Adv.* **2015**, *5*, 52642–52657. [CrossRef]
139. Abrishami, M.E.; Zahabi, V. Reinforcing graphene oxide/cement composite with NH2 functionalizing group. *Bull. Mater. Sci.* **2016**, *39*, 1073–1078. [CrossRef]
140. Lv, S.; Deng, L.; Yang, W.; Zhou, Q.; Cui, Y. Fabrication of polycarboxylate/graphene oxide nanosheet composites by copolymerization for reinforcing and toughening cement composites. *Cem. Concr. Compos.* **2016**, *66*, 1–9. [CrossRef]
141. Farooq, F.; Rahman, S.K.U.; Akbar, A.; Khushnood, R.A.; Javed, M.F.; alyousef, R.; alabduljabbar, H.; aslam, F. A comparative study on performance evaluation of hybrid GNPs/CNTs in conventional and self-compacting mortar. *Alex. Eng. J.* **2020**, *59*, 369–379. [CrossRef]
142. Farooq, F.; Akbar, A.; Khushnood, R.A.; Muhammad, W.L.; Rehman, S.K.; Javed, M.F. Experimental Investigation of Hybrid Carbon Nanotubes and Graphite Nanoplatelets on Rheology, Shrinkage, Mechanical, and Microstructure of SCCM. *Materials* **2020**, *13*, 230. [CrossRef] [PubMed]
143. Hou, D.; Lu, Z.; Li, X.; Ma, H.; Li, Z. Reactive molecular dynamics and experimental study of graphene-cement composites: Structure, dynamics and reinforcement mechanisms. *Carbon* **2017**, *115*, 188–208. [CrossRef]
144. Lv, S.; Liu, J.; Sun, T.; Ma, Y.; Zhou, Q. Effect of GO nanosheets on shapes of cement hydration crystals and their formation process. *Constr. Build. Mater.* **2014**, *64*, 231–239. [CrossRef]
145. Lv, S.; Ting, S.; Liu, J.; Zhou, Q. Use of graphene oxide nanosheets to regulate the microstructure of hardened cement paste to increase its strength and toughness. *CrystEngComm* **2014**, *16*, 8508–8516. [CrossRef]
146. Wang, Q.; Wang, J.; Lu, C.-X.; Liu, B.-W.; Zhang, K.; Li, C.-Z. Influence of graphene oxide additions on the microstructure and mechanical strength of cement. *New Carbon Mater.* **2015**, *30*, 349–356. [CrossRef]
147. Metaxa, Z.S. Exfoliated graphene nanoplatelet cement-based nanocomposites as piezoresistive sensors: Influence of nanoreinforcement lateral size on monitoring capability. *Cienc. Tecnol. Dos Mater.* **2016**, *28*, 73–79. [CrossRef]
148. Zhao, L.; Guo, X.; Ge, C.; Li, Q.; Guo, L.; Shu, X.; Liu, J. Investigation of the effectiveness of PC@ GO on the reinforcement for cement composites. *Constr. Build. Mater.* **2016**, *113*, 470–478. [CrossRef]
149. Lu, Z.; Hou, D.; Meng, L.; Sun, G.; Lu, C.; Li, Z. Mechanism of cement paste reinforced by graphene oxide/carbon nanotubes composites with enhanced mechanical properties. *RSC Adv.* **2015**, *5*, 100598–100605. [CrossRef]
150. Sun, X.; Wu, Q.; Zhang, J.; Qing, Y.; Wu, Y.; Lee, S. Rheology, curing temperature and mechanical performance of oil well cement: Combined effect of cellulose nanofibers and graphene nano-platelets. *Mater. Des.* **2017**, *114*, 92–101. [CrossRef]

151. Zhou, C.; Li, F.; Hu, J.; Ren, M.; Wei, J.; Yu, Q. Enhanced mechanical properties of cement paste by hybrid graphene oxide/carbon nanotubes. *Constr. Build. Mater.* **2017**, *134*, 336–345. [CrossRef]
152. Gong, K.; Pan, Z.; Korayem, A.H.; Qiu, L.; Li, D.; Collins, F.; Wang, C.M.; Duan, W.H. Reinforcing effects of graphene oxide on portland cement paste. *J. Mater. Civ. Eng.* **2014**, *27*, A4014010. [CrossRef]
153. Zhao, L.; Guo, X.; Ge, C.; Li, Q.; Guo, L.; Shu, X.; Liu, J. Mechanical behavior and toughening mechanism of polycarboxylate superplasticizer modified graphene oxide reinforced cement composites. *Compos. Part B Eng.* **2017**, *113*, 308–316. [CrossRef]
154. Qian, Y.; Abdallah, M.Y.; Kawashima, S. Characterization of Cement-Based Materials Modified with Graphene-Oxide. In *Nanotechnology in Construction*; Springer: Berlin/Heidelberg, Germany, 2015; pp. 259–264.
155. Sun, H.; Ling, L.; Ren, Z.; Memon, S.A.; Xing, F. Effect of graphene oxide/graphene hybrid on mechanical properties of cement mortar and mechanism investigation. *Nanomaterials* **2020**, *10*, 113. [CrossRef] [PubMed]
156. Zhao, L.; Guo, X.; Liu, Y.; Ge, C.; Guo, L.; Shu, X.; Liu, J. Synergistic effects of silica nanoparticles/polycarboxylate superplasticizer modified graphene oxide on mechanical behavior and hydration process of cement composites. *RSC Adv.* **2017**, *7*, 16688–16702. [CrossRef]
157. Mohammed, A.; Sanjayan, J.; Duan, W.; Nazari, A. Graphene Oxide Impact on Hardened Cement Expressed in Enhanced Freeze–Thaw Resistance. *J. Mater. Civ. Eng.* **2016**, *28*, 04016072. [CrossRef]
158. Kaur, R.; Kothiyal, N.C.; Arora, H. Studies on combined effect of superplasticizer modified graphene oxide and carbon nanotubes on the physico-mechanical strength and electrical resistivity of fly ash blended cement mortar. *J. Build. Eng.* **2020**, *30*, 101304. [CrossRef]
159. Tong, T.; Fan, Z.; Liu, Q.; Wang, S.; Tan, S.; Yu, Q. Investigation of the effects of graphene and graphene oxide nanoplatelets on the micro-and macro-properties of cementitious materials. *Constr. Build. Mater.* **2016**, *106*, 102–114. [CrossRef]
160. Rhee, I.; Lee, J.S.; Kim, Y.A.; Kim, J.H.; Kim, J.H. Electrically conductive cement mortar: Incorporating rice husk-derived high-surface-area graphene. *Constr. Build. Mater.* **2016**, *125*, 632–642. [CrossRef]
161. Rhee, I.; Kim, Y.A.; Shin, G.-O.; Kim, J.H.; Muramatsu, H. Compressive strength sensitivity of cement mortar using rice husk-derived graphene with a high specific surface area. *Constr. Build. Mater.* **2015**, *96*, 189–197. [CrossRef]
162. Devasena, M.; Karthikeyan, J. Investigation on strength properties of graphene oxide concrete. *Int. J. Eng. Sci. Invent. Res. Dev.* **2015**, *1*, 307–310.
163. Duarte, F.; Ferreira, A. Energy harvesting on road pavements: State of the art. *Proc. Inst. Civ. Eng.-Energy* **2016**, *169*, 79–90. [CrossRef]
164. Ghosh, S.; Harish, S.; Rocky, K.A.; Ohtaki, M.; Saha, B.B. Graphene enhanced thermoelectric properties of cement based composites for building energy harvesting. *Energy Build.* **2019**, *202*, 109419. [CrossRef]
165. Balandin, A.A. Thermal properties of graphene and nanostructured carbon materials. *Nat. Mater.* **2011**, *10*, 569–581. [CrossRef]
166. Wei, J.; Zhao, L.; Zhang, Q.; Nie, Z.; Hao, L. Enhanced thermoelectric properties of cement-based composites with expanded graphite for climate adaptation and large-scale energy harvesting. *Energy Build.* **2018**, *159*, 66–74. [CrossRef]
167. Wei, J.; Fan, Y.; Zhao, L.; Xue, F.; Hao, L.; Zhang, Q. Thermoelectric properties of carbon nanotube reinforced cement-based composites fabricated by compression shear. *Ceram. Int.* **2018**, *44*, 5829–5833. [CrossRef]
168. Tzounis, L.; Liebscher, M.; Fuge, R.; Leonhardt, A.; Mechtcherine, V. P- and n-type thermoelectric cement composites with CVD grown p- and n-doped carbon nanotubes: Demonstration of a structural thermoelectric generator. *Energy Build.* **2019**, *191*, 151–163. [CrossRef]
169. Cooper, D.R.; D'Anjou, B.; Ghattamaneni, N.; Harack, B.; Hilke, M.; Horth, A.; Majlis, N.; Massicotte, M.; Vandsburger, L.; Whiteway, E. Experimental review of graphene. *ISRN Condens. Matter Phys.* **2012**, *2012*, 1–56. [CrossRef]
170. Zhao, H.; Bai, J. Highly sensitive piezo-resistive graphite nanoplatelet–carbon nanotube hybrids/polydimethylsilicone composites with improved conductive network construction. *ACS Appl. Mater. Interfaces* **2015**, *7*, 9652–9659. [CrossRef]
171. Han, B.G.; Han, B.Z.; Ou, J.P. Experimental study on use of nickel powder-filled Portland cement-based composite for fabrication of piezoresistive sensors with high sensitivity. *Sens. Actuators A Phys.* **2009**, *149*, 51–55. [CrossRef]

172. Han, B.; Guan, X.; Ou, J. Electrode design, measuring method and data acquisition system of carbon fiber cement paste piezoresistive sensors. *Sens. Actuators A Phys.* **2007**, *135*, 360–369. [CrossRef]
173. Li, G.Y.; Wang, P.M.; Zhao, X. Pressure-sensitive properties and microstructure of carbon nanotube reinforced cement composites. *Cem. Concr. Compos.* **2007**, *29*, 377–382. [CrossRef]
174. Le, J.-L.; Du, H.; Dai Pang, S. Use of 2D Graphene Nanoplatelets (GNP) in cement composites for structural health evaluation. *Compos. Part B Eng.* **2014**, *67*, 555–563. [CrossRef]
175. Valdes, L.B. Resistivity measurements on germanium for transistors. *Proc. IRE* **1954**, *42*, 420–427. [CrossRef]
176. Chung, D.D.L. Piezoresistive cement-based materials for strain sensing. *J. Intell. Mater. Syst. Struct.* **2002**, *13*, 599–609. [CrossRef]
177. Qu, J.; Han, B. Piezoresistive cement-based strain sensors and self-sensing concrete components. *J. Intell. Mater. Syst. Struct.* **2008**, *20*, 329–336.
178. Li, H.; Xiao, H.-G.; Ou, J.-P. A study on mechanical and pressure-sensitive properties of cement mortar with nanophase materials. *Cem. Concr. Res.* **2004**, *34*, 435–438. [CrossRef]
179. Yu, X.; Kwon, E. A carbon nanotube/cement composite with piezoresistive properties. *Smart Mater. Struct.* **2009**, *18*, 055010. [CrossRef]
180. Du, H.; Quek, S.T.; Dai Pang, S. Smart multifunctional cement mortar containing graphite nanoplatelet. In Proceedings of the SPIE Smart Structures and Materials+ Nondestructive Evaluation and Health Monitoring, San Diego, CA, USA, 10–14 March 2013; Volume 8692, p. 869238.
181. Lahiri, I.; Verma, V.P.; Choi, W. An all-graphene based transparent and flexible field emission device. *Carbon* **2011**, *49*, 1614–1619. [CrossRef]
182. Endo, M.; Hayashi, T.; Ahm Kim, Y.; Terrones, M.; Dresselhaus, M.S. Applications of carbon nanotubes in the twenty–first century. *Philos. Trans. R. Soc. Lond. Ser. A Math. Phys. Eng. Sci.* **2004**, *362*, 2223–2238. [CrossRef]
183. Horszczaruk, E.; Mijowska, E.; Kalenczuk, R.J.; Aleksandrzak, M.; Mijowska, S. Nanocomposite of cement/graphene oxide–Impact on hydration kinetics and Young's modulus. *Constr. Build. Mater.* **2015**, *78*, 234–242. [CrossRef]
184. Raidongia, K.; Tan, A.T.; Huang, J. Graphene oxide: Some new insights into an old material. In *Carbon Nanotubes and Graphene*; Elsevier: Amsterdam, The Netherlands, 2014; pp. 341–374.
185. Iris, K.; Xiong, X.; Tsang, D.C.; Ng, Y.H.; Clark, J.H.; Fan, J.; Zhang, S.; Hu, C.; Ok, Y.S. Graphite oxide-and graphene oxide-supported catalysts for microwave-assisted glucose isomerisation in water. *Green Chem.* **2019**, *21*, 4341–4353.
186. Stankovich, S.; Dikin, D.A.; Piner, R.D.; Kohlhaas, K.A.; Kleinhammes, A.; Jia, Y.; Wu, Y.; Nguyen, S.T.; Ruoff, R.S. Synthesis of graphene-based nanosheets via chemical reduction of exfoliated graphite oxide. *Carbon* **2007**, *45*, 1558–1565. [CrossRef]
187. Chandrasekaran, S.; Sato, N.; Tölle, F.; Mülhaupt, R.; Fiedler, B.; Schulte, K. Fracture toughness and failure mechanism of graphene based epoxy composites. *Compos. Sci. Technol.* **2014**, *97*, 90–99. [CrossRef]
188. Harris, P.J. New perspectives on the structure of graphitic carbons. *Crit. Rev. Solid State Mater. Sci.* **2005**, *30*, 235–253. [CrossRef]
189. Huang, X.; Yin, Z.; Wu, S.; Qi, X.; He, Q.; Zhang, Q.; Yan, Q.; Boey, F.; Zhang, H. Graphene-based materials: Synthesis, characterization, properties, and applications. *Small* **2011**, *7*, 1876–1902. [CrossRef]
190. Roni Peleg. Available online: https://www.ossila.com/pages/introduction-to-graphene (accessed on 1 August 2018).
191. Cataldi, P.; Athanassiou, A.; Bayer, I.S. Graphene nanoplatelets-based advanced materials and recent progress in sustainable applications. *Appl. Sci.* **2018**, *8*, 1438. [CrossRef]

Publisher's Note: MDPI stays neutral with regard to jurisdictional claims in published maps and institutional affiliations.

© 2020 by the authors. Licensee MDPI, Basel, Switzerland. This article is an open access article distributed under the terms and conditions of the Creative Commons Attribution (CC BY) license (http://creativecommons.org/licenses/by/4.0/).

Article

Reduced Graphene Oxide Sheets as Inhibitors of the Photochemical Reactions of α-Lipoic Acid in the Presence of Ag and Au Nanoparticles

N'ghaya Toulbe [1,2], Malvina S. Stroe [1], Monica Daescu [1], Radu Cercel [1], Alin Mogos [3], Daniela Dragoman [4], Marcela Socol [1], Ionel Mercioniu [5] and Mihaela Baibarac [1,*]

1. National Institute of Materials Physics, Laboratory of Optical Processes in Nanostructure Materials, Atomistilor str. 405 A, 77125 Bucharest, Romania; toulbe.nghaya@infim.ro (N.T.); malvina@infim.ro (M.S.S.); monica.daescu@infim.ro (M.D.); radu.cercel@infim.ro (R.C.); marcela.socol@infim.ro (M.S.)
2. Interdisciplinary School of Doctoral Studies, University of Bucharest, Șoseaua Panduri 90, 050663 Bucharest, Romania
3. S.C. Agilrom Scientific S.R.L., 77190 Bucharest, Romania; alin.mogos@agilrom.ro
4. Faculty of Physics, University of Bucharest, Șoseaua Panduri 90, București 050663, Bucharest, P.O. Box MG-11, 077125 Bucharest-Magurele, Romania; danieladragoman@yahoo.com or daniela@solid.fizica.unibuc.ro
5. National Institute of Materials Physics, Atomic Structures and Defects in Advanced Materials Laboratory, Atomistilor str. 405 A, 77125 Bucharest, Romania; imercioniu@infim.ro
* Correspondence: barac@infim.ro; Tel.: + 40-21-3690170

Received: 12 October 2020; Accepted: 7 November 2020; Published: 11 November 2020

Abstract: The influence of Ag and Au nanoparticles and reduced graphene oxide (RGO) sheets on the photodegradation of α-lipoic acid (ALA) was determined by UV-VIS spectroscopy. The ALA photodegradation was explained by considering the affinity of thiol groups for the metallic nanoparticles synthesized in the presence of trisodium citrate. The presence of excipients did not induce further changes when ALA interacts with Ag and Au nanoparticles with sizes of 5 and 10 nm by exposure to UV light. Compared to the Raman spectrum of ALA powder, changes in Raman lines' position and relative intensities when ALA has interacted with films obtained from Au nanoparticles with sizes between 5 and 50 nm were significant. These changes were explained by considering the chemical mechanism of surface-enhanced Raman scattering (SERS) spectroscopy. The photodegradation of ALA that had interacted with metallic nanoparticles was inhibited in the presence of RGO sheets.

Keywords: α-lipoic acid; UV-VIS spectroscopy; SERS spectroscopy

1. Introduction

α-Lipoic acid (ALA) is present in various organisms of the human body, e.g., the kidney, heart, liver, and foods such as broccoli, spinach, and yeast extract [1]. The molecular structure of ALA contains both a carboxylic group and a dithiolane ring, and, depending on the chirality of the substituted carbon atom of the dithiolane ring, two enantiomers are known at present for this compound, i.e., RLA and SLA. ALA is used as a drug in diabetic neuropathy [2], as a dietary supplement to slow down the aging process of the body [3], and in obesity treatment [4]. The most important effect of ALA is its antioxidative capacity [5]. An improvement in ALA's antioxidative effects has been reported in the presence of the Coenzyme Q10 (CQ10) [5]. Pharmaceutical compounds containing ALA are marketed under the name Alpha Lipoic Sustain, Tiolin, Alanerv, Alasod 600, and Cerebinox, the last of which also contains CQ10.

Until now, the main methods used in the characterization of ALA were (i) FTIR spectroscopy [6]; (ii) Raman scattering [6], (iii) UV-VIS spectroscopy [7], and (iv) cyclic voltammetry or square-wave voltammetry [8]. The detection of ALA was achieved using (i) electrochemistry methods [9], (ii) liquid chromatography [10], (iii) mass spectrometry [11], and optical methods based on surface plasmon resonance [12]. A recent application of metallic nanoparticles that have interacted with ALA was used as a sensorial platform for nerve agents [13].

Considering the applications of ALA in the health domain, in this work, special attention is paid to the photodegradation process of ALA in the absence and the presence of metallic nanoparticles. The dependence of the photodegradation process of ALA in the presence of Ag and Au nanoparticles with various sizes is studied, too. The photodegradation process is also analyzed in pharmaceutical compounds such as Alpha Lipoic Sustain and Cerebinox. Using UV-VIS spectroscopy, we demonstrate that inhibition of ALA's photodegradation that has interacted with Ag and Au nanoparticles can be induced in the presence of reduced graphene oxide (RGO) sheets. The differences between the Raman spectrum of ALA in the powder state and the surface-enhanced Raman scattering (SERS) spectrum of ALA as a thin layer deposited on rough films, obtained from the colloidal solutions of Au nanoparticles with sizes of 5, 10, and 50 nm, are also reported. A sustained effort was made to develop SERS supports by various methods such as (i) the oxidation/electrochemical reduction of metal electrodes, (ii) evaporation by the vacuum deposition of shaved metal films, (iii) lithography, and (iv) synthesis of colloidal metallic structures [14,15]. The two mechanisms that underlie the SERS effect, i.e., electromagnetic and chemical mechanisms, allow us to understand these differences. As is well known, the electromagnetic enhancement process of the Raman spectra is induced by the excitation of localized or delocalized surface plasmons generated at the dielectric/metal interface, while the chemical process involves the generation of new chemical bonds at the adsorbed molecule/metal interface [14]. Generally, SERS supports are characterized by rough structures with sizes in the range of 10–100 nm [15]. In order to highlight this property for the films obtained from the colloidal solution of Au nanoparticles, atomic force microscopy (AFM) studies are reported. The changes induced on the roughness parameters by the adsorption of RGO sheets onto Au films are also shown.

2. Materials and Methods

ALA ($C_8H_{14}O_2S_2$), sodium borohydride ($NaBH_4$), trisodium citrate ($Na_3C_6H_5O_7$), silver nitrate ($AgNO_3$), sodium hydroxide (NaOH), sodium citrate ($C_6H_5Na_3O_7$), tannic acid ($C_{76}H_{52}O_{46}$), tetrachloroauric acid ($HAuCl_4$), potassium carbonate (K_2CO_3), and dimethylformamide (DMF, with 0.1% water) were purchased from the Aldrich-Sigma company (St. Louis, MO, USA). Pharmaceutical compounds, Alpha Lipoic Sustain and Cerebinox, were purchased from a local pharmacy (Jarrow Formulas, Inc., Los Angeles, CA, USA and Polisano Pharmaceuticals, Sibiu, Romania). The Alpha Lipoic Sustain pharmaceutical product contains, as active compounds, 300 mg of ALA and 333 mg of Biotin and, as additional ingredients, polyacrylic acid, $Ca_3(PO_4)_2$, $C_{18}H_{36}O_2$, $C_{36}H_{70}MgO_4$, SiO_2, cellulose, beetroot powder (for color), natural vanilla flavor, and a food-grade coating. The Cerebinox pharmaceutical product contains 300 mg of ALA, 10 mg of CQ10, 100 mg of wheat germ extract, caking, and SiO_2.

The Ag nanoparticles with an average size of cca. 5 and 10 nm were prepared according to Reference [16]. Briefly, to obtain Ag nanoparticles with a size of 5 nm, mixing by vigorous stirring in the dark of a 48 mL aqueous solution containing 2 mM $NaBH_4$ and 4.28 mM $Na_3C_6H_5O_7$ at 60 °C, for 30 min, was carried out. After that, 2 mL of 1 mM $AgNO_3$ was added to the above solution. Further, the reaction mixture was preserved at 90 °C for another 20 min, the pH of the reaction mixture was adjusted to 10.5 using a 0.1 M NaOH solution. The yellow suspension of Ag nanoparticles was cooled to 25 °C and then centrifuged at 12,000 rpm for 15 min in order to remove all unreacted compounds. This step was followed by washing and redispersion in distilled water. A similar protocol was used to obtain Ag nanoparticles with a size of 10 nm, the only difference being the addition of 2 mL of 1.17 mM $AgNO_3$ to the 48 mL solution containing 2 mM $NaBH_4$ and 4.28 mM $Na_3C_6H_5O_7$. According to Reference [16], the particle concentration in the colloidal solutions of Ag with particle sizes of 5 and 10 nm was equal to 1.03×10^{15} and 1.53×10^{14} particles/mL, respectively.

Au nanoparticles with average sizes of cca. 5 and 10 nm were synthesized using the protocol described in Reference [17]. The synthesis protocol involves (i) the preparation of a solution containing 150 mL of 2.2 mM $Na_3C_6H_5O_7$ and 0.01 mL of 2.5 mM $C_{76}H_{50}O_{46}$, which was heated at 70 °C, and (ii) the addition of 1 mL of 25 mM $HAuCl_4$ to the above solution, when one observes that, after cca. 10 s, the uncolored solution becomes orange-red. The addition of 1 mL of 150 mM K_2CO_3 to the above reaction mixture induced a change in pH from 10.4 to 8.3. This reaction mixture was preserved at 100 °C, for a period of 5 min, until a suspension of Au nanoparticles with a size of around 5 nm resulted [17]. In order to obtain Au nanoparticles with a size of cca. 10 nm, the only difference in the above protocol needed is the use of 0.01 mL of 2.5 mM tannic acid [17]. In the case of the colloidal solutions of Au with particle sizes of 5 and 10 nm, the particle concentration was equal to 4.4×10^{13} and 1.2×10^{13} particles/mL [17]. In References [16] and [17], TEM studies demonstrated that these protocols for the preparation of the metallic nanoparticles reported led to (i) Ag nanoparticles with size ranges of 5 ± 0.7 nm and 10 ± 2.0 nm [16] and (ii) Au nanoparticles with size ranges of 5.1 ± 0.5 nm and 10.5 ± 0.9 nm [17].

The Au nanoparticles with a size of cca. 50 nm were synthesized by the addition of 1 mL of 1wt.% $HAuCl_4$ to 47 mL of water, and the mixture was heated to 100 °C when 1.5 mL of $Na_3C_6H_5O_7$ were injected [18,19]. The reaction mixture was preserved at 100 °C for 45 min, until its color became red, and was then cooled at 25 °C [18,19]. The colloidal solution of Au, with a particle size of 50 nm, had a particle concentration equal to 1.05×10^{11} particles/mL [19]. For the SERS studies, the rough Au films were obtained by the evaporation of a 0.12 mL colloidal solution of Au with a particle size of 50 nm. The roughness parameters of these Au films before and after the deposition of RGO sheets were evaluated by atomic force microscopy (AFM).

The AFM images were collected (in phase feedback) with a Nanonics Multiview 4000 microscope (Nanonics Imaging Ltd., Jerusalem, Israel) working in tapping mode. We used a cantilever with a 20 nm diameter, a 1730 factor of merit, and a 34.7 kHz vibration frequency.

The photodegradation processes were monitored by UV-VIS spectroscopy. In this order, solutions of ALA in (i) DMF, (ii) a stabilized suspension in a citrate buffer of Au nanoparticles with diameters equal to 5 and 10 nm, respectively, and (iii) the dispersions of the citrate functionalized Ag nanospheres with sizes equal to 5 and 10 nm, respectively, in aqueous sodium citrate, were used. The solutions of ALA or pharmaceutical products used in the UV-VIS spectroscopy studies had a 1 mg/mL concentration. All spectra were recorded with a Perkin Elmer UV-VIS-NIR spectrophotometer, Lambda 950 model (PerkinElmer, Inc., Waltham, MA, USA), in the 200–600 nm spectral range, the scan rate being 96 nm/min. The UV-VIS spectra of each sample were recorded at intervals of 5 min of UV irradiation for a period of 60 min. UV irradiation was performed using a Hg-vapors lamp with a power of 350 W.

The Raman spectra of ALA were recorded with an FT Raman spectrophotometer, model RFS100S (Bruker Optik GmbH, Ettlingen, Germany). The concentration of ALA solutions was 10^{-2} M for the Raman scattering studies. The films obtained from the colloidal solution of Au, with a particle size of 50 nm, were prepared on quartz substrates with an area of 1 cm^2.

Samples for transmission electron microscopy (TEM) and high-resolution transmission electron microscopy (HRTEM) were prepared by suspending them in ethanol and transferring them to a copper grid coated with amorphous carbon support. TEM and HRTEM images were recorded with a JEOL JEM ARM 200 F electron microscope (JEOL (Europe) SAS) operated at 200 keV.

A solution consisting of 0.5 mg of RGO in 1 mL of DMF was prepared to highlight the influence of the RGO sheets on the roughness parameters of Au films.

3. Results and Discussion

3.1. The Photodegradation of ALA Highlighted by UV-VIS Spectroscopy

Figure 1a shows the UV-VIS spectrum of ALA in DMF in the initial state (black curve), characterized by a band with a maximum at 338 nm. The exposure of the sample at the UV light led to a decrease in the absorbance of this band, simultaneous with the appearance and the increase in the band's absorbance at 272 nm, a modification that involved the appearance of an isosbestic point at 300 nm. This suggests that new chemical compounds are generated under UV light in a solution of ALA in DMF containing 0.1 wt.% water, according to Scheme 1. In the case of a semi-aqueous ALA solution, the exposure to UV light induced only a decrease in the band's absorbance at 335 nm (Figure 1b).

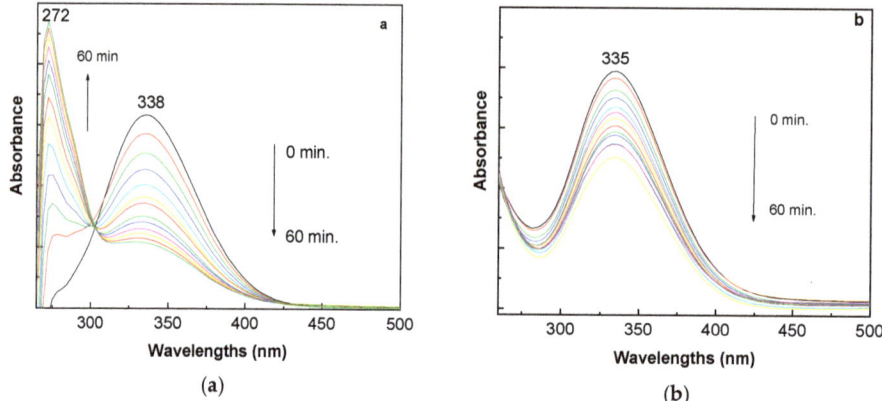

Figure 1. UV-VIS spectra of α-lipoic acid (ALA) in dimethyl formamide (DMF) (**a**) and in the mixture of DMF–H$_2$O with the volume ratio equal to 1:1 (**b**). The concentration of ALA is 1 mg/mL. The two samples' UV-VIS spectra were recorded at intervals of 5 min of exposure to UV light, for 60 min.

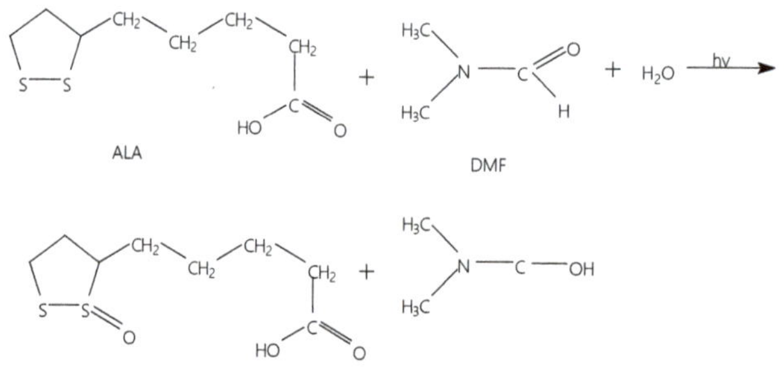

Scheme 1. The photochemical reaction of ALA in the presence of DMF and H$_2$O.

The UV-VIS spectra of Ag and Au nanoparticles must be shown to assess the influence of metallic nanoparticles on ALA. In this context, Figure 2 highlights that the plasmonic bands of the Ag nanoparticles with sizes of 5 nm and 10 nm peaked at 398 nm (Figure 2a) and 400 nm (Figure 2b), while those of the Au nanoparticles with sizes of 5 nm and 10 nm were situated at 522 nm (Figure 2c) and 524 nm (Figure 2d). Figure 3 shows the HRTEM images of the metallic nanoparticles. According to Figure 3, regardless of the Ag and Au nanoparticles' size, they show a spherical shape. The UV-VIS spectra show relatively broad bands, which indicates a broad size distribution. This fact is confirmed by the TEM images shown in Figure 3.

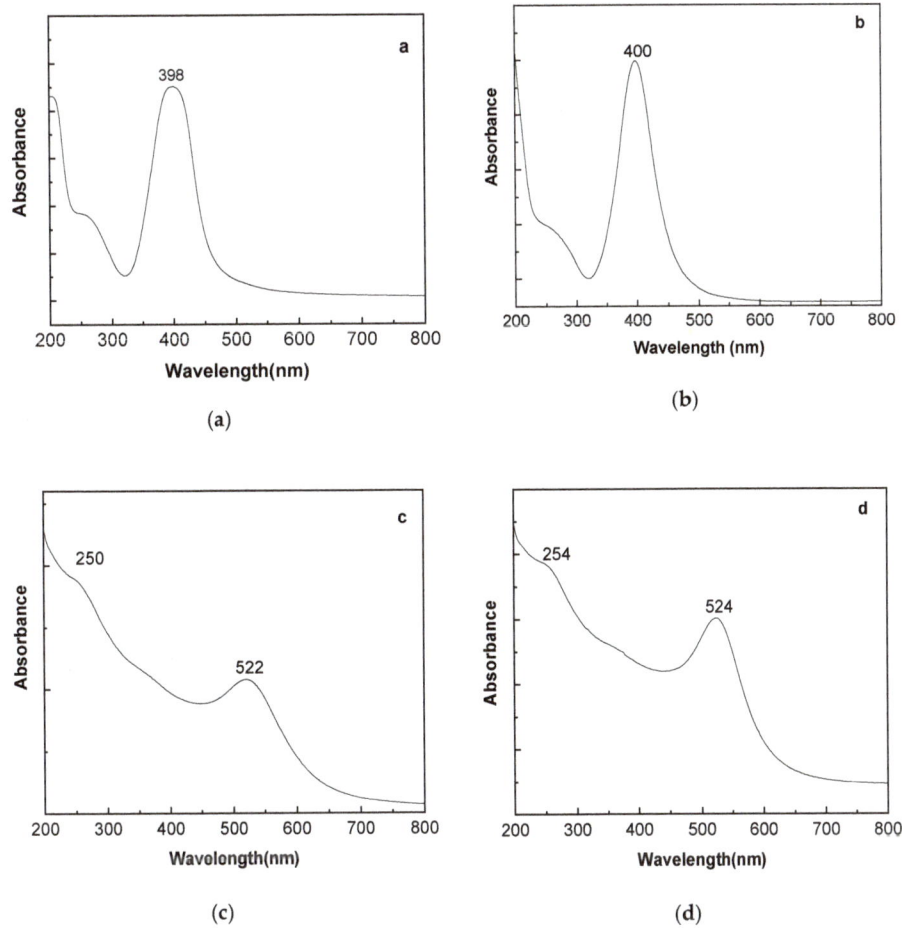

Figure 2. UV-VIS spectra of nanoparticles: (**a**) Ag with a particle size of 5 nm, (**b**) Ag with a particle size of 10 nm, (**c**) Au with a particle size of 5 nm, and (**d**) Au with a particle size of 10 nm.

Figure 4 shows the UV-VIS spectra of ALA that has interacted with Au and Ag nanoparticles, both with sizes of 5 and 10 nm, the ALA concentration in the aqueous solution of metallic nanoparticles being equal to 1 mg/mL. According to Figure 4 (red curves), in the initial state, i.e., before exposure to UV light, the UV-VIS spectra were characterized by absorption bands with maxima at 322 nm and 542 nm in the case of ALA that has interacted with Au nanoparticles with a size of 10 nm (Figure 4d). These bands were situated at 320 nm and 530 nm in the case of ALA that has interacted with Au nanoparticles with a size of 5 nm (Figure 4c). In the case of ALA that has interacted with Ag nanoparticles with sizes of 10 nm and 5 nm, the UV-VIS spectra show absorption bands with maxima at 424 nm (Figure 4b) and 401 nm (Figure 4a), respectively. The exposure of these samples to UV light for 60 min led to a gradual decrease in the absorbance of the band localized in the spectral range 300–700 nm (Figure 4). Supplementarily, in the case of ALA that has interacted with Ag nanoparticles with sizes of 10 and 5 nm, one observes the following: (i) an up-shift of the absorption band at 454 nm (Figure 4b) and 431 nm (Figure 4a), respectively; (ii) the appearance of isosbestic points at 339 nm and 568 nm in the case of ALA that has interacted with Ag nanoparticles with a size of 10 nm (Figure 4b); and (iii) the appearance of isosbestic points at 335 nm and 512 nm in the case of ALA that has interacted with Ag nanoparticles with a size of 5 nm (Figure 4a). Regardless of the type of nanoparticles, i.e., Ag or Au, these results suggest development of photochemical reactions between ALA and metallic nanostructures. Considering the decreased absorbance of UV-VIS spectra, the photochemical processes between ALA and metallic nanoparticles were more intense in Ag nanoparticles, compared to Au nanoparticles. The more pronounced shift of the absorption band in the case of Ag nanoparticles that have interacted with ALA (Figure 4a,b) is believed to have its origin in the process of the chemical adsorption of ALA onto the surface of metal particles. Such a process would involve obtaining a more stable electronic state, i.e., the transformation of Ag into Ag^+ (from [Kr] $4d^{10}5s^1$ to [Kr] $4d^{10}$) and of Au into Au^+ (from [Xe]$4f^{14}5d^{10}6s^1$ to [Xe]$4f^{14}5d^{10}$]), for which it is known that the ionization energy is equal to 7.57 eV and 9.22 eV, respectively [20]. This fact was considered for the higher affinity of thiol groups of ALA for Ag nanoparticles compared with that for Au nanoparticles. Ag nanoparticles' transformation into Ag_2S was reported in 2017 when the appearance of non-stoichiometry in Ag sublattices was envisaged [21,22]. The formation of the thiolate-gold clusters is explained by several models reported by (i) Walter et al., who took into account the magic numbers of free valence electrons [23], and (ii) Cheng et al., who considered the superatom network and the super valence bond mode [24].

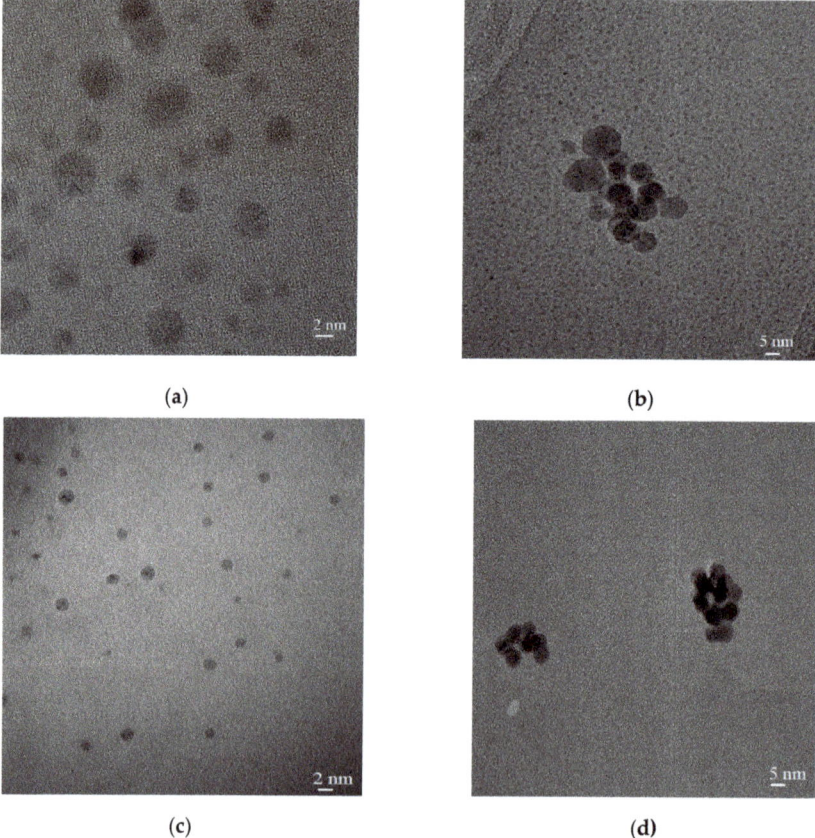

Figure 3. HRTEM images of the nanoparticles: (**a**) Ag with a particle size of 5 nm, (**b**) Ag with a particle size of 10 nm, (**c**) Au with a particle size of 5 nm, and (**d**) Au with a particle size of 10 nm.

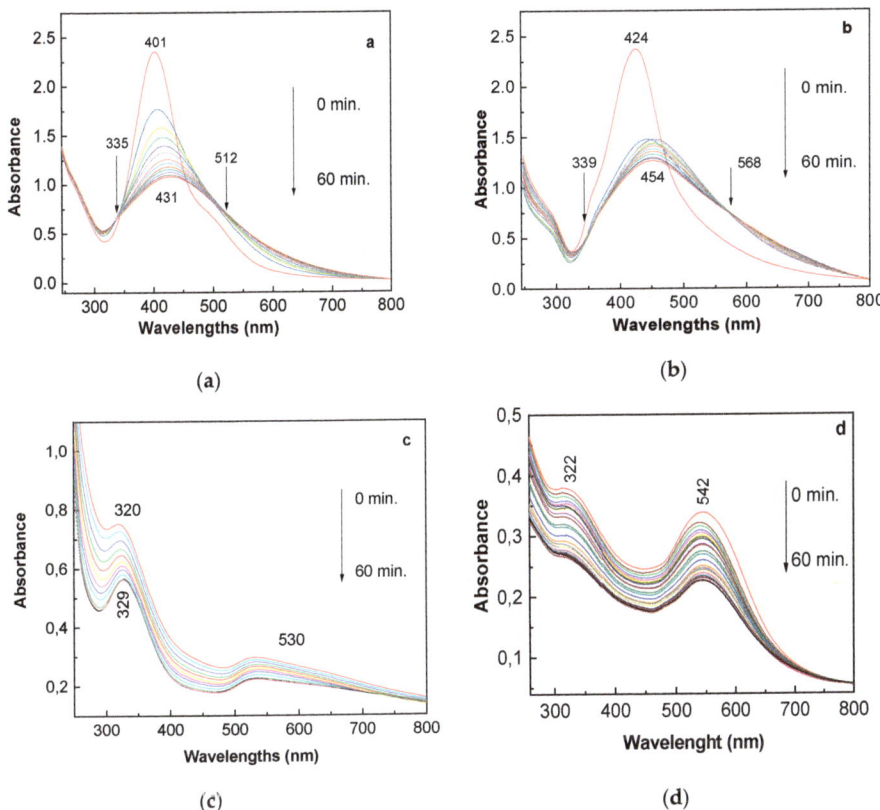

Figure 4. UV-VIS spectra of ALA in the presence of Ag with a particle size of 5 nm (**a**), Ag with a particle size of 10 nm (**b**), Au with a particle size of 5 nm (**c**), and Au with a particle size of 10 nm (**d**). The concentration of ALA is 1 mg/mL. The four samples' UV-VIS spectra were recorded at intervals of 5 min of exposure to UV light, for 60 min.

Figures 5 and 6 show the influence of the additional active compounds to ALA and excipients. In this context, Figure 5 highlights the influence of the pharmaceutical compound concentration and the Ag nanoparticle size on the photochemical processes found in UV-VIS spectroscopy studies.

Figure 5b highlights that, in the case of the Cerebinox solution with a concentration of 1 mg/mL, prepared in the presence of Ag nanoparticles with a size of 5 nm, the UV-VIS spectrum shows an intense band with a maximum at 401 nm, whose absorbance decreases as the time of exposure to UV light increases, up to 60 min. This change was accompanied by the appearance of isosbestic points at 344 and 635 nm. The decrease in the concentration of Cerebinox in the solution of Ag nanoparticles with a size of 5 nm, at 0.25 mg/mL, induced an up-shift of the absorption band from 401 nm to 419 nm and the presence of isosbestic points situated at 329 and 566 nm (Figure 5a). The exposure of the Cerebinox solution with a concentration of 1 mg/mL to UV light for 60 min, prepared in the presence of Ag nanoparticles with a size of 10 nm, induced a shift in the band from 431 nm to 443 nm, which was simultaneous with the appearance of isosbestic points at 371 and 533 nm (Figure 5c). Similar behavior is evidenced in the pharmaceutical compound marketed under the name of Alpha Lipoic Sustain (Figure 6).

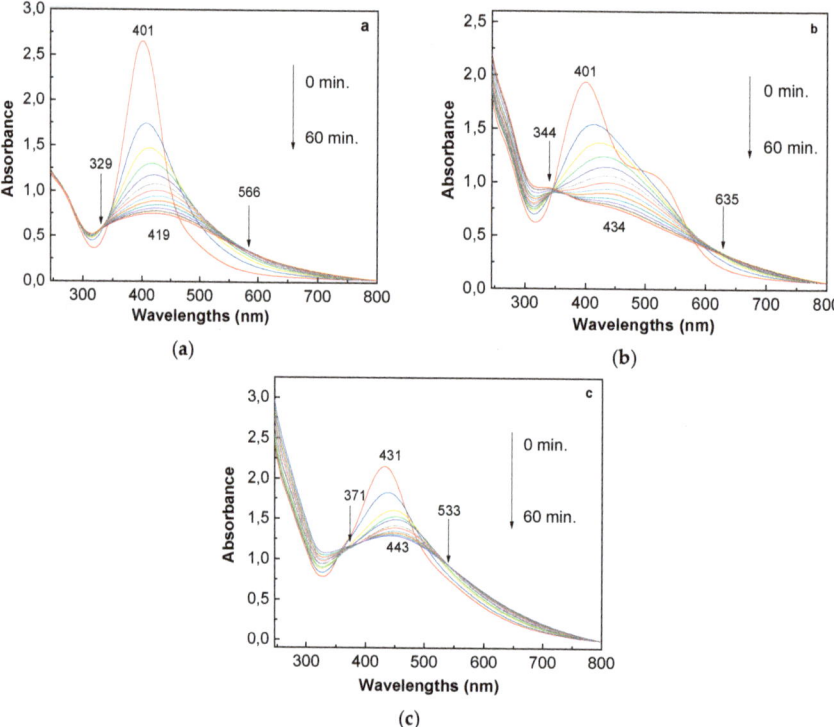

Figure 5. UV-VIS spectra of the ALA solution in the presence of excipients (commercial product marketed under the name Cerebinox) and Ag nanoparticles with a size of 5 nm, having an analyte concentration equal to 0.25 mg/mL (**a**) and 1 mg/mL (**b**), and their changes at exposure to UV light. Figure (**c**) shows the UV-VIS spectra of the Cerebinox solution with a concentration of 1 mg/mL in the presence of Ag nanoparticles with a size 10 nm and its change at exposure to UV light. The three samples' UV-VIS spectra were recorded at intervals of 5 min of exposure to UV light, for 60 min.

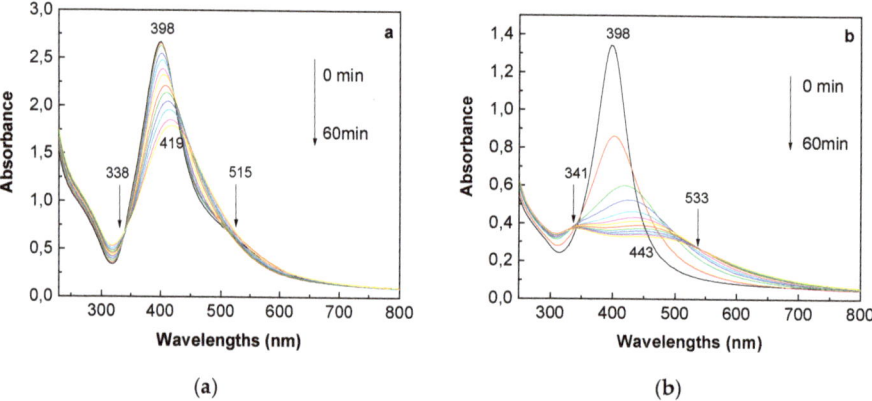

Figure 6. UV-VIS spectra of the ALA solution in the presence of excipients (commercial product marketed under the name of Alpha Lipoic Sustain) in the presence of Ag nanoparticles with sizes of 5 nm (**a**) and 10 nm (**b**), with a concentration of 1 mg/mL, and their changes at the exposure to UV light. The two samples' UV-VIS spectra were recorded at intervals of 5 min of exposure to UV light for 60 min.

Thus, in the case of Alpha Lipoic Sustain solution, with a concentration of 1 mg/mL, (i) the exposure to UV light in the presence of Ag nanoparticles with a size of 5 nm induced a shift of the band from 398 nm to 419 nm, which was simultaneous with its decrease in absorbance and the appearance of isosbestic points at 338 nm and 515 nm. (ii) The exposure to UV light in Ag nanoparticles' presence with a size of 10 nm led to a more significant decrease in the absorbance of the UV-VIS spectrum accompanied by a shift from 398 nm to 443 nm and the presence of isosbestic points at 341 nm and 533 nm. The results shown in Figures 5 and 6 demonstrate that the presence of additional active compounds to ALA and the excipients do not influence the photochemical reaction of ALA with metallic (Me) nanoparticles. The origin of these spectral variations can be explained by taking into account the formation of new compounds in the presence of UV light, according to Scheme 2.

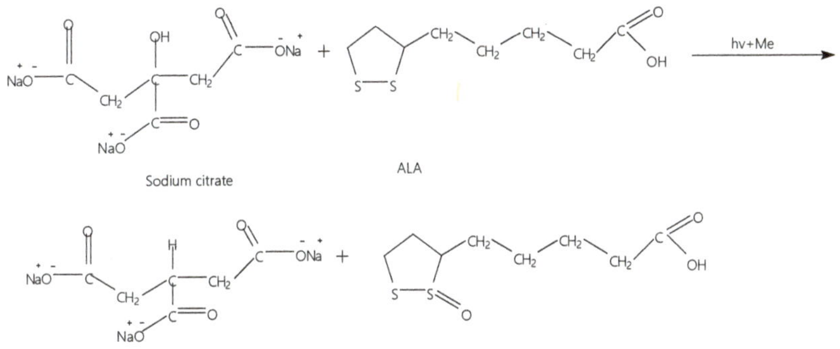

Scheme 2. The photochemical reaction of ALA in the presence of metallic (Me) nanoparticles.

3.2. The RGO Sheets as Inhibitors of the ALA Photodegradation Highlighted by UV-VIS Spectroscopy

Figures 7 and 8 highlight the influence of RGO sheets on ALA's photochemical reactions in the presence of metallic nanoparticles. According to Figure 7, the absorption band of the RGO sheets decorated with Ag nanoparticles with a size of 5 nm has a maximum at 407 nm. The UV-VIS spectrum of ALA in the presence of 0.5 mg RGO sheets in 1 mL of DMF and 1 mL of Ag nanoparticles with a size of 5 nm, before exposure to UV light, is characterized by a band with a maximum at 389 nm. According to Figure 7a, the exposure to UV light induced the following: (i) in the first 5 min, a shift of the band from 389 nm to 398 nm and the appearance of a new band observed as a shoulder at 338 nm; (ii) a gradual shift of the bands from 398 nm and 338 nm to 425 nm and 323 nm, respectively, when the time of UV irradiation varied from 5 min to 60 min; (iii) a decrease in the absorbance of the two bands at 398–425 nm and 338 nm, which involved a change in the ratio between their absorbance ($A_{398-425}/A_{338}$) from 1.05 to 0.83, when the exposure time at UV light increases from 5 to 60 min; and (iv) the appearance of an isosbestic point at 512 nm.

According to Figure 7b, the exposure of the ALA to UV light in the presence of 0.5 mg RGO sheets in 1.5 mL of DMF and 0.5 mL of Ag nanoparticles with a size of 5 nm induced a decrease in the absorbance of the band at 332 nm. This was simultaneous with the increase in the absorbance of the band at 263 nm and the appearance of an isosbestic point at 305 nm. A careful analysis of the band's absorbance variation in the spectral range 300–500 nm highlights a decrease from (i) 0.92 to 0.6 in Figure 7a and (ii) 2.36 to 1.08 in Figure 3d. These results demonstrate the inhibitory role of RGO sheets on the photochemical reactions between ALA and Ag nanoparticles. Such behavior is also shown in the case of Au nanoparticles. Figure 8 is relevant in this context.

The black curve in Figure 8a shows the UV-VIS spectrum of the ALA solution resulting from the dissolution of 0.5 mg of ALA in the presence of 0.5 mg RGO in 0.5 mL of DMF and 1.5 mL of Au nanoparticles with a size of 5 nm. Three bands with maxima at 269 nm, 317 nm, and 530 nm are shown in Figure 8a. According to Figure 8a, the exposure to UV light induced a decrease in the absorbance of bands at (i) 269 nm, from 0.3 to 0.26, (ii) 317 nm, from 0.36 to 0.27, and (iii) 530 nm, from 0.25 to

0.24. In the ALA solution resulting from the dissolution of 0.5 mg of ALA in the presence of 0.5 mg RGO in 1 mL of DMF and 1 mL of Au nanoparticles with a size of 5 nm, no change was observed in the absorbance of the band with a maximum at 515 nm (Figure 8b). The only change observed in Figure 8b regards the band's shift from 311 nm to 323 nm, a variation accompanied by an increase in its absorbance from 0.41 to 0.47. An increase in the absorbance of the band at 260 nm from 0.54 to 0.62 was also reported. These values were significantly lower than those reported in Figure 5a. These results proved once again the inhibitory role of RGO sheets on the photochemical reaction between ALA and Au nanoparticles. An explanation for the inhibitory role of the RGO sheets on the photochemical reaction between ALA and Me nanoparticles is shown in Scheme 3.

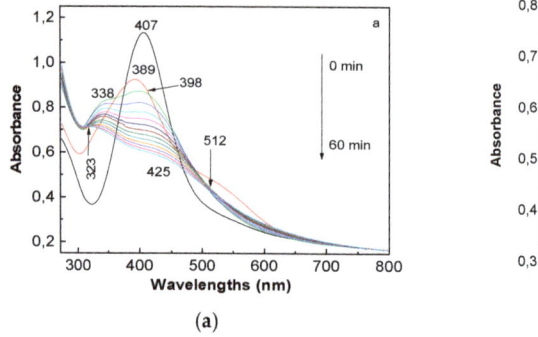
(a)

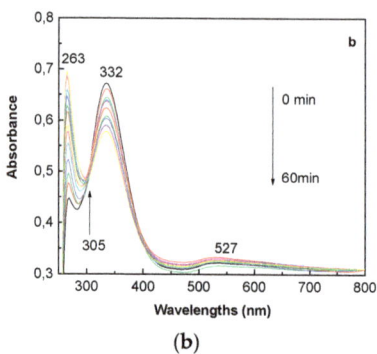
(b)

Figure 7. The influence of the UV light on the ALA solution resulting from the dissolution of 1 mg of ALA in the presence of (**a**) 0.5 mg RGO sheets in 1 mL of DMF and 1 mL of Ag nanoparticles with a size of 5 nm and (**b**) 0.5 mg RGO sheets in 1.5 mL of DMF and 0.5 mL of Ag nanoparticles with a size of 5 nm. The black curve in Figure **a** shows the UV-VIS spectrum of 0.5 mg RGO sheets in 1 mL of DMF and 1 mL of Ag nanoparticles with a size of 5 nm in the absence of ALA. The two samples' UV-VIS spectra were recorded at intervals of 5 min of exposure to UV light, for 60 min.

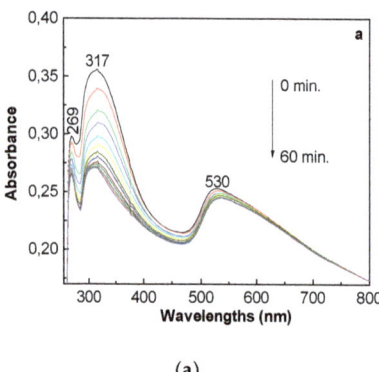
(a)

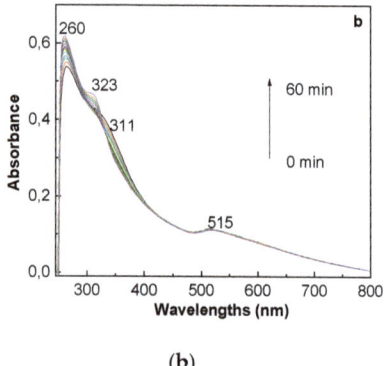
(b)

Figure 8. UV-VIS spectra of the ALA solution resulting from the dissolution of 0.5 mg in the presence of (**a**) 0.5 mg RGO in 0.5 mL of DMF and 1.5 mL of Au nanoparticles with a size of 5 nm and (**b**) 0.5 mg RGO in 1 mL of DMF and 1 mL of Au nanoparticles with a size of 5 nm, as well as their variations at exposure to UV light. The two samples' UV-VIS spectra were recorded at intervals of 5 min of exposure to UV light, for 60 min.

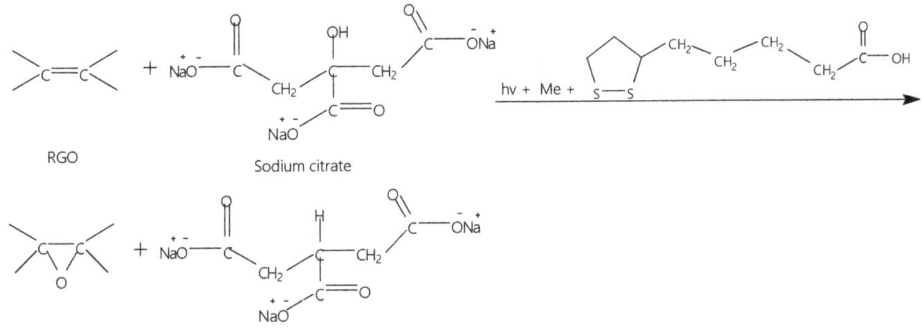

Scheme 3. The interaction of RGO sheets with ALA and Me nanoparticles stabilized with sodium citrate.

3.3. Raman Scattering and SERS Spectroscopy Studies on ALA

In order to prove the adsorption of ALA onto metallic nanoparticles functionalized with sodium citrate, Figure 9 shows the differences between the Raman spectra of ALA when this organic compound is in the form of a powder (Figure 9a) and a layer deposited onto Ag and Au films resulting from colloidal solutions of Ag and Au with particle sizes of 5 and 10 nm (Figure 9b–e). According to Figure 9a, the main Raman lines of ALA peaked at 243, 372–418–457, 513, 565, 636, 684, 902, 1024, 1055, 1084–1230, 1307, 1442, 1649, 2846–2860, and 2916–2931–2974 cm^{-1}, assigned to the following vibrational modes: deformation ring, deformation CSSC, stretching S-S, stretching 1, 2-dithiolane ring, stretching C-S out-of-phase, stretching C-S in-phase, stretching C-C in 1, 2-dithiolane ring, stretching C-C, stretching C-C trans, in-plane deformation C-H in a heteroatomic ring, twisting of CH_2 group, deformation scissoring of CH_2 group, stretching C=O in amide, and anti-symmetrical stretching CH_2 in alkyl chain/anti-symmetrical stretching CH_2 in a 1,2-dithiolane ring. [2,25] The values of the ratio between Raman lines' intensities peaked at 513, 1442, 1024, and 2916 cm^{-1}, i.e., I_{513}/I_{2916}, I_{1442}/I_{2916}, and I_{1024}/I_{2916} are equal to 0.57, 0.35, and 0.09, respectively. In the context of the discussion concerning the interaction of ALA with metallic nanoparticles, it is important to know that the ratio I_{513}/I_{684} has a value of 2.36. In contrast to Figure 9a, the following vibrational changes are observed in Figure 9b,d, i.e., in the case of ALA layers deposited onto metallic films resulting from a colloidal solution of Ag nanoparticles with sizes of 5 and 10 nm: (i) a down-shift of the Raman lines from 513, 684, 1024, 1442, and 2916 cm^{-1} (Figure 9a) to 504, 684, 1018, 1430, and 2908 cm^{-1}, respectively (Figure 9b,d); and (ii) a change in the value of the ratios I_{513}/I_{2916}, I_{1442}/I_{2916}, I_{1024}/I_{2916}, and I_{513}/I_{684} from 2, 0.28, 0.4, and 8.35 (Figure 9b) to 3.94, 0.56, 0.79, and 8.2, respectively (Figure 9d). The changes in the intensities and the positions of Raman lines when the ALA layer is deposited onto metallic films resulting from the colloidal solutions of Au nanoparticles with sizes of 5 and 10 nm are noted in Figure 9c,e as follows: (i) the Raman lines from 513, 680, 1024, 1442, and 2916 cm^{-1} (Figure 9a) shifted to 504, 673, 1014, 1430, and 2909 cm^{-1}, respectively (Figure 9c,e); (ii) in the 3000–3500 cm^{-1} spectral range, a new Raman line with a maximum at 3075 cm^{-1} appeared; (iii) values of the ratios I_{504}/I_{2909}, I_{1430}/I_{2909}, I_{1014}/I_{2909}, and I_{504}/I_{673} changed from 1.01, 0.29, 0.2, and 3.7 (Figure 9c) to 2, 0.57, 0.4 and 3.67 (Figure 9e).

The variations above indicate a chemical interaction between metallic nanoparticles functionalized with sodium citrate and ALA, as shown in Scheme 2. Regardless of the metallic nanoparticle type, i.e., Ag or Au, the increase in the Raman line's intensity with a maximum at 504 cm^{-1} suggests preferential chemical adsorption of ALA onto metallic nanoparticles, as reported in the case of other thiols [25]. Evidence of the chemical adsorption of ALA onto the metallic nanoparticle surface is shown by the Raman line with a maximum at 234 cm^{-1}, observed in Figure 9b–e, which was assigned to the stretching vibration mode of the metal-adsorbate bond [2]. The presence of these Raman lines in Figure 9b–e can be explained by considering the chemical mechanism of the SERS effect that involves a charge transfer at the metal/dielectric interface—in our case, the Ag or Au nanoparticles and ALA, when new bonds between the metal and adsorbate of the type Ag-S or Au-S are generated [25].

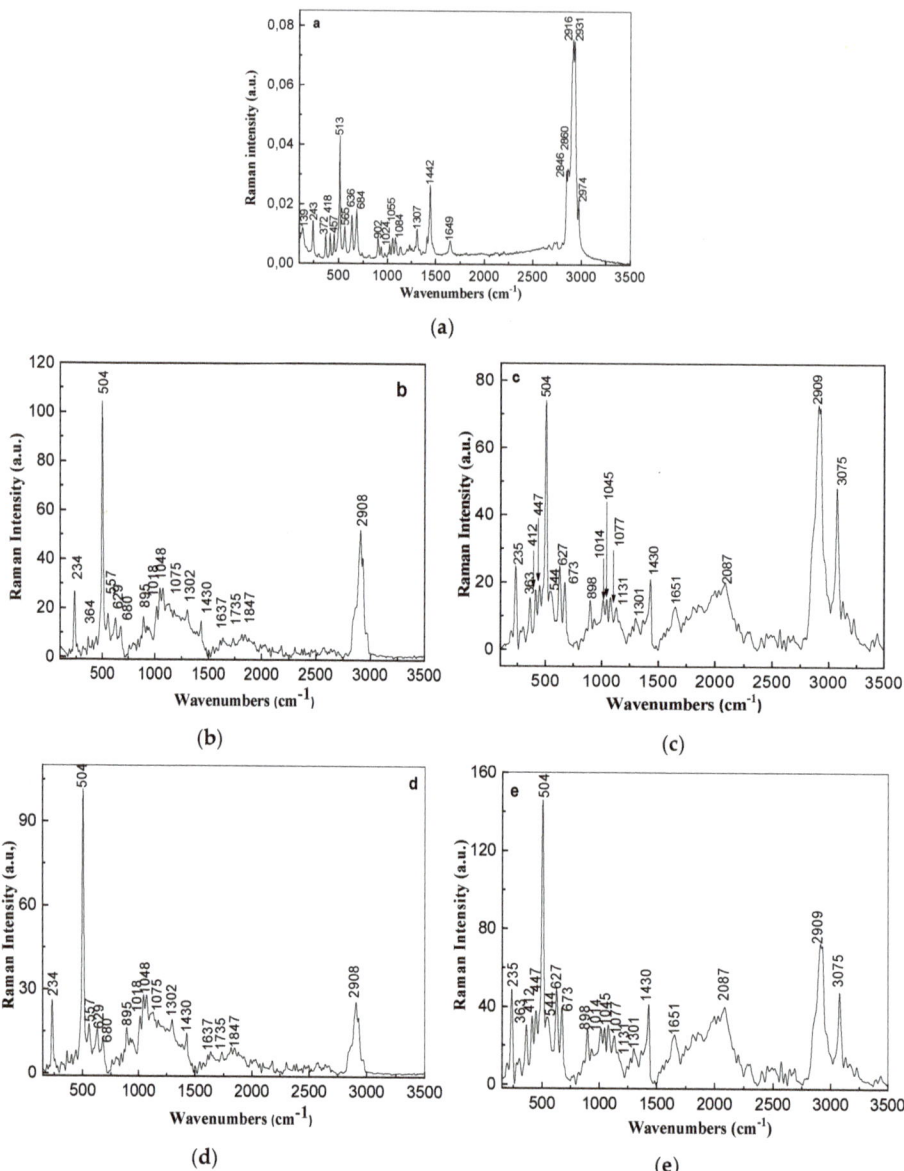

Figure 9. Raman spectra of ALA in a solid state (**a**) and as layers deposited onto metallic films prepared using colloidal solutions of Au nanoparticles with sizes of 5 nm (**b**) and 10 nm (**c**) and of Ag nanoparticles with sizes of 5 nm (**d**) and 10 nm (**e**). The ALA layers were obtained by using ALA solutions with a concentration of 10^{-2} M.

Using a protocol proposed by Laurent et al. [26] to assess the enhancement factor (EF) of the SERS process and the experimental results shown in Figure 9, the following EFs of ALA were calculated for the Raman line situated in the spectral range 500–520 cm^{-1}: (i) 1.15×10^7, in the case of Ag nanoparticles with sizes of 5 nm and 10 nm, (ii) 8.16×10^6, in the case of Au nanoparticles with a size of 5 nm, and (iii) 1.6×10^7, in the case of Au nanoparticles with a size of 10 nm.

Figure 10 highlights the dependence of the Raman spectrum intensity of ALA on the Au nanoparticles weight used to prepare metallic films. According to Figure 10a, the Raman spectrum intensity of ALA is enhanced as Au nanoparticles' weight with a size of 50 nm increases. This fact is a consequence of the electromagnetic mechanism of the SERS effect, when the generation of localized surface plasmons at the metal/dielectric interface occurs [14]. According to Figure 10b, for the same weight of Au nanoparticles, a higher Raman intensity of ALA is found when the size of metallic nanoparticles is equal to 50 nm compared to those with a size of 10 nm. In the case of Au nanoparticles with a size of 50 nm, the EF of the ALA Raman line intensity situated in the spectral range 500–520 cm^{-1} was equal to 5.18×10^7, a value superior to that reported in the case of Au nanoparticles with a size of 10 or 5 nm (Figure 9c,e).

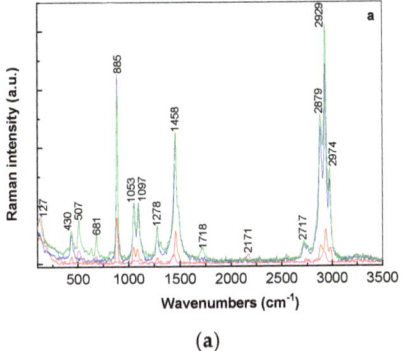
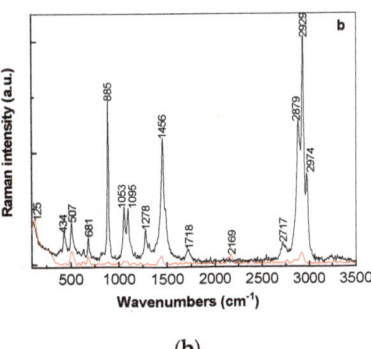

Figure 10. Raman spectra of the ALA layer deposited onto metallic films prepared using 1 mL (red curve), 2 mL (blue curve), 4 mL (magenta curve), and 6 mL (green curve) colloidal solutions of Au nanoparticles with a size of 50 nm (**a**). The dependence of the intensity of Raman spectra on the ALA layers deposited onto metallic films prepared from 6 mL colloidal solutions of Au nanoparticles with sizes of 50 nm (black curve) and 10 nm (red curve) (**b**). In all samples, the ALA solution concentration was 10^{-2} M.

3.4. HRTEM and AFM Studies

A short characterization of the RGO sheets by HRTEM and the film of Au obtained from the colloidal solutions of Au nanoparticles with a size of 50 nm before and after the adsorption of the RGO sheets by AFM is shown in Figure 11. Figure 11a shows the TEM image of the RGO sheets stacked with various folding. The HRTEM image of the RGO sheets (Figure 11b) highlights an interplanar distance d_{002} equal to 3.85 Å.

The AFM image of the films obtained by the evaporation of a 0.12 mL colloidal solution of Au, with a particle size of 50 nm, onto quartz substrates with an area of 1 cm^2, are shown in Figure 11c$_1$. As a consequence of the agglomerate and aggregate processes of the Au nanoparticles, the AFM images revealed that the gold nanoparticle diameter is ~60 nm (Figure 11c$_2$). Two roughness parameters were calculated from the AFM scans for the Au film, i.e., the root mean square (RMS) and roughness average (Ra), with values equal to 18 nm and 13 nm, respectively. The deposition of the RGO sheets onto the Au film induced RMS and Ra values of 91 nm and 71 nm, respectively. Ra's higher value in the case of the RGO sheets deposited onto the Au film caused their random orientation onto the metallic film surface (Figure 11d).

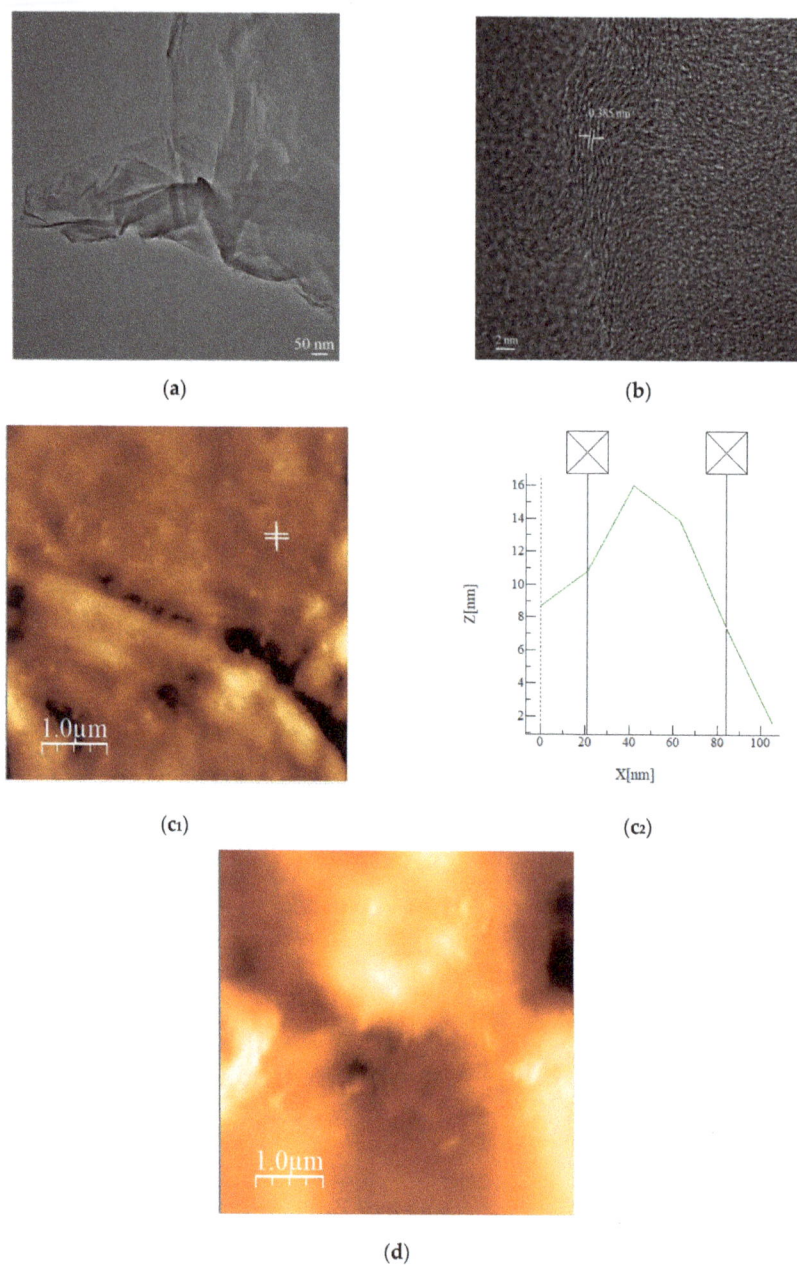

Figure 11. TEM (**a**) and HRTEM (**b**) images of the RGO sheets. The AFM image of the metallic film prepared from Au nanoparticles' colloidal solutions with a size of 50 nm before (**c₁**, **c₂**) and after the deposition of the RGO sheets (**d**).

4. Conclusions

Using UV-VIS spectroscopy, new results are reported in this article concerning the photodegradation of ALA in the presence of Ag and Au nanoparticles. We have demonstrated that (i) the presence of excipients and additional active compounds to ALA, by exposure to UV light for 1 h, do not induce further changes once ALA has interacted with a colloidal solution of Ag and Au nanoparticles with sizes of 5 and 10 nm. (ii) The changes reported in the position and the relative intensities of Raman lines of ALA in the presence of Ag and Au nanoparticles with sizes between 5 and 10 nm, in contrast to the Raman spectrum of ALA powder, are explained by taking into account the chemical mechanism of surface-enhanced Raman scattering (SERS) spectroscopy. (iii) The photodegradation process of ALA that has interacted with metallic nanoparticles was inhibited in the presence of RGO sheets.

Author Contributions: Conceptualization, M.B. and A.M.; methodology, M.B.; investigation by UV-VIS spectroscopy, N.T., M.D., M.S.S., and R.C.; AFM investigation, M.S.; TEM and HRTEM investigations, I.M.; Investigations by Raman scattering and preparation of metallic nanoparticles M.B.; preparation of gold films, M.D.; writing—original draft preparation, M.B.; writing—review and editing, M.B., A.M., and D.D.; visualization, N.T., M.D., M.S., and R.C.; supervision, M.B.; project administration, M.B.; funding acquisition, M.B. All authors have read and agreed to the published version of the manuscript.

Funding: This research was co-funded by the European Regional Development Fund under the Competitiveness Operational Program 2014–2020, project number 58/05.09.2016, signed between the National Institute of Materials Physics and the National Authority for Scientific Research and Innovation as an Intermediate Body, on behalf of the Ministry of European Funds as Managing Authority for Operational Program Competitiveness (POC).

Acknowledgments: This work was achieved in the frame of sub-contract of type D, no. 1965/20.09.2017 of the POC project number 58/05.09.2019.

Conflicts of Interest: The authors declare no conflict of interest. The funders had no role in the design of the study; in the collection, analyses, or interpretation of data; in the writing of the manuscript, or in the decision to publish the results.

References

1. Durrani, A.; Schwartz, H.; Nagl, M.; Sontag, G. Determination of free [alpha]-lipoic acid in foodstuffs by HPLC coupled with CEAD and ESI-MS. *Food Chem.* **2010**, *120*, 38329–38336. [CrossRef]
2. Ziegle, D.; Reljanovic, M.; Mehnert, H.; Gries, F.A. α-lipoic acid in treatment of diabetic polyneuropathy in Germany. *Exp. Clin. Endocr. Diab.* **1999**, *107*, 421–430. [CrossRef]
3. Shay, K.P.; Moreau, R.F.; Smith., E.J.; Smith, A.R.; Hagen, T.M. Alpha-lipoic acid as a dietary supplement. Molecular mechanism and therapeutic potential. *Biochim. Biophys. Acta (BBA)-Gen. Subj.* **2009**, *1790*, 1149–1160. [CrossRef] [PubMed]
4. Namazi, N.; Larijani, B.; Azadbakht, L. Alpha-lipoic acid supplement in obesity treatment: A systematic review and meta-analysis of clinical trials. *Clin. Nutr.* **2018**, *37*, 419–428. [CrossRef]
5. Packer, L.; Witt, E.H.; Tritschler, H.J. Alpha-lipoic acid as a biological antioxidant. *Free Radical Bio. Med.* **1995**, *19*, 227–250. [CrossRef]
6. Ikuta, N.; Tanaka, A.; Otsubo, A.; Ogawa, N.; Yamamoto, H.; Mizukami, T.; Arai, S.; Okuno, M.; Terao, K.; Matsugo, S. Spectroscopic studies of R (+)-α-lipoic acid-cyclodextrin complex. *Int. J. Mol. Sci.* **2014**, *15*, 20469–20485. [CrossRef]
7. Busby, R.W.; Schelvis, J.P.M.; Yu, D.S.; Babcock, G.T.; Marletta, M.A. Lipoic acid biosynthesis: Lip A is an iron-sulfur protein. *J. Am. Chem. Soc.* **1999**, *121*, 4706–4707. [CrossRef]
8. Miranda, M.P.; de Rio, R.; del Valle, M.A.; Faundez, M.; Armijo, F. Use of fluorine-doped tin oxide electrodes for lipoic acid determination in dietary supplements. *J. Electroanaly. Chem.* **2012**, *668*, 1–6. [CrossRef]
9. Charoenkitamorn, K.; Chaiyo, S.; Chailapakul, W.; Siangproh, W. Low-cost and disposable sensors for the simultaneous determination of coenzyme Q10 and α-lipoic acid using manganese (IV) oxide – modified screen-printed graphene electrodes. *Anal. Chem. Acta* **2018**, *104*, 22–31. [CrossRef]
10. Inoue, T.; Sudo, M.; Yoshida, H.; Todoroki, K.; Nohta, H.; Yamaguchi, M. Liquid chromatographic determination of polythiols based on pre-colum excimer fluorescence derivatization and its application to α-lipoic acid analysis. *J. Chromatogr. A* **2009**, *1216*, 7564–7569. [CrossRef]
11. Ruiz-Jimez, J.; Friego-capote, F.; Mata-Granados, J.M.; Quesda, J.M.; Luque de Castro, M.D. Determination of the ubiquinol-10 and ubiquinone-10 (coenzyme A10) in human serum by liquid chromatography tandem mass spectrometry to evaluate the oxidative stress. *J. Chromatogr. A* **2007**, *1175*, 242–248. [CrossRef] [PubMed]

12. Vaishnav, S.K.; Patel, K.; Chandraker, K.; Korram, J.; Nagwanshi, R.; Ghosh, K.K.; Satnami, M.L. Surface plasmons resonance based spectrophotometric determination of medicinally important thiol compounds using unmodified silver nanoparticles. *Spectroc. Acta Pt. A Molec. Biomolec. Spectr.* **2017**, *179*, 155–164. [CrossRef] [PubMed]
13. Sun, J.; Guo, L.; Bao, J.; Xie, J. A simple, label-free AuNPs—Based colorimetric ultrasensitive detection of nerve agents and highly toxic organophosphate pesticide. *Biosens. Bioelectron.* **2011**, *28*, 152–157. [CrossRef] [PubMed]
14. Raether, H. *Surface Plasmons on Smooth and Rough Surfaces and Gratings*; Springer Tracts in Modern Physics: Spinger, Berlin, 1988.
15. Lefrant, S.; Baltog, I.; Lamy de la Chapelle, M.; Baibarac, M.; Louarn, G.; Journet, C.; Bernier, P. Structural properties of some conducting polymers and carbon nanotubes investigated by SERS spectroscopy. *Synth. Met.* **1999**, *100*, 13–27. [CrossRef]
16. Agnihotri, S.; Mukherje, S.; Mukherji, S. Size-controlled silver nanoparticles synthesized over the range 5-100 nm using the same protocol and their antibacterial efficacy. *RSC Adv.* **2014**, *4*, 3974–3983. [CrossRef]
17. Piella, J.; Bastus, N.G.; Puntes, V. Size-controlled synthesis of sub-10-nanometer citrate-stabilized gold nanoparticles and related optical properties. *Chem. Mater.* **2016**, *28*, 1066–1075. [CrossRef]
18. Li, M.; Guo, X.; Wang, H.; Wen, Y.; Yang, H. Rapid and label-free Raman detection of azadicarbonamide with asthma risk. *Sens. Actuator B-Chem.* **2015**, *216*, 535–541. [CrossRef]
19. Perrault, S.D.; Chan, W.C.W. Synthesis and surface modified of highly monodispersed spherical gold nanoparticles of 50-200 nm. *J. Am. Chem. Soc.* **2009**, *131*, 17042–17043. [CrossRef]
20. Baibarac, M.; Lapkowski, M.; Pron, A.; Lefrant, S.; Baltog, I. SERS spectra of poly(3-hexylthiophene) in oxidized and unoxidized states. *J. Raman Spectrosc.* **1998**, *29*, 825–832. [CrossRef]
21. Sadovnikov, S.I.; Gusev, A.I. Recent progress in nanostructured silver sulfide: from synthesis and nonstoichiometry to properties. *J. Mater. Chem. A* **2017**, *5*, 17676–17704. [CrossRef]
22. Potter, P.M.; Navratilova, J.; Rogers, K.R.; Al-Abed, S.R. Transofrmation of silver nanoparticle consumer products during simulated usage and disposal. *Environ. Sci. Nano* **2019**, *6*, 592–598. [CrossRef] [PubMed]
23. Walter, M.; Akola, J.; Lopez-Acevedo, O.; Jadzinsky, P.D.; Calero, G.; Ackerson, C.J.; Whetten, R.L.; Grönbeck, H.; Häkkinen, H. A unified view of ligand-protected gold clusters as superatom complexes. *Proc. Natl. Acad. Sci. USA* **2008**, *105*, 9157–9162. [CrossRef] [PubMed]
24. Cheng, L.; Ren, C.; Zhang, X.; Yang, J. New insight into the electronic shell of $Au_{38}(SR)_{24}$: A superatomic molecule. *Nanoscale* **2013**, *5*, 1475–1478. [CrossRef] [PubMed]
25. Bryant, M.A.; Pemberton, J.E. Surface Raman scattering of self-assembled monolayers formed from 1-alkanethiols: behavior of films at Au and comparison to films at Ag. *J. Am. Chem. Soc.* **1991**, *113*, 8284–8293. [CrossRef]
26. Laurent, G.; Felidj, N.; Aubard, J.; Levi, G. Evidence of multipolar excitations in surface enhanced Raman scattering. *Phys. Rev. B* **2005**, *71*, 045430. [CrossRef]

Publisher's Note: MDPI stays neutral with regard to jurisdictional claims in published maps and institutional affiliations.

© 2020 by the authors. Licensee MDPI, Basel, Switzerland. This article is an open access article distributed under the terms and conditions of the Creative Commons Attribution (CC BY) license (http://creativecommons.org/licenses/by/4.0/).

Article

Highly Multifunctional GNP/Epoxy Nanocomposites: From Strain-Sensing to Joule Heating Applications

Xoan F. Sánchez-Romate *, Alejandro Sans, Alberto Jiménez-Suárez *, Mónica Campo, Alejandro Ureña and Silvia G. Prolongo

Materials Science and Engineering Area, Escuela Superior de Ciencias Experimentales y Tecnología, Universidad Rey Juan Carlos, Calle Tulipán s/n, 28933 Móstoles, Madrid, Spain; a.sans@alumnos.urjc.es (A.S.); monica.campo@urjc.es (M.C.); alejandro.urena@urjc.es (A.U.); silvia.gonzalez@urjc.es (S.G.P.)
* Correspondence: xoan.fernandez.sanchezromate@urjc.es (X.F.S.-R.); alberto.jimenez.suarez@urjc.es (A.J.-S.); Tel.: +34-914-884-771 (X.F.S.-R.); +34-914-887-141 (A.J.-S.)

Received: 16 November 2020; Accepted: 3 December 2020; Published: 5 December 2020

Abstract: A performance mapping of GNP/epoxy composites was developed according to their electromechanical and electrothermal properties for applications as strain sensors and Joule heaters. To achieve this purpose, a deep theoretical and experimental study of the thermal and electrical conductivity of nanocomposites has been carried out, determining the influence of both nanofiller content and sonication time. Concerning dispersion procedure, at lower contents, higher sonication times induce a decrease of thermal and electrical conductivity due to a more prevalent GNP breakage effect. However, at higher GNP contents, sonication time implies an enhancement of both electrical and thermal properties due to a prevalence of exfoliating mechanisms. Strain monitoring tests indicate that electrical sensitivity increases in an opposite way than electrical conductivity, due to a higher prevalence of tunneling mechanisms, with the 5 wt.% specimens being those with the best results. Moreover, Joule heating tests showed the dominant role of electrical mechanisms on the effectiveness of resistive heating, with the 8 wt.% GNP samples being those with the best capabilities. By taking the different functionalities into account, it can be concluded that 5 wt.% samples with 1 h sonication time are the most balanced for electrothermal applications, as shown in a radar chart.

Keywords: carbon nanotubes; thermal properties; electrical properties; strain sensing; joule heating

1. Introduction

Nowadays, polymeric materials are gaining much attention. They present some interesting properties that make them suitable for use in a wide variety of applications, including as coatings for environmental and corrosion protection [1–3] or as a matrix in composite materials due to their high compatibility with most widely used reinforcements [4,5].

In this regard, the use of carbon nanoparticles such as graphene nanoplatelets (GNPs) or carbon nanotubes (CNTs) is now of interest. This can be explained by their excellent mechanical, thermal and electrical properties. In fact, they can reach values of Young's Modulus over 1 TPa, thermal conductivity around 5000 W/mK and electrical conductivities of 10^7 S/m [6–10]. These superior properties make them highly suitable for multiple applications. More specifically, they are commonly used as reinforcement for polymeric materials. Furthermore, when added to an insulator matrix, both electrical and thermal conductivity grow several orders of magnitude, becoming conductive materials, because of the creation of percolating networks inside the material [11–13]. These facts promote their use, for example, in Structural Health Monitoring (SHM), electromagnetic interference shield and as thermal interface materials (TIMs) [14–19].

This work is focused on the effect of GNP content and sonication time on several properties such as electrical and thermal conductivity, strain-sensing and Joule heating capabilities. Here, GNP nanocomposites have demonstrated good capabilities as strain sensors with gauge factors, that is, the correlation between the variation of the normalized electrical resistance divided by the applied strain is much superior to that of CNT-based ones, especially at higher strain levels [20–25]. This is explained by the 2D disposition of the GNPs within the material allowing a higher interparticle distance between adjacent nanoparticles, thus leading to a more prominent tunneling effect [26–28]. In addition, GNPs can be added into the resin in contents superior to those possible with CNTs without inducing a drastic degradation of mechanical properties. Therefore, these materials present much higher values of thermal conductivity, which mainly depends on nanofiller content [14,29,30].

However, the correlation among the different properties is sometimes not well understood, as there are multiple factors affecting the final properties of the nanocomposite, including content, dispersion and geometry of the nanofiller [11,24,31,32]. In this regard, several dispersion techniques are commonly used to ensure a proper homogenization of the nanofillers inside the polymer. Among others, three roll milling and sonication have proved to be the most effective techniques due to the higher shear or cavitation forces induced during the process, which lead to an adequate breakage of larger agglomerates along with some exfoliating mechanisms [33,34]. Moreover, the enhancement of one property can lead to the degradation of another, as observed in highly conductive nanocomposites that present low gauge factors, as the interparticle distance between adjacent nanoparticles is much lower [20]. For these reasons, this study aims to better understand the role of nanoparticle content and dispersion state in the final properties of the nanocomposites.

First, the electrical and thermal conductivity of GNP nanocomposites are determined for different combinations of GNP content and sonication time. Then, two examples of specific applications are also measured and deeply explored: their use as strain sensors by means of electrical measurements with applied strain and their capacity for Joule's effect resistive heating. Finally, a summary of the obtained results is shown by balancing the final properties of each material in order to select the optimum one as a function of the desired application.

2. Experimental Procedure

2.1. Materials

The nanocomposite is based on a GNP reinforced epoxy matrix. The resin is an *Araldite LY 556* from *Hunstman* supplied by *Antala* (Barcelona, Spain) with an amino hardener *XB 3473* in a stoichiometry proportion of 100:23 monomer to hardener from the same supplier.

GNPs are *M25* supplied by *XG Sciences* (Lansing, MI, USA) with a lateral size of 25 μm and a thickness of 6 to 10 nm.

2.2. Nanocomposite Manufacturing

First, GNPs were manually dispersed into the epoxy resin. The mixture was then subjected to an ultrasonication process by using a horn sonicator *UP400S* supplied by *Hielscher* (Teltow, Germany) at an amplitude of 80% and a pulse period of 0.5 s. Sonication time and GNP content were varied in order to analyze their effects in the final properties of the nanocomposites. Nanofiller content and sonication time were varied accordingly to that shown in Table 1.

Once the dispersion was made, the mixture was degassed at 80 °C for 15 min in order to remove the possible entrapped air. Then, it was subjected to a curing cycle at 140 °C for 8 h.

Finally, the plates obtained were demolded and machined to the dimensions required by the different tests which are explained below.

Table 1. Nomenclature used for materials manufactured and tested.

GNP Content (wt.%)	Sonication Time (h)	Designation
5	1	5GNP-1 h
	2	5GNP-2 h
	3	5GNP-3 h
8	1	8GNP-1 h
	2	8GNP-2 h
	3	8GNP-3 h

2.3. Electrical, Thermal and Microstructural Characterization

Four-probe DC volume conductivity tests were carried out for 4 different samples of $10 \times 10 \times 1$ mm^3 dimensions for each condition (Figure 1a). Electrical resistance was determined as the slope of I-V curve, and electrical resistivity was determined accordingly to the geometry of the samples. The voltage range was set at 0–200 V for low conductive samples and 0–25 V for high conductive samples. The tests were performed in a *SMU, Keithley Instrument Inc. mod. 2410* (Cleveland, OH, USA).

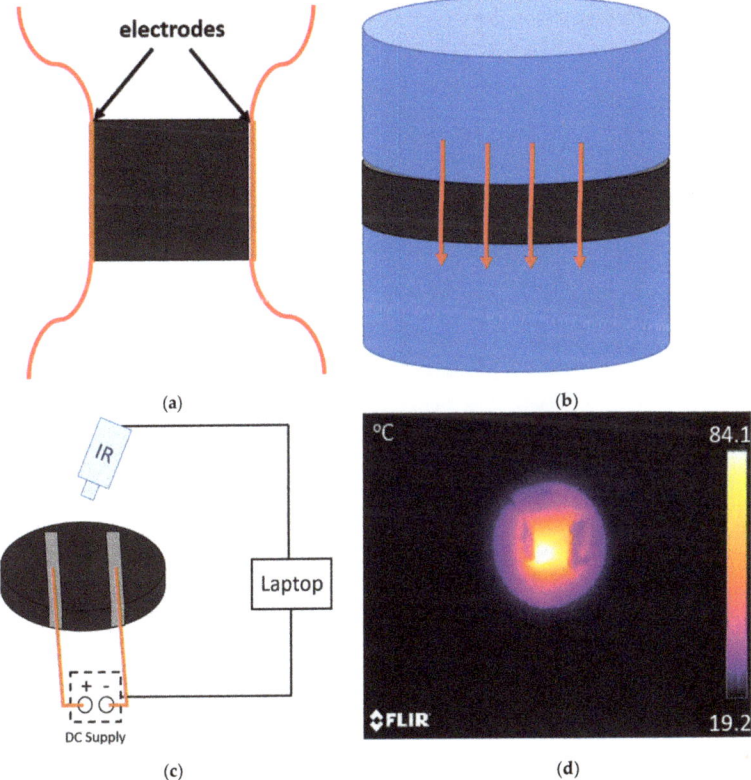

Figure 1. Schematics of (**a**) electrode disposition on the electrical conductivity tests, (**b**) set-up of thermal conductivity measurements (where the red arrows indicate the sense of the heat flow), (**c**) Joule heating tests indicating the electrode's disposition in the sample and (**d**) an example of thermal image of a 5GNP-1 h sample (the dark shapes around the central region correspond to the silver paint coating of the electrodes).

Thermal conductivity was measured by estimating the heat flow through 50 mm diameter round samples by using a Heat Flow Meter (*FOX 50 Heat Flow Meter 190_C VHS 220VAC*) from *TA Instruments* (New Castle, DE, USA) as shown in the schematic of Figure 1b. The thickness of the samples varied in the range of 4–5 mm. Two samples were tested for each condition and thermal conductivity was determined at 30, 90 and 180 °C.

The GNP distribution was determined by means of Scanning Electron Microscopy analysis. For this purpose, GNPs were filtrated after sonication process in an acetone bath using a 0.22 μm porous paper. The obtained powder was then analyzed by using a *Hitachi S 3400N* apparatus from *Hitachi Global* (Tokyo, Japan).

2.4. Strain Monitoring Tests

Tensile tests were performed according to standard ASTM D638 at a test rate of 1 mm/min. Simultaneously, the electrical resistance was measured in order to characterize the strain monitoring capabilities of the manufactured materials by using an *Agilent* hardware *34410 A* (Agilent Technologies, Santa Clara, CA, USA).

To achieve this purpose, the electrical resistance was recorded between two electrodes made of copper wire and silver ink to ensure a good electrical contact with the substrate. Here, the sensitivity, also called, gauge factor (*GF*) of the materials has been determined.

GF is given by the ratio between the change of the normalized resistance divided by the applied strain:

$$GF = \frac{\frac{\Delta R}{R_0}}{\varepsilon} \tag{1}$$

where $\Delta R/R_0$ denotes the change of the electrical resistance divided by the initial resistance of the specimen.

In the tests conducted, *GF* was determined at low strain levels where crack mechanisms are not supposed to taking place.

2.5. Joule Effect Heating Tests

Electrothermal properties were determined by resistive heating. In this experiment, thermal conductivity samples were subjected to a varying applied voltage. The temperature of the samples was measured by using a FTIR thermal camera (FLIR E50) (FLIR Systems, Wilsonville, OR, USA) as shown in the schematics of Figure 1c,d. The electrodes were also made with copper wire and silver ink with a distance of 30 mm between them. The voltage was applied by steps of 50 V until a temperature of around 180 °C was reached in the sample as it is near the degradation of the epoxy matrix.

3. Results and Discussion

The electromechanical and thermal properties of GNP nanocomposites are discussed in this section. First, electrical conductivity measurements are shown, while thermal properties are also explored. Finally, the electromechanical characteristics of the proposed materials are given on the basis of strain monitoring tests.

3.1. Electrical Properties of GNP/Epoxy Nanocomposites

Figure 2 shows the values of the electrical conductivity for GNP nanocomposites. It can be observed that an increase of GNP content from 5 to 8 wt.% leads to a significant increase in the electrical conductivity, from values of 10^{-4}–10^{-3} S/m to values of 0.1–1 S/m. This is easily explained by the effect of the higher volume fraction of the nanofillers that induces the creation of a higher number of percolating networks inside the material as has been widely explored in other studies [35,36].

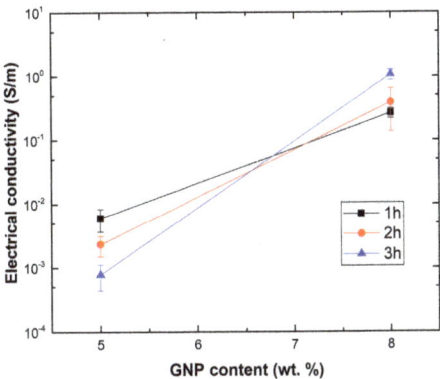

Figure 2. Electrical conductivity measurements for the different samples.

However, the effect of sonication time is quite more complex. Here, at lower GNP contents, it is observed that the increasing sonication time leads to a reduction of the electrical conductivity. This is explained by the effect that sonication has on the GNP mixture. On one hand, there is the prevalence of exfoliating mechanisms of graphene layers during the sonication process [37], leading to a reduction of GNP thickness and, thus, to an increase of the aspect ratio of the nanoparticles as well as to an enhancement of GNP dispersion [38]. However, it has been also widely investigated that very large sonication times lead to a significant breakage of the nanofillers due to the higher cavitation forces induced during the sonication process [35]. This breakage of GNPs leads to a reduction in the lateral size. At lower contents, the viscosity of the media is low, so the cavitation forces are more effective [39]. This means that the optimum sonication time to achieve the best electrical performance is lower and this fact explains that increasing this sonication time too much could result in a detriment of the electrical properties, because of a very aggressive rupture of GNPs that leads to a reduction of the effective aspect ratio.

On the other hand, when increasing the GNP content, the viscosity of the mixture is much higher, so that cavitation process is not so efficient and the optimum sonication time to achieve the desired properties is increased. For this reason, the highest electrical conductivity for 8 wt.% GNP nanocomposites is achieved at 3 h of sonication time.

In this regard, the SEM analysis of GNP powder after sonication process can confirm the previous statements. On one hand, when comparing 5GNP-1 h to 5GNP-3 h samples, an evident reduction of the lateral size can be pointed out (Figure 3a,b), which is more prevalent than the reduction of GNP stacking, which is qualitatively similar in both cases (Figure 3c,d) as, at low times, the sonication is effective at this GNP content. However, in the case of 8GNP samples, the increase of the sonication time from 1 h to 3 h promotes a very efficient reduction of the GNP stacking (Figure 3e,f) due to the previously commented higher efficiency of the sonication process at higher times explained by the higher viscosity of the mixture.

These statements are of high novelty, as sonication time can have a positive effect on GNP properties depending on the viscosity of the mixture, as observed for 8 wt.% samples whereas longer sonication times will have a negative effect at lower contents due to an initial higher efficiency of the process that trends to rapidly break the nanoplatelets.

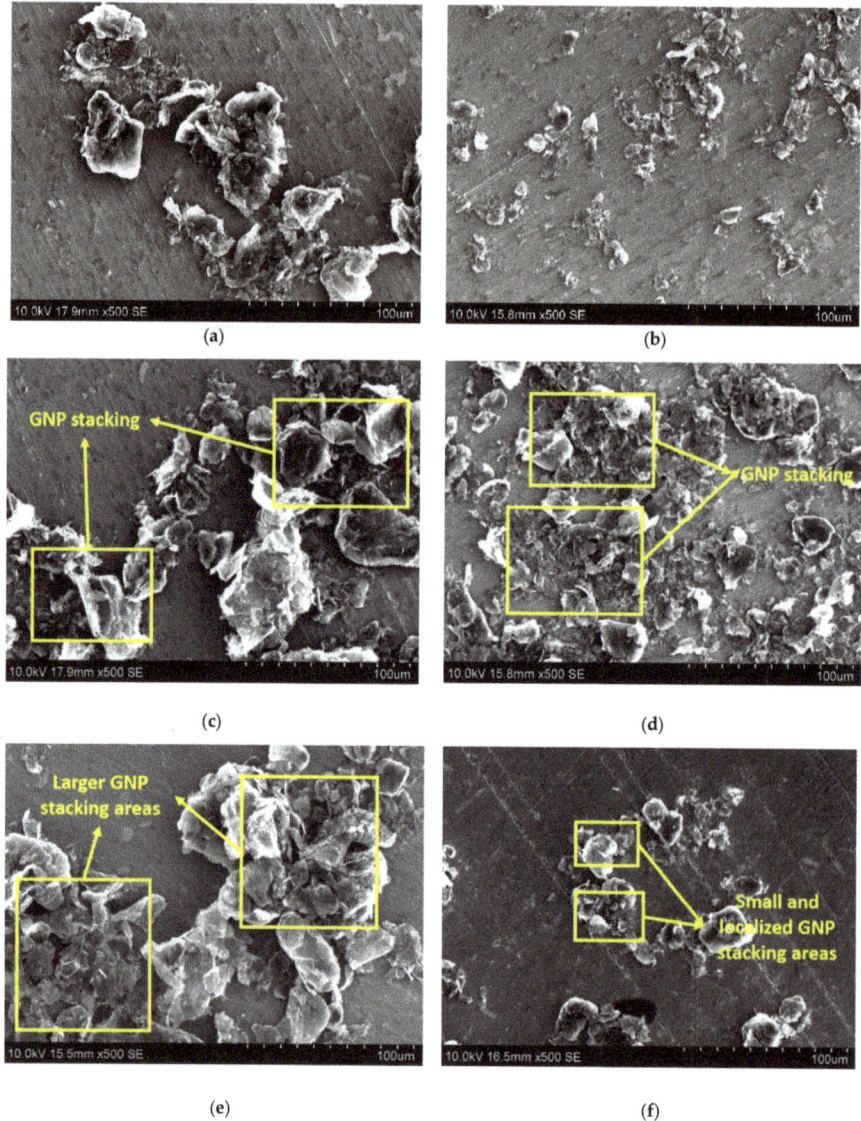

Figure 3. SEM images of GNP powder after the sonication process for (**a**,**c**) 5GNP-1 h, (**b**,**d**) 5GNP-3 h, (**e**) 8GNP-1 h, and (**f**) 8GNP-3 h samples.

3.2. Thermal Properties of GNP/Epoxy Nanocomposites

Figure 4 summarizes the values of the thermal conductivity for GNP nanocomposites. It is observed that an increase of GNP content induces an enhancement of the thermal conductivity, as expected, due to a higher presence of nanofiller. Here, the effect of sonication time is not so prevalent, and only induces slight differences in the thermal properties of the nanocomposites. This can be explained on the basis of the role of GNP geometry and distribution inside the material.

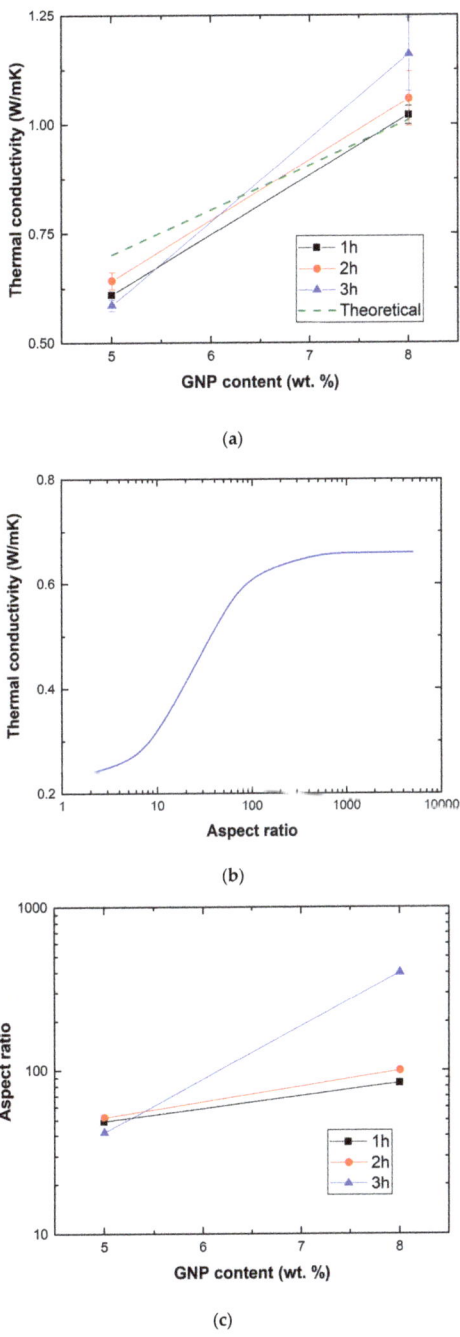

Figure 4. (**a**) Thermal conductivity measurements and theoretical estimations (dashed line), (**b**) variation of thermal conductivity for a 5 wt.% GNP sample accordingly to Hatta et al. model [40] as a function of GNP aspect ratio, and (**c**) estimation of aspect ratio accordingly to Hatta model for the different conditions.

Thermal conductivity can be estimated from Hatta et al. model [40] knowing the thermal conductivity of the epoxy and the GNPs:

$$S_{11} = S_{22} = \frac{\beta}{2 \times (\beta^2 - 1)^{\frac{3}{2}}} \times \left[\beta \times (\beta^2 - 1)^{\frac{1}{2}} - \cosh^{-1} \times \beta\right] S_{33} = 1 - 2 \times S_{11} \quad (2)$$

where β is the aspect ratio of GNPs, and S_{11}, S_{22} and S_{33} are the thermal tensors in the principal axis. In the case of a 3D randomly distribution of nanofillers, the thermal conductivity of nanocomposite k_c can be estimated from the thermal conductivity of matrix, k_m (set as 0.22 W/mK) and from the nanoreinforcement k_f (set as 100 W/mK), as well as from its volume fraction, ϕ:

$$\frac{k_c}{k_m} = 1 + \phi \times \left[\left(k_f - k_m\right) \times (2 \times S_{33} + S_{11}) + 3 \times k_m\right]/J$$
$$J = 3 \times (1 - \phi) \times \left(k_f - k_m\right) \times S_{11} \times S_{33} + k_m \cdot [3 \times (S_{11} + S_{33}) - \phi \times (2 \times S_{11} + S_{33})] + \frac{3 \times k_m^2}{(k_f - k_m)} \quad (3)$$

Therefore, the aspect ratio of the nanofillers also plays a significant role. In this context, the dashed green line in Figure 4a indicates the estimation of the thermal conductivity when supposing that the aspect ratio of GNPs is the same for every condition. It can be observed that at lower contents the model generally overestimates the value of the thermal conductivity while at higher contents the estimations are below the measured values. This can be attributed to the differences in the geometry between GNPs for each condition. In this regard, the influence of the aspect ratio on the thermal conductivity is analyzed in the graph of Figure 4b. Here, an increase of the aspect ratio leads to an increase of the thermal conductivity, which is more prevalent in a range of l/d from 10 to 1000.

Therefore, the increasing thermal conductivity with sonication time in the case of 8 wt.% GNP nanocomposites is explained by the increase of the aspect ratio due to a better correlation between the exfoliation induced by cavitation forces and the breakage of GNPs and, thus, a reduction on the lateral size. The opposite effect is observed at lower contents, as explained previously, as the sonication process is much more aggressive and, thus, at higher sonication time there is a more prevalent breakage of GNPs in comparison to exfoliating effect. In this context, the graph of Figure 4c shows the prediction of the aspect ratio of GNPs by adjusting the theoretical model to the experimental measurements. Here, the reduction of the aspect ratio due to a very aggressive breakage of GNPs can be stated when increasing the sonication time at lower contents, whereas the opposite effect is clearly seen at higher contents, validating the previous statements.

3.3. Analysis of Strain Monitoring Capabilities

Figure 5 summarizes the measured gauge factor at a low strain level ($\varepsilon \sim 0.0025$) for the different GNP nanocomposites. It can be noticed that GF values show the opposite trend when compared to the electrical conductivity measurements in Figure 2. This is explained by understanding the role of tunneling mechanisms inside the material. According to Simmons [41], the electrical resistance associated with tunneling mechanisms, R_{tunnel}, follows an exponential trend with the distance between adjacent nanoparticles, also called tunneling distance, t:

$$R_{tunnel} = \frac{h^2 t}{Ae^2 \sqrt{2m\varphi}} exp\left(\frac{4\pi t}{h} \sqrt{2m\Phi}\right) \quad (4)$$

where h is Planck's constant, m and e are the electron mass and charge, A the cross-sectional area of GNPs, and φ the height barrier of the matrix.

Therefore, the higher the tunneling distance, the more prevalent the exponential effect of tunneling resistance is. For this reason, lower values of conductivity, which imply higher values of tunneling resistance, usually lead to higher values of sensitivity, as seen in several studies [20,42].

Moreover, the electromechanical response of the samples shows a very prevalent exponential behavior, as can be seen in the graphs of Figure 5b. This is in good agreement with previous studies, where a prevalence of contact mechanisms takes places at a low strain level whereas the breakage of electrical pathways is dominant at higher strain levels, thus leading to a sharper increase of electrical resistance that is reflected in a higher GF at higher strain levels and, thus, to a very prevalent exponential response [24,42].

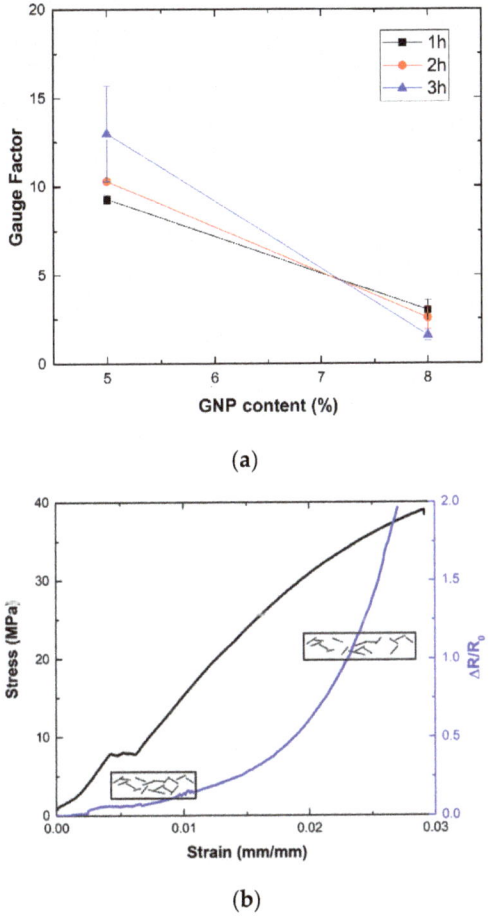

Figure 5. Electromechanic response of GNP nanocomposites showing (**a**) the measured GF and (**b**) an example of a strain-sensing curve.

In this case, the effect of sonication time can significantly affect the sensing properties of these materials. At lower contents and due to a higher efficiency of the dispersion method, there is a reduction of the aspect ratio, as commented, that leads to an increase of percolation threshold [43,44]. This higher percolation threshold implies a higher distance between adjacent nanoparticles and thus, a higher sensitivity when increasing this time. However, the opposite effect can be clearly seen at higher contents, where the highest sensitivities are observed at the lowest sonication time. Therefore, to achieve the best sensing response, the system with lower GNP content and higher sonication time will be selected.

3.4. Joule Effect Heating Analysis

Figure 6 summarizes the results of the Joule effect resistive heating tests, where the applied voltage and its corresponding average temperature reached in the sample are correlated. Here, it can be observed that both GNP content and sonication time have a significant influence in the resistive heating capacities of the samples.

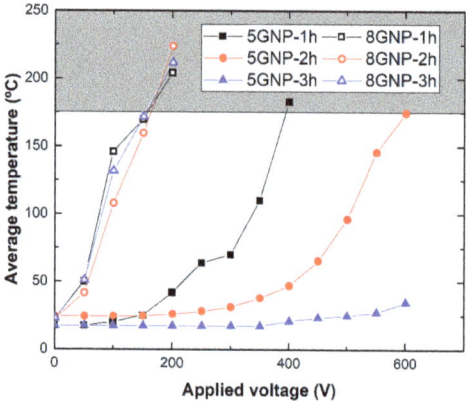

Figure 6. Average temperature reached as a function of the applied voltage for each tested condition (grey-colored area indicates the degradation zone of the epoxy resin).

On one side, by increasing the GNP content, the average temperature reached on the samples increases drastically. More specifically, the maximum allowable temperature, given by the degradation temperature of the epoxy resin (around 180–200 °C), is reached at 400–600 V under the 5GNP-1 h and 5GNP-2 h conditions. However, for 8 wt.% GNP samples, the applied voltage needed is around 150–200 V.

On the other hand, the sonication time also affects the Joule heating properties of the nanocomposites. For 5 wt.% GNP samples, an increase in the sonication time implies a drastic decrease of resistive heating capabilities. In fact, samples with a sonication time of 1 h present a limit voltage of 400 V while samples with a sonication time of 3 h do not reached the maximum allowable temperature at the range of the voltage tested. Nevertheless, sonication time does not have a prevalent effect for the samples with an 8 wt.% GNP content, where the maximum allowable temperature is reached at a similar applied voltage.

These results can be explained accordingly to Joule's Law:

$$Q = i^2 \times R \times t \tag{5}$$

where Q is the generated heat during the test, i, the current flow, R, the electrical resistance of the specimen and t the time that the specimen is subjected to resistive heating.

Therefore, the electrical properties of these materials play a crucial role in their resistive heating capabilities. Here, it can be concluded that the higher the electrical conductivity of the samples, the higher the current flow, i, and thus the heat generated during the Joule's effect tests. This is in good agreement with the previously mentioned electrical conductivity results shown in Figure 2. For 5 wt.% GNP samples, there is a significant variation of electrical conductivity with sonication time which is reflected in a poor heating capability for the samples at 3 h of sonication. Moreover, in the case of 8 wt.% GNP samples, their higher electrical conductivity thus leads to higher heating properties. Here, the differences observed among the different sonication times are less prevalent as the electrical network formed inside the material is good enough to ensure proper electrical connections between adjacent nanoparticles.

Furthermore, the Joule heating tests show very good heating capabilities in comparison to other studies with similar reinforcements and equivalent geometries [45]. Here, the main difference is correlated with the dispersion technique which, in the case of sonication, tends to form a more homogeneous dispersion inside the material without seriously affect the electrical and thermal properties of the GNPs themselves than in three roll milling process, where there is a prevalent breakage of GNPs due to the high shear forces involved in the dispersion process, leading to lower values of electrical and thermal conductivity and, thus, lower resistive heating capabilities.

3.5. Analysis of Optimum Conditions for Application

In this section, the behavior of the different manufactured samples is analyzed depending on the property tested. The aim is to select the optimum conditions depending on the desired application. In this context, Figure 7 shows a head-to-head comparison between the thermal and electrical conductivity (Figure 7a) and between SHM and Joule heating capabilities of the different samples (Figure 7b).

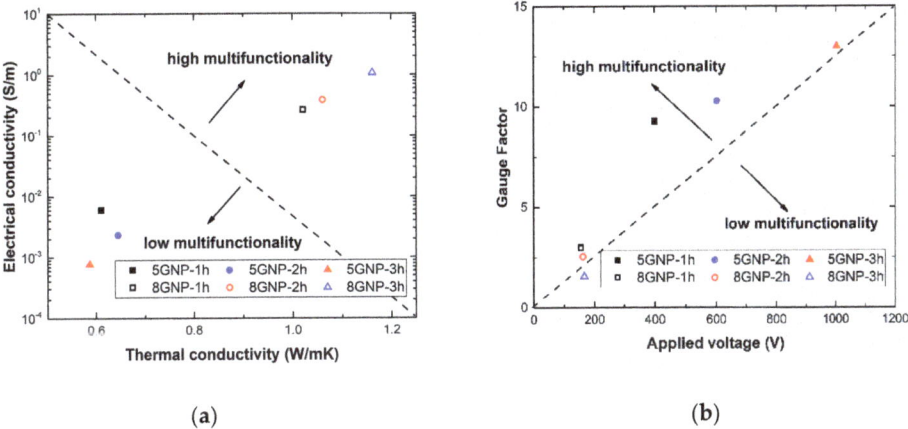

Figure 7. Graphs representing a comparison (**a**) between electrical and thermal conductivity and (**b**) between SHM capabilities, given by the Gauge Factor and Joule heating properties, given by the applied voltage to reach the maximum allowable temperature.

In the first case, specimens with higher GNP contents and sonication times show the optimum combination of properties (hollow symbols of Figure 7a). This is explained by the prevalent role of the nanofiller content along with the selection of a higher sonication time that allows a more significant exfoliating effect without any substantial detriment on electrical and thermal properties of the GNPs themselves. However, when comparing the capability for SHM applications and Joule heating ones, the selection of an optimum condition is quite a bit more complex. This is explained by the opposite effect of Joule heating capabilities, which are mainly governed by the creation of a highly conductive electrical network inside the material and SHM ones, which are dominated by tunneling mechanisms that are more prevalent in less conductive networks. Here, 5GNP-1 h samples are very competitive (black solid symbol of Figure 7b), as they have a very high electrical sensitivity to strain due to a higher prevalence of tunneling mechanisms. In addition, their electrical conductivity is high enough to allow a relatively good Joule heating effect in comparison to 2 and 3 h samples because of a better GNP dispersion inside the material without affecting the intrinsic thermal and electrical properties, as previously explained.

The selection of an optimum condition will therefore depend on the desired functionality. For these reasons, a radar chart was constructed to obtain a complete overview.

In this chart, each measured property or functionality has been rescaled from 0 to 1, where 1 denotes the highest performance for this property. Therefore, the "best" material will have a factor of 1, whereas the rest of conditions were rescaled accordingly to their value of this property.

This re-scalation follows a linear trend for Joule Effect, Gauge Factor and thermal conductivity. However, due to the highest sensitivity to small variations of electrical conductivity, it has been rescaled following a logarithmic trend, where 1 denotes again the highest measured conductivity and 0, the value of conductivity at percolation threshold, fixed at 10^{-6} S/m as observed in other studies [46,47].

Figure 8 shows the calculated values of the factors for each property and condition tested. Here, it can be observed that there is a high correspondence among electrical, thermal and Joule heating properties, whereas electrical sensitivity follows an opposite trend due to the previously commented factors. Accordingly, 5GNP-1 h seems to be a very promising solution for accomplishing all the analyzed functionalities, due to the good balance conferred by a good GNP dispersion without any detriment on nanoparticle intrinsic properties. More specifically, when compared to other works with similar nanoreinforcements, they show much higher Joule heating capabilities [45,48] and similar gauge factors at low strain levels [20,42], showing a high potential for diverse applications.

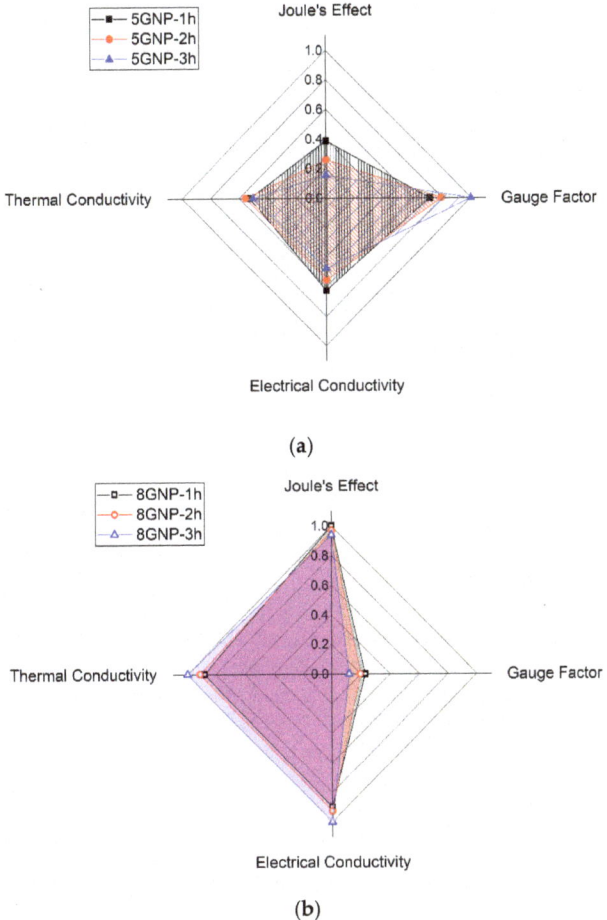

Figure 8. Radar chart of the different tested properties for (**a**) 5GNP and (**b**) 8GNP samples, scaled from 0 to 1.

4. Conclusions

Thermo-electrical and strain-sensing capabilities of GNP nanocomposites were deeply studied by varying GNP content and sonication time.

It was observed that the strain-sensing capabilities and the electrical conductivity follow an opposite trend. Here, the highest strain-sensing gauge factors have been achieved for the samples with the lower GNP content and higher sonication time, which show the lowest electrical conductivity. Furthermore, the effect of the dispersion procedure by means of sonication time on electrical and electromechanical properties was also explored. At lower GNP contents, higher sonication times induced a higher breakage of GNPs, with this effect being more prevalent than the exfoliating effect. However, at higher GNP contents and due to the higher viscosity of the mixture, the exfoliating effect is more prevalent at higher sonication times, explaining the higher values of electrical conductivity reached for these samples.

Concerning the thermal and electrothermal properties, a similar trend to that of electrical conductivity is noticed. Here, the samples with higher GNP content show the highest thermal conductivities and Joule heating capabilities. However, dispersion procedure at higher contents does not play a crucial role, as there is a high enough percolating network to ensure good resistive heating responses.

Therefore, by comparing the measured properties, it is possible to select the optimum manufacturing conditions as a function of the desired application. In this regard, 5GNP-1 h samples show a good balance among properties, as their Joule heating capabilities are much higher than 2 and 3 h sonication samples and their sensitivity is also much higher than 8 wt.% GNP specimens. Furthermore, they are very competitive when compared to similar nanocomposites of the literature.

Author Contributions: X.F.S.-R. conceptualization, formal anaylisis, writing—original draft preparation, writing-review; A.S. methodology, formal analysis, A.J.-S. conceptualization, supervision, writing—review, funding acquisition; M.C. supervision, writing—review; A.U., funding acquisition, S.G.P. writing—review, supervision, funding acquisition; All authors have read and agreed to the published version of the manuscript.

Funding: This research was funded by the Ministerio de Economía y Competitividad of Spanish Government [PROJECT PID2019-106703RB-I00], Comunidad de Madrid Regional Government [PROJECT ADITIMAT-CM (S2018/NMT-4411)] and Young Researchers R&D Project (Ref. M2183, SMART-MULTICOAT) financed by Universidad Rey Juan Carlos and Comunidad de Madrid.

Conflicts of Interest: The authors declare no conflict of interest.

References

1. Zhu, G.; Cui, X.; Zhang, Y.; Chen, S.; Dong, M.; Liu, H.; Shao, Q.; Ding, T.; Wu, S.; Guo, Z. Poly (vinyl butyral)/Graphene oxide/poly (methylhydrosiloxane) nanocomposite coating for improved aluminum alloy anticorrosion. *Polymer* **2019**, *172*, 415–422. [CrossRef]
2. Khan, A.; Sliem, M.H.; Arif, A.; Salih, M.A.; Shakoor, R.A.; Montemor, M.F.; Kahraman, R.; Mansour, S.; Abdullah, A.M.; Hasan, A. Designing and performance evaluation of polyelectrolyte multilayered composite smart coatings. *Prog. Org. Coat.* **2019**, *137*, 105319. [CrossRef]
3. Mohammadkhani, R.; Ramezanzadeh, M.; Saadatmandi, S.; Ramezanzadeh, B. Designing a dual-functional epoxy composite system with self-healing/barrier anti-corrosion performance using graphene oxide nano-scale platforms decorated with zinc doped-conductive polypyrrole nanoparticles with great environmental stability and non-toxicity. *Chem. Eng. J.* **2020**, *382*, 122819.
4. Hintze, W.; Hartmann, D.; Schütte, C. Occurrence and propagation of delamination during the machining of carbon fibre reinforced plastics (CFRPs)—An experimental study. *Compos. Sci. Technol.* **2011**, *71*, 1719–1726. [CrossRef]
5. Pervaiz, M.; Panthapulakkal, S.; Birat, K.; Sain, M.; Tjong, J. Emerging trends in automotive lightweighting through novel composite materials. *Mater. Sci. Appl.* **2016**, *7*, 26. [CrossRef]
6. Ruoff, R.S.; Lorents, D.C. Mechanical and thermal properties of carbon nanotubes. *Carbon* **1995**, *33*, 925–930. [CrossRef]
7. Popov, V.N. Carbon nanotubes: Properties and application. *Mater. Sci. Eng.* **2004**, *43*, 61–102. [CrossRef]
8. Kuang, Y.; He, X. Young's moduli of functionalized single-wall carbon nanotubes under tensile loading. *Compos. Sci. Technol.* **2009**, *69*, 169–175. [CrossRef]
9. Wang, H.; Gao, E.; Liu, P.; Zhou, D.; Geng, D.; Xue, X.; Wang, L.; Jiang, K.; Xu, Z.; Yu, G. Facile growth of vertically-aligned graphene nanosheets via thermal CVD: The experimental and theoretical investigations. *Carbon* **2017**, *121*, 1–9. [CrossRef]
10. Patel, K.D.; Singh, R.K.; Kim, H. Carbon-based nanomaterials as an emerging platform for theranostics. *Mater. Horizons* **2019**, *6*, 434–469. [CrossRef]
11. Prolongo, S.; Moriche, R.; Jiménez-Suárez, A.; Sánchez, M.; Ureña, A. Advantages and disadvantages of the addition of graphene nanoplatelets to epoxy resins. *Eur. Polym. J.* **2014**, *61*, 206–214. [CrossRef]
12. Nistal, A.; Garcia, E.; Pérez-Coll, D.; Prieto, C.; Belmonte, M.; Osendi, M.I.; Miranzo, P. Low percolation threshold in highly conducting graphene nanoplatelets/glass composite coatings. *Carbon* **2018**, *139*, 556–563. [CrossRef]
13. Chen, J.; Han, J.; Xu, D. Thermal and electrical properties of the epoxy nanocomposites reinforced with purified carbon nanotubes. *Mater. Lett.* **2019**, *246*, 20–23. [CrossRef]
14. Chu, K.; Jia, C.; Li, W. Effective thermal conductivity of graphene-based composites. *Appl. Phys. Lett.* **2012**, *101*, 121916. [CrossRef]

15. Vertuccio, L.; Guadagno, L.; Spinelli, G.; Lamberti, P.; Zarrelli, M.; Russo, S.; Iannuzzo, G. Smart coatings of epoxy based CNTs designed to meet practical expectations in aeronautics. *Compos. Part B Eng.* **2018**, *147*, 42–46. [CrossRef]
16. Abbasi, H.; Antunes, M.; Ignacio Velasco, J. Recent advances in carbon-based polymer nanocomposites for electromagnetic interference shielding. *Prog. Mater. Sci.* **2019**, *103*, 319–373. [CrossRef]
17. Rafiee, M.; Nitzsche, F.; Laliberte, J.; Hind, S.; Robitaille, F.; Labrosse, M.R. Thermal properties of doubly reinforced fiberglass/epoxy composites with graphene nanoplatelets, graphene oxide and reduced-graphene oxide. *Compos. Part B Eng.* **2019**, *164*, 1–9. [CrossRef]
18. Yang, Y.; Luo, C.; Jia, J.; Sun, Y.; Fu, Q.; Pan, C. A wrinkled Ag/CNTs-PDMS composite film for a high-performance flexible sensor and its applications in human-body single monitoring. *Nanomaterials* **2019**, *9*, 850. [CrossRef]
19. Coppola, B.; Di Maio, L.; Incarnato, L.; Tulliani, J. Preparation and characterization of polypropylene/carbon nanotubes (PP/CNTs) nanocomposites as potential strain gauges for structural health monitoring. *Nanomaterials* **2020**, *10*, 814. [CrossRef]
20. Moriche, R.; Sanchez, M.; Jimenez-Suarez, A.; Prolongo, S.G.; Urena, A. Strain monitoring mechanisms of sensors based on the addition of graphene nanoplatelets into an epoxy matrix. *Compos. Sci. Technol.* **2016**, *123*, 65–70. [CrossRef]
21. Zha, J.; Zhang, B.; Li, R.K.Y.; Dang, Z. High-performance strain sensors based on functionalized graphene nanoplates for damage monitoring. *Compos. Sci. Technol.* **2016**, *123*, 32–38. [CrossRef]
22. Ke, K.; Bonab, V.S.; Yuan, D.; Manas-Zloczower, I. Piezoresistive thermoplastic polyurethane nanocomposites with carbon nanostructures. *Carbon* **2018**, *139*, 52–58. [CrossRef]
23. Sánchez-Romate, X.F.; Artigas, J.; Jiménez-Suárez, A.; Sánchez, M.; Güemes, A.; Ureña, A. Critical parameters of carbon nanotube reinforced composites for structural health monitoring applications: Empirical results versus theoretical predictions. *Compos. Sci. Technol.* **2019**, *171*, 44–53. [CrossRef]
24. Sánchez, M.; Moriche, R.; Sánchez-Romate, X.F.; Prolongo, S.G.; Rams, J.; Ureña, A. Effect of graphene nanoplatelets thickness on strain sensitivity of nanocomposites: A deeper theoretical to experimental analysis. *Compos. Sci. Technol.* **2019**, *181*, 107697. [CrossRef]
25. Han, S.; Chand, A.; Araby, S.; Car, R.; Chen, S.; Kang, H.; Cheng, R.; Meng, Q. Thermally and electrically conductive multifunctional sensor based on epoxy/graphene composite. *Nanotechnology* **2020**, *31*, 075702. [CrossRef]
26. Hu, N.; Karube, Y.; Yan, C.; Masuda, Z.; Fukunaga, H. Tunneling effect in a polymer/carbon nanotube nanocomposite strain sensor. *Acta Mater.* **2008**, *56*, 2929–2936. [CrossRef]
27. Hashemi, R.; Weng, G.J. A theoretical treatment of graphene nanocomposites with percolation threshold, tunneling-assisted conductivity and microcapacitor effect in AC and DC electrical settings. *Carbon* **2016**, *96*, 474–490. [CrossRef]
28. Spinelli, G.; Lamberti, P.; Tucci, V.; Guadagno, L.; Vertuccio, L. Damage Monitoring of Structural Resins Loaded with Carbon Fillers: Experimental and Theoretical Study. *Nanomaterials* **2020**, *10*, 434. [CrossRef]
29. Nan, C.; Liu, G.; Lin, Y.; Li, M. Interface effect on thermal conductivity of carbon nanotube composites. *Appl. Phys. Lett.* **2004**, *85*, 3549–3551. [CrossRef]
30. Wang, T.; Tsai, J. Investigating thermal conductivities of functionalized graphene and graphene/epoxy nanocomposites. *Comput. Mater. Sci.* **2016**, *122*, 272–280. [CrossRef]
31. Chandrasekaran, S.; Seidel, C.; Schulte, K. Preparation and characterization of graphite nano-platelet (GNP)/epoxy nano-composite: Mechanical, electrical and thermal properties. *Eur. Polym. J.* **2013**, *49*, 3878–3888. [CrossRef]
32. Kashi, S.; Gupta, R.K.; Kao, N.; Hadigheh, S.A.; Bhattacharya, S.N. Influence of graphene nanoplatelet incorporation and dispersion state on thermal, mechanical and electrical properties of biodegradable matrices. *J. Mater. Sci. Technol.* **2018**, *34*, 1026–1034. [CrossRef]
33. Ahmadi-Moghadam, B.; Taheri, F. Effect of processing parameters on the structure and multi-functional performance of epoxy/GNP-nanocomposites. *J. Mater. Sci.* **2014**, *49*, 6180–6190. [CrossRef]
34. Sánchez-Romate, X.F.; Sáiz, V.; Jiménez-Suárez, A.; Campo, M.; Ureña, A. The role of graphene interactions and geometry on thermal and electrical properties of epoxy nanocomposites: A theoretical to experimental approach. *Polym. Test.* **2020**, *90*, 106638.

35. Ghaleb, Z.; Mariatti, M.; Ariff, Z. Properties of graphene nanopowder and multi-walled carbon nanotube-filled epoxy thin-film nanocomposites for electronic applications: The effect of sonication time and filler loading. *Compos. Part A Appl. Sci. Manuf.* **2014**, *58*, 77–83. [CrossRef]
36. Hamidinejad, M.; Zhao, B.; Zandieh, A.; Moghimian, N.; Filleter, T.; Park, C.B. Enhanced Electrical and Electromagnetic Interference Shielding Properties of Polymer-Graphene Nanoplatelet Composites Fabricated via Supercritical-Fluid Treatment and Physical Foaming. *ACS Appl. Mater. Interfaces* **2018**, *10*, 30752–30761. [CrossRef]
37. Moriche, R.; Prolongo, S.G.; Sánchez, M.; Jiménez-Suárez, A.; Sayagués, M.J.; Ureña, A. Morphological changes on graphene nanoplatelets induced during dispersion into an epoxy resin by different methods. *Compos. Part B Eng.* **2015**, *72*, 199–205. [CrossRef]
38. Atif, R.; Wei, J.; Shyha, I.; Inam, F. Use of morphological features of carbonaceous materials for improved mechanical properties of epoxy nanocomposites. *RSC Adv.* **2016**, *6*, 1351–1359. [CrossRef]
39. Frømyr, T.R.; Hansen, F.K.; Olsen, T. The optimum dispersion of carbon nanotubes for epoxy nanocomposites: Evolution of the particle size distribution by ultrasonic treatment. *J. Nanotechnol.* **2012**, *2012*, 1–14. [CrossRef]
40. Hatta, H.; Taya, M.; Kulacki, F.; Harder, J. Thermal diffusivities of composites with various types of filler. *J. Compos. Mater.* **1992**, *26*, 612–625. [CrossRef]
41. Simmons, J.G. Generalized formula for the electric tunnel effect between similar electrodes separated by a thin insulating film. *J. Appl. Phys.* **1963**, *34*, 1793–1803. [CrossRef]
42. Moriche, R.; Jiménez-Suárez, A.; Sánchez, M.; Prolongo, S.G.; Ureña, A. Sensitivity, influence of the strain rate and reversibility of GNPs based multiscale composite materials for high sensitive strain sensors. *Compos. Sci. Technol.* **2018**, *155*, 100–107. [CrossRef]
43. Li, J.; Kim, J. Percolation threshold of conducting polymer composites containing 3D randomly distributed graphite nanoplatelets. *Compos. Sci. Technol.* **2007**, *67*, 2114–2120. [CrossRef]
44. Sanchez-Romate, X.F.; Jimenez-Suarez, A.; Sanchez, M.; Guemes, A.; Urena, A. Novel approach to percolation threshold on electrical conductivity of carbon nanotube reinforced nanocomposites. *RSC Adv.* **2016**, *6*, 43418–43428. [CrossRef]
45. Redondo, O.; Prolongo, S.G.; Campo, M.; Sbarufatti, C.; Giglio, M. Anti-icing and de-icing coatings based Joule's heating of graphene nanoplatelets. *Compos. Sci. Technol.* **2018**, *164*, 65–73. [CrossRef]
46. Bauhofer, W.; Kovacs, J.Z. A review and analysis of electrical percolation in carbon nanotube polymer composites. *Compos. Sci. Technol.* **2009**, *69*, 1486–1498. [CrossRef]
47. Nan, C.; Shen, Y.; Ma, J. Physical properties of composites near percolation. *Annu. Rev. Mater. Res.* **2010**, *40*, 131–151. [CrossRef]
48. Prolongo, S.G.; Moriche, R.; Jimenez-Suarez, A.; Delgado, A.; Urena, A. Printable self-heating coatings based on the use of carbon nanoreinforcements. *Polym. Compos.* **2020**, *41*, 271–278. [CrossRef]

Publisher's Note: MDPI stays neutral with regard to jurisdictional claims in published maps and institutional affiliations.

© 2020 by the authors. Licensee MDPI, Basel, Switzerland. This article is an open access article distributed under the terms and conditions of the Creative Commons Attribution (CC BY) license (http://creativecommons.org/licenses/by/4.0/).

Review

Graphene-Based Scaffolds for Regenerative Medicine

Pietro Bellet [1,†], Matteo Gasparotto [1,†], Samuel Pressi [2,†], Anna Fortunato [2,†], Giorgia Scapin [3,*], Miriam Mba [2,*], Enzo Menna [2,*] and Francesco Filippini [1,*]

1. Department of Biology, University of Padua, 35131 Padua, Italy; pietro.bellet@studenti.unipd.it (P.B.); matteo.gasparotto.1@phd.unipd.it (M.G.)
2. Department of Chemical Sciences, University of Padua & INSTM, 35131 Padua, Italy; samuel.pressi@unipd.it (S.P.); anna.fortunato.1@phd.unipd.it (A.F.)
3. Department of Medicine, Harvard Medical School, Boston, MA 02115, USA
* Correspondence: giorgia.scapin17@gmail.com (G.S.); miriam.mba@unipd.it (M.M.); enzo.menna@unipd.it (E.M.); francesco.filippini@unipd.it (F.F.)
† Equal contributions.

Abstract: Leading-edge regenerative medicine can take advantage of improved knowledge of key roles played, both in stem cell fate determination and in cell growth/differentiation, by mechanotransduction and other physicochemical stimuli from the tissue environment. This prompted advanced nanomaterials research to provide tissue engineers with next-generation scaffolds consisting of smart nanocomposites and/or hydrogels with nanofillers, where balanced combinations of specific matrices and nanomaterials can mediate and finely tune such stimuli and cues. In this review, we focus on graphene-based nanomaterials as, in addition to modulating nanotopography, elastic modulus and viscoelastic features of the scaffold, they can also regulate its conductivity. This feature is crucial to the determination and differentiation of some cell lineages and is of special interest to neural regenerative medicine. Hereafter we depict relevant properties of such nanofillers, illustrate how problems related to their eventual cytotoxicity are solved via enhanced synthesis, purification and derivatization protocols, and finally provide examples of successful applications in regenerative medicine on a number of tissues.

Keywords: graphene; graphene oxide; reduced graphene oxide; tissue regeneration; 2D-scaffolds; hydrogels; fibers; stem cell differentiation

1. Introduction

Graphene consists of an atomic honeycomb lattice composed of carbon atoms that can be considered as an indefinite large polycyclic aromatic hydrocarbon with an infinite number of condensed benzene rings. Graphene family is constituted by several derivatives such as graphene oxide (GO), reduced graphene oxide (RGO), graphene quantum dots (GQDs), graphene nanosheets, monolayer graphene, and few-layer graphene [1]. A schematic representation of graphene-based materials (GBMs) taken into account in this review is shown in Figure 1. Although an accurate description of the state of the art in GBM synthesis is out of the scope of this review, a brief outline is provided in Section 2.1. It is vital to stress out that GBMs are highly heterogenous, especially when considering biological properties and applications. Therefore, careful choice of the synthetic method is required to obtain a material with the desired properties (i.e., dimensions, conductivity and eventual functional groups).

Due its high electrical conductivity, mechanical properties and aspect ratio, graphene has become attractive in many fields. In addition to being a rising star in scientific fields other than biology and medicine, graphene, GBMs and composites are widely used for important biotechnological and biomedical applications. Almost all graphene derivatives and composites are being used and tuned to develop special delivery carriers for theranostics [2], gene therapy and drug delivery, and a huge number of examples have been reviewed in recent years [3–6]. Therefore, we can just list here a few examples of applications in biosensing and bioimaging, before moving to the focus of this review, which is regenerative medicine.

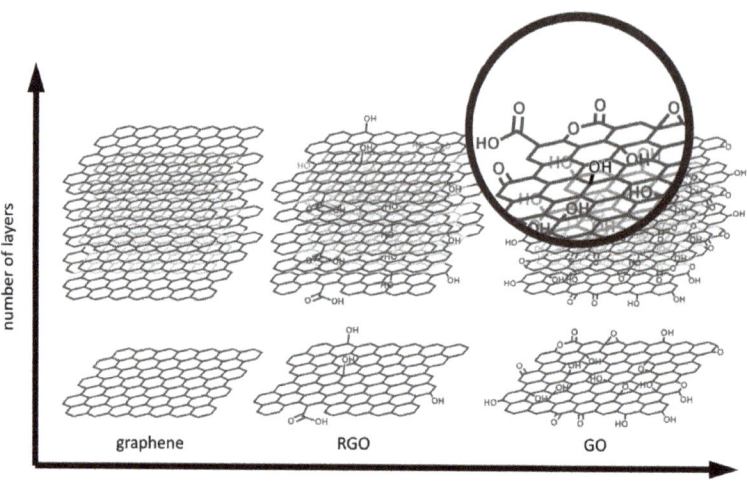

Figure 1. Structural overview of graphene-based materials.

Conductivity and high transporter capability of graphene allow for tuning biosensor surface features and outperforming many other biosensor types in terms of speed, accuracy, specificity, selectivity and sensitivity. In general, proteins (either catalysts or receptors/ligands) are associated to the graphene-based biosensor surface via electrostatic interaction, covalent bond or by polymer mediated capture. To avoid electrostatic interaction may alter the protein conformation, GBMs are combined with polymers in nanocomposites where mild electrostatics combines efficient binding and maintenance of the original conformation [7]. When instead the conformational dynamics of proteins has to be studied, multilayer graphene nanopore sensors can be used [8]. Several compounds can be detected electrochemically using graphene and GBMs as electrochemical sensor, e.g., cancer markers and cells, ATP, DNA, glucose, toxins, or even proteins [6]. Graphene-based field-effect transistor (FET) biosensors, which can be integrated with electronic chips, easing compatibility with industry standards, are especially applicable in detection of charged molecules such as DNA. Graphene-based fluorescence resonance energy transfer (FRET) biosensors are also widely used with small molecules, nucleic acids and proteins, as reviewed by Zhao et al. [9]. Some biosensors integrate graphene in the surface plasmon resonance (SPR) technology, showing improved sensitivity and detection range [10]. Graphene quantum dots (GQDs) are of special interest to bioimaging in vitro and in vivo because of their biocompatibility, tunable fluorescence with excellent photostability, ultra-small size and hydrophilicity [11]. Stable photoluminescence makes GQDs suitable for cancer bioimaging and has led to biofunctionalization for specific cancer cell imaging and real-time imaging in living cells [12]. GO and RGO are used in bioimaging as well, as their combination with different polymers (e.g. PGA), metal ions or biomolecules can modulate emissions in three main fluorescence regions (blue, green and red), making (R)GO-derived platforms suitable for multiple tracing and bio-imaging purposes [13,14].

GBMs, and especially GO and RGO find plenty of applications in tissue engineering, where they are employed as scaffolds for tissue regeneration. Tissue engineering is an interdisciplinary technology that gains insights from material chemistry, engineering, cell biology, and immunology to develop biomaterials capable of restoring, maintaining, or improving tissue function or a whole organ [15]. Scaffolds act as biological substitutes that enhance cellular interactions and are able to stimulate the differentiation of stem cells or precursor cells into the desired lineage. The extracellular environment provides biochemical, biophysical, and electrical signals, which all together define tissue-specific niches for proper tissue function and homeostasis. By recapitulating such features in biomimetic scaffolds, the goal of tissue engineering is to guide stem cell development and differentiation to resemble cell organization and behavior in the natural, tissue-specific environment. Such approach offers an interesting translational perspective for tissue repair and regeneration [16,17]. However, successfully reproducing a tissue is extremely challenging since a number of different aspects must be taken into account. In this scenario, nanocomposite materials have proven to be effective in mimicking the required characteristics.

Graphene-based scaffolds (GBSs) are a particular class of scaffold made from graphene, GO and/or RGO nanocomposites. Among the plethora of nanomaterials available, graphene and its derivatives are attractive candidates for developing tissue engineering scaffolds thanks to their tuneable electrical conductivity, excellent mechanical properties, biocompatibility, chemically modifiable surface, and nanoscale dimension matching cell surface receptors and extracellular matrix (ECM) nanoroughness/nanotopography. Morover, they display good capacity to adsorb proteins from the serum (e.g., fibronectin, laminin and albumin), favoring cell adhesion, proliferation and differentiation.

Graphene structural features and dimensions resemble many components of the extracellular environment such as proteins of the ECM (e.g. collagen), ion channels, signalling proteins and cytoskeletal elements [18]. Therefore, the introduction of graphene or its derivatives into polymeric scaffolds endows them with features that can be tailored to match the ones of the natural tissue of interest. For instance, each tissue has specific mechanical and electrical properties that should be matched by artificial scaffolds. Intuitively, scaffolds for bone regenerative medicine should be stiffer ($E > 10^9$ Pa), whereas nervous tissue requires much softer supports ($E < 4 \cdot 10^2$ Pa) and muscles need substrate with intermediate stiffness ($E > 10^4$ Pa) [19].

Being one of the toughest and strongest nanomaterials discovered so far, graphene incorporation into polymeric scaffolds enhances their mechanical properties, toughness and tensile strength [20]. Therefore, graphene percentage within the scaffold can be modulated in order to better mimic the ECM mechanical properties of the tissue of interest. Moreover, graphene nanocomposite scaffolds are endowed with nanoroughness, which contributes to cell anchoring while modulating cell morphology [18]. This property is particularly important for the differentiation of neuronal cells as graphene establishes tight contact with the growth cone and guides the spreading of developing neurites [21]. Lastly, empirical evidence suggests that engineering the electrical conductivity of the scaffold plays a crucial role in producing a functional electroactive tissue. Since graphene is electrically conductive and its conductivity is stable in biological environments, its incorporation in polymeric scaffolds can reduce the polymer electrical resistance. As a result, graphene-based scaffolds can be used to mimic and regenerate the electroactive tissues like the cardiac and neural ones, but also to boost the repair of non-excitable cells that are subjected to electrical field after an injury, like during bone repair and wound healing [22]. However, Burnstine-Townley and co-workers pointed out that the actual role of scaffold conductivity in cell differentiation is not completely clear. Specifically, disentangling the effect of a single scaffold feature on cell fate can be challenging, as varying graphene content has effect on several properties, such as surface roughness, cellular adhesion and interaction with nutrients, growth factors and wastes [23].

At a glance, graphene ability to mimic the natural extracellular environment nanotopography, to retain signalling molecules, to be easily incorporated in both natural and synthetic polymers, and to modulate stiffness and conductivity of the scaffold make it the ideal nanomaterial to provide cues needed to guide cell behaviour and hence an invaluable tool for regenerative medicine applications. Nevertheless, the toxicology profile of graphene and its derivative has not been completely elucidated yet.

Several drawbacks of GBMs employment for regenerative medicine approaches have been reported which might include membrane damage, hydrophobic interaction, oxidative stress, genotoxicity, mitochondrial disorders and autophagy. However, safety risks should be evaluated case by case based on the intrinsic properties of GBMs, such as purity, surface functional groups, lateral size, stiffness, hydrophobicity and structural defects. Moreover, several reports showed that graphene cytotoxicity is influenced by multiple parameters such as cell population tested as well as graphene dispersibility and functionalization [24,25].

2. Graphene-Based Scaffolds
2.1. Methods for GBM Synthesis

As thoroughly reviewed by Wu and co-workers graphene synthesis can be performed through a plethora of bottom-up or top-down approaches [26]. Among the most common, Chemical Vapor Deposition (CVD), Physical Vapor Deposition (PVD), spin coating, laser ablation and arch discharge needs to be mentioned. Although a systematic review of graphene synthetic methods is out of the scope of this review, it needs to be stressed that the final properties, complexity and cost of a nanomaterial are strictly related to its procedure of synthesis. Each protocol has its advantages and drawbacks, thus the choice should be done taking into consideration the final application of the product. For the sake of clarity, a brief overview of standard approaches is provided in this section.

CVD is often exploited to produce graphene for 2D composites and graphenic foams, described in the next sections. In CVD, gaseous precursors (typically hydrocarbons) are flowed at high temperatures over a metal surface, which acts as a catalyst for their decomposition and leads to the condensation of carbon atoms, forming a graphene sheet. In a typical process, graphene is grown onto a metal surface, supported with a polymer (e.g., poly(methyl methacrylate)—PMMA), and the catalyst is etched by acidic treatment. Subsequently, the graphene foil is transferred on a substrate and the supporting polymer is appropriately dissolved. The choice of metal or alloy for deposition changes process thermodynamics and kinetics, and allows to finely tune the number of graphene layers of the resulting material. The most common metal catalysts for CVD are nickel [27,28] and copper [29], with a preference for the latter due to its capability to produce single- and bi-layered graphene.

However, bulk production of graphene is more conveniently achieved starting from graphite and weakening the van der Waals forces between its stacked monoatomic carbon layers. Examples for such top-down approaches are liquid-phase exfoliation, surfactant-assisted liquid-phase exfoliation and chemical functionalization. In the first two methods [30–34], exfoliation is achieved through different combinations of factors such as (i) the choice of a solvent with proper surface tension (e.g, $\gamma = 40$ mJ m^{-2}) [35]; (ii) the use of surfactants, to minimize the interfacial tension between solvent and graphene; (iii) sonication or other external mechanical driving forces; (iv) centrifugation stages to remove thicker graphitic flakes. The principal shortcomings of these methods are the generation of defects and the reduced size attributed to sonication-induced cavitation [36,37].

Among the most common chemical top-down methods, there is the oxidation and subsequent exfoliation of graphite to GO, followed by either chemical reduction or thermal cleavage of oxidized groups to obtain RGO. In a typical procedure, graphite is mixed with sulfuric acid and oxidizing agents in an iterative and synergic action of intercalation and oxidation [38]. Subsequent exfoliation in water then easily yield GO, which can be further modified due its large amount of different oxygen functional groups (such as epoxy, hydroxyl, carbonyl and carboxyl groups). Even if the production of GO induces a large number of defects in the graphenic sp^2 network, the enhanced hydrophilicity that results from the oxidation can be beneficial for its compatibility in different types of matrixes.

Oxidation is usually carried out at 40–50 °C. However, as demonstrated by Eigler and co-workers, working at lower temperatures could reduce damages to the basal plane. They demonstrated the possibility to synthesize a minimally damaged GO with an almost intact σ framework of C atoms [39] and superior thermal properties [40] while maintaining the oxidation temperature below 10 °C, and effectively controlling kinetics of process.

RGO is obtained from GO with different synthetic methods, yielding materials with different properties. Indeed, thermal treatment (often improperly called "thermal reduction") and chemical reduction of GO to RGO do not have the same effect on graphene structure, hence on the properties of the resulting materials. The disproportionation induced by thermal treatment of highly oxidized GO brings defective holes in the plane.

Chemical reduction, on the other hand, can be achieved with different reactants [41] leading to different results: as an example, hydrazine leads to N-doping in plane, while reduction with L-ascorbic acid leaves adsorbates on RGO that are not easily removed by washing procedures [42]. Since complete reduction of GO is not achieved, RGO differs from graphene due to the presence of residual functional groups; however, O/C ratio of RGO is much lower than that of GO. Even if the sp^2 network is partially restored, the performances are still lower than those of CVD graphene.

Stability and reactivity of GO are also affected by other parameters such as pH of the dispersion, which is often neglected or underestimated. Indeed, Hirsch and coworkers [43] found evidence that the carbon lattice is damaged by treatments with a base at 40 °C while at 10 °C the partial cleavage of epoxy groups is observed. According to the above-mentioned observations, assessing the O/C ratio, which is often the only parameter considered to assess the successful synthesis of RGO, is clearly not enough to describe the obtained material. It must also be emphasized that the choice of starting materials, different methods of synthesis, and purification procedures have a direct impact on the presence of impurities, that can have a biological effect and can lead to controversial results when materials are used for bio-applications.

In the next sections, graphene-based nanocomposites will be considered. In these materials, GBMs act as fillers while the matrix is typically an organic polymer (natural or synthetic FDA-approved polymer), though bioglasses and ceramics are also used. In the first part, two-dimensional (2D) scaffolds are discussed, whereas the second part is devoted to three-dimensional (3D) scaffolds (Figure 2). In particular, three types of 3D scaffolds are considered: porous foams, fibrous scaffolds, and hydrogels.

2.2. Two-Dimensional Scaffolds

Two-dimensional scaffolds are relatively low cost and easy to fabricate, thus they are often used in preliminary studies to investigate the effect of a specific substrate on cell behavior.

The simplest example of a graphenic 2D-scaffold is represented by CVD-grown graphene (one or more layers) on a PMMA-supported metal catalyst, and then transferred onto a substrate after etching of the metal catalyst. Jangho et al. used this technique to transfer the monolayer graphene onto glass to study its effects on the reciprocal interactions between cells and substrate and to test the possible promotion of human mesenchymal stem cell (hMSC) neurogenesis and neurite outgrowth [44]. In a similar way Nayak et al. transferred CVD-grown graphene on different polymeric substrates to verify the effect of nanotopography induced by interactions between graphene and polymers. Differently from the glass control, their 2D scaffold exhibited nanoripples due to a weaker adhesion, and boosted hMSCs differentiation similarly to treatment with bone morphogenic protein BMP2 [45].

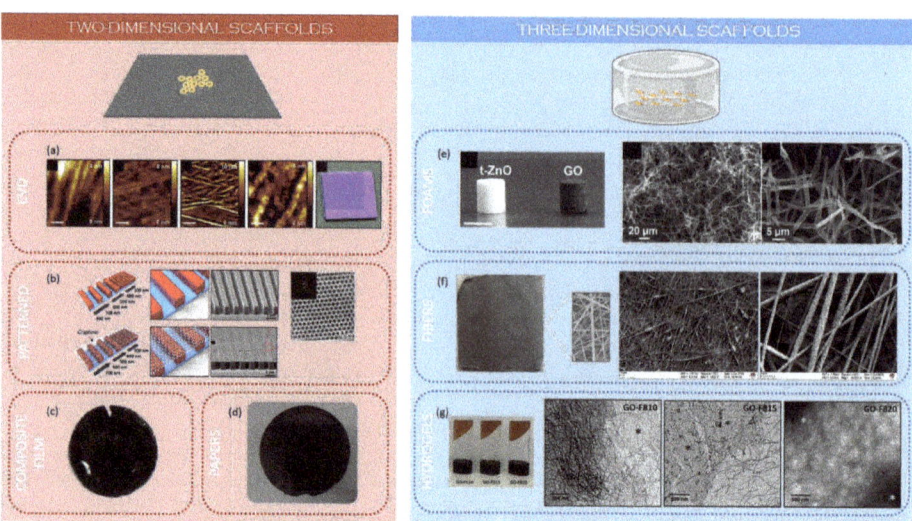

Figure 2. Examples of two-dimensional and three-dimensional scaffolds. (**a**) CVD graphene on Si/SiO$_2$ chip and AFM images of graphene transferred on different polymeric substrates [45]; (**b**) graphene transferred on nanopatterned substrate and AFM image [46]; (**c**) PLLA-RGO film obtained as reported in [47]; (**d**) graphite oxide paper [48]; (**e**) GO foams and SEM images [49]; (**f**) PLLA-RGO electrospun fibers obtained as reported in [50]; (**g**) peptide–GO hybrid hydrogels and TEM images [51].

Another method to obtain a graphene-coated surface is based on the chemical modification of a substrate to enable specific interactions with graphene-based materials (GBMs). Ryoo et al. used (3-aminopropyl)triethoxylane (APTES) to decorate the surface of glass coverslips with aminic groups. As a result, they obtained a positively charged surface which could effectively interact with negatively charged GO. Similarly, they exploited (3-glycidyloxypropyl)trimethoxylane (GPTMS) to promote glass interaction with aminated carbon nanotubes (CNT). In vitro tests proved carbon nanomaterial-coated glass to be better at promoting the number and dimensionality of focal adhesions, suggesting good biocompatibility [52].

Two-dimensional graphene-based scaffolds can also be obtained by vacuum filtration of material suspensions. For instance, Jasin and co-workers fabricated graphene-based paper as a substrate for cell growth, air drying vacuum filtrated dispersions of three different starting materials: (i) graphite oxide and graphene oxide with (ii) small and (iii) large average lateral dimensions. Although they did not observe any significant difference on cell adhesion, morphology or proliferation, the smaller release of lactate dehydrogenase (LDH) enzyme compared to control samples, suggested that their scaffold can enhance cell viability [48].

A higher degree of versatility is achieved with hybrid or composite scaffolds, where graphene is used as a filler or coating for polymeric matrices. As an example, Pandele et al. prepared chitosan/GO composites by solution blending, obtaining films with a rough surface useful for cell adhesion. The homogeneous dispersion of GO in a polymeric matrix led to an enhancement of the mechanical properties due to the large aspect ratio of the nanomaterial and its interaction with the polymer chains [53]. Furthermore, Jin et al. tested the viability of a free-standing film composed of GO and bacterial cellulose (BC) obtained from *Gluconacetobacter intermedius*. GO was added to the growth media and *G. intermedius* bio-reduction capabilities were exploited to obtain BC-RGO composites. hMSCs seeded onto these materials showed higher proliferation compared to ones seeded onto films of RGO without the fibrous structure of cellulose [54]. Li et al. fabricated RGO-cellulose paper by drop-casting GO dispersions on cellulose paper, subsequently reducing it with L-ascorbic acid (Figure 3).

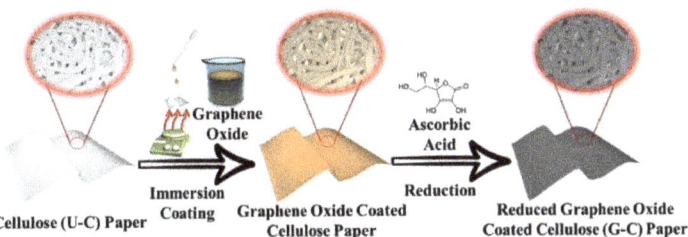

Figure 3. Assembly of RGO-cellulose hybrid paper through deposition of GO followed by in situ reduction [55].

These scaffolds showed low resistivity (~300 Ω/sq), increased mechanical strength and a specific surface micro-topography induced by RGO, which led to improved stem cell adhesion and osteogenic induction. Furthermore, their 2D-scaffolds could be employed with pseudo-3D stacked multilayered constructs that can be configured by rolling or folding, allowing designing a large number of different setups [55].

To enhance their biological effects, two-dimensional scaffolds can be micro- or nanopatterned with specific topographical cues that can direct cell growth and differentiation. Different methods have been developed to this aim, and a pattern can be drawn with either the help of a positive photoresists spin-coated on graphene oxide surface [56], or by transferring CVD graphene on a polymeric nanopatterned substrate [46,57]. This latter approach was adopted by Jangho and co-workers. They transferred a graphene layer on a poly(urethane acrylate)-patterned surface featuring regular parallel nanogrooves, thus obtaining a chemically homogeneous but mechanically heterogeneous substrate. In fact, graphene has lower mechanical properties in regions where it is suspended between nanoridges. Indeed, alignment of hMSCs along the nanotopographical cues of the substrate was observed [46].

Among the plethora of chemical studies presenting new kinds of scaffolds, there is a modest number of works specifically focused on specific GBM functionalization strategies to improve biocompatibility or differentiation capabilities. As an example, Qi et al. functionalized GO with L-theanine, an amino acid that promotes neuronal differentiation. Its presence in a poly(lactic-co-glycolic acid) (PLGA) film increased its hydrophilicity and enhanced neuronal differentiation of neuronal stem cells (NSCs) [58]. In our lab we [47,59] designed composite poly-L-lactic acid (PLLA) scaffolds with different carbon nanostructures (CNS) as filler—namely RGO, carbon nanohorns (CNH) and CNT—covalently functionalized with p-methoxyphenyl (PhOMe) groups in order to improve biocompatibility, and the electrical and mechanical properties of materials. RGO- and CNH-based scaffolds (RGO-PhOMe and CNH-PhOMe respectively) showed promising activity in enhancing the expression of myogenic markers during human circulating multipotent stem cell (hCMCs) differentiation. Moreover, electric percolation was found to take place within the considered range of RGO concentration, tough with lower performances compared to CNT-based samples. This difference is likely due to the influence of aspect ratios on electrical behavior.

Despite the aforementioned potentialities, 2D scaffolds have limitations. First of all, a two-dimensional environment is not suited to reproduce natural ECM. Then, nutrients are directly available to cells and wastes can diffuse to a limited extent. Lastly, altered cell–cell interactions may result in unpredictable cell responses. Therefore, in recent years the focus has shifted towards the study and design of 3D-scaffolds in order to overcome these limitations.

2.3. Three-Dimensional Scaffolds

As already mentioned, 3D scaffolds recapitulate tissue biophysical features thus are better candidates for in vivo applications. Scaffolds with a three-dimensional architecture should be endowed with a highly interconnected porous network. Recently, Lutzweiler and co-workers reviewed the effects of porosity, pore size and shape, interconnectivity and curvature in scaffolds used for tissue regeneration: not only these properties directly influence migration of nutrients and wastes inside the scaffold, but also the permeation and communication between cells [60]. Recent evidence suggests that scaffolds with pore diameters between 100 and 750 μm are generally beneficial while larger pores make cells experience a planar pseudo-2D environment, which differs from their natural environment [61,62].

2.3.1. Foams

The easiest method to fabricate porous scaffolds involve freeze-drying filtrates or suspensions. For example, Domínguez-Bajo et al. produced RGO foams by drying GO slurries, obtaining structures with 43% of porosity and 30 μm of pore size after thermal reduction. In addition, these scaffolds had a relatively low Young's modulus (~1.3 kPa) and made a good candidate for nervous tissue engineering. When their applicability on neural repair after spinal cord injury was tested in vivo, not only scaffolds were populated by nerve cells, but the authors also observed full vascularization [63]. In another instance, the same group exploited ice segregation-induced self-assembly, based on unidirectional freezing of dipped suspensions and lyophilization, to fabricate hierarchically channeled RGO scaffolds with controlled porosity and pore size (80% and 150 μm respectively) [64]. Liao et al. exploited a freeze-drying approach to produce a porous hybrid scaffold based on a copolymer composite of methacrylated chondroitin sulfate (CSMA) and poly(ethylene glycol) methyl ether-ε-caprolactone-acryloyl chloride (PECA) with GO, synthesized by heat initiated free radical polymerization. Not only scaffolds pore size could be tuned by CSMA:PECA ratio, but the compressive strength increased with PECA content, with values consistent with cartilage tissue. The plateau limit of conductivity (1.84 S/m) resulted at 3% GO content [65].

In a similar way, Hermenean et al. fabricated a porous chitosan/GO scaffold with improved mechanical performance—i.e., increased compressive strength and tunable Young's modulus while keeping scaffold flexibility—observing that the incorporation of 3% of GO significantly enhanced bone regeneration in vivo, compared to pure chitosan scaffolds, even in the absence of additional differentiating agents, confirming the active action of GO in facilitating cell infiltration and differentiation [66].

Graphene foams are the first porous structures composed of single layer graphene, applied in tissue engineering. Besides porosity, these scaffolds are endowed with a wrinkled topography induced by the synthetic process, which is beneficial for cell adhesion and proliferation since it better mimics the ECM [67,68]. Li et al. compared NSCs differentiation performance on 2D CVD graphene scaffolds and 3D graphene foam and observed improved proliferation and differentiation towards mature phenotypes on the latter substrate [69].

In the techniques described so far, pore size and interconnectivity depended on the Ni foam features. However, Xiao et al. recently managed to finely tailor these properties, fabricating an ordered architecture of Ni: they used photolithography to define a mask in which Ni was deposited by electroplating and aligned. Graphene was then grown through CVD on the resulting Ni template (Figure 4). Thanks to this procedure, they managed to design a scaffold with defined features by tuning pore and skeleton size (10–50 μm range), orientation angles (45° or 90°), electrical conductivity (60–80 S cm^{-1} range) and density (around 3–4 mg cm^{-3}). Such a scaffold was able to direct neuronal growth and align neurons along a defined path to form a network [70].

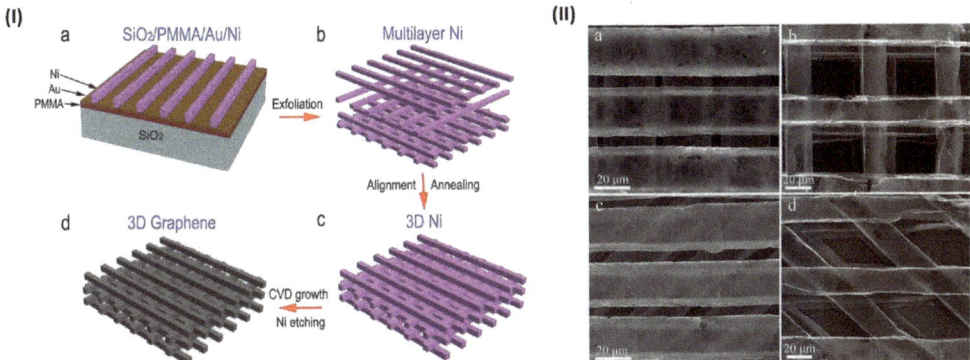

Figure 4. (**I**) (**a**–**d**) Schematic illustration of the procedure used to fabricate a 3D-CG. (**II**) (**a**–**d**) SEM images of four layer freestanding 3D-CGs with different patterns, pores, and skeleton sizes [70].

Another method to fabricate porous structures has been employed by Rasch et al. [49]. Starting from tetrapod-shaped ZnO, pressed and annealed in a mold, they were able to synthesize templates with high porosity (50 to 98%). GBM deposition was obtained by infiltrating a GO suspension in the templates, followed by chemical etching with hydrochloric acid. Their protocol allowed easy, versatile and cost-effective deposition of nanomaterials. Moreover, biological evaluations of these scaffolds by Schmitt et al. showed they could be promising for nervous tissue engineering [71].

An alternative approach to induce porosity in scaffolds is supercritical foaming which allows to control scaffold morphology through a careful choice of experimental parameters, such as chamber pressure, temperature and decompression rate. Evlashin et al. exploited this process to manufacture RGO-reinforced polycaprolactone (PCL/RGO) and PCL/GO foams in a carbon dioxide atmosphere. Although the presence of RGO in the polymer matrix led to an increase of pore size, those foams showed poor cell adhesion properties. Conversely, they found PCL/GO scaffolds to enhance cell adhesion. However, both composites displayed lack of interconnected porosity, resulting in cells attaching only on scaffold surface [72]. Polymer-enriched hybrids can also be obtained starting from CVD graphene foams, by depositing the polymer from a solution through spin or dip coating. Resulting scaffolds show improved mechanical performances and cellular responses. In order to retain porosity, it is crucial to avoid pore saturation through fine optimization of dip coating time and by choosing a polymer with a favorable, near zero contact angle. Nieto et al. exploited this technique with a copolymer of polylactic acid (PLA) and poly-ε-caprolactone (PCL) and achieved improved tensile strength due to filling of the pre-existing microcracks in pristine G foams. In vitro tests demonstrated these materials are able to support hMSCs viability and differentiation, making them suitable for musculoskeletal tissue engineering [73].

A layer-by-layer (LBL) technique was followed by Song and co-workers who deposited a positively charged polymer, poly(diallyl dimethylammonium) chloride (PDDA) on a negatively charged Ni template and subsequently placed negatively charged GO onto its surface, which was then thermally converted to RGO. Electrochemical deposition of polypyrrole (PPY) and hydroxyapatite (HA) on top, increased scaffold roughness and surface area, favoring cell adhesion and proliferation as confirmed by *in vitro* tests on the pre-osteoblast cell line MC3T3-E1 [74].

Besides metallic templates, polymeric organic foams are used for polymer replication technique, especially in the inorganic scaffold field. Deliormanlı et al. used polyurethane foam to fabricate HA scaffolds, eliminating the template and sintering HA by heat treatment. PCL/GO was added by dip coating, leading to a scaffold with improved mechanical performance and higher bioactivity [75].

The same procedure can be applied to bioactive glass, another important class of useful scaffolds in tissue engineering. As an example, Turk et al. incorporated 10% graphene directly in the glass matrix before sintering borate-based porous scaffolds, doubling the compressive strength and obtaining an electrical conductivity (0.060 S/cm) which could be exploited to electrically stimulate cell growth [76]. Moreover, Deliormanlı et al. fabricated more chemically stable and biocompatible silicate-based scaffolds coated with PCL/graphene with pore size between 100–500 μm, without detrimental effects of polymer coating on pore structure [77].

An alternative approach to porous structure design is 3D printing, which allows to accurately control scaffold geometrical features without the limitation of using a template. Jakus et al. exploited a PLGA-based ink where they incorporated graphene with the use of surfactants and plasticizers. The mechanical properties of composites are affected by graphenic particles, with an increase of elastic modulus to a value of 16 MPa at 20 vol% loading of graphene, but with detrimental effects at higher loadings (40–60 vol% of graphene). In addition, they observed an anisotropic alignment of graphene flakes, enhancing electrical conductivity due to shear forces produced during the 3D printing extrusion process, which increased with the decrease of the nozzle diameter [78].

Cabral et al. used extrusion 3D printing to produce multicomponent scaffolds, based on tricalcium phosphate chitosan and gelatin, which mimicked the inorganic and organic components of bones, respectively. GO was added to this blend and reduced to RGO in situ by L-ascorbic acid treatment. When comparing mechanical properties of scaffolds incorporating GO or RGO they found the latter to better mimic bone Young modulus, thus their scaffold might be useful as a temporary support for bone regeneration [79].

2.3.2. Electrospun Fibers

Fiber-based scaffolds are largely employed in tissue engineering because they intrinsically resemble the microstructure of natural tissues. Fiber diameter, porosity, and orientation are the main features that influence cell growth and tissue regeneration [80]. One of the most common techniques to produce continuous fibers is electrospinning. Electrospinning offers several advantages, including (i) ease of processing, (ii) possibility of large-scale production (iii) availability of advanced modes [81]. Moreover, it is highly versatile and electrospun fibers can be deposited in a random orientation or in an aligned fashion which enhances cell alignment and elongation along the contacted fiber direction [82]. Most thermoplastic materials can be electrospun by fine-tuning the properties of the polymeric solution and the electrospinning parameters such as voltage, electrodes distance and flow rate. The American Food and Drug Administration (FDA) approved several thermoplastic biomaterials for in vivo implantation. Nevertheless, their applications are restricted by the high hydrophobicity, low mechanical properties, lack of specific interactions with cells and sometimes relative slow in vivo degradation rate. Luckily, these limits can be easily overcome by introducing proper nanofillers, and several examples of electrospun thermoplastic materials reinforced with GBMs have been reported [83–85]. As highlighted by Song et al., the solubility of the filler strongly influences the mechanical properties of the final material: a poor dispersibility or a too-high loading leads to aggregation, which results in fractures and disconnections along the nanofibers. It has been observed that electrospinning graphene-based composites yields thinner fibers (Figure 5), but on the other hand, even a small amount of GO or RGO inside the electrospun fibers reinforces their structure and overcomes the detrimental effect of a reduced diameter on mechanical properties. Therefore, it is crucial to finely tune the CNS content in order to find the right balance between a uniform CNS dispersion, nanofiber diameter and reinforcement effect. Besides, it has been widely demonstrated that incorporation of CNS in fibrous scaffolds results in an improved biomimetic microenvironment that enhances cell adhesion and proliferation on different cell types [84].

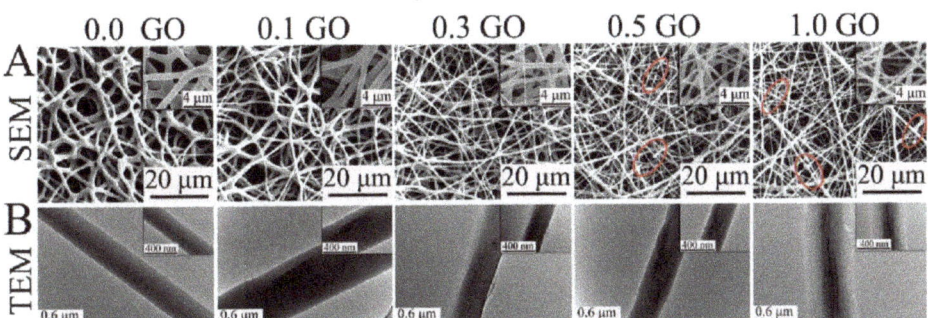

Figure 5. Surface morphological and constructional images of PCL/GO composite nanofibers with different GO concentrations (wt%): (**A**) SEM images and (**B**) TEM images. The red ellipses in SEM images are the fractures [84].

Generally, the smaller thickness induced by graphene-based nanofillers on electrospun fibers allows mimicking the structure of ECM even better. It is believed that the effect of GBMs on fiber diameter is due to the electrical conductivity of the feeder solution, which is a key factor in determining the diameter and size distribution of the electrospun fibers. Moreover, it is reported that fiber diameter is highly correlated to the viscosity of the feeder solution. Scaffaro and colleagues pointed out that the decrease of viscosity of a PCL solution by addition of GO induces electrospinning of thinner fibers. On the other hand, they observed an opposite effect on viscosity (and fiber diameter) with GO-grafted-PEG (GO-g-PEG). Functionalization of the filler not only increased fiber diameter, but also improved dispersion of the filler and maximized the filler/matrix interfaced area, making GO-g-PEG more effective than GO in reinforcing composite fibers, in particular at low concentration [86].

In 2019, Basar and co-workers developed a PCL/GO composite scaffold [87] by functionalizing GO with either an RGD-peptide (GRGDSP), thiophene (Th) or both. Besides having the aforementioned effect on fiber diameter, GO functionalization yielded an enhanced electrical behavior to the scaffold, with conductivities reaching 15.06 $\mu S\ cm^{-1}$ in PCL/GO-GRGDSP-Th (2% of GO), a 15-fold increase compared to neat PCL (0.95 $\mu S\ cm^{-1}$). However, while scaffolds with higher content of GO (2%) showed higher electrical performances, the elastic modulus and tensile strength of 0.5% GO-scaffolds were found to be higher. Once again, this result was associated with the uniform dispersion of GO in the polymer matrix. Interestingly, this modification resulted in an increment of both electric conductivity and mechanical stability due to the ability of sulfur moieties to enable the crosslink between GO and PCL [87].

Scaffold properties can also be altered by combining different organic or inorganic fillers. Lui et al. developed electrospun PLA scaffolds reinforced with GO (1–3 wt %) and/or nano-HA (15 wt %). Interestingly, addition of 15 wt % nano-HA improved both elastic modulus and tensile strength, whereas concentrations of GO above 2% diminished them due to filler aggregation. Nanofiller addition slightly increased scaffold glass transition temperature and modified the hydrophobicity of PLA, enhancing the polymer water uptake, which in turn assisted cell adhesion and proliferation [88].

Different strategies have been developed to obtain polymeric nanofiber scaffolds based on graphene and its derivatives. However, nanocomposites fail to provide a pure graphene interface. An alternative approach aims to immobilize nanostructures on the surface of polymeric nanofibers. The surface of aliphatic polyesters such as PCL and PLA can easily be functionalized with hydroxyl and amino groups by treating the polymeric scaffold with a diamine solution. In tissue engineering, aminolysis of polyesters improves their interactions with cells and allows them to form a stable graphenic coating [89]. Recently, Jalili-Firoozinezhad et al. reported an easy method to generate electrically conductive nanofibers by coating a PCL nanofibrous mat with GO liquid crystals, which were then reduced to RGO to form PCL-templated graphene nanofibers [90]. Proper electrical conductivity and nanofibrous topography of these constructs make them an ideal platform for cell culture, tissue engineering, drug delivery, and biosensor applications. Preliminary in vitro analyses using hMSCs revealed no induced cytotoxicity and confirmed an enhanced cellular metabolism and proliferation rate compared to standard culture plates and PCL nanofibers.

Indeed, coated fibers can be obtained without any surface treatment. Wang et al. developed a conductive graphene-based fibrous scaffold by coating RGO via an in situ redox reaction of GO on the surface of silk fibroin/poly(L-lactic acid-co-caprolactone) (ApF/PLCL) composite nanofibers [91]. The authors highlighted that the coating did not affect the nanoscale topography of the scaffold and enhanced its mechanical properties, electroactivity and biocompatibility. They then investigated how these conductive scaffolds regulated in vitro and in vivo cell behavior and differentiation under electrical stimulation. RGO-coated ApF/PLCL scaffolds boosted cell migration, proliferation and myelin gene expression of Schwann cells (SCs), whereas pheochromocytoma-derived PC12 cells cultured on these scaffolds exhibited enhanced differentiation. In vivo implantation of the constructs promoted peripheral nerve regeneration in rats.

Polymer core-CNS shell fibers can be obtained by electrospinning the polymer into a solution of graphene or one of its derivatives. Subsequently, it is possible to further functionalize or reduce the shell. Jin et al. exploited this principle to develop an RGO core-shell nanofiber (RGO-CSNFM) [92]. The RGO core-shell structure displayed high mechanical, electrical conductivity (10.0 S cm^{-1}) and a charge carrying capacity. This property is likely due to both RGO-CSNFM large surface areas and the extended π–π conjugated bond network generated over the surface of the RGO shell layer. Wu et al. developed an LBL method to coat electrospun nanofibers that mimic vascular ECM and enhance proliferation of endothelial cells. PLLA surface modification was achieved via electrostatic LBL self-assembly by alternately immersing PLLA fibers in a positively charged solution of 0.1 wt% chitosan and a negatively charged solution of 0.1 wt% heparin (PLLA-CS/Hep) or 0.1 wt% heparin/graphite oxide (PLLA-CS/Hep/GO). After the LBL coating, the hydrophilicity and mechanical properties of the modified PLLA nanofibers were greatly enhanced. Moreover, the CS/Hep/GO coating positively influenced cell attachment, viability, and proliferation of endothelial cells [93].

The versatility of electrospinning allows to obtain complex and ordered structures. A compelling example has been reported by Shao and co-workers [94] who used electrospinning to develop a 3D scaffold with multiple orthogonal aligned fibers. This peculiar architecture improved mechanical properties and decreased issues that may arise when working with parallel fibers or random networks. Moreover, a 3D structure better mimics the natural cellular environment. They developed an electrospun PLGA/silk fibroin/GO/hydroxyapatite (PLGA/TSF/GO/HA) 3D scaffold. hMSCs seeded onto these scaffolds showed enhanced proliferation and elongated morphology along the long axis of the nanofibers. Lastly, biological assays indicated that composite scaffolds enhanced osteogenesis and alkaline phosphatase activity.

In another work, Zhang and co-workers combined GO nanosheets and aligned aminolyzed PLLA nanofibers which favored nerve regeneration. The aminolysis of PLLA nanofibers allowed to form a stable GO coating. Schwann cells (SCs) cultured on these nanocomposite scaffolds displayed improved proliferation and elongation along the fiber direction compared to those grown on the aligned PLLA and aminolyzed-PLLA. The coated structure was also able to improve differentiation and neurite outgrowth of pheochromocytoma derived PC12 cell line. The authors suggest that these results may arise from the modification of surface chemistry and roughness induced by the GO coating [95].

2.3.3. Hydrogels

Hydrogels are three-dimensional entangled networks able to retain large amounts of water. Despite being mostly liquid, they display a solid-like rheological behavior and recently they have been employed as scaffolds for tissue engineering [96,97]. Hydrogels can be categorized into two main classes based on the forces involved in building the network: (i) chemical and (ii) physical gels. The network of chemical hydrogels is obtained through covalent cross-linking of its components, which generates a permanent structure. On the other hand, the structure of physical hydrogel is characterized by reversible non-covalent interactions which make these gels suitable for cell encapsulation but highly susceptible to environmental conditions (i.e., ionic strength, pH, temperature), such that even minor changes can cause the network to collapse. Indeed, physical hydrogels exhibit lower mechanical properties than their chemical counterpart. However, even chemical gels generally cannot withstand high mechanical stress despite the covalent cross-links [98].

Graphene and graphene derivatives in hydrogels may play the role of (i) self-assembling gelator molecule or (ii) filler in order to prepare multi-functional nanocomposite hydrogels. Self-assembly has been recognized as one of the most effective "bottom-up" strategies for building structured networks. Driven by non-covalent π-π interactions that arise from their 2D structure, graphene and graphene derivatives spontaneously re-organize into a 3D structure. Self-assembled hydrogels can be prepared through a one-step hydrothermal method starting from a graphene-based solution [99]. For example, Yang and colleagues have demonstrated the jellification of GO at the solution–filter membrane interface, creating highly conductive and anisotropic films [100].

The employment of pure graphene and/or graphene derivatives hydrogels is quite restricted, thus they are mainly used as high-quality nanofillers for composite hydrogels [101]. Different synthetic and natural polymers able to form hydrogels are suitable for tissue engineering scaffolds. Among synthetic polymers we may mention polyethylene glycol (PEG), poly(acrylamide), poly(lactic acid) or synthetic peptides. Natural-derived polymers such as alginate, chitosan, collagen, silk or gelatin are also widely used to fabricate hydrogel scaffolds for tissue engineering. Polymeric scaffolds display good biocompability and biodegradability but lack, for example, the ability tolerate strong mechanical forces [102,103].

Alginate is a natural polysaccharide composed of β-D-mannuroic acid (M) and α-L-guluronic acid (G) typically obtained from brown seaweed. In the presence of various divalent cation (Ca^{2+}, Mg^{2+}), alginate polymers form gels via non-covalent cross-linking of the carboxylate groups of the G blocks on the polymer backbone. Even if the concentration of crosslinker, percentage of G content and jellification time allows to tune the properties of alginate hydrogels, other limitations cannot be overcome without the use of specific fillers. Particularly, alginate-based hydrogels do not permit good control over their internal architecture, they lack cell receptors adhesion sites and suffer from low protein adsorption capability [104]. As independently highlighted by Losic et al. and Chen et al. [105,106], the introduction of GO and RGO in an alginate matrix allows to modify and control the porosity of the gel (ca. 99%±0.3%), making the pores size uniform from surface to its inner core and fostering cellular activity. GO and RGO composite gels also allow to reach the optimal swelling index required for an efficient scaffold. Investigation of mechanical and electrical properties revealed an optimum GO content of 0.1 wt%. Above this concentration a detrimental effect was observed due to an imperfect dispersion of GO within the alginate matrix.

Chitosan, as well its derivatives, is a widely available natural polymer characterized by excellent biological properties (i.e., biocompatibility, coagulation activity, biodegradability). Agarose (AG), on the other hand, is a polysaccharide obtained from red algae, displays a thermo-sensitive behavior and exhibits mechanical properties similar to that of soft tissues. However, its employment is limited by the lack of cell recognition sites.

Sivashankari and Prabaharan used GO as a nanofiller for the fabrication of agarose/chitosan (AG/CS)-based scaffold [107]. Through a freeze-drying method, they prepared 3D AG/CS/GO scaffolds with different concentrations of GO (0–1.5 wt %). GO introduced changes in the scaffold morphology, in their swelling behavior and in their water retention ability. In particular, AG/CS/GO scaffolds with 1 and 1.5 wt % of GO exhibited the highest porosity (Figure 6), with an average pore size (237–274 µm) matching the demands for bone tissue regeneration [108,109]. Even with the increase in porosity, GO likewise enhanced the mechanical properties due to interactions established between fillers and polymer matrix and favored cell attachment and proliferation. Freeze-drying techniques are widely employed for scaffold generation and also allows to obtain anisotropic scaffolds. Liu et al. developed a highly oriented hydrogel through directional freezing of CS/GO suspension on a copper plate cooled with liquid nitrogen [110]. This method produced micro-sized ice rods within the suspension, which act as template for a honeycomb-like structure resembling a bone lamellae structure. The resulting hydrogel displayed anisotropic mechanical behavior improved by the incorporation of GO and were able to guide the growth of mouse osteoblastic MC3T3-E1 cells along the longitudinal direction of the honeycomb structure.

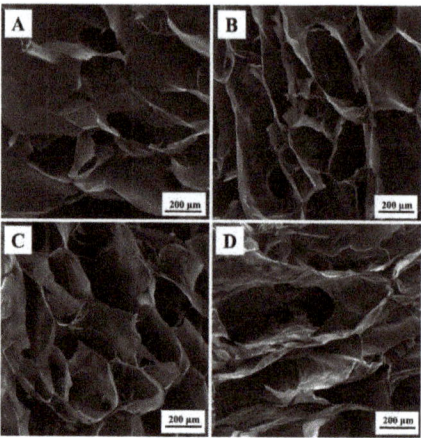

Figure 6. SEM images of (**A**) AG/CS (0% GO), (**B**) AG/CS/GO (0.5% GO), (**C**) AG/CS/GO (1% GO), and (**D**) AG/CS/GO (1.5% GO) composite scaffolds [107].

Self-assembling peptide-based hydrogels (SAPHs) have been widely employed as vehicle for drug delivery, but they can also be employed in tissue engineering due to their biocompatibility and non-immunogenic nature [104,111,112]. Ligorio et al. used GO as nanofiller in a peptide (FEFKFEFK) hydrogel for tissue engineering [51]. After conditioning with cell culture media (i.e., pH 7.4), all gels displayed an enhanced storage modulus. Bovine nucleus polposus (NP) cells were cultured on these hydrogels to assess cell viability and GO-hydrogels with shear modulus similar to the native NP showed higher viability and constant metabolic activity throughout the culture period.

Wang et al. prepared a silk fibroin scaffold incorporating exfoliated graphene [113]. An aligned silk fibroin hybrid hydrogel was obtained by application of an electric field. Even if aligned silk nanofiber gels were previously proven to be able to influence behavior such as cell orientation and migration [114], they failed to actively induce neural differentiation. The nanocomposite hydrogels displayed anisotropic mechanical properties, and the one with the highest content of graphene showed doubled parallel and orthogonal compressive moduli compared to graphene-free samples, making them suitable for nerve tissue engineering. After the addition of graphene, cell proliferation was further enhanced, indicating that graphene sheets effectively induced neurite differentiation.

In recent years, injectable hydrogels have drawn major attention since they need minimal invasive procedure to be administered and have reduced therapeutic costs. The hydrogel precursor should be injected as a controllable liquid (i.e., characterized by low viscosity) and must jellify into a robust hydrogel as quickly as possible *in situ* [115]. Finally, it is uttermost important that gelation occurs after injection and at physiological conditions (temperature and pH). Recently, Lee et al. developed an injectable GO-incorporated glycolchitosan-oxidized hyaluronic acid (gCS/oHA) hydrogel [116]. Gelation of gCS/oHA was obtained through the cross-link between the aldehyde group on oHA and the amine groups of gCS (Schiff-base reaction). Frequency sweep experiments were used to investigate the mechanical properties in a plate–plate geometry. The results showed that when the GO content increased, the G' value gradually increased too, suggesting a more robust hydrogel formation. GO may enhance polymer cross-linking through hydrogen bonding interactions [117,118]. GO-incorporated hydrogels displayed lower cytotoxicity and higher osteogenic activity compared to control both in vitro and in vivo. High levels of COL1 expression observed in cultures hinted that these injectable gels could be suitable for treating bone injuries. Saravanan et al. explored chitosan-glycerophosphate-based injectable hydrogels for treatment of bone defects [119]. Due to newly introduced non-covalent interactions, GO (0.5% w/v) composite hydrogels significantly increased swelling, protein adsorption and cell interaction compared to their GO-free counterparts. Moreover, GO introduction reduced gelation times and controlled degradation rates.

Poor dispersion of GBMs within the polymer matrix causes aggregation [105], which may be detrimental for scaffold properties. To achieve homogeneous and stable dispersions of GBMs, covalent and/or non-covalent functionalization may be required. Díez-Pascual et al. [120] fabricated poly(propylene fumarate) (PPF)-based nanocomposites reinforced with GO, non-covalently functionalized with PEG (PEG-GO). PEG functionalization reduces the aggregation tendency and cytotoxicity of GO without impairing its unique features. The presence of PEG-GO leads to a threefold increase of Young's modulus at 3% loading of filler and improved cell adhesion and growth. The results have been ascribed to the roughness of the scaffold, the hydrogen-bonded network established between GO and the polymer and the good GO dispersion inside the matrix. Polymers may be also covalently bonded to GO, for example by esterification. Noh et al. designed a graphene oxide GO covalently functionalized with acrylated polyethylene glycol (PEGA-GO) through ester formation [121]. The PEGA-GO was photopolymerized with polyethylene glycol diacrylate (PEGDA) leading to gel formation. GO-doped hydrogels boosted cell adhesion and osteogenic differentiation, though no changes were observable in swelling and mechanical properties.

Wu et al. took advantage of the abundant functional groups on the surface of nanosized GO to link starch chains via esterification [122]. Starch is a widely available and cost-effective polysaccharide, which does not release degradation products that induce inflammations in vivo. Nanosized GO was synthesized from starch through microwave-assisted degradation and then covalently bonded to the polysaccharide itself to improve its mechanical features and bioactivity. In another example, Ruan et al. crosslinked carboxymethyl chitosan (CMC) to GO by amide bond formation [123]. The obtained GO-CMC scaffolds appeared rougher than their GO-free counterparts and showed better retention properties and slower degradation rates thanks to the higher cross-linking degree compared to the GO-free and CS/GO-CS samples. Water uptake and retention rates are important parameters, since the scaffold is the vessel for nutrients and metabolites for cell activity. The authors highlighted that GO introduction also deals with the poor mechanical strength typical of bare CMC [124].

3. Stem Cell Differentiation and Mechano-Transduction
3.1. Tissue Engineering and Stem Cells

Stem cells are non-specialized cells with self-renewal potential and the ability to differentiate into various cell types if directed with appropriate stimuli, making them a powerful tool for the regeneration of injured tissues [125]. Embryonic Stem Cells (ESC) are pluripotent stem cells able to originate all the cell types of the body [126]. Despite their ideal self-renewing capabilities and differentiation potential, they are not widely used for tissue engineering studies due to the ethical restrictions of human embryo use in research. As a valid alternative, tissue engineering switched the focus to adult stem cells, which are stem cells residing throughout the body whose role is to maintain and repair the tissue in which there are found. Such cells have a limited differentiation potential compared to ESCs but offer the advantage of being isolated directly from the patient for autologous regenerative therapies. Good examples of adult stem cells are the Mesenchymal Stem Cells (MSC) and Hematopoietic Stem Cells (HSC). They both can be isolated from patient bone marrow and can regenerate bone, cartilage, and adipose tissue (MSCs), as well as the entire immune system (HSCs) [127,128]. However, some adult stem cells, like neuronal stem cells (NSCs), can be isolated only with very invasive procedures and in small quantities [129]. As of now, the most promising stem cell type for regenerative applications are the induced Pluripotent Stem Cells (iPSCs). iPSCs are generated from the "reprogramming" of somatic cells back to the pluripotent "embryonic" state [130]. Therefore, they show the same "unlimited" self-renewal and differentiation capabilities of the ESCs, with the advantage of being free from ethical issues as reprogrammed from patient or donor-matched somatic cells [131]. Challenges associated with the iPSC clinical use are (i) the difficulties in finding HLA-matched donors (especially for mixed-race patients) and (ii) the time and costs for the development of patient-derived iPSCs, particularly considering the extensive validation and stringent regulatory processes that would require each patient-derived cell line [132,133]. However, recent works proposed new strategies to engineer such iPSCs to make them "invisible" to the recipient immune system, showing that we are very close to the generation of off-the-shelf, universally compatible iPSCs for the allogenic treatment of a myriad of diseases [134–136]. The interaction between stem cells and the extracellular microenvironment is critical in controlling stem cell differentiation, as depicted Figure 7.

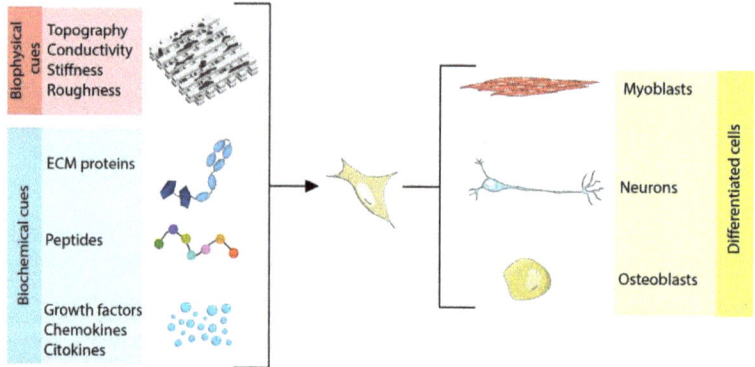

Figure 7. Schematic representation of biophysical and biochemical cues on cell differentiation.

3.2. Cues Controlling Stem Cell Behavior

Biochemical cues are provided by reciprocal interactions between the cell, soluble bioactive agents, and the ECM. Soluble molecules, such as growth factors, chemokines, and cytokines, diffuse to bind the cell surface receptors and have potent effects on cell growth, proliferation, and differentiation. Insoluble ECM macromolecules (e.g., collagens, elastin, and laminin), glycoproteins (e.g., fibronectin and vitronectin), and polysaccharides (e.g., heparan sulfate and hyaluronic acid) form a meshwork of fibers or fibrils with ECM glycoproteins incorporated into them. The resulting matrix is tissue-specific and functions as both a structural and signaling scaffold to cells [137].

Many works showed that some of the aforementioned molecules—if administrated both in vitro and in vivo—are able to elicit specific cell responses [138]; moreover, different strategies have been developed to link such proteins to biomaterial scaffolds in order to help delivery at the injured sites [139]. However, coating surfaces with recombinant proteins or native matrix macromolecules extracted from animal tissues encounters the problem of eliciting immune responses, in particular when using proteins from different species. Furthermore, their isolation and purification from native tissues or their production as recombinant proteins at a larger scale for tissue engineering purposes is expensive and subject to batch-to-batch variability [140]. For these reasons, the production of specific motifs known to mediate regulatory signals as synthetic peptides presents significant advantages compared to using entire recombinant/native tissue proteins: (i) low immunogenic activity; (ii) increased stability; (iii) low production costs; and (iv) simplified preparation and immobilization onto substrates. Moreover, peptides can be: (v) presented to cells at surface densities significantly higher than those possibly achieved with entire proteins or domains; and (vi) tailored in composition for each tissue-specific application [141]. The biomimetic peptides most used for scaffold functionalization are the ones representing the ECM protein epitopes for integrin binding and therefore promoting cell adhesion [142]. In addition, tissue-specific peptides, resembling active motifs of growth factors and transmembrane proteins, have also been used to tune the cell differentiation [143,144].

Cells are capable of sensing and responding to biophysical cues, over a wide range of length scales. Many of these cues are provided by the ECM, which acts as a cellular scaffold and is the primary extracellular component in tissues. In vivo, the ECM, through its structure and molecular composition, presents a variety of geometrically defined, three-dimensional (3D) physical cues in the submicron to micron scale, referred to as topographies. Cell response to topographies is mediated by a phenomenon called contact guidance, which is known to affect cell adhesion, morphology, migration, and differentiation [145]. Another physical cue displayed by the ECM is mechanical stiffness through which, similar to topography, a diverse set of cellular functions can be modulated. Matrix sensing requires the ability of cells to pull against the matrix and cellular mechanotransducers to generate signals based on the force that the cell must generate to deform the matrix. Mechano-sensitive pathways subsequently convert these biophysical cues into biochemical signals that commit the cells to a specific lineage [145]. For example, MSC differentiation can be modulated by substrate stiffness [146], while developing neurons are able to transduce topographical stimuli through the interaction of the growth cone with the immediate environment. Such mechanical cues direct neurite extension, ensuring appropriate and regulated connectivity within the overall neural circuitry [147].

ECM mimicry can be achieved using either natural or synthetic polymers interconnected by physical and ionic interactions and even covalent linkages [148]. Electrospun polymer fibrous substrates with controlled fiber architectures and diameter provide topographical cues to cells by presenting geometries mimetic of the scale and 3D arrangement of the collagen and laminin fibrils of the ECM. Such polymer fibers present a high surface-to-volume ratio and porosity and are hence well-suited for promoting cell adhesion, growth, and differentiation and enable growth factor/drug loading; such properties are inherent to bioactive matrix microniches [50,149]. Recent advances in 3D bioprinting strongly improved our ability to imitate natural features of ECM. As an additive manufacture technology, the 3D bioprinting allows deposition of polymers, hydrogels, cells, growth factors, and peptide active motifs by using a layer-by-layer approach to build up arrangements favorable to tissue-like structure formation, which are endowed with superior differentiating properties compared to the conventional 2D culture vessels [148].

Endogenous electrical signals are present in many developing systems and influence crucial cellular behaviors—such as cell division, cell migration, and cell differentiation [150]. Some cell types, like osteoblasts, neurons, and cardiomyocytes, are especially sensitive to electrical signals as they activate membrane receptors and downstream intracellular signaling elements leading to specific cell responses [151]. Not only cells, but also extracellular matrix proteins, such as collagen, fibrin, and keratin, can generate electrical currents upon mechanical stress, a phenomenon known as piezoelectricity [152].

Electrically conductive scaffolds not only enhance path finding of growing axons [153], but also improve cell survival and functional integration after transplantation in vivo by providing structural support for transplanted cells and facilitating synaptogenesis with host cells by restoring the neuronal network activity [154,155].

Since electroactive myocytes are responsible for heart and muscle contraction, electrically conductive materials found applications in cardiac and muscle tissue engineering as they support and maintain cell electrophysiology [22].

Even though bone cells are non-excitable cells, stress-generated piezoelectricity has been shown to stimulate bone precursor cell proliferation and differentiation to restore the injured site, making electroactive materials and electrical stimulation a valid tool for bone regeneration strategies [156].

Stem cells are also sensitive to electrical cues and their differentiation can be modulated by electrical stimulation and culture on electroactive materials. NSC ability to undergo neuronal and glial differentiation is boosted by electroactive material and exogenous electrical field [157,158]. The use of electrically conductive material have also been shown to promote the neuronal differentiation of adult stem cells derived from non-neural tissue without the addition of neuron-specific growth factors and cytokines [159,160]. Cardiomyocyte differentiation of iPSCs, ESCs, and MSCs is possible with the use of chemically defined media, but it dramatically increases if coupled with electroactive materials, showing protein expression, cell morphology, and contractility of the natural tissue [161,162]. Similarly, MSC differentiation into osteoblasts can be achieved with specific osteogenic media; but it is further supported by the aid of electroactive scaffolds [163].

3.3. The Importance of the "Nanoscale"

Cells have micro and nanoscale sensitivity because the extracellular environment presents a variety of spatially defined cues in the sub-micron to micron scale (Figure 8).

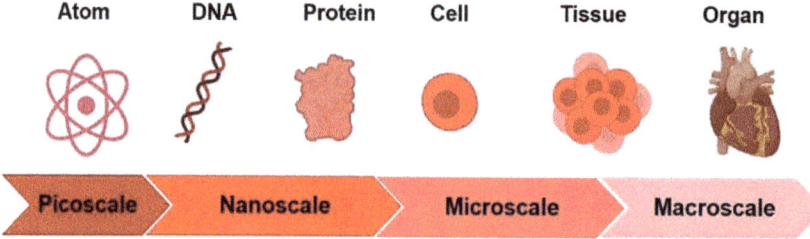

Figure 8. The relative scale of biological molecules and structures.

At the nanometer level, the extracellular environment affects sub-cellular behaviors such as the organization of cell adhesion molecule receptors. At the micron level, the extracellular environment affects cellular and supracellular characteristics such as cell morphology and [163]. The nanoscale physical features of the scaffolds can affect cell behavior. Natural tissues have indeed a hierarchical structure ranging from the macroscale (>1 mm) to the microscale (1 μm–1 mm), and the nanoscale (<1 μm). As a result, individual cells (typically in the size range 10–50 μm) respond in different ways to structures at different length scales. It has been shown that integrin receptors possess characteristic dimensions on the order of 10 nm [164]. The basement membrane of organs consists of nanoscale fibers (line topography) and pores (holes) that range in diameter from a few nanometers to several hundred nanometers [165]. The tubular fibers of collagen also have nanoscale dimensions [166] and laminin shows a nanoscale texture as well [167]. Given that cell ECM is patterned down to the nanoscale, cell-biomaterial interactions in scaffolds can be optimized by incorporating features of nanoscale dimensions. Indeed, surfaces topographically structured at the submicron scale can affect a wide variety of growth parameters, such as cell adhesion, morphology, viability, genic regulation, apoptosis, motility, and differentiation [168]. Evidence from nanoscale topography analysis suggests that nanoscale features eliciting a cell response are in the same size range (50–70 nm) that is associated with integrin cluster formation [169]. Further studies showed that scaffold nanotopography can control cell fate by altering cell and nucleus shapes, hence activating intracellular signal transduction and silent gene expression [125,170]. This is particularly true for neurons that, thanks to their growth cones, sense and actively respond to the surface nanotopography with a surprising sensitivity to variations of few nanometers [171].

Molecular deposition and lithographic techniques allow the patterning of tissue-specific molecules with nanometer resolutions. For example, the deposition of molecules that promote and support neuronal adhesion, growth, and differentiation on regenerating scaffolds enables the selective adhesion and growth of neural cells and a controlled neurite extension along the geometric pattern [172]. Apart from peptide/protein nanopatterns, nanomaterials and nanotechnology tools can also be used to develop special scaffolds able to recapitulate the architecture of structural proteins within ECM and the nanoscale features that model native ECM nanotopography [142]. Nanomaterials take advantage of their unique molecular features to induce, with high specificity, a number of desired physiological responses in target cells and tissues, while minimizing undesirable effects [173]. The peculiar mechanical and chemical properties of nanomaterials can be exploited for integration with native tissue in long-term implants; moreover, their nanoscale features have the potential to interact with the biological system at the molecular scale, while offering elevated levels of control [174].

Combinations of stimulatory cues may be used to incorporate nanoscale topographical, biochemical, and electrical cues in the same scaffold to provide an environment for tissue regeneration that is superior to inert scaffolds. This approach—able to precisely regulate cell differentiation, morphology, and polarization—is fundamental for the development of next-generation scaffolds suitable for clinical applications.

3.4. Role of the Biomolecular Corona

Interactions between the surface chemistry of nanomaterials and surfaces of biological components (proteins, phospholipids, organelles, DNA etc.) are crucial to determine the effects on cells and tissues. As soon as a nanomaterial comes in contact with a biological fluid (i.e., cell culture media, blood or interstitial fluid) it is coated with ions and proteins and develop a new interface which is often referred to as the protein corona or biomolecular corona (BC). This layer at the nanobio interface defines the biological identity of the nanomaterial, determining cell interactions, uptake and clearance [175]. Protein adsorption by GBMs has been reported in numerous studies. Umadevi and Sastry [176] analyzed non-covalent interactions on the surface of carbon nanostructures and highlighted that the graphitic lattice of graphene allowed hydrophobic interactions and strong π-π stacking with aromatic amino acids (Phe, Tyr, Trp) with binding energies between 15 and 20 kcal mol^{-1}. Surface chemistry is key in tuning the strength and type of interactions. Epoxide, hydroxyl and carboxyl groups on the surfaces favor hydrogen and electrostatic bonding with proteins, facilitating adsorption on GO compared to pristine graphene or RGO. GBMs have been shown to strongly bind to different serum proteins such as albumin, fibronectin (Fn), collagen, and laminin [177]. Therefore, when cells grow onto a graphene-based scaffold, they show an enhanced capacity to form focal adhesions by clustering integrin molecules and favoring cell adhesion [125]. In addition, GBMs capacity to adsorb proteins results in trapping growth factors produced by the cells during their differentiation. Such growth factors can progressively be released during cell maturation, allowing a continuous supply, which is suitable for long-term cell differentiation [177].

Not only graphene physical properties favor the adsorption of proteins, but they also offer tremendous opportunities for the covalent functionalization of protein active motifs and chemical groups [178]. Such approach allows the stable attachment of signalling molecules to the graphene structure to influence cell behaviour, but it also simplifies combination of graphene with both natural and synthetic polymers for the development of superior scaffolds combining multiple cues for cell growth and differentiation.

4. Nanotoxicology and Functionalization

4.1. In Vitro Cytotoxicity

The use of graphene-based nanomaterials (GBMs) does not come without possible concerns about in vitro cytotoxicity and in vivo biocompatibility. As anticipated in Section 3.4, the biomolecular corona (BC) plays an important role in regulating the fate and toxicity of nanomaterials that interface with a biological environment. Due to the unique and distinct physico-chemical properties of graphene and its derivatives, there is an enormous variability at the nano-bio interface which leads to different intrinsic toxicological effects. Moreover, nanomaterials are often pre-bound to chemical moieties that originate from the manufacturing process, from stabilizers used in their preparation or from exposure to gasses or buffers, all of which might further impact biocompatibility. Therefore, any generalization would be inaccurate, possibly misleading and must be avoided [25,179].

Pristine GBMs have been shown to have a dose- and time-dependent in vitro toxicity in both procaryotic [180–183] and eucaryotic cells [184–187]. Graphene has a hydrophobic nature that often causes irreversible aggregation in cell culture media and it has been reported to agglomerate on cell membranes causing physical damage [188]. Conversely, oxidized derivatives of graphene, such as graphene oxide (GO) and reduced graphene oxide (RGO), are more hydrophilic and show little aggregation in biological buffers resulting in lesser cytotoxicity [179]. According to Chatterjee and co-workers [189], who performed a comprehensive study about biological interaction of oxidized graphene derivatives, GO and RGO had similar toxic responses with different dose-dependency and distinct molecular mechanisms which were attributed to their peculiar surface oxidation status. However, the presence of oxidative functional groups on the surface can lead to the generation of reactive oxygen species (ROS). In addition, if they are not correctly washed, graphene nanomaterials might retain residual chemicals applied to separate the graphitic layers or during the fabrication of oxidized derivatives.

To solve these problems, novel green approaches for nanoparticle synthesis and modification have been developed, involving the use of biocompatible surfactants and reducing agents. According to Askari et al. [190], graphene nanosheets can be synthetized in the presence of Herceptin, a natural antibody, using an ultrasonic-assisted exfoliation method. The toxicity of graphene was tested in 3D spheroid cultures of human breast adenocarcinoma cell line (SKBR-3) to better mimic the natural tissue micro-environment. The authors concluded that that presence of Herceptin and its residues on graphene nanoparticles created a biocompatible platform suitable for cell growth. In another study, Narayanan et al. [191] described a facile and green synthesis of reduced graphene oxide by the deoxygenation of GO under aqueous alkaline conditions in the presence of soluble starch as a reducing agent (SRGO). The cytotoxicity of SRGO on skin fibroblasts was evaluated using a Wst-1 assay and showed that SRGO showed a substantial increase in cell viability at high concentrations (200 μg mL^{-1}) compared to non-reduced GO. The authors also investigated the hemocompatibility profiles of the nanomaterials and revealed that both caused a hemolysis effects compared to negative controls. However, SRGO did not exhibit a direct proportionality between hemolytic activity and concentration, with hemolysis staying as low as ~4.9% in maximum concentration samples.

4.2. Hemocompatibility and Interaction with Immune System Cells

Understanding interactions between nanomaterials and blood is key to determining in vivo biocompatibility due to the unavoidable contact between the two. Thanks to the protein corona effect, nanoparticles that touch blood or enter the bloodstream are coated by a milieu of proteins that may undergo conformational changes, exposing new epitopes and promoting phagocytosis or elimination from the circulation [192]. Nanomaterials can cause hemolysis and activate or interfere with clotting and coagulation cascades [193], seriously hindering the health of the organism.

The hemolytic property of nanoparticles is influenced by their distribution size, shape, surface charge and chemical composition [194]. Jaworski et al. [195] studied pristine graphene, RGO and GO effects on chicken embryo red blood cells (RBCs) and reported altered RBC morphology with loss of biconcavity. All of the nanomaterials exhibited dose-dependent hemolytic activity towards RBCs, with highest hemolysis rates observed at 5 mg mL^{-1}. Pristine graphene showed the highest hemolysis (73%), followed by RGO (42%) and GO (27%), correlating with the degree of surface oxidation. Lower hemolytic concentrations and activity have been reported by other groups [193]. However, according to Duan et al. [196] the hemolytic potential of GO can be largely reduced by pre-incubating it with BSA or FBS, exploiting their extremely high protein adsorption ability. In another work, Sasidharan et al. [197] provided evidence that pristine graphene and GO have excellent hemocompatibility showing no hemolysis, platelet activation or plasma coagulation up to a relatively high concentration (75 µg mL^{-1}) and under in vitro conditions. The authors also highlighted that pristine graphene had the potential to upregulate the production under sterile conditions of pro-inflammatory cytokines, such as IL-6 and IL-8. Cytokines are soluble glycoproteins released during an inflammatory response that recruit immune cells in order to tackle foreign bodies that have entered the organism.

Understanding interactions of GBMs with the immune system is of considerable relevance both from a toxicological and biomedical perspective. The BC of carbon-based nanomaterials is abundant in complement proteins. These proteins play a central role in modulating the immune and inflammatory responses towards intruders and may be a key factor in generating chronic ailments (such as allergy and sterile inflammation) by recruiting neutrophils and macrophages [198,199]. In addition, complement activation can promote cell-mediated immunity by enhancing generation of antigen-specific immunoglobulins by B-cells, activation of T-cells and uptake by dendritic cells [200]. Neutrophils and macrophages are part of the reticuloendothelial system (RES) which is responsible for the uptake and clearance of foreign bodies that have entered the organisms: the former are normally the first to intervene in an inflammation reaction, whereas the latter arrive later and promote tissue healing. It has been reported that macrophages better uptake hydrophilic systems compared to hydrophobic graphene since it is poorly dispersible in water and remains blocked on the cell surface [197]. Similarly, neutrophils are involved in nanoparticle clearance and it has been shown that exposure to carbon-based nanomaterials may upregulate neutrophils infiltration in tissues [201]. Carbon nanomaterials are also known to trigger apoptosis and/or cell death in macrophages, causing significant impairment in the immune resistance of subjects if used in vivo. However, Lin et al. [202] reported in a recent study that macrophage viability and activation are found to be mainly unaffected by few-layered graphene (FLG) at doses up to 50 µg mL^{-1} and therefore it is of little toxicity for M1 and M2 human macrophages, even though it triggers cell stress, ROS and inflammatory cytokines. Notably, neutrophils and macrophages are cleared from the circulation via the liver, spleen and kidneys and there is evidence that bone marrow may also play a major role in their clearance [203]. Therefore, nanomaterials carried by these cells can accumulate in those districts, causing unexpected issues and altering their fate.

4.3. In Vivo Biocompatibility

In vivo biodistribution and pharmacokinetics of GBMs have been studied in small and large animal models [204–209] in order to investigate the adsorption, distribution, metabolism and excretion (ADME). The fate of nanomaterials in organisms is influenced not only by their properties, but also from the pathway of exposure. Thanks to the wide range of potential applications of GBMs in biomedicine, exposure can occur in a number of ways including inhalation, intratracheal instillation, oral gauge, injection (intraperitoneal, intravenous or subcutaneous) or through debris generated from worn or biodegraded implants [204,210]. Once inside the organism, nanomaterials can make their way into the bloodstream even if not directly injected there and spread throughout the body. In addition, there is evidence that GBMs can diffuse across biological barriers such as the blood-air, blood–brain, blood-testis or blood-placental barrier, and accumulate in organs causing acute and chronic inflammation, tissue lesions and necrosis [211,212]. Krajnak and coworkers [213] examined graphene nanoparticles of different sizes and different forms (carbon black, graphene, GO and RGO) to determine if pulmonary exposure resulted in changes in vascular function and expression of acute response markers in mice. It was observed that while graphene altered gene expression in cardiovascular system, no changes were produced in the peripheral vascular function. On the other hand, pulmonary exposure to the oxidized forms of graphene had a more acute effect on heart and kidneys and repeated exposure might lead to injury or dysfunctions. Another study reported that GO provokes severe and persistent injury in mice lungs including granulomas persisting for up to 90 days [214]. Biodistribution experiments on intratracheally instilled carbon-14 labeled FLG showed that even if it was mainly retained in lungs, it was also redistributed to the liver and spleen passing through the air-blood barrier. However, no detectable absorption of FLG was observed when administered orally [212]. Conversely, radioactive-labelled RGO given through an oral gauge was rapidly absorbed in the intestine, metabolized by the kidneys and then excreted via urine [215]. Intravenous injection of GO in mice elicited blood platelets aggregation and extensive pulmonary thromboembolism, while low uptake was observed in the RES [216]. Surprisingly, a recent study by Newman et al. [217] highlighted that GO sheets accumulate preferentially in the spleen and progressively biodegrade over nine months. They evaluated the potential consequences of this prolonged accumulation and found limited effects on spleen histopathology and splenic function. Cell-mediated immune response was measured by determining the populations of T lymphocytes, specifically CD4+ and CD8+ cells as the major immune component of the splenic white pulp. Moderate changes were seen in both cell populations in mice injected with GO (2.5, 5, and 10 mg/kg) at both 24 h and one month after administration and no significant differences in the levels of the proinflammatory cytokines IL-6 and TNF-α were detected at any time point compared to control. However, they registered a significant drop in anti-inflammatory cytokines expression at 24 h and at the one-month time point for all tested GO doses. The authors concluded that reduction in cytokine expression after GO treatment may indicate the involvement of the innate immune system in regulating the effects of GO.

4.4. Minimizing GBM Toxicity

Although the inherent toxicity of graphene and its derivatives is a major drawback for their biomedical applications, it is a well-known problem, and different strategies have been developed to overcome it. In an attempt to enhance their overall safety and minimize the risks for adverse reactions in humans from exposure, Bussy et al. [218] offered a set of rules for the development of graphene and its derivatives: (1) use small, individual graphene sheets which are more efficiently internalized by macrophages in the body and removed from the site of deposition; (2) minimize aggregation using hydrophilic, stable, colloidal dispersions of graphene sheets; (3) use excretable graphene material or chemically-modified graphene that can be degraded effectively. Biological responses to these nanomaterials depend on various properties and it has been reported that smaller particles and higher oxidation improve biocompatibility [219]. Most importantly, variations in surface chemistry play a major role determining their toxicity and pharmacokinetic profile [220]. Highly hydrophobic graphene tends to aggregate in aqueous solvents thanks to intermolecular attractive Van der Waals forces, π-π stacking, hydrogen bonds and electrostatic interactions [221]. This tendency makes it hard to manipulate and characterize their biocompatibility and it has been suggested that the high percentage of controversial results in toxicity statistics could be owing to the dissimilarities in GBMs solubility [222]. Therefore, improving the dispersion of graphene-based nanomaterials in various solvents is a prerequisite for their further applications. Recent strategies include sonication, stabilization with surfactant and surface functionalization.

Wojtoniszak et al. [223] showed that GO and RGO exhibit a surfactant-dependent toxicity by comparing the homogeneity of GO and RGO dispersions in phosphate buffered saline (PBS) and cell viability on mice fibroblasts L929 cells. Three different dispersants were used, namely PEG, poly(ethylene glycol)-*block*-poly(propylene glycol-*block*-poly(ethylene glycol) (Pluronic P123), and sodium deoxycholate (DOC). The authors concluded that both materials had relatively good cytocompatibility in the 3.125–12.5 µg mL^{-1} range, with lowest toxicity detected in PEG-stabilized GO.

The BC can be tuned and modified by exploiting the ability of GBMs to adsorb moieties from the culture medium. As previously discussed, biological and bioactive species (DNA, carbohydrates and proteins) can be used as surfactants to stabilize graphene nanomaterials in aqueous solution, paving the way for different biomedical applications [191,224,225]. Pre-incubation in protein solutions was shown to form a thin coating on nanomaterials in suspension, minimizing cytotoxicity by limiting their direct interaction with cells. It has been reported that GO and RGO coated with BSA [226], FBS [227] or serum proteins [228] showed attenuate cytotoxicity and could improve biocompatibility. In addition, another interesting method of surface modification is functionalization through exposure to a specific enzyme, peptide or antibody [190,229]. Bussy et al. [230] exposed human lung carcinoma (A549) and bronchial epithelial (BEAS-2B) cell lines to GO and analyzed its effect in serum-free HEPES-buffered salt solution (BSS), Dulbecco's phosphate-buffered saline (PBS), and the normal media recommended for these cell lines (F12 for A549 and RPMI for BEAS-2B). Surprisingly, they reported more pronounced cellular responses in both BSS and PBS, but not in F12 or RPMI, and concluded that the interaction between GO and cells may differ depending on the concentration of salts and ions present in the aqueous environment. These charged moieties could influence nanofiller aggregation, bundling, stacking, or other colloidal properties of the negative surface-charged nanoparticles. In summary, the presence of proteins and other moieties in the cell culture medium influences the results on cytotoxicity and we could consider that GO and RGO might not be hemolytic in vivo where an abundant BC forms on their surfaces and protects the nanomaterial. These types of non-covalent surface functionalization are not stable in prolonged circulation and it is important to consider the dynamic changes of the BC as the nanoparticles translocate from one biological compartment to another or from the ECM intracellular locations [231].

Covalent functionalization is another strategy for enhancing solubility in different matrices and is frequently used to obtain nanocomposites, as previously discussed in Section 2. However, covalently bonding molecules to the surface leads to the disruption of the graphitic lattice changing its electronic and transport properties [232]. A number of surface modifications allow to obtain more hydrophilic GBMs with remarkably reduced toxicity. According to Kiew and coworkers [233], graphene-based nanomaterials with hydrophilic surfaces weaken the opsonin–protein interaction and could avoid being recognized by macrophage, thus an inflammatory response. Among the different strategies, the combination with polymeric materials represents a commonly used approach to overcome the limitations of graphene-based nanomaterials in biomedicine [234]. Several studies have reported that covalent modification with polyethilenglycole (PEGylation) can reduce cytotoxicity resulting in increased biocompatibility and stability in physiological buffers [235]. PEG is known to prolong particle circulation in the blood due to its ability to camouflage particle surfaces, sterically shielding against opsonization and uptake by the RES cells [192]. Other approaches involve the covalent attachment of conductive polymers (such as poly(pyrrole), poly(aniline), poly(allylamine)) or biodegradable synthetic polymers (e.g. poly(lactic) acid, poly(glycolic) acid, poly(lactide-co-glycolide)) or natural polysaccharides such as chitosan [236], alginate, hyaluronic acid and dextran (DEX). Like PEGylation, dextran coating reduces the adsorption of proteins the surface and improves biocompatibility. Compared with non-functionalized GO, GO–DEX conjugates showed improved stability in physiological solutions, accumulation in liver and spleen after intravenous injection, and most importantly clearance from body within a week without causing noticeable short-term toxicity [237].

Finally, graphene and its derivatives have been used in combination with biocompatible polymeric matrices to obtain conductive nanocomposites with enhanced cell adhesion, differentiation and biocompatibility [238]. These materials trigger reduced biological responses without the impairment of the GBMs capability to cross cell membranes and deliver therapeutic species [239]. As our lab has recently shown, organic-functionalized carbon nanofillers dispersed in a polymeric poly(L-lactic) acid exhibited enhanced cell viability (~90%) and supported cell growth [59], while having interest effects on the differentiation of neuronal precursors [50] and human circulating multipotent stem cells [47,159].

5. Examples of Tissue Regeneration
5.1. Bone Regeneration

Bones possess a remarkable regenerative capacity, as they maintain the ability to remodel themselves throughout adult life and they can repair fractures spontaneously [240]. After bone damage, soluble factors accumulate at the injury site and recruit mesenchymal stem cells, which, in turn, proliferate and differentiate toward osteoblasts. Subsequent calcification of the region results in a woven bone, which is finally remodeled by the renewing and resorptive actions of osteoblasts and osteoclasts [241]. Despite this regenerative process, there are instances where injuries may require clinical intervention to be completely healed. Autologous bone graft is a standard medical procedure for the treatment of bone-related diseases. Unfortunately, it is mostly limited by the availability of appropriate donor tissue [242]. Several scaffolds have been developed to enhance bone regeneration to overcome this issue and some representative examples are reviewed in this section.

For instance, graphene was used to coat three-dimensional hydroxyapatite scaffolds to support the growth and osteogenic differentiation of hMSCs. Scaffolds were found to be self-standing, as hMSCs differentiation did not require common differentiative molecules (i.e., dexamethasone or the bone morphogenetic protein 2) [160]. Moreover, graphene oxide was covalently linked to chitosan (CS), an animal-derived polymer already known to support cell adhesion and proliferation. The resulting polymer had better elastic modulus and hardness, which resulted in an increase in cell adhesion, spreading, proliferation, and formation of the extracellular matrix. Most importantly, cells grown onto GO-CS scaffolds showed an enhancement in calcium and phosphate deposition levels, a hallmark for osteoblastic differentiation [243].

Arnold and co-workers managed to enhance hMSCs osteogenic differentiation by directly functionalizing GO. Expressly, they set up an elegant universal synthetic procedure to covalently tether polyphosphates onto GO, generating a new phosphate-graphene material (CaPG) [244]. Their approach allowed them to obtain scaffolds with hydroxyapatite-like functionality at the interface, loaded with osteoinductive ions. They developed a 3D scaffold and assessed that its mechanical properties were comparable with bones (Young's modulus up to 1.8 GPa, compressive storage modulus up to 291 MPa, shear storage modulus up to 545 MPa, and ultimate compressive strengths up to 300 MPa). When hMSCs were seeded onto those scaffolds, a significant increase in the osteogenic marker alkaline phosphatase (ALP) and increased calcium deposits were observed, even when cells were cultured in growth medium (designed to maintain multipotency). Histological analyses of mice tissue after scaffold implantation showed no apparent damage, toxicological effects, or inflammation up to 8 weeks after treatment. More importantly, CaPG scaffolds enhanced donor cells' retention and provided differentiative signals favoring bone regeneration without using growth factors to direct osteogenesis.

Li and co-workers employed graphene oxide and lysozyme films to favor bone regeneration while minimizing the possibility of infection. Precisely by depositing overlapping layers of GO and lysozyme onto a chitosan base they obtained a construct not only able to support dental pulp stem cell growth and differentiation but also with improved antimicrobial activity. While GO is responsible for scaffolds stiffness and roughness, lysozyme improves the antimicrobial activity of GO by degrading the bacterial cell wall [245].

Among 3D scaffolds, Li and co-workers [246] provided an interesting proof of concept of the usage of 3D-printed alginate hydrogels as scaffolds for bone engineering. They used 3D-bioprinting to obtain gelatin-alginate scaffolds with defined porosity, then coated them with RGO. Although hydrogels are much less stiff than other composites, the authors observed a significant increase in adipose-derived stem cell (ADSC) differentiation toward the osteogenic lineage, as proven by the increase in ALP expression and calcification of the substrate.

Graphene oxide osteogenic potential was further investigated by Wu and co-workers, which grafted it with a peptide derived from the bone morphogenetic protein 2 [247]. GO-BMP2 was then bonded to silk-fibroin electrospun fibers to obtain biocompatible scaffolds that favor MSC adhesion and differentiation in vitro and in vivo and is are able to repair mice bone defects in less than 14 days.

5.2. Muscle Regeneration

Skeletal muscles made up most of the mass of the human body and are essential for motion and support. They are composed of multinucleated myofibers, which developed from mononucleated stem cell precursors during embryonic development. Satellite cells are unipotent stem cells that remain associated with adult myofibers and are responsible for muscle growth and regeneration [248,249]. Because of them, muscle tissue is endowed with a remarkable regenerative capacity, and most injuries sustained during everyday life fully recover via well-characterized processes [250]. However, severe injuries such as volumetric muscle loss and neuromuscular degenerative diseases, or aging, can result in significant muscular impairment, severely dampening life quality. In recent years, the possibility to produce scaffold recapitulating features of adult muscle tissue to enhance regeneration has drawn much attention, and several features that can enhance muscle regeneration have been identified. Among those, Gilbert and co-workers found substrate elasticity to be pivotal for muscle regeneration, as substrates mimicking tissue elasticity (~12 kPa) were able to sustain muscle stem cell self-renewal in vitro and differentiation in vivo [251]. Moreover, it was found that electroconductive scaffolds can enhance myoblasts fusion into myotubes in vitro, possibly by mimicking neuromuscular activity [252,253]. Starting from the observation that scaffold elastic properties are pivotal to resist the dynamic condition of the muscle tissue environment, Jo and co-workers [254] developed polyurethane/graphene oxide nanocomposite fibrous scaffolds to form a flexible and myogenic stimulating matrix for tissue engineering. They found nanocomposite to have better tensile strength, hydrophilicity, and biocompatibility than pristine materials. When they seeded mouse skeletal muscle cells C2C12 (a standard model for muscle differentiation studies) onto their scaffolds, they found an enhancement in cell adhesion and spreading, as demonstrated by the increase in the expression of actin and vinculin. Scaffolds were also capable of inducing muscle differentiation, as immunocytochemistry against myosin heavy chain (MHC, a marker for mature muscle cells) and RT-PCR against MyoG, α-actinin, and MyoD (markers for differentiating muscle cells) showed an increase directly proportional to GO concentration. Most importantly, they also found that scaffolds were able to sustain dynamic tensional stimuli, which, in turn, further increased the expression of differentiative markers.

As muscle cells are aligned along the fiber axis, materials patterned with surface features resembling native extracellular environment can influence mechanotransduction and favor cell differentiation. Park and co-workers [255] employed femtosecond laser ablation (FLA) to produce GO and RGO-based micropatterned conductive PAAm-hydrogels, which can support muscle differentiation in vitro and proved to have good stability in vivo. All scaffolds resembled muscle tissue Young's modulus, but only rGO-based ones possessed enough conductivity to deliver signals to cells. FLA allowed them to pattern scaffolds with 20 μm wide, 10 μm deep canals, and only scaffolds with a pattern distance comparable to cell dimension (50–80 μm) proved to affect differentiation. Specifically, when fusion index (i.e., the ratio of nuclei inside myotubes to all nuclei) and nuclear shape (which becomes less rounded during differentiation) were considered, it was found they could be improved by 50 and 80 μm patterned scaffolds independently on their conductivity. Morphological analyses were confirmed by immunocytochemical and qRT-PCR analyses, which demonstrated an increase in the expression of differentiative and mature myoblast markers (i.e., MHC, MyoG, and MyoD). In spite of this, conductivity proved to be pivotal to enhance cell aspect ratio, and electrical stimulation (2V, 10ms duration, 1 Hz) enhanced myotube formation with respect to untreated control. Hydrogels were also found to be suitable for implantation, as they remained intact for 4 weeks after subcutaneous implantation in mice, proving they can be a good platform for tissue implantation.

Besides skeletal muscle, cardiac muscle regeneration has drawn much attention because of the severity of heart diseases. Cardiomyocytes are specialized muscle cells which have a crucial role in the propagation of electric signal throughout the heart. Unlike skeletal muscle cells, cardiomyocytes have a reduced regenerative potential, and, after damage, are often replaced by scar tissue, which may lead to pathological heart failure. In an elegant comparative study, Lee and co-workers [256] compared the effects of gelatin methacrylic (GelMA) functionalized with either CNTs, GO, or RGO on the structural organization and functionality of rat primary cardiomyocytes. Even though all scaffolds resembled the elastic modulus of the heart, GO functionalized scaffolds exhibited low conductivity. Moreover, GO and RGO functionalized scaffolds displayed higher surface roughness compared to the GelMA and CNT-GelMA ones. Despite those differences, all scaffolds proved to support cell attachment and proliferation; however, they had different effects on cell differentiation. Specifically, when cells were stained against Cx43 (indicating electrical and metabolic coupling between cells), troponin-I and sarcomeric α-actin (both involved in muscle contraction), where enhanced only on RGO and CNT-GelMA but not in GO-GelMa. Moreover, even RGO failed to enhance the expression of troponin-I. Cells were further analyzed by patch-clamp to determine the extent and shape of the membrane action potential. Based on results, they found that CNT-GelMA led to the formation of ventricular like cardiomyocytes, whereas GO-GelMA resulted in an atrial-like phenotype. Instead, RGO-GelMA led to cells with a mixed phenotype. This finding suggests that different properties of the graphene derivative in the scaffold can be exploited to fine-tune cardiomyocyte phenotype.

In the context of injectable gels, Choe et al. developed an RGO-modified alginate gel and studied its antioxidant activity for cardiac tissue repair post myocardial infarction (MI) [257]. One of the hallmarks of MI is the high oxidative stress of heart tissues due to the formation of reactive oxygen species. Mesenchymal stem cell transplantation is a promising treatment for repairing heart tissues post MI, but after transplantation, their survival is compromise by the oxidative stress of the tissue. In their study, Choe et al. encapsulated hMSC in alginate microgels with a spherical shape (235 ± 11 μm diameter) suitable for easy injection. Nanocomposite microgels displayed higher cell viability than GO- and RGO- beads. To further improve survival, hMSCs were first enclosed in GO/alginate hydrogels and then GO was reduced. Nanocomposite microgels showed greater scavenging activity in all assays, while the graphene-free counterpart had a negligible antioxidant activity. These injectable anti-oxidizing nanomaterial-embedded microgels were able to scavenge radicals and lower the oxidative stress post MI, support MSC viability and maturation, thus increasing therapeutic activities and regeneration of infarcted tissues.

5.3. Nerve Regeneration

The nervous system represents the most intricate and vulnerable system in the human body, as, despite its pivotal importance, it is substantially unable to regenerate itself after injury. Because of its vital role, its organization is extraordinarily complex. Briefly, the nervous system comprises two main classes of cells: the glial cells and neurons. Neurons act as functional units, as they are characterized by peculiar electrophysiological features which allow them to rapidly transmit information between each other. Connections are established during neuritogenesis by the sprouting of dendrites and axons from the cell body. Specifically, each growing axon is tipped by the growth cone, a complex molecular machinery that senses environmental stimuli to guide growth toward the proper target [258]. On the other hand, glial cells consist of various specialized cell types (including Schwann cells, oligodendrocytes, and astrocytes) that regulate homeostasis, form myelin sheets around axons, and provide support and protection for neurons by maintaining a proper microenvironment [259]. Anatomically, the nervous system has been divided into the central (CNS) and peripheral (PNS) nervous system. Besides their different physiological role, they also respond differently to damages. Central nervous system regeneration is made more challenging, mainly because adult CNS is naturally resilient to cell repair and differentiation. For instance, after axotomy, glial cells of the CNS secrete inhibitory cues and form a physical and chemical barrier, the glial scar, which prevents regenerating axons to cross the injury site and reach their new target. Moreover, the basal expression of anti-regenerative cues such as chondroitin sulfate proteoglycans, Nogo-A, and myelin-associated glycoproteins, semaphorin 4D, and ephrin, is upregulated, further suppressing the capacity of the axonal growth cone to elongate [260–262]. Conversely, PNS neurons are endowed with a higher regenerative capacity due to the lack of CNS inhibitory factors [263].

Because of the inability of central neurons to regenerate, traumatic brain injury, and spinal cord injury have profound adverse effects on life quality and are a significant cause of mortality [264]. Efforts from the scientific community to address this issue resulted in several pharmacological and surgical therapeutic strategies [265]. However, in recent years graphene and its derivatives emerged as intriguing tools to design biomaterials mimicking tissue properties, encapsulate biomolecules and favor stem cell differentiation or tissue regeneration [266–269]. Recently, Quian and co-workers [266] used 3D printing and layer-by-layer casting methods to produce graphene and polycaprolactone scaffolds, which improved axonal regrowth and remyelination. Their technique allowed them to optimize quality control, mechanical strength, drug delivery distribution, and achieve the ideal electric conductivity for nerve growth. To increase scaffold biocompatibility, they coated it with polydopamine (PDA) and arginylglycylaspartic acid (RGD), which can encapsulate small molecules and favor cell adhesion, respectively. When tested with rat-immortalized Schwann cells, they found the optimal proliferation and viability rates on scaffolds at 1% graphene in PCL and that those scaffolds were able to support cell proliferation or up to seven days. Moreover, they found a higher expression of vinculin and N-cadherin on PDA/RGD-G/PCL scaffolds rather than on control scaffolds, indicating that graphene can have a role in promoting cell adhesion. Western blotting and qRT-PCR analyses indicated that not only scaffolds were able to induce expression of neural markers (such as glial fibrillary acidic protein, Class III ß-tubulin, and S100) but also they increase the expression of neurotrophic factors (NGF, BDNF, GDNF, and CNTF), which are vital to establishing a permissive environment for nerve regeneration. Moreover, when Schwann cell-loaded PDA/RGD-G/PCL scaffolds were grafted onto Sprague Dawley rats, histological and immunohistochemical observations 18 weeks after surgery suggested that regenerated nerves were well organized, lacked scar tissue and, most importantly, functional recovery was comparable to autograft implants.

As neurons require network formation to acquire proper function, tools to build 3D neuronal networks are required to enhance their function. An elegant method to encapsulate neurons onto a self-assembled micro-roll made of a bilayer of graphene and parylene-C [268], provided a proof of concept for designing a 3D neuronal network, which might also serve as a platform for modeling neurodegenerative diseases or producing cells suitable for transplantation. Their approach allowed them to create a support that allows neurons to interact with their surroundings without mixing with the external population, thus keeping a precisely controlled cell distribution. They exploited a sacrificial layer of calcium alginate to support a graphene layer, which was then coated with a parylene-C layer. Finally, the bilayer was patterned with an array of microscale pores to allow axons, but not cell bodies, to contact surrounding cells. Self-assembly into a tubular structure was induced by treating the sandwich with ethylenediaminetetraacetic acid (EDTA) to de-polymerize the alginate layer. Accessibility of reagents to the internal of the micro-roll was assessed by Ca^{2+} imaging in response to the addition of glutamate: encapsulated hippocampal neurons showed a coherent and coordinated response, and no delay with the response of external neurons was observed. Moreover, the formation of functional synapses between neurons was demonstrated both by immunocytochemistry against synapsin I, which is expressed by neurons at the synapse puncta, and by monitoring the synchronization of spontaneous Ca^{2+} waves. Besides serving as support for cell growth, the authors claim the graphene in their scaffold might serve as an electrode for electrophysiological recording and neuronal activity stimulation.

In order to study the role of substrate conductivity in neuronal network formation and alignment, Wang and colleagues developed a 3D conductive GO-coated scaffold based on printed PLCL microfibers using a near-field electrostatic printing (NFEP) [270]. NFEP is a technique that combines electrostatic spinning and 3D printing that allows to obtain fiber sizes of a few micrometers and complex architectures [271]. By manipulating the motion of the collection surface along X-Y-Z axes, NEFP easily generates arbitrary patterns (2D or 3D). PLCL scaffolds with different fiber overlay angles, diameters, and spatial organization were coated with GO, which was then reduced to RGO in situ without damaging the architecture. Depending on the layer thickness, RGO coating improved electrical conductivity while increasing surface roughness. The scaffolds were then used to assess the correlation between electrical stimulation (ES) and neurite outgrowth of the pheochromocytoma-derived PC12 cell line and primary neurons from hippocampal tissue of embryonic mice. ES enhanced neurite outgrowth and alignment with respect to control without ES stimulation. Strikingly, while neurite outgrowth resulted in being strictly correlated with the strength of the electric field, its directionality did not seem to influence neurite alignment. However, it was found that neurite outgrowth tightly followed the orientation of the smaller microfiber pattern and a more dispersive distribution of neurites was observed on fibers with higher diameters, where neurites had a higher tendency to branch out and lose their directional orientation (Figure 9).

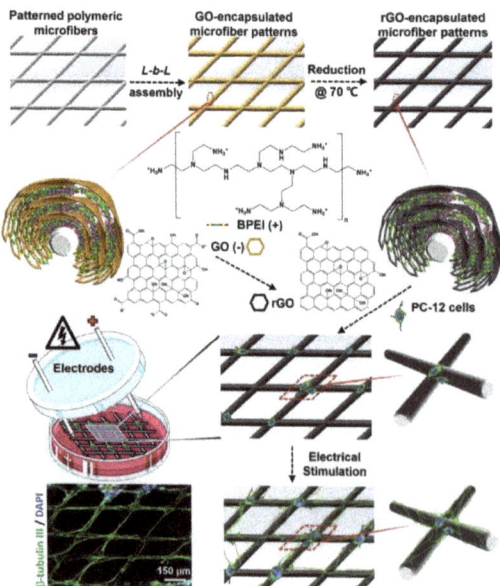

Figure 9. A diagram of the key procedures toward neuronal-like network formation with the guidance of conductive microfiber patterns under electrical stimulation [272].

Glial cells are as necessary as neurons to ensure proper nervous system functionality, therefore their regeneration after injury is as crucial as neuronal restoration [272]. Specifically, oligodendrocytes are responsible for myelination of central neurons and must be restored to ensure proper neuronal connectivity. The most common way to obtain oligodendrocytes is to differentiate multipotent NSCs or induced pluripotent stem cells (iPSCs). However, the process has proven to be challenging, as it requires long culture periods (up to 150 days) and has a limited yield [273]. Shah and co-workers [267] developed a nanocomposite PCL-GO scaffold, which allowed for NSCs differentiation into oligodendrocytes in just 6 days of culture. They treated electrospun PCL nanofibers of 200–300 nm diameter with oxygen plasma to render their surface hydrophilic, then GO was deposited on their surface at either 0.1, 0.5, or 1 mg/mL. Finally, laminin, an ECM protein essential for adhesion, growth, and differentiation of NSCs, was used to coat scaffolds. Rat hippocampal NSCs displayed significant differences in cell morphology after just 6 days of culture. Moreover, concentrations of GO as low as 0.1 mg/mL were able to strongly enhance the expression of the myelin basic protein (MBP), a marker specific for oligodendrocyte differentiation. The absence of an effect on the expression of the neuronal marker Tubβ3 and the astrocytic marker GFAP further suggested those scaffolds were able to selectively direct differentiation toward the oligodendrocytic lineage. When they treated cells grown on PCL-GO with integrin signaling inhibitors, they observed a steep decrease in oligodendrocyte markers. This observation suggested that the GO-coating of the nanofiber scaffolds might promote differentiation through specific microenvironmental interactions that activate integrin-related intracellular signaling.

Besides rigid scaffolds, biocompatible conductive hydrogels have attracted much attention because of their ability to better reproduce the mechanical properties of host tissues. Javadi and co-workers [269] developed a biocompatible hydrogel, based on polyurethane (PU), poly(3,4-ethylenedioxythiophene) (PEDOT) doped with poly(4-styrenesulfonate) (PSS) and liquid crystal graphene oxide (LCGO). They obtained a formulation with excellent conductivity, tensile modulus, and yield strength to support neuronal stem cells differentiation toward neurons and glial cells (as proven by the increase of the neuronal marker Tubβ-3 and the astrocyte marker GFAP). The authors claim LCGO liquid crystal nature synergistically combined with the properties of PEDOT:PSS to increase hydrogel mechanical and electrical properties.

Starting from evidence that the cholinergic system is involved in several neuron protective processes, cortical plasticity, and functional recovery after brain injury, Pradhan and co-workers developed a choline-graphene oxide functionalized (CFGO) injectable hydrogel based on poly(acrylic acid). Not only their hydrogels were able to support neuronal cell growth and differentiation, but they also stabilized the actin cytoskeleton. As choline is involved in enhancing neural recovery in TBI treatment, they injected their hydrogels in mice with parietal cortex brain injuries. They found scaffolds were able to restore cortical loss in just 7 days of treatment [274].

5.4. Wound Healing

GBMs have also been employed as fillers for wound healing hydrogels. Rehman and co-workers [275] developed RGO-GelMA hydrogels which enhance migration of fibroblasts, keratinocytes and endothelial cells in vitro and favor angiogenesis, in vivo, in chicken embryos. The authors speculate that this property could be due to an increase in intracellular ROS levels caused by RGO. In another study, Li and co-workers [276] developed N-acetyl cysteine (NAC) loaded GO-collagen membranes. In this formulation, GO has been reported to enhance mechanical properties and water retention of the collagen scaffolds, whereas NAC is used to lower ROS levels in the damaged tissue. The membrane accelerated cell migration, maturation and angiogenesis, leading to rapid skin regeneration. Moreover, the expression of profibrotic factors was found to be downregulated, indicating those scaffolds could promote scarless wound healing. A common problem of wound healing hydrogel is their vulnerability to bacterial infection [277]. To solve this problem, Yan and co-workers [278] developed an Ag reduced GO sodium alginate film which not only is able to inhibit bacterial growth but also to stimulate rapid wound healing in vivo.

6. Conclusions

Graphene-based scaffolds have been proven to be versatile tools in mediating tissue regeneration, as highlighted by the examples of in vitro and in vivo applications that have been discussed in this review. However, much more effort is required from the scientific community to clarify and rationalize their mechanism of action. It is clear that different composites can be employed to obtain similar results and yet subtle changes in scaffold formulation may result in completely different results. Therefore, a systematic analysis of the effects of scaffold composition on differentiation is required in order to disentangle the role of each scaffold component on cell fate. This would allow for a direct comparison between different scaffolds and a finer rational design. Moreover, it needs to be stressed that most applications rely on qRT-PCR data to prove successful differentiation. However, this approach reveals only an average trend in cell differentiation, without detecting potentially significant and biologically important differences between cells seeded onto different spots of the scaffold. Even when immunocytochemical data are provided, most authors fail to mention whether or not those data are representative of the whole sample or are just isolate cases. Coupling scaffold engineering with single cell RNA sequencing would overcome those limitations, allowing both a better understanding of scaffold effects on cell physiology and comparisons between the in vitro culture and the in vivo reference.

To date, scaffold engineering has focused on regeneration of a single tissue. However, clinical applications often require grafting of whole organs. Obtaining a scaffold that is able to efficiently reproduce a whole organ, or even multiple tissues (e.g., innervate muscles), has proven challenging and further studies are required before any viable clinical usage. In our opinion, finding the rationale behind graphene regulation of cell fates will allow us to obtain scaffolds that can reliably support and differentiate cells in a number of pre-determined types at the same time.

Although a molecular rationale for graphene-mediated effects is still lacking, it is remarkable that graphene-based scaffolds are able to determine cell fate more rapidly and efficiently than any other differentiation protocol, even without the addition of exogenous pro-differentiation factors. Indeed, in recent years research on mechanotransduction has unveiled several details on how nanotopography and stiffness stimuli are perceived and transduced by cells, whereas further efforts are needed to elucidate the contribution of other chemical and physical stimuli from the tissue environment. In particular, conductivity is of special interest to neuronal cell fate and differentiation, as it is specific to the nervous system. In addition to the aforementioned nanotopographic features, graphene and GBMs are endowed with tunable conductivity. Therefore, graphene-based nanomaterials represent a useful and cost-effective tool to enhance neuronal differentiation and tissue repair.

Author Contributions: All authors contributed to writing manuscript parts, to the overall draft preparation, review and editing. All authors have read and agreed to the published version of the manuscript.

Funding: This research received no external funding. The APC was funded by F.F.

Institutional Review Board Statement: Not applicable.

Informed Consent Statement: Not applicable.

Data Availability Statement: Not applicable.

Conflicts of Interest: The authors declare no conflict of interest.

References

1. Smith, A.T.; LaChance, A.M.; Zeng, S.; Liu, B.; Sun, L. Synthesis, properties, and applications of graphene oxide/reduced graphene oxide and their nanocomposites. *Nano Mater. Sci.* **2019**, *1*, 31–47. [CrossRef]
2. Patel, K.D.; Singh, R.K.; Kim, H.-W. Carbon-based nanomaterials as an emerging platform for theranostics. *Mater. Horizons* **2019**, *6*, 434–469. [CrossRef]
3. Sharma, H.; Mondal, S. Functionalized Graphene Oxide for Chemotherapeutic Drug Delivery and Cancer Treatment: A Promising Material in Nanomedicine. *Int. J. Mol. Sci.* **2020**, *21*, 6280. [CrossRef]
4. Gazzi, A.; Fusco, L.; Khan, A.; Bedognetti, D.; Zavan, B.; Vitale, F.; Yilmazer, A.; Delogu, L.G. Photodynamic Therapy Based on Graphene and MXene in Cancer Theranostics. *Front. Bioeng. Biotechnol.* **2019**, *7*, 295. [CrossRef]

5. Kim, N.Y.; Blake, S.; De, D.; Ouyang, J.; Shi, J.; Kong, N. Two-Dimensional Nanosheet-Based Photonic Nanomedicine for Combined Gene and Photothermal Therapy. *Front. Pharmacol.* **2020**, *10*, 1573. [CrossRef]
6. Priya Swetha, P.D.; Manisha, H.; Sudhakaraprasad, K. Graphene and Graphene-Based Materials in Biomedical Science. *Part. Part. Syst. Charact.* **2018**, *35*, 1800105. [CrossRef]
7. Weng, Y.; Jiang, B.; Yang, K.; Sui, Z.; Zhang, L.; Zhang, Y. Polyethyleneimine-modified graphene oxide nanocomposites for effective protein functionalization. *Nanoscale* **2015**, *7*, 14284–14291. [CrossRef] [PubMed]
8. Qiu, W.; Skafidas, E. Detection of Protein Conformational Changes with Multilayer Graphene Nanopore Sensors. *ACS Appl. Mater. Interfaces* **2014**, *6*, 16777–16781. [CrossRef] [PubMed]
9. Zhao, Y.; Li, X.; Zhou, X.; Zhang, Y. Review on the graphene based optical fiber chemical and biological sensors. *Sensors Actuators B Chem.* **2016**, *231*, 324–340. [CrossRef]
10. Omar, N.A.; Fen, Y.W.; Saleviter, S.; Daniyal, W.M.; Anas, N.A.; Ramdzan, N.S.; Roshidi, M.D. Development of a Graphene-Based Surface Plasmon Resonance Optical Sensor Chip for Potential Biomedical Application. *Materials* **2019**, *12*, 1928. [CrossRef]
11. Younis, M.R.; He, G.; Lin, J.; Huang, P. Recent Advances on Graphene Quantum Dots for Bioimaging Applications. *Front. Chem.* **2020**, *8*, 424. [CrossRef] [PubMed]
12. Li, K.; Zhao, X.; Wei, G.; Su, Z. Recent Advances in the Cancer Bioimaging with Graphene Quantum Dots. *Curr. Med. Chem.* **2018**, *25*, 2876–2893. [CrossRef] [PubMed]
13. Esmaeili, Y.; Bidram, E.; Zarrabi, A.; Amini, A.; Cheng, C. Graphene oxide and its derivatives as promising In-vitro bio-imaging platforms. *Sci. Rep.* **2020**, *10*, 18052. [CrossRef]
14. Jagiełło, J.; Chlanda, A.; Baran, M.; Gwiazda, M.; Lipińska, L. Synthesis and Characterization of Graphene Oxide and Reduced Graphene Oxide Composites with Inorganic Nanoparticles for Biomedical Applications. *Nanomaterials* **2020**, *10*, 1846. [CrossRef]
15. Langer, R.; Vacanti, J.P. Tissue engineering. *Science* **1993**, *260*, 920–926. [CrossRef]
16. Dvir, T.; Timko, B.P.; Kohane, D.S.; Langer, R. Nanotechnological strategies for engineering complex tissues. *Nat. Nanotechnol.* **2011**, *6*, 13–22. [CrossRef] [PubMed]
17. Chao, T.-I.; Xiang, S.; Chen, C.-S.; Chin, W.-C.; Nelson, A.J.; Wang, C.; Lu, J. Carbon nanotubes promote neuron differentiation from human embryonic stem cells. *Biochem. Biophys. Res. Commun.* **2009**, *384*, 426–430. [CrossRef] [PubMed]
18. Lee, T.-J.; Park, S.; Bhang, S.H.; Yoon, J.-K.; Jo, I.; Jeong, G.-J.; Hong, B.H.; Kim, B.-S. Graphene enhances the cardiomyogenic differentiation of human embryonic stem cells. *Biochem. Biophys. Res. Commun.* **2014**, *452*, 174–180. [CrossRef]
19. Barnes, J.M.; Przybyla, L.; Weaver, V.M. Tissue mechanics regulate brain development, homeostasis and disease. *J. Cell Sci.* **2017**, *130*, 71–82. [CrossRef] [PubMed]
20. Ege, D.; Kamali, A.R.; Boccaccini, A.R. Graphene Oxide/Polymer-Based Biomaterials. *Adv. Eng. Mater.* **2017**, *19*, 1700627. [CrossRef]
21. He, Z.; Zhang, S.; Song, Q.; Li, W.; Liu, D.; Li, H.; Tang, M.; Chai, R. The structural development of primary cultured hippocampal neurons on a graphene substrate. *Colloids Surf. B. Biointerfaces* **2016**, *146*, 442–451. [CrossRef]
22. Da Silva, L.P.; Kundu, S.C.; Reis, R.L.; Correlo, V.M. Electric Phenomenon: A Disregarded Tool in Tissue Engineering and Regenerative Medicine. *Trends Biotechnol.* **2020**, *38*, 24–49. [CrossRef]
23. Burnstine-Townley, A.; Eshel, Y.; Amdursky, N. Conductive Scaffolds for Cardiac and Neuronal Tissue Engineering: Governing Factors and Mechanisms. *Adv. Funct. Mater.* **2020**, *30*, 1901369. [CrossRef]
24. Plachá, D.; Jampilek, J. Graphenic Materials for Biomedical Applications. *Nanomaterials* **2019**, *9*, 1758. [CrossRef]
25. Seabra, A.B.; Paula, A.J.; De Lima, R.; Alves, O.L.; Durán, N. Nanotoxicity of Graphene and Graphene Oxide. *Chem. Res. Toxicol.* **2014**, *27*, 159–168. [CrossRef]
26. Wu, Y.; Wang, S.; Komvopoulos, K. A review of graphene synthesis by indirect and direct deposition methods. *J. Mater. Res.* **2020**, *35*, 76–89. [CrossRef]
27. Yu, Q.; Lian, J.; Siriponglert, S.; Li, H.; Chen, Y.P.; Pei, S.-S. Graphene segregated on Ni surfaces and transferred to insulators. *Appl. Phys. Lett.* **2008**, *93*, 113103. [CrossRef]
28. Reina, A.; Thiele, S.; Jia, X.; Bhaviripudi, S.; Dresselhaus, M.S.; Schaefer, J.A.; Kong, J. Growth of large-area single- and Bi-layer graphene by controlled carbon precipitation on polycrystalline Ni surfaces. *Nano Res.* **2009**, *2*, 509–516. [CrossRef]
29. Liu, W.; Li, H.; Xu, C.; Khatami, Y.; Banerjee, K. Synthesis of high-quality monolayer and bilayer graphene on copper using chemical vapor deposition. *Carbon N. Y.* **2011**, *49*, 4122–4130. [CrossRef]
30. Ciesielski, A.; Samorì, P. Graphene via sonication assisted liquid-phase exfoliation. *Chem. Soc. Rev.* **2014**, *43*, 381–398. [CrossRef]
31. Yi, M.; Shen, Z. A review on mechanical exfoliation for the scalable production of graphene. *J. Mater. Chem. A* **2015**, *3*, 11700–11715. [CrossRef]
32. Narayan, R.; Kim, S.O. Surfactant mediated liquid phase exfoliation of graphene. *Nano Converg.* **2015**, *2*, 20. [CrossRef] [PubMed]
33. Xu, Y.; Cao, H.; Xue, Y.; Li, B.; Cai, W. Liquid-Phase Exfoliation of Graphene: An Overview on Exfoliation Media, Techniques, and Challenges. *Nanomaterials* **2018**, *8*, 942. [CrossRef]
34. Le Ba, T.; Mahian, O.; Wongwises, S.; Szilágyi, I.M. Review on the recent progress in the preparation and stability of graphene-based nanofluids. *J. Therm. Anal. Calorim.* **2020**, *142*, 1145–1172. [CrossRef]
35. Hernandez, Y.; Nicolosi, V.; Lotya, M.; Blighe, F.M.; Sun, Z.; De, S.; McGovern, I.T.; Holland, B.; Byrne, M.; Gun'Ko, Y.K.; et al. High-yield production of graphene by liquid-phase exfoliation of graphite. *Nat. Nanotechnol.* **2008**, *3*, 563–568. [CrossRef]

36. Polyakova, E.Y.; Rim, K.T.; Eom, D.; Douglass, K.; Opila, R.L.; Heinz, T.F.; Teplyakov, A.V.; Flynn, G.W. Scanning Tunneling Microscopy and X-ray Photoelectron Spectroscopy Studies of Graphene Films Prepared by Sonication-Assisted Dispersion. *ACS Nano* **2011**, *5*, 6102–6108. [CrossRef]
37. Skaltsas, T.; Ke, X.; Bittencourt, C.; Tagmatarchis, N. Ultrasonication Induces Oxygenated Species and Defects onto Exfoliated Graphene. *J. Phys. Chem. C* **2013**, *117*, 23272–23278. [CrossRef]
38. Seiler, S.; Halbig, C.E.; Grote, F.; Rietsch, P.; Börrnert, F.; Kaiser, U.; Meyer, B.; Eigler, S. Effect of friction on oxidative graphite intercalation and high-quality graphene formation. *Nat. Commun.* **2018**, *9*, 836. [CrossRef]
39. Eigler, S.; Enzelberger-Heim, M.; Grimm, S.; Hofmann, P.; Kroener, W.; Geworski, A.; Dotzer, C.; Röckert, M.; Xiao, J.; Papp, C.; et al. Wet Chemical Synthesis of Graphene. *Adv. Mater.* **2013**, *25*, 3583–3587. [CrossRef]
40. Eigler, S.; Grimm, S.; Hirsch, A. Investigation of the Thermal Stability of the Carbon Framework of Graphene Oxide. *Chem. A Eur. J.* **2014**, *20*, 984–989. [CrossRef]
41. Chua, C.K.; Pumera, M. Chemical reduction of graphene oxide: A synthetic chemistry viewpoint. *Chem. Soc. Rev.* **2014**, *43*, 291–312. [CrossRef]
42. Eigler, S.; Grimm, S.; Enzelberger-Heim, M.; Müller, P.; Hirsch, A. Graphene oxide: Efficiency of reducing agents. *Chem. Commun.* **2013**, *49*, 7391–7393. [CrossRef]
43. Eigler, S.; Grimm, S.; Hof, F.; Hirsch, A. Graphene oxide: A stable carbon framework for functionalization. *J. Mater. Chem. A* **2013**, *1*, 11559–11562. [CrossRef]
44. Kim, J.; Park, S.; Kim, Y.J.; Chang Su, J.; Lim, K.T.; Seonwoo, H.; Cho, S.P.; Chung, T.D.; Choung, P.H.; Choung, Y.H.; et al. Monolayer graphene-directed growth and neuronal differentiation of mesenchymal stem cells. *J. Biomed. Nanotechnol.* **2015**, *11*, 2024–2033. [CrossRef] [PubMed]
45. Nayak, T.R.; Andersen, H.; Makam, V.S.; Khaw, C.; Bae, S.; Xu, X.; Ee, P.L.R.; Ahn, J.H.; Hong, B.H.; Pastorin, G.; et al. Graphene for controlled and accelerated osteogenic differentiation of human mesenchymal stem cells. *ACS Nano* **2011**, *5*, 4670–4678. [CrossRef] [PubMed]
46. Kim, J.; Bae, W.G.; Park, S.; Kim, Y.J.; Jo, I.; Park, S.; Jeon, N.L.; Kwak, W.; Cho, S.; Park, J.; et al. Engineering structures and functions of mesenchymal stem cells by suspended large-area graphene nanopatterns. *2D Mater.* **2016**, *3*, 35013. [CrossRef]
47. Tonellato, M.; Piccione, M.; Gasparotto, M.; Bellet, P.; Tibaudo, L.; Vicentini, N.; Bergantino, E.; Menna, E.; Vitiello, L.; Di Liddo, R.; et al. Commitment of autologous human multipotent stem cells on biomimetic poly-l-lactic acid-based scaffolds is strongly influenced by structure and concentration of carbon nanomaterial. *Nanomaterials* **2020**, *10*, 415. [CrossRef]
48. Jasim, D.A.; Lozano, N.; Bussy, C.; Barbolina, I.; Rodrigues, A.F.; Novoselov, K.S.; Kostarelos, K. Graphene-based papers as substrates for cell growth: Characterisation and impact on mammalian cells. *FlatChem* **2018**, *12*, 17–25. [CrossRef]
49. Rasch, F.; Schütt, F.; Saure, L.M.; Kaps, S.; Strobel, J.; Polonskyi, O.; Nia, A.S.; Lohe, M.R.; Mishra, Y.K.; Faupel, F.; et al. Wet-Chemical Assembly of 2D Nanomaterials into Lightweight, Microtube-Shaped, and Macroscopic 3D Networks. *ACS Appl. Mater. Interfaces* **2019**, *11*, 44652–44663. [CrossRef]
50. Vicentini, N.; Gatti, T.; Salice, P.; Scapin, G.; Marega, C.; Filippini, F.; Menna, E. Covalent functionalization enables good dispersion and anisotropic orientation of multi-walled carbon nanotubes in a poly(l-lactic acid) electrospun nanofibrous matrix boosting neuronal differentiation. *Carbon N. Y.* **2015**, *95*, 725–730. [CrossRef]
51. Ligorio, C.; Zhou, M.; Wychowaniec, J.K.; Zhu, X.; Bartlam, C.; Miller, A.F.; Vijayaraghavan, A.; Hoyland, J.A.; Saiani, A. Graphene oxide containing self-assembling peptide hybrid hydrogels as a potential 3D injectable cell delivery platform for intervertebral disc repair applications. *Acta Biomater.* **2019**, *92*, 92–103. [CrossRef] [PubMed]
52. Ryoo, S.R.; Kim, Y.K.; Kim, M.H.; Min, D.H. Behaviors of NIH-3T3 fibroblasts on graphene/carbon nanotubes: Proliferation, focal adhesion, and gene transfection studies. *ACS Nano* **2010**, *4*, 6587–6598. [CrossRef]
53. Pandele, A.M.; Dinescu, S.; Costache, M.; Vasile, E.; Obreja, C.; Iovu, H.; Ionita, M. Preparation and in vitro, bulk, and surface investigation of chitosan/graphene oxide composite films. *Polym. Compos.* **2013**, *34*, 2116–2124. [CrossRef]
54. Jin, L.; Zeng, Z.; Kuddannaya, S.; Wu, D.; Zhang, Y.; Wang, Z. Biocompatible, Free-Standing Film Composed of Bacterial Cellulose Nanofibers-Graphene Composite. *ACS Appl. Mater. Interfaces* **2016**, *8*, 1011–1018. [CrossRef] [PubMed]
55. Li, J.; Liu, X.; Tomaskovic-Crook, E.; Crook, J.M.; Wallace, G.G. Smart graphene-cellulose paper for 2D or 3D "origami-inspired" human stem cell support and differentiation. *Colloids Surfaces B Biointerfaces* **2019**, *176*, 87–95. [CrossRef] [PubMed]
56. Kim, S.E.; Kim, M.S.; Shin, Y.C.; Eom, S.U.; Lee, J.H.; Shin, D.M.; Hong, S.W.; Kim, B.; Park, J.C.; Shin, B.S.; et al. Cell migration according to shape of graphene oxide micropatterns. *Micromachines* **2016**, *7*, 186. [CrossRef] [PubMed]
57. Smith, A.S.T.; Yoo, H.; Yi, H.; Ahn, E.H.; Lee, J.H.; Shao, G.; Nagornyak, E.; Laflamme, M.A.; Murry, C.E.; Kim, D.H. Micro-and nano-patterned conductive graphene-PEG hybrid scaffolds for cardiac tissue engineering. *Chem. Commun.* **2017**, *53*, 7412–7415. [CrossRef]
58. Qi, Z.; Chen, X.; Guo, W.; Fu, C.; Pan, S. Theanine-Modified Graphene Oxide Composite Films for Neural Stem Cells Proliferation and Differentiation. *J. Nanomater.* **2020**, *2020*, 3068173. [CrossRef]
59. Vicentini, N.; Gatti, T.; Salerno, M.; Hernandez Gomez, Y.S.; Bellon, M.; Gallio, S.; Marega, C.; Filippini, F.; Menna, E. Effect of different functionalized carbon nanostructures as fillers on the physical properties of biocompatible poly(L-lactic acid) composites. *Mater. Chem. Phys.* **2018**, *214*, 265–276. [CrossRef]
60. Lutzweiler, G.; Halili, A.N.; Vrana, N.E. The overview of porous, bioactive scaffolds as instructive biomaterials for tissue regeneration and their clinical translation. *Pharmaceutics* **2020**, *12*, 602. [CrossRef]

61. Bružauskaitė, I.; Bironaitė, D.; Bagdonas, E.; Bernotienė, E. Scaffolds and cells for tissue regeneration: Different scaffold pore sizes—different cell effects. *Cytotechnology* **2016**, *68*, 355–369. [CrossRef]
62. Reilly, G.C.; Engler, A.J. Intrinsic extracellular matrix properties regulate stem cell differentiation. *J. Biomech.* **2010**, *43*, 55–62. [CrossRef] [PubMed]
63. Domínguez-Bajo, A.; González-Mayorga, A.; Guerrero, C.R.; Palomares, F.J.; García, R.; López-Dolado, E.; Serrano, M.C. Myelinated axons and functional blood vessels populate mechanically compliant rGO foams in chronic cervical hemisected rats. *Biomaterials* **2019**, *192*, 461–474. [CrossRef]
64. Serrano, M.C.; Patiño, J.; García-Rama, C.; Ferrer, M.L.; Fierro, J.L.G.; Tamayo, A.; Collazos-Castro, J.E.; Del Monte, F.; Gutiérrez, M.C. 3D free-standing porous scaffolds made of graphene oxide as substrates for neural cell growth. *J. Mater. Chem. B* **2014**, *2*, 5698–5706. [CrossRef]
65. Liao, J.F.; Qu, Y.; Chu, B.; Zhang, X.; Qian, Z. Biodegradable CSMA/PECA/graphene porous hybrid scaffold for cartilage tissue engineering. *Sci. Rep.* **2015**, *5*, 9879. [CrossRef]
66. Hermenean, A.; Codreanu, A.; Herman, H.; Balta, C.; Rosu, M.; Mihali, C.V.; Ivan, A.; Dinescu, S.; Ionita, M.; Costache, M. Chitosan-Graphene Oxide 3D scaffolds as Promising Tools for Bone Regeneration in Critical-Size Mouse Calvarial Defects. *Sci. Rep.* **2017**, *7*, 16641. [CrossRef] [PubMed]
67. Chae, S.J.; Güneş, F.; Kim, K.K.; Kim, E.S.; Han, G.H.; Kim, S.M.; Shin, H.-J.; Yoon, S.-M.; Choi, J.-Y.; Park, M.H.; et al. Synthesis of Large-Area Graphene Layers on Poly-Nickel Substrate by Chemical Vapor Deposition: Wrinkle Formation. *Adv. Mater.* **2009**, *21*, 2328–2333. [CrossRef]
68. Ma, Q.; Yang, L.; Jiang, Z.; Song, Q.; Xiao, M.; Zhang, D.; Ma, X.; Wen, T.; Cheng, G. Three-Dimensional Stiff Graphene Scaffold on Neural Stem Cells Behavior. *ACS Appl. Mater. Interfaces* **2016**, *8*, 34227–34233. [CrossRef]
69. Li, N.; Zhang, Q.; Gao, S.; Song, Q.; Huang, R.; Wang, L.; Liu, L.; Dai, J.; Tang, M.; Cheng, G. Three-dimensional graphene foam as a biocompatible and conductive scaffold for neural stem cells. *Sci. Rep.* **2013**, *3*, 1604. [CrossRef] [PubMed]
70. Xiao, M.; Kong, T.; Wang, W.; Song, Q.; Zhang, D.; Ma, Q.; Cheng, G. Interconnected Graphene Networks with Uniform Geometry for Flexible Conductors. *Adv. Funct. Mater.* **2015**, *25*, 6165–6172. [CrossRef]
71. Schmitt, C.; Rasch, F.; Cossais, F.; Held-Feindt, J.; Lucius, R.; Vazquez, A.R.; Nia, A.S.; Lohe, M.R.; Feng, X.; Mishra, Y.K.; et al. Glial cell responses on tetrapod-shaped graphene oxide and reduced graphene oxide 3D scaffolds in brain in vitro and ex vivo models of indirect contact. *Biomed. Mater.* **2020**. [CrossRef]
72. Evlashin, S.; Dyakonov, P.; Tarkhov, M.; Dagesyan, S.; Rodionov, S.; Shpichka, A.; Kostenko, M.; Konev, S.; Sergeichev, I.; Timashev, P.; et al. Flexible polycaprolactone and polycaprolactone/graphene scaffolds for tissue engineering. *Materials* **2019**, *12*, 2991. [CrossRef] [PubMed]
73. Nieto, A.; Dua, R.; Zhang, C.; Boesl, B.; Ramaswamy, S.; Agarwal, A. Three Dimensional Graphene Foam/Polymer Hybrid as a High Strength Biocompatible Scaffold. *Adv. Funct. Mater.* **2015**, *25*, 3916–3924. [CrossRef]
74. Song, F.; Jie, W.; Zhang, T.; Li, W.; Jiang, Y.; Wan, L.; Liu, W.; Li, X.; Liu, B. Room-temperature fabrication of a three-dimensional reduced-graphene oxide/polypyrrole/hydroxyapatite composite scaffold for bone tissue engineering. *RSC Adv.* **2016**, *6*, 92804–92812. [CrossRef]
75. Deliormanlı, A.M.; Türk, M.; Atmaca, H. Preparation and characterization of PCL-coated porous hydroxyapatite scaffolds in the presence of MWCNTs and graphene for orthopedic applications. *J. Porous Mater.* **2019**, *26*, 247–259. [CrossRef]
76. Turk, M.; Deliormanll, A.M. Electrically conductive borate-based bioactive glass scaffolds for bone tissue engineering applications. *J. Biomater. Appl.* **2017**, *32*, 28–39. [CrossRef] [PubMed]
77. Deliormanlı, A.M.; Türk, M. Investigation the in vitro biological performance of graphene/bioactive glass scaffolds using MC3T3-E1 and ATDC5 cells. *Mater. Technol.* **2018**, *33*, 854–864. [CrossRef]
78. Jakus, A.E.; Secor, E.B.; Rutz, A.L.; Jordan, S.W.; Hersam, M.C.; Shah, R.N. Three-dimensional printing of high-content graphene scaffolds for electronic and biomedical applications. *ACS Nano* **2015**, *9*, 4636–4648. [CrossRef]
79. Cabral, C.S.D.; Miguel, S.P.; de Melo-Diogo, D.; Louro, R.O.; Correia, I.J. In situ green reduced graphene oxide functionalized 3D printed scaffolds for bone tissue regeneration. *Carbon N. Y.* **2019**, *146*, 513–523. [CrossRef]
80. Jun, I.; Han, H.-S.; Edwards, J.R.; Jeon, H. Electrospun Fibrous Scaffolds for Tissue Engineering: Viewpoints on Architecture and Fabrication. *Int. J. Mol. Sci.* **2018**, *19*, 749. [CrossRef]
81. Agarwal, S.; Wendorff, J.H.; Greiner, A. Progress in the field of electrospinning for tissue engineering applications. *Adv. Mater.* **2009**, *21*, 3343–3351. [CrossRef] [PubMed]
82. Gnavi, S.; Fornasari, B.E.; Tonda-Turo, C.; Laurano, R.; Zanetti, M.; Ciardelli, G.; Geuna, S. The Effect of Electrospun Gelatin Fibers Alignment on Schwann Cell and Axon Behavior and Organization in the Perspective of Artificial Nerve Design. *Int. J. Mol. Sci.* **2015**, *16*, 12925–12942. [CrossRef] [PubMed]
83. Mohammadi, S.; Shafiei, S.S.; Asadi-Eydivand, M.; Ardeshir, M.; Solati-Hashjin, M. Graphene oxide-enriched poly(ε-caprolactone) electrospun nanocomposite scaffold for bone tissue engineering applications. *J. Bioact. Compat. Polym.* **2017**, *32*, 325–342. [CrossRef]
84. Song, J.; Gao, H.; Zhu, G.; Cao, X.; Shi, X.; Wang, Y. The preparation and characterization of polycaprolactone/graphene oxide biocomposite nanofiber scaffolds and their application for directing cell behaviors. *Carbon N. Y.* **2015**, *95*, 1039–1050. [CrossRef]

85. Luo, Y.; Shen, H.; Fang, Y.; Cao, Y.; Huang, J.; Zhang, M.; Dai, J.; Shi, X.; Zhang, Z. Enhanced proliferation and osteogenic differentiation of mesenchymal stem cells on graphene oxide-incorporated electrospun poly(lactic-co-glycolic acid) nanofibrous mats. *ACS Appl. Mater. Interfaces* **2015**, *7*, 6331–6339. [CrossRef]
86. Scaffaro, R.; Lopresti, F.; Maio, A.; Botta, L.; Rigogliuso, S.; Ghersi, G. Electrospun PCL/GO-g-PEG structures: Processing-morphology-properties relationships. *Compos. Part A Appl. Sci. Manuf.* **2017**, *92*, 97–107. [CrossRef]
87. Basar, A.O.; Sadhu, V.; Turkoglu Sasmazel, H. Preparation of electrospun PCL-based scaffolds by mono/multi-functionalized GO. *Biomed. Mater.* **2019**, *14*, 45012. [CrossRef] [PubMed]
88. Liu, C.; Wong, H.M.; Yeung, K.W.K.; Tjong, S.C. Novel electrospun polylactic acid nanocomposite fiber mats with hybrid graphene oxide and nanohydroxyapatite reinforcements having enhanced biocompatibility. *Polymers* **2016**, *8*, 287. [CrossRef]
89. Jeznach, O.; Kolbuk, D.; Sajkiewicz, P. Aminolysis of Various Aliphatic Polyesters in a Form of Nanofibers and Films. *Polymers* **2019**, *11*, 1669. [CrossRef] [PubMed]
90. Jalili, S.; Mohamadzadeh, M.; Ghanian, M.; Ashtiani, M.; Alimadadi, H.; Baharvand, H.; Martin, I.; Scherberich, A. Polycaprolactone-templated reduced-graphene oxide liquid crystal nanofibers towards biomedical applications. *RSC Adv.* **2017**, *7*, 39628–39634. [CrossRef]
91. Wang, J.; Cheng, Y.; Chen, L.; Zhu, T.; Ye, K.; Jia, C.; Wang, H.; Zhu, M.; Fan, C.; Mo, X. In vitro and in vivo studies of electroactive reduced graphene oxide-modified nanofiber scaffolds for peripheral nerve regeneration. *Acta Biomater.* **2019**, *84*, 98–113. [CrossRef] [PubMed]
92. Jin, L.; Wu, D.; Kuddannaya, S.; Zhang, Y.; Wang, Z. Fabrication, Characterization, and Biocompatibility of Polymer Cored Reduced Graphene Oxide Nanofibers. *ACS Appl. Mater. Interfaces* **2016**, *8*, 5170–5177. [CrossRef] [PubMed]
93. Wu, K.; Zhang, X.; Yang, W.; Liu, X.; Jiao, Y.; Zhou, C. Influence of layer-by-layer assembled electrospun poly (l-lactic acid) nanofiber mats on the bioactivity of endothelial cells. *Appl. Surf. Sci.* **2016**, *390*, 838–846. [CrossRef]
94. Shao, W.; He, J.; Wang, Q.; Cui, S.; Ding, B. Biomineralized Poly(l -lactic-co-glycolic acid)/Graphene Oxide/Tussah Silk Fibroin Nanofiber Scaffolds with Multiple Orthogonal Layers Enhance Osteoblastic Differentiation of Mesenchymal Stem Cells. *ACS Biomater. Sci. Eng.* **2017**, *3*, 1370–1380. [CrossRef] [PubMed]
95. Zhang, K.; Zheng, H.; Liang, S.; Gao, C. Aligned PLLA nanofibrous scaffolds coated with graphene oxide for promoting neural cell growth. *Acta Biomater.* **2016**, *37*, 131–142. [CrossRef]
96. Peppas, N.A.; Hilt, J.Z.; Khademhosseini, A.; Langer, R. Hydrogels in biology and medicine: From molecular principles to bionanotechnology. *Adv. Mater.* **2006**, *18*, 1345–1360. [CrossRef]
97. Schäfer, M.K.E.; Altevogt, P. L1CAM malfunction in the nervous system and human carcinomas. *Cell. Mol. Life Sci.* **2010**, *67*, 2425–2437. [CrossRef] [PubMed]
98. Haraguchi, K. Stimuli-responsive nanocomposite gels. *Colloid Polym. Sci.* **2011**, *289*, 455–473. [CrossRef]
99. Xu, Y.; Sheng, K.; Li, C.; Shi, G. Self-Assembled Graphene Hydrogel via a One-Step Hydrothermal Process. *ACS Nano* **2010**, *4*, 4324–4330. [CrossRef]
100. Yang, X.; Qiu, L.; Cheng, C.; Wu, Y.; Ma, Z.-F.; Li, D. Ordered Gelation of Chemically Converted Graphene for Next-Generation Electroconductive Hydrogel Films. *Angew. Chemie Int. Ed.* **2011**, *50*, 7325–7328. [CrossRef]
101. Liao, G.; Hu, J.; Chen, Z.; Zhang, R.; Wang, G.; Kuang, T. Preparation, Properties, and Applications of Graphene-Based Hydrogels. *Front. Chem.* **2018**, *6*, 450. [CrossRef]
102. Prasadh, S.; Suresh, S.; Wong, R. Osteogenic Potential of Graphene in Bone Tissue Engineering Scaffolds. *Materials* **2018**, *11*, 143. [CrossRef]
103. Spicer, C.D. Hydrogel scaffolds for tissue engineering: The importance of polymer choice. *Polym. Chem.* **2020**, *11*, 184–219. [CrossRef]
104. Moradali, M.F.; Ghods, S.; Rehm, B.H.A. *Alginate Biosynthesis and Biotechnological Production BT—Alginates and Their Biomedical Applications*; Rehm, B.H.A., Moradali, M.F., Eds.; Springer: Singapore, 2018; pp. 1–25. ISBN 978-981-10-6910-9.
105. Mansouri, N.; Al-Sarawi, S.F.; Mazumdar, J.; Losic, D. Advancing fabrication and properties of three-dimensional graphene–alginate scaffolds for application in neural tissue engineering. *RSC Adv.* **2019**, *9*, 36838–36848. [CrossRef]
106. Jiao, C.; Xiong, J.; Tao, J.; Xu, S.; Zhang, D.; Lin, H.; Chen, Y. Sodium alginate/graphene oxide aerogel with enhanced strength–toughness and its heavy metal adsorption study. *Int. J. Biol. Macromol.* **2016**, *83*, 133–141. [CrossRef] [PubMed]
107. Sivashankari, P.R.; Prabaharan, M. Three-dimensional porous scaffolds based on agarose/chitosan/graphene oxide composite for tissue engineering. *Int. J. Biol. Macromol.* **2020**, *146*, 222–231. [CrossRef] [PubMed]
108. Balagangadharan, K.; Dhivya, S.; Selvamurugan, N. Chitosan based nanofibers in bone tissue engineering. *Int. J. Biol. Macromol.* **2017**, *104*, 1372–1382. [CrossRef] [PubMed]
109. Wubneh, A.; Tsekoura, E.K.; Ayranci, C.; Uludağ, H. Current state of fabrication technologies and materials for bone tissue engineering. *Acta Biomater.* **2018**, *80*, 1–30. [CrossRef]
110. Liu, Y.; Fang, N.; Liu, B.; Song, L.; Wen, B.; Yang, D. Aligned porous chitosan/graphene oxide scaffold for bone tissue engineering. *Mater. Lett.* **2018**, *233*, 78–81. [CrossRef]
111. Ulijn, R.V.; Smith, A.M. Designing peptide based nanomaterials. *Chem. Soc. Rev.* **2008**, *37*, 664–675. [CrossRef] [PubMed]
112. Zhao, X.; Zhang, S. Designer Self-Assembling Peptide Materials. *Macromol. Biosci.* **2007**, *7*, 13–22. [CrossRef]
113. Wang, L.; Song, D.; Zhang, X.; Ding, Z.; Kong, X.; Lu, Q.; Kaplan, D.L. Silk–Graphene Hybrid Hydrogels with Multiple Cues to Induce Nerve Cell Behavior. *ACS Biomater. Sci. Eng.* **2019**, *5*, 613–622. [CrossRef] [PubMed]

114. Wang, L.; Lu, G.; Lu, Q.; Kaplan, D.L. Controlling Cell Behavior on Silk Nanofiber Hydrogels with Tunable Anisotropic Structures. *ACS Biomater. Sci. Eng.* **2018**, *4*, 933–941. [CrossRef]
115. Piantanida, E.; Alonci, G.; Bertucci, A.; De Cola, L. Design of Nanocomposite Injectable Hydrogels for Minimally Invasive Surgery. *Acc. Chem. Res.* **2019**, *52*, 2101–2112. [CrossRef] [PubMed]
116. Lee, S.J.; Nah, H.; Heo, D.N.; Kim, K.-H.; Seok, J.M.; Heo, M.; Moon, H.-J.; Lee, D.; Lee, J.S.; An, S.Y.; et al. Induction of osteogenic differentiation in a rat calvarial bone defect model using an In situ forming graphene oxide incorporated glycol chitosan/oxidized hyaluronic acid injectable hydrogel. *Carbon N. Y.* **2020**, *168*, 264–277. [CrossRef]
117. Sayyar, S.; Gambhir, S.; Chung, J.; Officer, D.L.; Wallace, G.G. 3D printable conducting hydrogels containing chemically converted graphene. *Nanoscale* **2017**, *9*, 2038–2050. [CrossRef]
118. Fan, L.; Yi, J.; Tong, J.; Zhou, X.; Ge, H.; Zou, S.; Wen, H.; Nie, M. Preparation and characterization of oxidized konjac glucomannan/carboxymethyl chitosan/graphene oxide hydrogel. *Int. J. Biol. Macromol.* **2016**, *91*, 358–367. [CrossRef] [PubMed]
119. Saravanan, S.; Vimalraj, S.; Anuradha, D. Chitosan based thermoresponsive hydrogel containing graphene oxide for bone tissue repair. *Biomed. Pharmacother.* **2018**, *107*, 908–917. [CrossRef]
120. Díez-Pascual, A.M.; Díez-Vicente, A.L. Poly(propylene fumarate)/Polyethylene Glycol-Modified Graphene Oxide Nanocomposites for Tissue Engineering. *ACS Appl. Mater. Interfaces* **2016**, *8*, 17902–17914. [CrossRef]
121. Noh, M.; Kim, S.H.; Kim, J.; Lee, J.R.; Jeong, G.J.; Yoon, J.K.; Kang, S.; Bhang, S.H.; Yoon, H.H.; Lee, J.C.; et al. Graphene oxide reinforced hydrogels for osteogenic differentiation of human adipose-derived stem cells. *RSC Adv.* **2017**, *7*, 20779–20788. [CrossRef]
122. Wu, D.; Bäckström, E.; Hakkarainen, M. Starch Derived Nanosized Graphene Oxide Functionalized Bioactive Porous Starch Scaffolds. *Macromol. Biosci.* **2017**, *17*, 1600397. [CrossRef] [PubMed]
123. Ruan, J.; Wang, X.; Yu, Z.; Wang, Z.; Xie, Q.; Zhang, D.; Huang, Y.; Zhou, H.; Bi, X.; Xiao, C.; et al. Enhanced Physiochemical and Mechanical Performance of Chitosan-Grafted Graphene Oxide for Superior Osteoinductivity. *Adv. Funct. Mater.* **2016**, *26*, 1085–1097. [CrossRef]
124. Zhao, L.; Wu, Y.; Chen, S.; Xing, T. Preparation and characterization of cross-linked carboxymethyl chitin porous membrane scaffold for biomedical applications. *Carbohydr. Polym.* **2015**, *126*, 150–155. [CrossRef]
125. Kim, J.A.; Jang, E.Y.; Kang, T.J.; Yoon, S.; Ovalle-Robles, R.; Rhee, W.J.; Kim, T.; Baughman, R.H.; Kim, Y.H.; Park, T.H. Regulation of morphogenesis and neural differentiation of human mesenchymal stem cells using carbon nanotube sheets. *Integr. Biol. (Camb).* **2012**, *4*, 587–594. [CrossRef]
126. Nishikawa, S.-I.; Jakt, L.M.; Era, T. Embryonic stem-cell culture as a tool for developmental cell biology. *Nat. Rev. Mol. Cell Biol.* **2007**, *8*, 502–507. [CrossRef]
127. Spangrude, G.J.; Heimfeld, S.; Weissman, I.L. Purification and characterization of mouse hematopoietic stem cells. *Science* **1988**, *241*, 58–62. [CrossRef]
128. Pittenger, M.F.; Mackay, A.M.; Beck, S.C.; Jaiswal, R.K.; Douglas, R.; Mosca, J.D.; Moorman, M.A.; Simonetti, D.W.; Craig, S.; Marshak, D.R. Multilineage potential of adult human mesenchymal stem cells. *Science* **1999**, *284*, 143–147. [CrossRef] [PubMed]
129. Nam, H.; Lee, K.-H.; Nam, D.-H.; Joo, K.M. Adult human neural stem cell therapeutics: Current developmental status and prospect. *World J. Stem Cells* **2015**, *7*, 126–136. [CrossRef]
130. Takahashi, K.; Tanabe, K.; Ohnuki, M.; Narita, M.; Ichisaka, T.; Tomoda, K.; Yamanaka, S. Induction of pluripotent stem cells from adult human fibroblasts by defined factors. *Cell* **2007**, *131*, 861–872. [CrossRef] [PubMed]
131. Rashid, S.T.; Alexander, G.J.M. Induced pluripotent stem cells: From Nobel Prizes to clinical applications. *J. Hepatol.* **2013**, *58*, 625–629. [CrossRef]
132. Lanza, R.; Russell, D.W.; Nagy, A. Engineering universal cells that evade immune detection. *Nat. Rev. Immunol.* **2019**, *19*, 723–733. [CrossRef] [PubMed]
133. Barker, J.N.; Boughan, K.; Dahi, P.B.; Devlin, S.M.; Maloy, M.A.; Naputo, K.; Mazis, C.M.; Davis, E.; Nhaissi, M.; Wells, D.; et al. Racial disparities in access to HLA-matched unrelated donor transplants: A prospective 1312-patient analysis. *Blood Adv.* **2019**, *3*, 939–944. [CrossRef] [PubMed]
134. Gornalusse, G.G.; Hirata, R.K.; Funk, S.E.; Riolobos, L.; Lopes, V.S.; Manske, G.; Prunkard, D.; Colunga, A.G.; Hanafi, L.-A.; Clegg, N.; et al. HLA-E-expressing pluripotent stem cells escape allogeneic responses and lysis by NK cells. *Nat. Biotechnol.* **2017**, *35*, 765–772. [CrossRef] [PubMed]
135. Deuse, T.; Hu, X.; Gravina, A.; Wang, D.; Tediashvili, G.; De, C.; Thayer, W.O.; Wahl, A.; Garcia, J.V.; Reichenspurner, H.; et al. Hypoimmunogenic derivatives of induced pluripotent stem cells evade immune rejection in fully immunocompetent allogeneic recipients. *Nat. Biotechnol.* **2019**, *37*, 252–258. [CrossRef]
136. Xu, H.; Wang, B.; Ono, M.; Kagita, A.; Fujii, K.; Sasakawa, N.; Ueda, T.; Gee, P.; Nishikawa, M.; Nomura, M.; et al. Targeted Disruption of HLA Genes via CRISPR-Cas9 Generates iPSCs with Enhanced Immune Compatibility. *Cell Stem Cell* **2019**, *24*, 566–578.e7. [CrossRef]
137. Mouw, J.K.; Ou, G.; Weaver, V.M. Extracellular matrix assembly: A multiscale deconstruction. *Nat. Rev. Mol. Cell Biol.* **2014**, *15*, 771–785. [CrossRef]
138. Alberts, P.; Rudge, R.; Hinners, I.; Muzerelle, A.; Martinez-Arca, S.; Irinopoulou, T.; Marthiens, V.; Tooze, S.; Rathjen, F.; Gaspar, P.; et al. Cross talk between tetanus neurotoxin-insensitive vesicle-associated membrane protein-mediated transport and L1-mediated adhesion. *Mol. Biol. Cell* **2003**, *14*, 4207–4220. [CrossRef] [PubMed]

139. Matsumoto, K.; Sato, C.; Naka, Y.; Kitazawa, A.; Whitby, R.L.D.; Shimizu, N. Neurite outgrowths of neurons with neurotrophin-coated carbon nanotubes. *J. Biosci. Bioeng.* **2007**, *103*, 216–220. [CrossRef]
140. Von der Mark, K.; Park, J.; Bauer, S.; Schmuki, P. Nanoscale engineering of biomimetic surfaces: Cues from the extracellular matrix. *Cell Tissue Res.* **2010**, *339*, 131–153. [CrossRef]
141. Chen, H.; Yuan, L.; Song, W.; Wu, Z.; Li, D. Biocompatible polymer materials: Role of protein–surface interactions. *Prog. Polym. Sci.* **2008**, *33*, 1059–1087. [CrossRef]
142. Shekaran, A.; Garcia, A.J. Nanoscale engineering of extracellular matrix-mimetic bioadhesive surfaces and implants for tissue engineering. *Biochim. Biophys. Acta* **2011**, *1810*, 350–360. [CrossRef]
143. Berezin, V.; Bock, E. NCAM mimetic peptides: Pharmacological and therapeutic potential. *J. Mol. Neurosci.* **2004**, *22*, 33–39. [CrossRef]
144. Scapin, G.; Salice, P.; Tescari, S.; Menna, E.; De Filippis, V.; Filippini, F. Enhanced neuronal cell differentiation combining biomimetic peptides and a carbon nanotube-polymer scaffold. *Nanomed. Nanotechnol. Biol. Med.* **2015**, *11*, 621–632. [CrossRef]
145. Nikkhah, M.; Edalat, F.; Manoucheri, S.; Khademhosseini, A. Engineering microscale topographies to control the cell-substrate interface. *Biomaterials* **2012**, *33*, 5230–5246. [CrossRef] [PubMed]
146. Engler, A.J.; Sen, S.; Sweeney, H.L.; Discher, D.E. Matrix elasticity directs stem cell lineage specification. *Cell* **2006**, *126*, 677–689. [CrossRef] [PubMed]
147. Pettikiriarachchi, J.T.S.; Parish, C.L.; Shoichet, M.S.; Forsythe, J.S.; Nisbet, D.R. Biomaterials for Brain Tissue Engineering. *Aust. J. Chem.* **2010**, *63*, 1143–1154. [CrossRef]
148. Nicolas, J.; Magli, S.; Rabbachin, L.; Sampaolesi, S.; Nicotra, F.; Russo, L. 3D Extracellular Matrix Mimics: Fundamental Concepts and Role of Materials Chemistry to Influence Stem Cell Fate. *Biomacromolecules* **2020**, *21*, 1968–1994. [CrossRef]
149. Landers, J.; Turner, J.T.; Heden, G.; Carlson, A.L.; Bennett, N.K.; Moghe, P.V.; Neimark, A. V Carbon nanotube composites as multifunctional substrates for in situ actuation of differentiation of human neural stem cells. *Adv. Healthc. Mater.* **2014**, *3*, 1745–1752. [CrossRef]
150. Yao, L.; Pandit, A.; Yao, S.; McCaig, C.D. Electric field-guided neuron migration: A novel approach in neurogenesis. *Tissue Eng. Part B. Rev.* **2011**, *17*, 143–153. [CrossRef]
151. Ning, C.; Zhou, Z.; Tan, G.; Zhu, Y.; Mao, C. Electroactive polymers for tissue regeneration: Developments and perspectives. *Prog. Polym. Sci.* **2018**, *81*, 144–162. [CrossRef] [PubMed]
152. Fukada, E. History and recent progress in piezoelectric polymers. *IEEE Trans. Ultrason. Ferroelectr. Freq. Control* **2000**, *47*, 1277–1290. [CrossRef]
153. Yao, L.; McCaig, C.D.; Zhao, M. Electrical signals polarize neuronal organelles, direct neuron migration, and orient cell division. *Hippocampus* **2009**, *19*, 855–868. [CrossRef] [PubMed]
154. Schmidt, C.E.; Shastri, V.R.; Vacanti, J.P.; Langer, R. Stimulation of neurite outgrowth using an electrically conducting polymer. *Proc. Natl. Acad. Sci. USA* **1997**, *94*, 8948–8953. [CrossRef]
155. Lovat, V.; Pantarotto, D.; Lagostena, L.; Cacciari, B.; Grandolfo, M.; Righi, M.; Spalluto, G.; Prato, M.; Ballerini, L. Carbon Nanotube Substrates Boost Neuronal Electrical Signaling. *Nano Lett.* **2005**, *5*, 1107–1110. [CrossRef]
156. Isaacson, B.M.; Bloebaum, R.D. Bone bioelectricity: What have we learned in the past 160 years? *J. Biomed. Mater. Res. A* **2010**, *95*, 1270–1279. [CrossRef]
157. Defterali, Ç.; Verdejo, R.; Majeed, S.; Boschetti-de-Fierro, A.; Méndez-Gómez, H.R.; Díaz-Guerra, E.; Fierro, D.; Buhr, K.; Abetz, C.; Martínez-Murillo, R.; et al. In Vitro Evaluation of Biocompatibility of Uncoated Thermally Reduced Graphene and Carbon Nanotube-Loaded PVDF Membranes with Adult Neural Stem Cell-Derived Neurons and Glia. *Front. Bioeng. Biotechnol.* **2016**, *4*, 94. [CrossRef] [PubMed]
158. Zhou, L.; Fan, L.; Yi, X.; Zhou, Z.; Liu, C.; Fu, R.; Dai, C.; Wang, Z.; Chen, X.; Yu, P.; et al. Soft Conducting Polymer Hydrogels Cross-Linked and Doped by Tannic Acid for Spinal Cord Injury Repair. *ACS Nano* **2018**, *12*, 10957–10967. [CrossRef]
159. Scapin, G.; Bertalot, T.; Vicentini, N.; Gatti, T.; Tescari, S.; De Filippis, V.; Marega, C.; Menna, E.; Gasparella, M.; Parnigotto, P.P.; et al. Neuronal commitment of human circulating multipotent cells by carbon nanotube-polymer scaffolds and biomimetic peptides. *Nanomedicine* **2016**, *11*, 1929–1946. [CrossRef] [PubMed]
160. Guo, W.; Zhang, X.; Yu, X.; Wang, S.; Qiu, J.; Tang, W.; Li, L.; Liu, H.; Wang, Z.L. Self-Powered Electrical Stimulation for Enhancing Neural Differentiation of Mesenchymal Stem Cells on Graphene-Poly(3,4-ethylenedioxythiophene) Hybrid Microfibers. *ACS Nano* **2016**, *10*, 5086–5095. [CrossRef] [PubMed]
161. Navaei, A.; Saini, H.; Christenson, W.; Sullivan, R.T.; Ros, R.; Nikkhah, M. Gold nanorod-incorporated gelatin-based conductive hydrogels for engineering cardiac tissue constructs. *Acta Biomater.* **2016**, *41*, 133–146. [CrossRef]
162. Liu, Y.; Lu, J.; Xu, G.; Wei, J.; Zhang, Z.; Li, X. Tuning the conductivity and inner structure of electrospun fibers to promote cardiomyocyte elongation and synchronous beating. *Mater. Sci. Eng. C. Mater. Biol. Appl.* **2016**, *69*, 865–874. [CrossRef]
163. Chen, J.; Yu, M.; Guo, B.; Ma, P.X.; Yin, Z. Conductive nanofibrous composite scaffolds based on in-situ formed polyaniline nanoparticle and polylactide for bone regeneration. *J. Colloid Interface Sci.* **2018**, *514*, 517–527. [CrossRef]
164. Comisar, W.A.; Hsiong, S.X.; Kong, H.-J.; Mooney, D.J.; Linderman, J.J. Multi-scale modeling to predict ligand presentation within RGD nanopatterned hydrogels. *Biomaterials* **2006**, *27*, 2322–2329. [CrossRef] [PubMed]
165. Abrams, G.A.; Goodman, S.L.; Nealey, P.F.; Franco, M.; Murphy, C.J. Nanoscale topography of the basement membrane underlying the corneal epithelium of the rhesus macaque. *Cell Tissue Res.* **2000**, *299*, 39–46. [CrossRef]

166. Curtis, A.S.; Casey, B.; Gallagher, J.O.; Pasqui, D.; Wood, M.A.; Wilkinson, C.D. Substratum nanotopography and the adhesion of biological cells. Are symmetry or regularity of nanotopography important? *Biophys. Chem.* **2001**, *94*, 275–283. [CrossRef]
167. Rodríguez Hernández, J.C.; Salmerón Sánchez, M.; Soria, J.M.; Gómez Ribelles, J.L.; Monleón Pradas, M. Substrate chemistry-dependent conformations of single laminin molecules on polymer surfaces are revealed by the phase signal of atomic force microscopy. *Biophys. J.* **2007**, *93*, 202–207. [CrossRef] [PubMed]
168. Kim, N.J.; Lee, S.J.; Atala, A. 1—Biomedical nanomaterials in tissue engineering. In *Woodhead Publishing Series in Biomaterials*; Gaharwar, A.K., Sant, S., Hancock, M.J., Hacking, S.A., Eds.; Woodhead Publishing: Cambridge, UK, 2013; pp. 1–25. ISBN 978-0-85709-596-1.
169. Arnold, M.; Cavalcanti-Adam, E.A.; Glass, R.; Blümmel, J.; Eck, W.; Kantlehner, M.; Kessler, H.; Spatz, J.P. Activation of integrin function by nanopatterned adhesive interfaces. *Chemphyschem* **2004**, *5*, 383–388. [CrossRef]
170. Yang, K.; Jung, K.; Ko, E.; Kim, J.; Park, K.I.; Kim, J.; Cho, S.-W. Nanotopographical manipulation of focal adhesion formation for enhanced differentiation of human neural stem cells. *ACS Appl. Mater. Interfaces* **2013**, *5*, 10529–10540. [CrossRef] [PubMed]
171. Brunetti, V.; Maiorano, G.; Rizzello, L.; Sorce, B.; Sabella, S.; Cingolani, R.; Pompa, P.P. Neurons sense nanoscale roughness with nanometer sensitivity. *Proc. Natl. Acad. Sci. USA* **2010**, *107*, 6264–6269. [CrossRef]
172. Staii, C.; Viesselmann, C.; Ballweg, J.; Williams, J.C.; Dent, E.W.; Coppersmith, S.N.; Eriksson, M.A. Distance dependence of neuronal growth on nanopatterned gold surfaces. *Langmuir* **2011**, *27*, 233–239. [CrossRef]
173. Silva, G.A. Nanotechnology approaches for drug and small molecule delivery across the blood brain barrier. *Surg. Neurol.* **2007**, *67*, 113–116. [CrossRef] [PubMed]
174. Kotov, N.A.; Winter, J.O.; Clements, I.P.; Jan, E.; Timko, B.P.; Campidelli, S.; Pathak, S.; Mazzatenta, A.; Lieber, C.M.; Prato, M.; et al. Nanomaterials for Neural Interfaces. *Adv. Mater.* **2009**, *21*, 3970–4004. [CrossRef]
175. Nel, A.E.; Mädler, L.; Velegol, D.; Xia, T.; Hoek, E.M.V.; Somasundaran, P.; Klaessig, F.; Castranova, V.; Thompson, M. Understanding biophysicochemical interactions at the nano–bio interface. *Nat. Mater.* **2009**, *8*, 543–557. [CrossRef]
176. Umadevi, D.; Sastry, G.N. Impact of the chirality and curvature of carbon nanostructures on their interaction with aromatics and amino acids. *ChemPhysChem* **2013**, *14*, 2570–2578. [CrossRef]
177. Kumar, S.; Parekh, S.H. Linking graphene-based material physicochemical properties with molecular adsorption, structure and cell fate. *Commun. Chem.* **2020**, *3*, 8. [CrossRef]
178. Georgakilas, V.; Otyepka, M.; Bourlinos, A.B.; Chandra, V.; Kim, N.; Kemp, K.C.; Hobza, P.; Zboril, R.; Kim, K.S. Functionalization of graphene: Covalent and non-covalent approaches, derivatives and applications. *Chem. Rev.* **2012**, *112*, 6156–6214. [CrossRef] [PubMed]
179. Lalwani, G.; D'Agati, M.; Khan, A.M.; Sitharaman, B. Toxicology of graphene-based nanomaterials. *Adv. Drug Deliv. Rev.* **2016**, *105*, 109–144. [CrossRef] [PubMed]
180. Hu, W.; Peng, C.; Luo, W.; Lv, M.; Li, X.; Li, D.; Huang, Q.; Fan, C. Graphene-based antibacterial paper. *ACS Nano* **2010**, *4*, 4317–4323. [CrossRef]
181. Liu, S.; Zeng, T.H.; Hofmann, M.; Burcombe, E.; Wei, J.; Jiang, R.; Kong, J.; Chen, Y. Antibacterial activity of graphite, graphite oxide, graphene oxide, and reduced graphene oxide: Membrane and oxidative stress. *ACS Nano* **2011**, *5*, 6971–6980. [CrossRef]
182. Kuo, W.-S.; Chen, H.-H.; Chen, S.-Y.; Chang, C.-Y.; Chen, P.-C.; Hou, Y.-I.; Shao, Y.-T.; Kao, H.-F.; Lilian Hsu, C.-L.; Chen, Y.-C.; et al. Graphene quantum dots with nitrogen-doped content dependence for highly efficient dual-modality photodynamic antimicrobial therapy and bioimaging. *Biomaterials* **2017**, *120*, 185–194. [CrossRef]
183. Prasad, K.; Lekshmi, G.S.; Ostrikov, K.; Lussini, V.; Blinco, J.; Mohandas, M.; Vasilev, K.; Bottle, S.; Bazaka, K.; Ostrikov, K. Synergic bactericidal effects of reduced graphene oxide and silver nanoparticles against Gram-positive and Gram-negative bacteria. *Sci. Rep.* **2017**, *7*. [CrossRef]
184. Vallabani, N.V.S.; Mittal, S.; Shukla, R.K.; Pandey, A.K.; Dhakate, S.R.; Pasricha, R.; Dhawan, A. Toxicity of graphene in normal human lung cells (BEAS-2B). *J. Biomed. Nanotechnol.* **2011**, *7*, 106–107. [CrossRef]
185. Nasirzadeh, N.; Azari, M.R.; Rasoulzadeh, Y.; Mohammadian, Y. An assessment of the cytotoxic effects of graphene nanoparticles on the epithelial cells of the human lung. *Toxicol. Ind. Health* **2019**, *35*, 79–87. [CrossRef] [PubMed]
186. Fujita, K.; Take, S.; Tani, R.; Maru, J.; Obara, S.; Endoh, S. Assessment of cytotoxicity and mutagenicity of exfoliated graphene. *Toxicol. Vitr.* **2018**, *52*. [CrossRef]
187. Yang, Z.; Pan, Y.; Chen, T.; Li, L.; Zou, W.; Liu, D.; Xue, D.; Wang, X.; Lin, G. Cytotoxicity and Immune Dysfunction of Dendritic Cells Caused by Graphene Oxide. *Front. Pharmacol.* **2020**, *11*. [CrossRef] [PubMed]
188. Dallavalle, M.; Calvaresi, M.; Bottoni, A.; Melle-Franco, M.; Zerbetto, F. Graphene can wreak havoc with cell membranes. *ACS Appl. Mater. Interfaces* **2015**, *7*, 4406–4414. [CrossRef]
189. Chatterjee, N.; Eom, H.J.; Choi, J. A systems toxicology approach to the surface functionality control of graphene-cell interactions. *Biomaterials* **2014**, *35*, 1109–1127. [CrossRef]
190. Askari, E.; Naghib, S.M.; Seyfoori, A.; Maleki, A.; Rahmanian, M. Ultrasonic-assisted synthesis and in vitro biological assessments of a novel herceptin-stabilized graphene using three dimensional cell spheroid. *Ultrason. Sonochem.* **2019**, *58*, 104615. [CrossRef]
191. Narayanan, K.B.; Kim, H.D.; Han, S.S. Biocompatibility and hemocompatibility of hydrothermally derived reduced graphene oxide using soluble starch as a reducing agent. *Colloids Surfaces B Biointerfaces* **2020**, *185*. [CrossRef]
192. Palmieri, V.; Perini, G.; De Spirito, M.; Papi, M. Graphene oxide touches blood: In vivo interactions of bio-coronated 2D materials. *Nanoscale Horizons* **2019**, *4*. [CrossRef] [PubMed]

193. Feng, R.; Yu, Y.; Shen, C.; Jiao, Y.; Zhou, C. Impact of graphene oxide on the structure and function of important multiple blood components by a dose-dependent pattern. *J. Biomed. Mater. Res. Part A* **2015**, *103*. [CrossRef] [PubMed]
194. Singh Bakshi, M. Nanotoxicity in Systemic Circulation and Wound Healing. *Chem. Res. Toxicol.* **2017**, *30*, 1253–1274. [CrossRef] [PubMed]
195. Jaworski, S.; Hinzmann, M.; Sawosz, E.; Grodzik, M.; Kutwin, M.; Wierzbicki, M.; Strojny, B.; Vadalasetty, K.P.; Lipińska, L.; Chwalibog, A. Interaction of different forms of graphene with chicken embryo red blood cells. *Environ. Sci. Pollut. Res.* **2017**, *24*. [CrossRef]
196. Duan, G.; Kang, S.G.; Tian, X.; Garate, J.A.; Zhao, L.; Ge, C.; Zhou, R. Protein corona mitigates the cytotoxicity of graphene oxide by reducing its physical interaction with cell membrane. *Nanoscale* **2015**, *7*, 15214–15224. [CrossRef]
197. Sasidharan, A.; Panchakarla, L.S.; Sadanandan, A.R.; Ashokan, A.; Chandran, P.; Girish, C.M.; Menon, D.; Nair, S.V.; Rao, C.N.R.; Koyakutty, M. Hemocompatibility and macrophage response of pristine and functionalized graphene. *Small* **2012**, *8*, 1251–1263. [CrossRef]
198. Sopotnik, M.; Leonardi, A.; Križaj, I.; Dušak, P.; Makovec, D.; Mesarič, T.; Ulrih, N.P.; Junkar, I.; Sepčić, K.; Drobne, D. Comparative study of serum protein binding to three different carbon-based nanomaterials. *Carbon N. Y.* **2015**, *95*, 560–572. [CrossRef]
199. Leso, V.; Fontana, L.; Iavicoli, I. Nanomaterial exposure and sterile inflammatory reactions. *Toxicol. Appl. Pharmacol.* **2018**, *355*, 80–92. [CrossRef] [PubMed]
200. Dobrovolskaia, M.A.; Aggarwal, P.; Hall, J.B.; McNeil, S.E. Preclinical Studies To Understand Nanoparticle Interaction with the Immune System and Its Potential Effects on Nanoparticle Biodistribution. *Mol. Pharm.* **2008**, *5*, 487–495. [CrossRef] [PubMed]
201. Keshavan, S.; Calligari, P.; Stella, L.; Fusco, L.; Delogu, L.G.; Fadeel, B. Nano-bio interactions: A neutrophil-centric view. *Cell Death Dis.* **2019**, *10*. [CrossRef]
202. Lin, H.; Ji, D.K.; Lucherelli, M.A.; Reina, G.; Ippolito, S.; Samorì, P.; Bianco, A. Comparative Effects of Graphene and Molybdenum Disulfide on Human Macrophage Toxicity. *Small* **2020**, *16*. [CrossRef] [PubMed]
203. Furze, R.C.; Rankin, S.M. The role of the bone marrow in neutrophil clearance under homeostatic conditions in the mouse. *FASEB J.* **2008**, *22*, 3111–3119. [CrossRef]
204. Chang, T.K.; Lu, Y.C.; Yeh, S.T.; Lin, T.C.; Huang, C.H.; Huang, C.H. In vitro and in vivo biological responses to graphene and graphene oxide: A murine calvarial animal study. *Int. J. Nanomed.* **2020**, *15*, 647–659. [CrossRef] [PubMed]
205. Dos Reis, S.R.R.; Pinto, S.R.; de Menezes, F.D.; Martinez-Manez, R.; Ricci-Junior, E.; Alencar, L.M.R.; Helal-Neto, E.; da Silva de Barros, A.O.; Lisboa, P.C.; Santos-Oliveira, R. Senescence and the Impact on Biodistribution of Different Nanosystems: The Discrepancy on Tissue Deposition of Graphene Quantum Dots, Polycaprolactone Nanoparticle and Magnetic Mesoporous Silica Nanoparticles in Young and Elder Animals. *Pharm. Res.* **2020**, *37*. [CrossRef] [PubMed]
206. Xiong, G.; Deng, Y.; Liao, X.; Zhang, J.; Cheng, B.; Cao, Z.; Lu, H. Graphene oxide nanoparticles induce hepatic dysfunction through the regulation of innate immune signaling in zebrafish (Danio rerio). *Nanotoxicology* **2020**, *14*, 667–682. [CrossRef] [PubMed]
207. Lin, Y.; Zhang, Y.; Li, J.; Kong, H.; Yan, Q.; Zhang, J.; Li, W.; Ren, N.; Cui, Y.; Zhang, T.; et al. Blood exposure to graphene oxide may cause anaphylactic death in non-human primates. *Nano Today* **2020**, *35*. [CrossRef]
208. El-Yamany, N.A.; Mohamed, F.F.; Salaheldin, T.A.; Tohamy, A.A.; Abd El-Mohsen, W.N.; Amin, A.S. Graphene oxide nanosheets induced genotoxicity and pulmonary injury in mice. *Exp. Toxicol. Pathol.* **2017**, *69*. [CrossRef]
209. Sasidharan, A.; Swaroop, S.; Koduri, C.K.; Girish, C.M.; Chandran, P.; Panchakarla, L.S.; Somasundaram, V.H.; Gowd, G.S.; Nair, S.; Koyakutty, M. Comparative in vivo toxicity, organ biodistribution and immune response of pristine, carboxylated and PEGylated few-layer graphene sheets in Swiss albino mice: A three month study. *Carbon N. Y.* **2015**, *95*. [CrossRef]
210. Xiaoli, F.; Qiyue, C.; Weihong, G.; Yaqing, Z.; Chen, H.; Junrong, W.; Longquan, S. Toxicology data of graphene-family nanomaterials: An update. *Arch. Toxicol.* **2020**, *94*. [CrossRef]
211. Mohamed, H.R.H.; Welson, M.; Yaseen, A.E.; El-Ghor, A. Induction of chromosomal and DNA damage and histological alterations by graphene oxide nanoparticles in Swiss mice. *Drug Chem. Toxicol.* **2019**, 1–11. [CrossRef]
212. Mao, L.; Hu, M.; Pan, B.; Xie, Y.; Petersen, E.J. Biodistribution and toxicity of radio-labeled few layer graphene in mice after intratracheal instillation. *Part. Fibre Toxicol.* **2016**, *13*. [CrossRef]
213. Krajnak, K.; Waugh, S.; Stefaniak, A.; Schwegler-Berry, D.; Roach, K.; Barger, M.; Roberts, J. Exposure to graphene nanoparticles induces changes in measures of vascular/renal function in a load and form-dependent manner in mice. *J. Toxicol. Environ. Heal. Part A* **2019**, *82*, 711–726. [CrossRef] [PubMed]
214. Rodrigues, A.F.; Newman, L.; Jasim, D.; Mukherjee, S.P.; Wang, J.; Vacchi, I.A.; Ménard-Moyon, C.; Bianco, A.; Fadeel, B.; Kostarelos, K.; et al. Size-Dependent Pulmonary Impact of Thin Graphene Oxide Sheets in Mice: Toward Safe-by-Design. *Adv. Sci.* **2020**, *7*. [CrossRef]
215. Zhang, D.; Zhang, Z.; Liu, Y.; Chu, M.; Yang, C.; Li, W.; Shao, Y.; Yue, Y.; Xu, R. The short- and long-term effects of orally administered high-dose reduced graphene oxide nanosheets on mouse behaviors. *Biomaterials* **2015**, *68*. [CrossRef] [PubMed]
216. Ferrari, A.C.; Bonaccorso, F.; Fal'ko, V.; Novoselov, K.S.; Roche, S.; Bøggild, P.; Borini, S.; Koppens, F.H.L.; Palermo, V.; Pugno, N.; et al. Science and technology roadmap for graphene, related two-dimensional crystals, and hybrid systems. *Nanoscale* **2015**, *7*. [CrossRef] [PubMed]

217. Newman, L.; Jasim, D.A.; Prestat, E.; Lozano, N.; De Lazaro, I.; Nam, Y.; Assas, B.M.; Pennock, J.; Haigh, S.J.; Bussy, C.; et al. Splenic Capture and in Vivo Intracellular Biodegradation of Biological-Grade Graphene Oxide Sheets. *ACS Nano* **2020**, *14*, 10168–10186. [CrossRef] [PubMed]
218. Bussy, C.; Ali-Boucetta, H.; Kostarelos, K. Safety considerations for graphene: Lessons learnt from carbon nanotubes. *Acc. Chem. Res.* **2013**, *46*, 692–701. [CrossRef]
219. Pinto, A.M.; Gonçalves, C.; Sousa, D.M.; Ferreira, A.R.; Moreira, J.A.; Gonçalves, I.C.; Magalhães, F.D. Smaller particle size and higher oxidation improves biocompatibility of graphene-based materials. *Carbon N. Y.* **2016**, *99*. [CrossRef]
220. Yan, L.; Zhao, F.; Li, S.; Hu, Z.; Zhao, Y. Low-toxic and safe nanomaterials by surface-chemical design, carbon nanotubes, fullerenes, metallofullerenes, and graphenes. *Nanoscale* **2011**, *3*. [CrossRef]
221. Qu, Y.; He, F.; Yu, C.; Liang, X.; Liang, D.; Ma, L.; Zhang, Q.; Lv, J.; Wu, J. Advances on graphene-based nanomaterials for biomedical applications. *Mater. Sci. Eng. C* **2018**, *90*. [CrossRef] [PubMed]
222. Rahmati, M.; Mozafari, M. Biological response to carbon-family nanomaterials: Interactions at the nano-bio interface. *Front. Bioeng. Biotechnol.* **2019**, *7*. [CrossRef]
223. Wojtoniszak, M.; Chen, X.; Kalenczuk, R.J.; Wajda, A.; Łapczuk, J.; Kurzewski, M.; Drozdzik, M.; Chu, P.K.; Borowiak-Palen, E. Synthesis, dispersion, and cytocompatibility of graphene oxide and reduced graphene oxide. *Colloids Surfaces B Biointerfaces* **2012**, *89*, 79–85. [CrossRef]
224. Di Luca, M.; Vittorio, O.; Cirillo, G.; Curcio, M.; Czuban, M.; Voli, F.; Farfalla, A.; Hampel, S.; Nicoletta, F.P.; Iemma, F. Electro-responsive graphene oxide hydrogels for skin bandages: The outcome of gelatin and trypsin immobilization. *Int. J. Pharm.* **2018**, *546*. [CrossRef]
225. Cirillo, G.; Curcio, M.; Spizzirri, U.G.; Vittorio, O.; Tucci, P.; Picci, N.; Iemma, F.; Hampel, S.; Nicoletta, F.P. Carbon nanotubes hybrid hydrogels for electrically tunable release of Curcumin. *Eur. Polym. J.* **2017**, *90*. [CrossRef]
226. Li, Y.; Feng, L.; Shi, X.; Wang, X.; Yang, Y.; Yang, K.; Liu, T.; Yang, G.; Liu, Z. Surface coating-dependent cytotoxicity and degradation of graphene derivatives: Towards the design of non-toxic, degradable nano-graphene. *Small* **2014**, *10*, 1544–1554. [CrossRef]
227. Hu, W.; Peng, C.; Lv, M.; Li, X.; Zhang, Y.; Chen, N.; Fan, C.; Huang, Q. Protein Corona-Mediated Mitigation of Cytotoxicity of Graphene Oxide. *ACS Nano* **2011**, *5*, 3693–3700. [CrossRef]
228. Chong, Y.; Ge, C.; Yang, Z.; Antonio Garate, J.; Gu, Z.K.; Weber, J.; Liu, J.; Zhou, R. Reduced Cytotoxicity of Graphene Nanosheets Mediated by Blood-Protein Coating. *ACS Nano* **2015**, *9*, 5713–5724. [CrossRef] [PubMed]
229. Kim, B.C.; Lee, I.; Kwon, S.J.; Wee, Y.; Kwon, K.Y.; Jeon, C.; An, H.J.; Jung, H.T.; Ha, S.; Dordick, J.S.; et al. Fabrication of enzyme-based coatings on intact multi-walled carbon nanotubes as highly effective electrodes in biofuel cells. *Sci. Rep.* **2017**, *7*. [CrossRef]
230. Bussy, C.; Kostarelos, K. Culture Media Critically Influence Graphene Oxide Effects on Plasma Membranes. *Chem* **2017**, *2*, 322–323. [CrossRef]
231. Bhattacharya, K.; Mukherjee, S.P.; Gallud, A.; Burkert, S.C.; Bistarelli, S.; Bellucci, S.; Bottini, M.; Star, A.; Fadeel, B. Biological interactions of carbon-based nanomaterials: From coronation to degradation. *Nanomed. Nanotechnol. Biol. Med.* **2016**, *12*. [CrossRef] [PubMed]
232. Punetha, V.D.; Rana, S.; Yoo, H.J.; Chaurasia, A.; McLeskey, J.T.; Ramasamy, M.S.; Sahoo, N.G.; Cho, J.W. Functionalization of carbon nanomaterials for advanced polymer nanocomposites: A comparison study between CNT and graphene. *Prog. Polym. Sci.* **2017**, *67*, 1–47. [CrossRef]
233. Kiew, S.F.; Kiew, L.V.; Lee, H.B.; Imae, T.; Chung, L.Y. Assessing biocompatibility of graphene oxide-based nanocarriers: A review. *J. Control. Release* **2016**, *226*, 217–228. [CrossRef]
234. Cirillo, G.; Peitzsch, C.; Vittorio, O.; Curcio, M.; Farfalla, A.; Voli, F.; Dubrovska, A.; Iemma, F.; Kavallaris, M.; Hampel, S. When polymers meet carbon nanostructures: Expanding horizons in cancer therapy. *Future Med. Chem.* **2019**, *11*, 2205–2231. [CrossRef]
235. Madni, A.; Noreen, S.; Maqbool, I.; Rehman, F.; Batool, A.; Kashif, P.M.; Rehman, M.; Tahir, N.; Khan, M.I. Graphene-based nanocomposites: Synthesis and their theranostic applications. *J. Drug Target.* **2018**, *26*, 858–883. [CrossRef]
236. Yang, H.; Bremner, D.H.; Tao, L.; Li, H.; Hu, Y.; Zhu, L. Carboxymethyl chitosan-mediated synthesis of hyaluronic acid-targeted graphene oxide for cancer drug delivery. *Carbohydr. Polym.* **2016**, *135*. [CrossRef] [PubMed]
237. Zhang, S.; Yang, K.; Feng, L.; Liu, Z. In vitro and in vivo behaviors of dextran functionalized graphene. *Carbon N. Y.* **2011**, *49*. [CrossRef]
238. Nezakati, T.; Seifalian, A.; Tan, A.M.; Seifalian, A. Conductive Polymers: Opportunities and Challenges in Biomedical Applications. *Chem. Rev.* **2018**, *118*, 6766–6843. [CrossRef]
239. Mahajan, S.; Patharkar, A.; Kuche, K.; Maheshwari, R.; Deb, P.K.; Kalia, K.; Tekade, R.K. Functionalized carbon nanotubes as emerging delivery system for the treatment of cancer. *Int. J. Pharm.* **2018**, *548*. [CrossRef]
240. Dimitriou, R.; Jones, E.; McGonagle, D.; Giannoudis, P.V. Bone regeneration: Current concepts and future directions. *BMC Med.* **2011**, *9*, 66. [CrossRef] [PubMed]
241. Majidinia, M.; Sadeghpour, A.; Yousefi, B. The roles of signaling pathways in bone repair and regeneration. *J. Cell. Physiol.* **2018**, *233*, 2937–2948. [CrossRef]
242. Lichte, P.; Pape, H.C.; Pufe, T.; Kobbe, P.; Fischer, H. Scaffolds for bone healing: Concepts, materials and evidence. *Injury* **2011**, *42*, 569–573. [CrossRef] [PubMed]

243. Depan, D.; Girase, B.; Shah, J.S.; Misra, R.D.K. Structure–process–property relationship of the polar graphene oxide-mediated cellular response and stimulated growth of osteoblasts on hybrid chitosan network structure nanocomposite scaffolds. *Acta Biomater.* **2011**, *7*, 3432–3445. [CrossRef]
244. Arnold, A.M.; Holt, B.D.; Daneshmandi, L.; Laurencin, C.T.; Sydlik, S.A. Phosphate graphene as an intrinsically osteoinductive scaffold for stem cell-driven bone regeneration. *Proc. Natl. Acad. Sci. USA* **2019**, *116*, 4855–4860. [CrossRef] [PubMed]
245. Li, M.; Li, H.; Pan, Q.; Gao, C.; Wang, Y.; Yang, S.; Zan, X.; Guan, Y. Graphene Oxide and Lysozyme Ultrathin Films with Strong Antibacterial and Enhanced Osteogenesis. *Langmuir* **2019**, *35*, 6752–6761. [CrossRef]
246. Li, J.; Liu, X.; Crook, J.M.; Wallace, G.G. 3D Printing of Cytocompatible Graphene/Alginate Scaffolds for Mimetic Tissue Constructs. *Front. Bioeng. Biotechnol.* **2020**, *8*, 824. [CrossRef]
247. Wu, J.; Zheng, A.; Liu, Y.; Jiao, D.; Zeng, D.; Wang, X.; Cao, L.; Jiang, X. Enhanced bone regeneration of the silk fibroin electrospun scaffolds through the modification of the graphene oxide functionalized by BMP-2 peptide. *Int. J. Nanomed.* **2019**, *14*, 733–751. [CrossRef]
248. Chal, J.; Pourquié, O. Making muscle: Skeletal myogenesis in vivo and in vitro. *Development* **2017**, *144*, 2104–2122. [CrossRef] [PubMed]
249. Ciciliot, S.; Schiaffino, S. Regeneration of mammalian skeletal muscle. Basic mechanisms and clinical implications. *Curr. Pharm. Des.* **2010**, *16*, 906–914. [CrossRef] [PubMed]
250. Corona, B.T.; Rivera, J.C.; Owens, J.G.; Wenke, J.C.; Rathbone, C.R. Volumetric muscle loss leads to permanent disability following extremity trauma. *J. Rehabil. Res. Dev.* **2015**, *52*, 785–792. [CrossRef]
251. Gilbert, P.M.; Havenstrite, K.L.; Magnusson, K.E.G.; Sacco, A.; Leonardi, N.A.; Kraft, P.; Nguyen, N.K.; Thrun, S.; Lutolf, M.P.; Blau, H.M. Substrate elasticity regulates skeletal muscle stem cell self-renewal in culture. *Science* **2010**, *329*, 1078–1081. [CrossRef]
252. Jun, I.; Jeong, S.; Shin, H. The stimulation of myoblast differentiation by electrically conductive sub-micron fibers. *Biomaterials* **2009**, *30*, 2038–2047. [CrossRef]
253. Nikolić, N.; Skaret Bakke, S.; Tranheim Kase, E.; Rudberg, I.; Flo Halle, I.; Rustan, A.C.; Thoresen, G.H.; Aas, V. Electrical Pulse Stimulation of Cultured Human Skeletal Muscle Cells as an In Vitro Model of Exercise. *PLoS ONE* **2012**, *7*, e33203. [CrossRef]
254. Jo, S.B.; Erdenebileg, U.; Dashnyam, K.; Jin, G.-Z.; Cha, J.-R.; El-Fiqi, A.; Knowles, J.C.; Patel, K.D.; Lee, H.-H.; Lee, J.-H.; et al. Nano-graphene oxide/polyurethane nanofibers: Mechanically flexible and myogenic stimulating matrix for skeletal tissue engineering. *J. Tissue Eng.* **2020**, *11*, 2041731419900424. [CrossRef]
255. Park, J.; Choi, J.H.; Kim, S.; Jang, I.; Jeong, S.; Lee, J.Y. Micropatterned conductive hydrogels as multifunctional muscle-mimicking biomaterials: Graphene-incorporated hydrogels directly patterned with femtosecond laser ablation. *Acta Biomater.* **2019**, *97*, 141–153. [CrossRef] [PubMed]
256. Lee, J.; Manoharan, V.; Cheung, L.; Lee, S.; Cha, B.-H.; Newman, P.; Farzad, R.; Mehrotra, S.; Zhang, K.; Khan, F.; et al. Nanoparticle-Based Hybrid Scaffolds for Deciphering the Role of Multimodal Cues in Cardiac Tissue Engineering. *ACS Nano* **2019**, *13*, 12525–12539. [CrossRef]
257. Choe, G.; Kim, S.-W.; Park, J.; Park, J.; Kim, S.; Kim, Y.S.; Ahn, Y.; Jung, D.-W.; Williams, D.R.; Lee, J.Y. Anti-oxidant activity reinforced reduced graphene oxide/alginate microgels: Mesenchymal stem cell encapsulation and regeneration of infarcted hearts. *Biomaterials* **2019**, *225*, 119513. [CrossRef]
258. McFarlane, S. Attraction vs. repulsion: The growth cone decides. *Biochem. Cell Biol.* **2000**, *78*, 563–568. [CrossRef]
259. Zuchero, J.B.; Barres, B.A. Glia in mammalian development and disease. *Development* **2015**, *142*, 3805–3809. [CrossRef] [PubMed]
260. Mietto, B.S.; Mostacada, K.; Martinez, A.M.B. Neurotrauma and inflammation: CNS and PNS responses. *Mediat. Inflamm.* **2015**, *2015*, 251204. [CrossRef]
261. Yiu, G.; He, Z. Glial inhibition of CNS axon regeneration. *Nat. Rev. Neurosci.* **2006**, *7*, 617–627. [CrossRef]
262. Rolls, A.; Shechter, R.; Schwartz, M. The bright side of the glial scar in CNS repair. *Nat. Rev. Neurosci.* **2009**, *10*, 235–241. [CrossRef] [PubMed]
263. Shim, S.; Ming, G. Roles of channels and receptors in the growth cone during PNS axonal regeneration. *Exp. Neurol.* **2010**, *223*, 38–44. [CrossRef] [PubMed]
264. Stein, D.M.; Knight, W.A. 4th Emergency Neurological Life Support: Traumatic Spine Injury. *Neurocrit. Care* **2017**, *27*, 170–180. [CrossRef]
265. Ahuja, C.S.; Nori, S.; Tetreault, L.; Wilson, J.; Kwon, B.; Harrop, J.; Choi, D.; Fehlings, M.G. Traumatic Spinal Cord Injury-Repair and Regeneration. *Neurosurgery* **2017**, *80*, S9–S22. [CrossRef]
266. Qian, Y.; Zhao, X.; Han, Q.; Chen, W.; Li, H.; Yuan, W. An integrated multi-layer 3D-fabrication of PDA/RGD coated graphene loaded PCL nanoscaffold for peripheral nerve restoration. *Nat. Commun.* **2018**, *9*, 323. [CrossRef]
267. Shah, S.; Yin, P.T.; Uehara, T.M.; Chueng, S.-T.D.; Yang, L.; Lee, K.-B. Guiding stem cell differentiation into oligodendrocytes using graphene-nanofiber hybrid scaffolds. *Adv. Mater.* **2014**, *26*, 3673–3680. [CrossRef] [PubMed]
268. Sakai, K.; Teshima, T.F.; Nakashima, H.; Ueno, Y. Graphene-based neuron encapsulation with controlled axonal outgrowth. *Nanoscale* **2019**, *11*, 13249–13259. [CrossRef]
269. Javadi, M.; Gu, Q.; Naficy, S.; Farajikhah, S.; Crook, J.M.; Wallace, G.G.; Beirne, S.; Moulton, S.E. Conductive Tough Hydrogel for Bioapplications. *Macromol. Biosci.* **2018**, *18*. [CrossRef] [PubMed]
270. Wang, J.; Wang, H.; Mo, X.; Wang, H. Reduced Graphene Oxide-Encapsulated Microfiber Patterns Enable Controllable Formation of Neuronal-Like Networks. *Adv. Mater.* **2020**, *32*, 2004555. [CrossRef]

271. Bisht, G.S.; Canton, G.; Mirsepassi, A.; Kulinsky, L.; Oh, S.; Dunn-Rankin, D.; Madou, M.J. Controlled continuous patterning of polymeric nanofibers on three-dimensional substrates using low-voltage near-field electrospinning. *Nano Lett.* **2011**, *11*, 1831–1837. [CrossRef]
272. Chamberlain, K.A.; Nanescu, S.E.; Psachoulia, K.; Huang, J.K. Oligodendrocyte regeneration: Its significance in myelin replacement and neuroprotection in multiple sclerosis. *Neuropharmacology* **2016**, *110*, 633–643. [CrossRef] [PubMed]
273. Ehrlich, M.; Mozafari, S.; Glatza, M.; Starost, L.; Velychko, S.; Hallmann, A.-L.; Cui, Q.-L.; Schambach, A.; Kim, K.-P.; Bachelin, C.; et al. Rapid and efficient generation of oligodendrocytes from human induced pluripotent stem cells using transcription factors. *Proc. Natl. Acad. Sci.* **2017**, *114*, E2243–E2252. [CrossRef] [PubMed]
274. Pradhan, K.; Das, G.; Khan, J.; Gupta, V.; Barman, S.; Adak, A.; Ghosh, S. Neuro-Regenerative Choline-Functionalized Injectable Graphene Oxide Hydrogel Repairs Focal Brain Injury. *ACS Chem. Neurosci.* **2019**, *10*, 1535–1543. [CrossRef] [PubMed]
275. Rehman, S.R.U.; Augustine, R.; Zahid, A.A.; Ahmed, R.; Tariq, M.; Hasan, A. Reduced Graphene Oxide Incorporated GelMA Hydrogel Promotes Angiogenesis For Wound Healing Applications. *Int. J. Nanomed.* **2019**, *14*, 9603–9617. [CrossRef]
276. Li, J.; Zhou, C.; Luo, C.; Qian, B.; Liu, S.; Zeng, Y.; Hou, J.; Deng, B.; Sun, Y.; Yang, J.; et al. N-acetyl cysteine-loaded graphene oxide-collagen hybrid membrane for scarless wound healing. *Theranostics* **2019**, *9*, 5839–5853. [CrossRef]
277. Kim, S.; Moon, J.-M.; Choi, J.S.; Cho, W.K.; Kang, S.M. Mussel-Inspired Approach to Constructing Robust Multilayered Alginate Films for Antibacterial Applications. *Adv. Funct. Mater.* **2016**, *26*, 4099–4105. [CrossRef]
278. Yan, X.; Li, F.; Hu, K.-D.; Xue, J.; Pan, X.-F.; He, T.; Dong, L.; Wang, X.-Y.; Wu, Y.-D.; Song, Y.-H.; et al. Nacre-mimic Reinforced Ag@reduced Graphene Oxide-Sodium Alginate Composite Film for Wound Healing. *Sci. Rep.* **2017**, *7*, 13851. [CrossRef] [PubMed]

Article

Reduced Graphene Oxide Inserted into PEDOT:PSS Layer to Enhance the Electrical Behaviour of Light-Emitting Diodes

Fernando Rodríguez-Mas *, Juan Carlos Ferrer, José Luis Alonso, Susana Fernández de Ávila and David Valiente

Communications Engineering Department, Universidad Miguel Hernández, 03202 Elche, Spain; jc.ferrer@umh.es (J.C.F.); j.l.alonso@umh.es (J.L.A.); s.fdezavila@umh.es (S.F.d.Á.); dvaliente@umh.es (D.V.)
* Correspondence: fernando.rodriguezm@umh.es

Abstract: In this study, poly(9-vinylcarbazole) (PVK)-based LEDs doped with reduced graphene oxide (rGO) and cadmium sulphide (CdS) nanocrystals were fabricated by spin-coating. The hybrid LED structure was a layer sequence of glass/indium tin oxide (ITO)/PEDOT:PSS | rGO/PVK/ Al. rGO was included in the poly(3,4-ethylenedioxythiophene)-poly(styrenesulfonate) (PEDOT:PSS) layer due to its energy bands being close to PEDOT:PSS bands, and the possibility of using water for dispersing both polymer and flakes. Optical properties such as photoluminescence and UV-Vis absorption were not affected by the addition of rGO to the PEDOT:PSS solution. However, PVK-based LEDs with rGO showed increased current density compared to those without rGO in the hole transporting layer. Higher electroluminescence intensities were observed for rGO-enriched LEDs, although the shape of the spectrum was not modified. LEDs including CdS nanocrystals in the poly(9-vinylcarbazole) emissive layer did not show such dependence on the rGO presence. Though the addition of rGO to PEDOT:PSS still produces a slightly higher current density in CdS doped LEDs, this growth is no longer proportional to the rGO load.

Keywords: cadmium sulphide; PVK; hybrid light-emitting device; electroluminescence; nanocrystals; reduced graphene oxide

Citation: Rodríguez-Mas, F.; Ferrer, J.C.; Alonso, J.L.; Fernández de Ávila, S.; Valiente, D. Reduced Graphene Oxide Inserted into PEDOT:PSS Layer to Enhance the Electrical Behaviour of Light-Emitting Diodes. *Nanomaterials* **2021**, *11*, 645. https://doi.org/10.3390/nano11030645

Academic Editor: José Miguel González-Domínguez

Received: 12 February 2021
Accepted: 2 March 2021
Published: 5 March 2021

Publisher's Note: MDPI stays neutral with regard to jurisdictional claims in published maps and institutional affiliations.

Copyright: © 2021 by the authors. Licensee MDPI, Basel, Switzerland. This article is an open access article distributed under the terms and conditions of the Creative Commons Attribution (CC BY) license (https:// creativecommons.org/licenses/by/ 4.0/).

1. Introduction

Organic-semiconductor devices have gained popularity in recent years due to several advantages they offer compared to their inorganic counterparts, namely mechanical flexibility, low cost [1], and a simple manufacturing processes [2,3]. Deposition of organic polymer layers can be achieved by means of simple techniques such as spin-coating. With this technique, uniform layers are applied onto a flexible or rigid substrate by means of centrifugal force. In these devices, specifically in light-emitting diodes (LEDs), the optical and electrical properties could be modified by the inclusion of different dopants, such as graphene and its derivatives, or semiconductor nanocrystals.

Since 2004 [4], graphene and its derivatives have generated high expectations in several areas, including as solar cells [5], field-effect transistors [6], electrocatalyst [7], transparent electrodes [8], graphene transistors [9], etc. Graphene has excellent properties, such as excellent thermal conductivity [10], electron mobility [11] and transparency [12], that make possible its involvement in the aforementioned areas. The simplest process for synthesizing graphene consists in the oxidation and exfoliation of graphite. This process transforms graphite into graphene oxide (GO). In a later step, GO is reduced, and either graphene or reduced graphene oxide (rGO) is synthesized depending on the reduction degree.

Not all the properties of graphene are extensible to graphene derivatives (GO or rGO). In the case of reduced graphene oxide, certain properties are maintained, but to a lesser extent, such as a conductivity of 200 S/m^1. In this paper, we will study the possibility of manufacturing hybrid LEDs with graphene derivatives. Normally, in these devices, the hole transport layer is formed by Poly(3,4-ethylenedioxythiophene)-poly(styrenesulfonate) (PEDOT:PSS), but in our analysis, this layer will be doped with different proportions of rGO. Reduced graphene oxide was chosen because the value of its energy bands ($-5.2 \sim -4.9$ eV) [13] is very close to PEDOT:PSS, ($-5.1 \sim -5$ eV) [13]. Additionally, rGO is aqueous dispersible, like the PEDOT:PSS, eliminating the possible problems that a mixture of different solvents might cause in the spin-coating technique [14].

On the other hand, Cadmium sulphide (CdS) nanocrystals (NCs) have been studied in recent years [15] due to their ability to modify the optical and electrical properties of the organic devices [16]. Cadmium sulphide nanoparticles have been used to improve or modify the properties of different devices, such as organic solar cells [17], light-emitting devices [18], hybrid memory devices as a compound of polymer composites and CdS nanocrystals [19,20], and memory devices with other nanoparticles, such as graphene oxide nanocrystals [21]. CdS is a semiconductor with a direct bandgap energy of 2.4 eV. This bandgap energy makes it suitable for applications in the visible band of the electromagnetic spectrum. A simple way to synthesize CdS nanocrystals could be through the process of thiolate decomposition [22]. This route is based on a two-step process: (i) synthesis of a thiolate (compound formed by the future surface ligand and the precursor metal of the nanoparticle) and (ii) reaction with a sulphur source, originating the nanocrystals. It should not be overlooked that a critical factor of nanocrystal behaviour is their size [23,24], since the optical properties depend on this parameter. To control the size and solubility, the nanocrystals are coated with ligands [25]. In this route, the nanocrystals are coated with thiophenol. The presence of thiophenol makes them soluble in dimethyl sulfoxide (DMSO). However, since the control of the exact mass of nanoparticles in solution is imprecise with this method, they were dried following reference [14]. Moreover, thiophenol was chosen as a ligand because thiophenol is formed by an aromatic ring [26]. This aromatic ring is expected to improve the charge transfer between the nanocrystals and the surrounding polymer because of the ring resonance. The influence of rGO on devices with their emission layer doped with CdS was also studied.

The aim of this paper is to analyse the influence of rGO and CdS nanoparticles embedded in PVK polymer on the electrical and optical properties of hybrid LEDs with active layers based on these materials.

2. Materials and Methods

2.1. Materials

Cadmium nitrate-tetrahydrate (Cd(NO$_3$)$_2 \cdot$4H$_2$O, 99.99%), thiophenol (99%), sulphur powder (99.98%), poly(9-vinylcarbazole) (PVK, 98%), poly(3,4-ethylenedi-oxythiophene): poly(styrene sulfonate) (PEDOT:PSS, 1.3% water solution), toluene, methanol and dimethyl sulfoxide (DMSO) were purchased from Sigma-Aldrich (Darmstadt, Germany) and used without further purification. Reduced graphene oxide (rGO) was obtained from Graphenano "nanotechnologies" (Yecla, Spain) and used without further purification.

2.2. Characterization

The measurements of optical absorption were carried out with a T92+ UV/VIS spectrophotometer from PG instruments Ltd. (Lutterworth, UK), and the measurements of photoluminescence (PL) were performed with a Modular Spectrofluorometer Fluorolog-3 from Horiba Scientific (Madrid, Spain). In all the photoluminescence measurements, the excitation wavelength was fixed at λ_{exc} = 365 nm.

Transmission electron microscopy (TEM) analysis was performed using a Jeol 2010 (Tokyo, Japan) operating at 200 kV. High-resolution TEM (HRTEM) images were obtained by phase contrast at Scherzer defocus in order to obtain easily interpretable images.

Current density vs. voltage (J–V) curves of the LEDs were measured using Keithley 2400 Sourcemeter equipment (Bracknell, UK).

Electroluminescence (EL) characterization was performed with a Triax 190 monochromator (Madrid, Spain) and a multichannel thermoelectrically cooled CCD Symphony detector by Horiba Jobin Yvon (Madrid, Spain).

2.3. Synthetic Pathway of PEDOT:PSS | rGo Dispersions

A rGO solution was prepared and mixed with PEDOT:PSS. A reduced graphene oxide flake solution was prepared with distilled water at 4 wt%. When the solution was completely dispersed, it was mixed with the PEDOT:PSS, 1.3% water solution, in accordance with the proportions in volume shown in Table 1. The rGO was redispersed in distilled water because PEDOT:PSS was dissolved in water, too. In this sense, possible problems resulting from the phase separation of different solvents were avoided. In the spin-coating process, the use of solution with different solvents could cause craters and agglomerations in the deposited layer, as we observed in our previous study [14].

Table 1. Summary of the quantities used in the PEDOT:PSS layer doped with rGO.

Solution	PEDOT:PSS (mL)	rGO (mL)
PEDOT:PSS	1.000	0.000
PEDOT:PSS	rGO [30:1]	1.000	0.033
PEDOT:PSS	rGO [15:1]	1.000	0.067
PEDOT:PSS	rGO [5:1]	1.000	0.200

2.4. Synthesis of CdS Nanocrystals Powder

The route used for the synthesis of CdS NCs was an extension of the thiolate decomposition method [14]. The main advantage of this method is that nanocrystals end up in a powder, making it simpler to embed them in the organic polymer. The first step in this process was the synthesis of cadmium thiolate, $Cd(C_6H_5S)_2$. For this, 1.54 g of $Cd(NO_3)_2 \cdot 4H_2O$ was dissolved in 25 mL of distilled water and 25 mL of methanol and the solution was stirred for 30 min. In another flask, 1.03 mL of thiophenol was dissolved in 50 mL of distilled water. After that, when both solutions were well dissolved, the solutions were mixed and the blend was stirred. When Reaction (1) evolved, the mixture turned a whitish colour and the cadmium thiolate precipitated. The white powder was filtered and dried obtaining the cadmium thiolate.

$$Cd(NO_3)_2 \cdot 4H_2O + C_6H_5\text{-SH} \rightarrow Cd(C_6H_5S)_2 \downarrow \qquad (1)$$

In a second step, two more solutions were prepared, 0.08 g of $Cd(C_6H_5S)_2$ was dissolved in 2 mL of DMSO and, in another vial, 0.17 g of sulphur was dissolved in 20 mL of toluene. Both solutions were stirred for thirty minutes and when the thiolate was dissolved, 0.4 mL of sulphur solution was added. The mixture changed and turned yellowish, indicating that the CdS NCs had been synthesized correctly. The nanocrystals ended up dissolved in DMSO and it was necessary to eliminate the solvent in order to avoid phase separation when mixed with the polymer solvent. To achieve this, the yellowish solution was heated to 200 °C for one hour. Once the solvent was evaporated, CdS NCs were obtained.

2.5. Hybrid LEDs Fabrication

Hybrid light-emitting diodes with a layer sequence of ITO/PEDOT:PSS | rGO/PVK:CdS/Al were fabricated by spin-coating. Commercial glass substrates covered with a semitransparent ITO layer were routinely cleaned by sequential sonication in 1,2,4-trichlorobenzene, acetone and isopropyl alcohol, and then dried with N_2.

Aqueous PEDOT-PSS dispersion with rGO was spin-coated onto the clean ITO surface and then annealed at 100 °C for 60 min. Then, the active layers were spin-coated and dried at 80 °C for 60 min.

Finally, the metallization of the cathodes was performed by evaporating aluminium in a high-vacuum chamber (10^{-6} mbar) until a thickness of 200 nm was achieved.

3. Results

3.1. CdS NCs Characterization

Absorbance and photoluminescence measurements were performed. The PL spectra are shown in Figure 1A. Therein, a narrow peak is observed at low wavelengths, the maximum of the narrow peak had a wavelength at 411 nm. This peak corresponded with the maximum photoluminescence peak of toluene (blue line). The solvent peaks are not usually observed in PL graphs, because the emission intensity of nanoparticles normally hides the emission intensity of solvents. In this case, toluene was present in the CdS curve because the concentration of measured solution was very low. This low concentration made the emission intensity of nanoparticles lower. The low concentration can also explain the noise that was observed in the CdS photoluminescence. In this PL spectrum, the main peak shows its maximum at 576 nm.

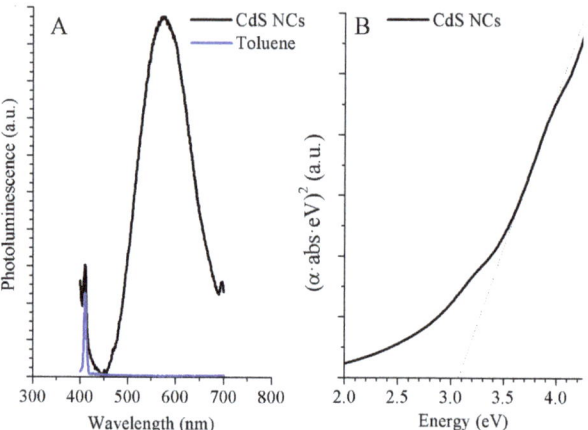

Figure 1. Photoluminescence spectra (**A**) of CdS nanocrystals (black line) and toluene (blue line). Optical absorption spectrum (**B**) of CdS NCs (black line).

To correctly observe the absorption edge and calculate the bandgap energy, the Tauc relation [27] was applied. In Figure 1B, $(\alpha h\nu)^2$ versus $h\nu$ are plotted, showing an excitonic shoulder, pointing to the absorption edge (dotted line). The resulting value of the band gap is 3.07 eV (403.7 nm). According to this, the wavelength corresponding to the maximum intensity at the CdS NCs photoluminescence spectrum (Figure 1A) should be at lower wavelengths, close to the absorption edge. This shift towards higher wavelengths is due to defects in the nanocrystal surface. The defects originate deep surface states, producing trap emission.

To characterize the size of the nanoparticles, two different methods were used, a theoretical method, whose calculations are based on the band gap energy, and the direct measurement from TEM images. As a theoretical method, the equation suggested by Brus was employed [28].

$$E_n = E_b + (\hbar^2\pi^2/2R^2) \times (1/m_e^* + 1/m_h^*) - 1.8e^2/(4\pi\varepsilon_0\varepsilon R), \tag{2}$$

where E_n is the nanoparticles band gap, E_b is the energy gap of the bulk material, m_e^* and m_h^* are the effective masses of electrons and holes, and ε is the dielectric constant. Solving the Brus equation (Equation (2)), these nanoparticles had an average size of 3.11 nm.

To prepare samples for TEM analysis, a drop of CdS NCs solution in toluene was deposited on a carbon grid and then dried at room temperature. A high-resolution image of the CdS nanocrystals is shown in Figure 2, as well as the histogram of the size distribution of more than 50 particles. Diameters range from 2.2 nm to 4.4 nm, and the average size is 3.20 ± 0.06 nm. When comparing the size calculated from the Brus equation (Table 2) to the image measurement, a minimal difference can be observed (0.09 nm) that can be neglected. It was verified that the Brus equation performs a good approximation of the real size.

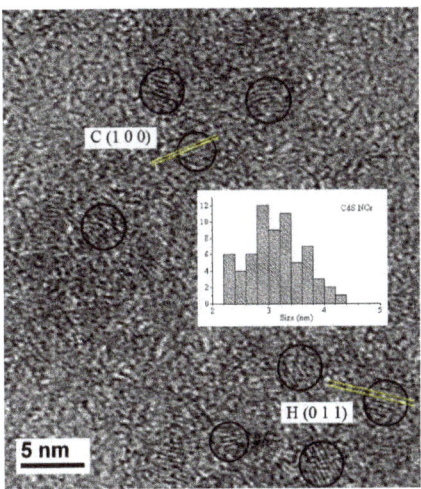

Figure 2. TEM images corresponding to CdS NCs and histogram of the size distribution for the nanocrystals.

Table 2. Summary of the CdS NCs characterization. Absorption edge, PL peak and size.

NCs	Absorption Edge (eV)	Emission Peak (nm)	Size (Brus) (nm)	Size (TEM) (nm)
CdS NCs	3.07	404 576	3.11	3.20 ± 0.06 nm

When measuring the interplanar distances of the CdS crystals in the HRTEM images, two different types of planes arise. The planes match the distances of the cubic zinc-blende-type structure and the hexagonal wurzite-type structure. In Figure 2, the {100} family planes of the cubic structure and the {011} family planes of the hexagonal phase are identified in two particles. The presence of both crystalline structures can justify the dispersion in the histogram of Figure 2. In addition, in Figure 1B, a second shoulder can be located at lower energy values. The dispersion in the histogram and the presence of the two shoulders in UV-Vis absorption, Figure 1B, confirm the two different structures [14]. The heat used to evaporate the solvent changes the NC structure. The energy could be enough to increase the size and transform the nanoparticles with cubic zinc blend structures into hexagonal wurzite-type structures [14,29,30].

3.2. Synthesis and Characterization of the Hybrid Solution

As discussed in the introduction, the active layer of the devices was doped with CdS NCs. The active layer was prepared from a solution of PVK mixed with CdS nanocrystals. CdS nanoparticle powder was weighed and added to a solution formed by PVK dissolved in toluene. This solution had a PVK:CdS mass ratio of 8:1 at 3 wt%. Moreover, another solution with pristine PVK was also prepared with a concentration of 3 wt%. Optical absorption and photoluminescence measurements of these solutions were performed with the results shown in Figure 3.

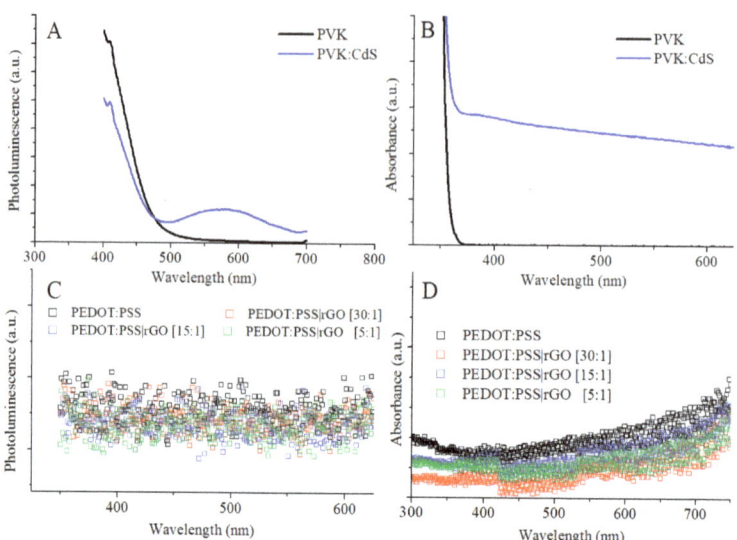

Figure 3. Photoluminescence (**A**) and absorption spectra (**B**) from pristine PVK and PVK doped with CdS NCs and photoluminescence (**C**) and absorption spectra (**D**) from PEDOT:PSS doped with different proportions of reduced graphene oxide.

The PL curves of pristine PVK and PVK with CdS NCs are plotted in Figure 3A. Both spectra exhibited a main peak corresponding to the PVK polymer located around 385 nm not shown in the figure. The influence of the nanoparticles was clearly visible in the blue line. The maximum of the secondary peak of the PVK solution with CdS NCs was localized at 577 nm. The same wavelength value that was presented in Figure 1A, where the CdS nanocrystals were measured without PVK. The photoluminescence of hybrid PVK shows a normal decrease in the PVK emission because the content of PVK polymer was reduced in the hybrid solution. Additionally, PVK photoluminescence quenching was observed [31]. In the hybrid solution, the excited electrons of the highest molecular orbital (HOMO) of the PVK polymer can either drift to the polymer's lowest unoccupied molecular orbital (LUMO) or go to the conduction band of CdS. This migration produces a charge transfer, lowering the PVK intensity [13]. As for the optical absorption, a constant background absorption was observed in the total range of wavelengths (Figure 3B) due to the high concentration of CdS NCs.

To study the influence of reduced graphene oxide on the devices, rGO was included in several hybrid LEDs with different proportions. For this purpose, the hole transport layer, represented by the PEDOT:PSS polymer, was doped with different rGO masses.

We performed optical absorption and photoluminescence measurements to check the influence of rGO in our devices. To observe if the addition of reduced graphene oxide had any influence on the optical properties of the hole transport layer solution, absorbance and PL measurements were performed on the samples described in Table 1. The excitation wavelength for the photoluminescence measurements was λ_{exc} = 365 nm. To perform UV-Vis absorption measurements, PEDOT:PSS and PEDOT:PSS doped with rGO solutions were diluted to prevent re-absorption. Diluted solutions were measured.

No difference was observed between the PL curves of PEDOT:PSS doped with rGO and those of PEDOT:PSS, as shown in Figure 3C. There was no significant difference in optical absorption measurements, either (Figure 3D). Since the introduction of reduced graphene oxide on PEDOT:PSS did not influence the optical characteristics of PEDOT:PSS, we assume that it should not influence the optical characteristics of hybrid LEDs.

3.3. Hybrid LEDs with rGO

Once the CdS nanocrystals and the different solutions were synthesized and characterized, hybrid light-emitting diodes were manufactured to check the influence of rGO on the devices. The hybrid LEDs consisted of a structure formed by the following stacked layers based on the solutions presented in the previous sections: glass/ITO/PEDOT:PSS(|rGO)/PVK(:CdS)/Al; Figure 4. Different devices were manufactured, employing PEDOT:PSS as a hole transport layer and PVK as the active layer. Excluding the changes in the concentrations of graphene and CdS nanoparticles, the same conditions were maintained for the fabricating process of LEDs.

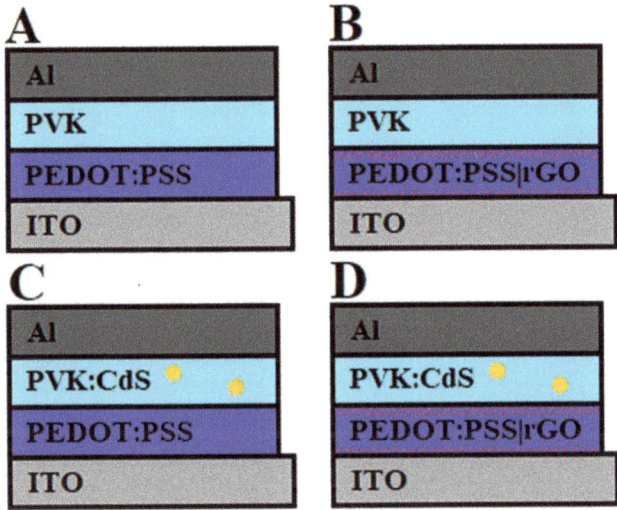

Figure 4. Structure of the different devices fabricated. ITO/PEDOT:PSS/PVK/Al (**A**), ITO/PEDOT:PSS|rGO/PVK/Al (**B**), ITO/PEDOT:PSS/PVK:CdS/Al (**C**), and ITO/PEDOT:PSS|rGO/PVK:CdS/Al (**D**).

Polymer and hybrid layers were deposited using the spin-coating technique. The hole transport layers were spin-coated at 2000 rpm on the substrate. When the hole transport layers were dried, the emission layers were deposited at 4000 rpm by spin-coating.

Two types of hybrid LEDs were produced, without or with CdS NCs embedded in the PVK active layer. In the first one, the solutions indicated in Table 1 were employed to spin cast the hole transport layers, and pristine PVK was used as the emissive layer. An LED without reduced graphene oxide was manufactured as a reference. The emission layer was composed of PVK at 3% wt. When the devices were fabricated, we collected the measurements of current density vs. voltage (J-V). The curves are shown in Figure 5A.

Hybrid LEDs doped with rGO exhibited changes in electrical properties. The current density increases with the inclusion of rGO. The maximum improvement was found for the hybrid LED with [PEDOT:PSS|rGO] ≡ [5:1], and this decreases with the ratio of rGO. The current density values at the threshold voltage are shown in Figure 5C. This current increases with rGO concentration. On the other hand, the threshold voltage is reduced by the presence of rGO in the PEDOT:PSS layer.

The variation of the electrical behaviour in the hybrid LEDs indicated that the rGO inclusion in the PEDOT:PSS layer improved the hole transport [32]. rGO is a two-dimensional structure with graphene distributions. In graphene distributions, the charge carriers tend to move. The reduction by means of which graphene transforms into reduced graphene oxide eliminates part of the graphene conducting regions present in rGO [33]. In these distributions, the π-bond (C=C) and σ-bond (C-H) alternate, generating optimal paths for the hole transport. With the increase of reduced graphene oxide, the number of graphene distributions increases in the layer, augmenting the optimal paths for transport.

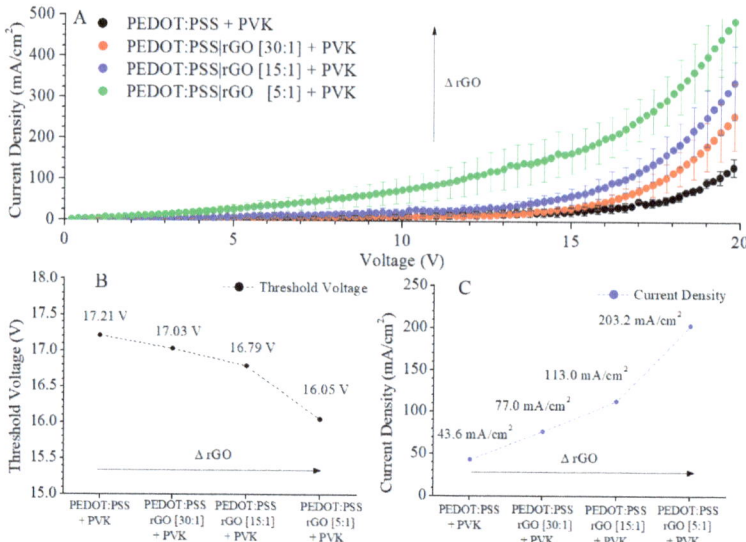

Figure 5. J-V curves (**A**), threshold voltage (**B**), and current density (**C**) at threshold voltage for the hybrid LEDs with rGO in the PEDOT:PSS layer.

Next, we studied the electroluminescence of the hybrid LEDs doped with rGO. Regarding the emission intensity, LEDs doped with reduced graphene oxide showed higher emission intensity than reference PVK-LEDs, as indicated in [34,35].

In Figure 6, the spectra were normalized to the maximum peak emission. All EL curves showed a similar morphology, with a narrow uppermost peak and two shoulders at longer wavelengths. To study these features, Gaussian deconvolutions were carried out for all spectra, indicated as non-solid lines in Figure 6 for PEDOT: PSS + PVK. Gaussian deconvolution is a statistical process where electroluminescence is decomposed by Gaussian curves, where the mean of each Gaussian curve coincides with the wavelength of the maximum emission peak, and the standard deviation is the difference of the wavelength of the maximum emission peak (mean) and the wavelength at which a $(1 - 1/\sqrt{e})$% decrease from the maximum emission peak is produced. We allowed three Gaussian curves to fit each spectrum. The wavelengths corresponding to the maximum of each Gaussian emission peak are detailed in Table 3.

Table 3. Positions of emission peaks for the hybrid LEDs with rGO and PVK LED.

Hybrid LEDs	Gaussian Emission Peaks (nm)		
PEDOT:PSS + PVK	428	495	596
PEDOT:PSS I rGO [30:1] + PVK	427	512	602
PEDOT:PSS I rGO [15:1] + PVK	425	506	594
PEDOT:PSS I rGO [5:1] + PVK	425	505	597

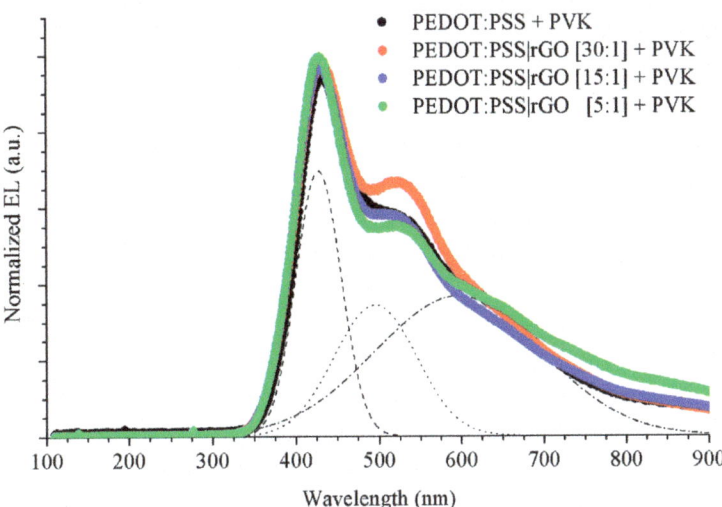

Figure 6. Normalized electroluminescence of PVK LED and hybrid LEDs using rGO in their hole transport layer. PEDOT:PSS + PVK (black), PEDOT:PSS | rGO [30:1] + PVK (red), PEDOT:PSS | rGO [15:1] + PVK (blue) and PEDOT:PSS | rGO [5:1] + PVK (green). Gaussian deconvolution of PEDOT:PSS + PVK (non-solid lines).

The EL spectra are different from the PL curves (Figure 3A). Like any polymer, PVK is a chain of monomers (a small molecule) bonded together. It is also known that photoluminescence of PVK has two peaks, one around 390 nm, produced by a small superposition of monomers, called p-PVK, because it is related to the phosphorescence of PVK. Another contribution is located around 410 nm, and is produced by the total superposition of the monomers, known as f-PVK, due to its relation to fluorescence of PVK [36]. In the EL curve of the PVK-reference, the peaks are not in the same positions as those observed for photoluminescence. The PVK electroluminescence presents a shift towards higher wavelengths. According to Ye et al. [36], with the increase of temperature, the emission intensity of p-PVK decreases and the emission intensity of the f-PVK increases and, at low temperatures, the PVK photoluminescence exhibits a broad peak at high wavelengths (~550 nm). This peak corresponds to the radiative transition of triplet states. In addition, Ye affirms that the PVK is a polar polymer, and that an electric current polarizes PVK. The polarization of PVK enhances the effect of f-PVK, since it reduces the intermolecular distance, and hence, the energy. Therefore, the peak of f-PVK is shifted towards higher wavelengths.

In PVK-reference, the maximum emission occurs for the narrowest peak, located at 428 nm, and the two shoulders that showed the EL curve were at 495 and 595 nm. The peak at 428 nm corresponds to the peak f-PVK enhanced by the electric field at room temperature. The two shoulders did not have any correspondence with the photoluminescence. Ye [36] stated that, for EL processes, the proportion of triplet states is increased compared to PL processes; this enables the observation of phosphorescent emissions at room temperature. This is the origin of the peak located at 495 nm. Finally, the lower shoulder observed in the EL spectrum close to a wavelength of 600 nm is attributed to the electromer of PVK. This emission is only visible in EL, not in PL, because photoexcitation does not normally generate the free charge carriers needed to form this complex.

For LEDs doped with rGO, the position of the f-PVK around 425 nm was not significantly modified. Additionally, the two shoulders that showed the EL curves of hybrid LEDs did not exhibit variations in wavelength with respect to the PVK-reference. Therefore, the rGO modified the emission intensity of the EL spectra, but this did not change the shape of the EL curves.

To verify that rGO does not modify the electroluminescence spectra, we compared the CIE 1931 chromatic coordinates of the manufactured LEDs. The colour coordinates for PVK-LED were (0.28, 0.28). The use of rGO did not qualitatively modify the chromatic coordinates. These were (0.28, 0.29) for PEDOT:PSS | rGO [30:1] + PVK; (0.27, 0.27) for PEDOT:PSS | rGO [15:1] + PVK and (0.28, 0.27) for PEDOT:PSS | rGO [5:1] + PVK.

3.4. Hybrid LEDs with rGO and CdS NCs

Having determined the electrical behaviour due to the inclusion of rGO, we studied the rGO influence on LEDs doped with CdS nanocrystals. In Figure 7A, the J-V curves are plotted. In these curves, it can be observed that at the same voltage values, the current density is higher in the doped LEDs (with any dopant, rGO or CdS NCs) than in the pristine PVK LED. In devices without rGO, the CdS inclusion considerably increased the electrical conduction. Since in PVK LED, the PVK lowest unoccupied molecular orbital (LUMO) is -2.3 eV, and for aluminium it is -4.3 eV, these values cause a potential barrier, the value of which is 2 eV. The inclusion of CdS nanocrystals in the active layer generates distributions where the potential barrier decreases, favouring the electron transport. The conduction band of CdS NCs is closer to that of aluminium than that of PVK LUMO, (the conduction band of CdS NCs is around -4 eV). Therefore, some electrons will occupy the CdS conduction band, producing improvements in electrical behaviour [14]. Additionally, LEDs doped with rGO and their active layer composed of PVK with CdS nanocrystals also increased the electrical conduction of the pristine PVK LED. With the same voltage, the CdS NCs augmented the current density of the LEDs manufactured previously. This modification of electrical behaviour can be justified by changing the hole transport. However, in this case, the improvement of the hole transport is not the only factor present. To the enhancement by rGO is added the influence of the nanoparticles. CdS nanoparticles reduce the potential barrier for electron injection.

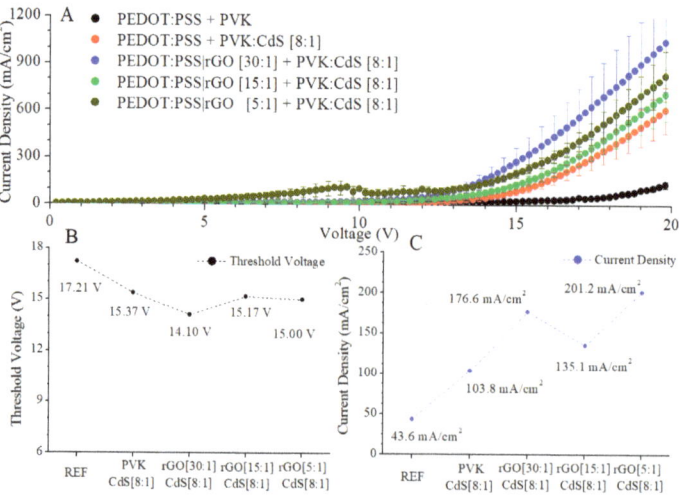

Figure 7. J-V curves (**A**), threshold voltage (**B**), and current density (**C**) at threshold voltage for the hybrid LEDs with rGO in the PEDOT:PSS layer and CdS NCs in the active layer.

In LEDs doped with reduced graphene oxide and CdS nanocrystals, a change was observed. These LEDs had a higher current density at threshold voltage than the reference PVK LED, Figure 7C. Additionally, the doped devices had a lower threshold voltage than the reference, Figure 7B. However, in this case, the direct relation between the rGO load and improvement of electronic transport was eliminated by the presence of the nanoparticles, and was not observed, as indicated in Figure 7B,C. Nevertheless, hybrid LEDs with rGO in the hole transport layer show higher currents and lower threshold voltages than LEDs without rGO. We expected that the electrical conduction would modify with the rGO increase, because more areas of rGO are present, but as shown in Figure 7A, this relation was not produced. The inclusion of CdS nanocrystals produced a higher electrical increase than the inclusion of rGO. In addition, the electrical conduction was higher in LEDs with two dopants than in the devices with a single dopant, but in these devices, the relation "electrical evolution–rGO" was not perceived.

A possible hypothesis to explain this behaviour is related to the morphology of the layers, mainly the interface between the hole transport layer (HTL) and the emissive layer.

Spin-coated layers of PEDOT:PSS are quite smooth, but their roughness increases when rGO is embedded, proportionally to the ratio. If the layer deposited on top of this HTL is pristine polymer, the roughness is corrected to some grade because of the ability of this soft material to coat the underlying surface. Thus, devices with a PVK emissive layer show an influence on the electrical characteristics correlated with the rGO load. On the other hand, if the layer deposited on top of the PEDOT:PSS|rGO layer contains inorganic nanoparticles that are rigid and non-deformable, the interface between the HTL and the emissive layer will have an abrupt profile with several imperfections acting, most probably, as charge carrier traps and recombination centres.

This might be the reason for CdS NCs enriching LEDs with higher rGO load in the HTL; the electrical characteristics do not show the same trend observed for LEDs without CdS nanocrystals.

The electroluminescence of hybrid LEDs doped with rGO and CdS nanocrystals was measured. To observe more clearly the influence of nanoparticles in the EL spectra, the proportion of PVK versus CdS NCs increased to the ratio [PVK:CdS] = [2:1] by weight. Thanks to this increase, the influence of CdS was more evident.

As in devices doped exclusively with rGO, all EL curves of hybrid LED doped with rGO and CdS NCs were normalized, and the average was calculated. All the results are plotted in Figure 8. Gaussian deconvolution was also performed to locate the peak positions of the spectra.

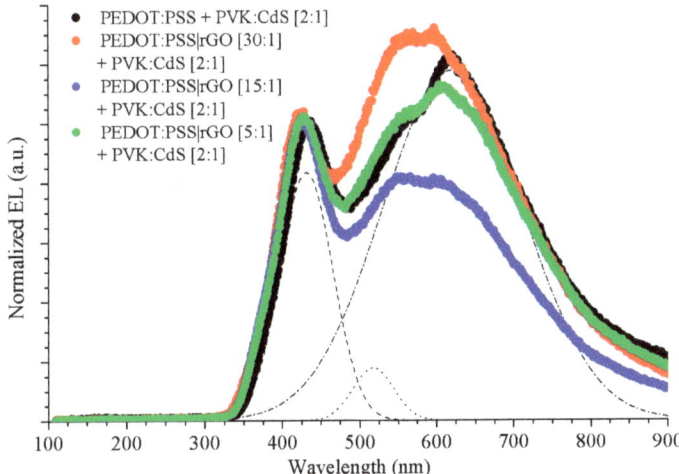

Figure 8. Normalized electroluminescence of PVK LED and hybrid LEDs using rGO in their hole transport layer and doped with CdS NCs. PEDOT:PSS + PVK:CdS [2:1] (black), PEDOT:PSS | rGO [30:1] + PVK:CdS [2:1] (red), PEDOT:PSS | rGO [15:1] + PVK:CdS [2:1] (blue), and PEDOT:PSS | rGO [5:1] + PVK:CdS [2:1] (green). Gaussian deconvolution of PEDOT:PSS + PVK:CdS [2:1] (non-solid lines).

The electroluminescence of hybrid PVK LEDs presented a narrow peak at shorter wavelengths and a broad emission for longer wavelengths. We carried out Gaussian analysis (Table 4) to determine the origin of the broad peak and the composition of the emission spectra.

Table 4. Positions of emission peaks for the hybrid LEDs with rGO and PVK LED doped with CdS NCs. As reference, pristine PVK LED is also listed.

Hybrid LEDs	Gaussian Emission Peaks (nm)		
PEDOT:PSS + PVK	428	495	596
PEDOT:PSS + PVK:CdS [2:1]	431	517	620
PEDOT:PSS	rGO [30:1] + PVK:CdS [2:1]	417	527	611
PEDOT:PSS	rGO [15:1] + PVK:CdS [2:1]	422	526	613
PEDOT:PSS	rGO [5:1] + PVK:CdS [2:1]	423	521	611

The narrow peak visible at lower wavelengths (Table 4) is due to the phosphorescence emission of PVK as in PVK-LED reference (428 nm) [14]. The broad peak of hybrid LEDs with CdS NCs was decomposed into two peaks. In Section 3.1, we indicated that two types of nanocrystals were present in the devices: cubic and hexagonal nanocrystals. These different structures are related to different sizes of CdS NCs, so the hybrid LEDs exhibit both contributions in their electroluminescence [14].

The addition of different amounts of rGO to hybrid LEDs with CdS NCs did not significantly shift the wavelength of the emission peaks' maximum intensity. In Figure 8, the highest peak is identified with CdS nanocrystals. In these devices, the charge carriers from the HOMO of PVK can migrate to the LUMO of PVK or to CdS. Due to this fact, the luminescence of the polymer and the nanoparticles are present in the electroluminescence. In our previous study [14], we demonstrated that with an increasing quantity of nanoparticles, the CdS luminescence is extended, because more charge carriers are able to migrate to CdS. At this point, the quantity of CdS nanoparticles is sufficient for it to account for the highest peak. As in Section 3.3, we studied the CIE 1931 colour coordinates, too. In our previous research, a shift towards the white colour was produced by the inclusion of CdS NCs in pristine PVK-LEDs [14]. The nanocrystals shifted the CIE coordinates to CdS NCs light emission [14]. Thus, we verified the relation of rGO and CdS NCs with PVK-LED, in their chromatic coordinates. The colour coordinates of PEDOT:PSS + PVK:CdS [2:1] were (0.37, 0.34). The nanocrystals produced the abovementioned shift towards the coordinates of CdS light emission (0.45, 0.50). As in the case of the previous study, the presence of rGO did not introduce qualitative variations to the chromatic coordinates of PVK-LEDs doped with CdS NCs. The CIE coordinates of all the manufactured LEDs are collected in Table 5. Apparently, hybrid LEDs with rGO [15:1] and CdS NCs are the closest to D65 white light emission.

Table 5. CIE 1931 colour coordinates of emission peaks for manufactured LEDs.

Hybrid LEDs	CIE 1931 Coordinates
PEDOT:PSS + PVK	(0.28, 0.28)
PEDOT:PSS	rGO [30:1] + PVK	(0.28, 0.29)
PEDOT:PSS	rGO [15:1] + PVK	(0.27, 0.27)
PEDOT:PSS	rGO [5:1] + PVK	(0.28, 0.27)
PEDOT:PSS	rGO [30:1] + PVK:CdS [2:1]	(0.36, 0.37)
PEDOT:PSS	rGO [15:1] + PVK:CdS [2:1]	(0.34, 0.32)
PEDOT:PSS	rGO [5:1] + PVK:CdS [2:1]	(0.36, 0.34)

4. Conclusions

Organic light-emitting diodes with the transport hole layer doped with reduced graphene oxide, and the active layer doped with CdS nanoparticles, were successfully fabricated.

Devices without nanoparticles showed an evolution in their electrical behaviour. This change was proportional to the amount of rGO. The rGO caused a decrease in the threshold voltage and an increase in the current density of the threshold voltage. Devices doped with rGO and CdS modified their electrical behaviour, too. However, the inclusion of CdS nanocrystals eliminated the dependence of the evolution on the amount of rGO.

Electroluminescence measurements were performed. As a result, it was demonstrated that rGO inclusion did not modify the position of the observed emissions in the spectra.

We did not find a clear influence of rGO on the electroluminescent emission in PVK-based LEDs, with or without CdS NCs. The CIE coordinates for LEDs doped with CdS NCs are quite close to white light sources; in particular, the LEDs including an intermediate load of rGO in the PEDOT:PSS layer emitted the light that was closest to the average midday light in Europe, known as D65.

Author Contributions: F.R.-M. designed the experiments; F.R.-M. performed the experiments; F.R.-M., J.C.F., J.L.A. and S.F.d.Á. analysed the data; F.R.-M. wrote the original paper draft; S.F.d.Á. supervised and performed critical revisions of the paper; J.C.F., J.L.A. and D.V. revised the manuscript. All authors have read and agreed to the published version of the manuscript.

Funding: This research received no external funding.

Acknowledgments: The authors gratefully acknowledge support from Graphenano "nanotechnology".

Conflicts of Interest: The authors declare no conflict of interest.

References

1. Ahmadian-Yazdi, M.; Eslamian, M. Toward scale-up of perovskite solar cells: Annealing-free perovskite layer by low-cost ultrasonic substrate vibration of wet films. *Mater. Today Commun.* **2018**, *15*, 151–159. [CrossRef]
2. O'Riordan, A.; O'Connor, E.; Moynihan, S.; Llinares, X.; van Deun, R.; Fias, P.; Nockemann, P.; Binnemans, K.; Redmond, G. Narrow bandwidth red electroluminescence from solution-processed lanthanide-doped polymer thin films. *Thin Solid Films* **2005**, *491*, 264–269. [CrossRef]
3. Han, J.; Kim, D.; Choi, K. Microcavity effect using nanoparticles to enhance the efficiency of organic light-emitting diodes. *Opt. Express* **2015**, *23*, 19863–19873. [CrossRef]
4. Novoselov, S.; Geim, A.; Morozov, S.; Jiang, D.; Zhang, Y.; Dubonos, S.; Grigorieva, I.; Firsov, A. Electric Field Effect in Atomically Thin Carbon Films. *Science* **2004**, *306*, 666–669. [CrossRef] [PubMed]
5. Kakavelakis, G.; Maksudov, T.; Konios, D.; Paradisanos, I.; Kioseoglou, G.; Stratakis, E.; Kymakis, E. Efficient and Highly Air Stable Planar Inverted Perovskite Solar Cells with Reduced Graphene Oxide Doped PCBM Electron Transporting Layer. *Adv. Energy Mater.* **2012**, *7*, 1602120. [CrossRef]
6. Wang, Q.; Kalantar-Zadeh, K.; Kis, A.; Coleman, J.; Strano, M. Electronics and optoelectronics of two-dimensional transition metal dichalcogenides. *Nat. Nanotechnol.* **2010**, *7*, 699–712. [CrossRef] [PubMed]
7. Qu, L.; Liu, Y.; Baek, J.; Dai, L. Nitrogen-Doped Graphene as Efficient Metal-Free Electrocatalyst for Oxygen Reduction in Fuel Cells. *ACS Nano* **2010**, *4*, 1321–1326. [CrossRef] [PubMed]
8. Bae, S.; Kim, H.; Lee, Y.; Xu, X.; Park, J.; Zheng, Y.; Balakrishnan, J.; Lei, T.; Kim, H.; Song, Y.; et al. Roll-to-roll production of 30-inch graphene films for transparent electrodes. *Nat. Nanotechnol.* **2010**, *5*, 574–578. [CrossRef]
9. Schwierz, F. Graphene transistors. *Nat. Nanotechnol.* **2010**, *5*, 487–496. [CrossRef] [PubMed]
10. Balandin, A.; Ghosh, S.; Bao, W.; Calizo, I.; Tewerdebrhan, D.; Miao, F.; Lau, C. Superior thermal conductivity of single-layer graphene. *Nano Lett.* **2008**, *8*, 902–907. [CrossRef]
11. Morozov, S.; Novoselov, K.; Katsnelson, M.; Schedin, F.; Elias, D.; Jaszezak, J.; Geim, A. Giant intrinsic carrier mobilities in graphene and its bilayer. *Phys. Rev. Lett.* **2008**, *100*, 016602. [CrossRef]
12. Nair, R.; Blake, P.; Grigorenko, A.; Novoselov, K.; Booth, T.; Stauber, T.; Peres, N.; Geim, A. Fine Structure Constant Defines Visual Transparency of Graphene. *Science* **2008**, *320*, 1308. [CrossRef]
13. Fan, X.; Zhang, M.; Wang, X.; Yang, F.; Meng, X. Recent progress in organic-inorganic hybrid solar cells. *J. Mater. Chem. A* **2013**, *1*, 8694. [CrossRef]
14. Rodríguez-Mas, F.; Ferrer, J.C.; Alonso, J.L.; Fernández de Ávila, S. Expanded electroluminescence in high load CdS nanocrystals PVK-based LEDs. *Nanomaterials* **2019**, *9*, 1212. [CrossRef]
15. Zang, Y.; Lei, J.; Ju, H. Principles and applications of photoelectrochemical sensing strategies based on biofunctionalized nanostructures. *Biosens. Bioelectron.* **2017**, *96*, 8–16. [CrossRef]
16. Vitukhnovskii, A.; Vaschenko, A.; Bychkovskii, D.; Dirin, D.; Tananaev, P.; Vakshtein, M.; Korzhonov, D. Photo-and electroluminescence from semiconductor colloidal quantum dots in organic matrices: QD-OLED. *Semiconductors* **2013**, *47*, 1567–1569. [CrossRef]

17. Seo, D.B.; Kim, S.; Gudala, R.; Challa, K.K.; Hong, K.; Kim, E.T. Synthesis and organic solar cell application of RNA-nucleobase-complexed CdS nanowires. *Sol. Energy* **2020**, *206*, 287–293. [CrossRef]
18. Ratnesh, R.K. Hot injection blended tuneable CdS quantum dots for production of blue LED and a selective detection of Cu2+ ions in aqueous medium. *Opt. Laser Technol.* **2019**, *116*, 103–111. [CrossRef]
19. Gogoi, K.K.; Chowdhurya, A. Performance improvement of organic resistive memories by exploiting synergistic layered nanohybrid dispersed polymer composites. *J. Appl. Phys.* **2020**, *126*, 065501. [CrossRef]
20. Gogoi, K.K.; Das, R.; Paul, T.; Ghosh, S.; Chowdhury, A. Tunning of bipolar resistive switching and memory characteristics of cadmium sulphide nanorods embedded in PMMA matrix. *Mater. Res. Express* **2019**, *6*, 115107. [CrossRef]
21. Das, R.C.; Gogoi, K.K.; Das, N.S.; Chowdhury, A. Optimization of quantum yield of highly luminescent graphene oxide quantum dots and their application in resistive memory devices. *Semicond. Sci. Technol.* **2019**, *34*, 125016. [CrossRef]
22. Ferrer, J.C.; Salinas-Castillo, A.; Alonso, J.L.; Fernández de Ávila, S.; Mallavia, R. Direct synthesis of PbS nanocrystal capped with 4-fluorothiophenol in semiconducting polymer. *Mater. Chem. Phys.* **2010**, *122*, 459–462. [CrossRef]
23. Swain, B.; Shin, D.; Joo, S.Y.; Ahn, N.K.; Lee, C.G.; Yoon, J.H. Synthesis of submicron silver powder from scrap low-temperature co-fired ceramic an e-waste: Understanding the leaching kinetics and wet chemistry. *Chemosphere* **2018**, *194*, 793–802. [CrossRef] [PubMed]
24. Wu, H.; Wu, C.P.; Zhang, N.; Zhu, X.N.; Ma, X.Q.; Zhigilei, L.V. Experimental and computational study of the effect of 1 atm background gas on nanoparticle generation in femtosecond laser ablation of metals. *Appl. Surf. Sci.* **2018**, *435*, 1114–1119. [CrossRef]
25. Herron, N.; Wang, Y.; Eckert, H. Synthesis and Characterization of Surface-Capped, Size-Quantized CdS Clusters. Chemical Control of Cluster Size. *J. Am. Chem. Soc.* **1990**, *112*, 1322–1326. [CrossRef]
26. Holze, R. The adsorption of thiophenol on gold—A spectroelectrochemical study. *Phys. Chem. Phys.* **2015**, *17*, 21364–21372. [CrossRef]
27. Osuwa, J.C.; Oriaku, C.I.; Kalu, I.A. Variation of optical band gap with post deposition annealing in CdS/PVA thin films. *Chalcogenide Lett.* **2009**, *6*, 433–436.
28. Brus, L. Electronic Wave Functions in Semiconductor Clusters: Experiment and Theory. *J. Phys. Chem.* **1986**, *90*, 2555–2560. [CrossRef]
29. Banerjee, R.; Jayakrishnan, R.; Ayyub, P. Effect of the size-induced structural transformation on the band gap in CdS nanoparticles. *J. Phys. Condens. Matter* **2000**, *12*, 10647. [CrossRef]
30. Wang, S.H.; Yang, S.H.; Yang, C.L.; Li, Z.Q.; Wang, J.N.; Ge, W.K. Poly(N-vinylcarbazole) (PVK) Photoconductivity enhancement induced by doping with CdS nanocrystals through chemical hybridization. *J. Phys. Chem.* **2000**, *104*, 11853. [CrossRef]
31. Masala, S.; Bizarro, V.; Re, M.; Nenna, G.; Villani, F.; Minarini, C.; Di Luccio, T. Photoluminescence quenching and conductivity enhancement of PVK induced by CdS quantum dots. *Phys. E Low Dimens Syst. Nanostruct.* **2012**, *44*, 1272–1277. [CrossRef]
32. Marsden, A.J.; Papageorgiou, D.G.; Vallés, C.; Liscio, A.; Palermo, V.; Bissett, M.A.; Young, R.J.; Kinloch, I.A. Electrical percolation in graphene-polymer composites. *2D Mater.* **2018**, *5*, 032003. [CrossRef]
33. Madsuha, A.F.; Pham, C.V.; Eck, M.; Neukom, M.; Krueger, M. Improved Hole Injection in Bulk Heterojunction (BHJ) Hybrid Solar Cells by Applying a Thermally Reduced Graphene Oxide Buffer Layer. *J. Nanomater.* **2019**, 6095863. [CrossRef]
34. Liu, Y.F.; Feng, J.; Zhang, Y.F.; Cui, H.F.; Yin, D.; Bi, Y.G.; Song, J.F.; Chen, Q.D.; Sun, H.B. Improved efficiency of indium-tin-oxide-free organic light-emitting devices using PEDOT:PSS/graphene oxide composite anode. *Org. Electron.* **2015**, *26*, 81–85. [CrossRef]
35. Jiang, X.Y.; Wang, Z.L.; Han, W.H.; Liu, Q.M.; Lu, S.Q. High performance silicon-organic hybrid solar cells via improving conductivity of PEDOT:PSS with reduced graphene oxide. *Appl. Surf. Sci.* **2017**, *407*, 398–404. [CrossRef]
36. Ye, T.; Chen, Y.; Ma, D. Electroluminescence of poly (N-vinylcarbazole) films: Fluorescence, phosphorescence and electromers. *Phys. Chem. Chem. Phys.* **2010**, *12*, 15410–15413. [CrossRef]

Communication
Drying-Time Study in Graphene Oxide

Talia Tene [1], Marco Guevara [2], Andrea Valarezo [3], Orlando Salguero [3], Fabian Arias Arias [4], Melvin Arias [3,5], Andrea Scarcello [3,6,7], Lorenzo S. Caputi [3,6] and Cristian Vacacela Gomez [2,3,*]

[1] Grupo de Fisicoquímica de Materiales, Universidad Técnica Particular de Loja, Loja EC-110160, Ecuador; tbtene@utpl.edu.ec
[2] CompNano, School of Physical Sciences and Nanotechnology, Yachay Tech University, Urcuquí EC-100119, Ecuador; marco.guevara@espoch.edu.ec
[3] UNICARIBE Research Center, University of Calabria, I-87036 Rende, Italy; andrea.valarezo@yachaytech.edu.ec (A.V.); orlando.salguero@yachaytech.edu.ec (O.S.); melvin.arias@intec.edu.do (M.A.); andrea.scarcello@unical.it (A.S.); lorenzo.caputi@fis.unical.it (L.S.C.)
[4] Facultad de Ciencias, Escuela Superior Politécnica de Chimborazo, Riobamba EC-060155, Ecuador; fabian.arias@espoch.edu.ec
[5] Laboratorio de Nanotecnología, Area de Ciencias Básicas y Ambientales, Instituto Tecnológico de Santo Do-mingo, Av. Los Próceres, Santo Domingo 10602, Dominican Republic
[6] Surface Nanoscience Group, Department of Physics, University of Calabria, Via P. Bucci, Cubo 33C, I-87036 Rende, Italy
[7] INFN, Sezione LNF, Gruppo Collegato di Cosenza, Via P. Bucci, I-87036 Rende, Italy
* Correspondence: cvacacela@yachaytech.edu.ec

Abstract: Graphene oxide (GO) exhibits different properties from those found in free-standing graphene, which mainly depend on the type of defects induced by the preparation method and post-processing. Although defects in graphene oxide are widely studied, we report the effect of drying time in GO and how this modifies the presence or absence of edge-, basal-, and sp^3-type defects. The effect of drying time is evaluated by Raman spectroscopy, UV-visible spectroscopy, and transmission electron microscopy (TEM). The traditional D, G, and 2D peaks are observed together with other less intense peaks called the D', D*, D**, D+G, and G+D. Remarkably, the D* peak is activated/deactivated as a direct consequence of drying time. Furthermore, the broad region of the 2D peak is discussed as a function of its deconvoluted $2D_{1A}$, $2D_{2A}$, and D+G bands. The main peak in UV-visible absorption spectra undergoes a redshift as drying time increases. Finally, TEM measurements demonstrate the stacking of exfoliated GO sheets as the intercalated (water) molecules are removed.

Keywords: graphite; few-layer graphene; graphene oxide; Raman; TEM; UV-vis; Lorentzian fitting

Citation: Tene, T.; Guevara, M.; Valarezo, A.; Salguero, O.; Arias Arias, F.; Arias, M.; Scarcello, A.; Caputi, L.S.; Vacacela Gomez, C. Drying-Time Study in Graphene Oxide. *Nanomaterials* **2021**, *11*, 1035. https://doi.org/10.3390/nano11041035

Academic Editor: José Miguel González-Domínguez

Received: 26 March 2021
Accepted: 12 April 2021
Published: 19 April 2021

Publisher's Note: MDPI stays neutral with regard to jurisdictional claims in published maps and institutional affiliations.

Copyright: © 2021 by the authors. Licensee MDPI, Basel, Switzerland. This article is an open access article distributed under the terms and conditions of the Creative Commons Attribution (CC BY) license (https://creativecommons.org/licenses/by/4.0/).

1. Introduction

Graphene, a two-dimensional carbon nanomaterial arranged in hexagonal symmetry, has already demonstrated excellent electronic, mechanical, electric, magnetic, and thermal properties [1–3] which guarantee exciting applications, including composites [4], energy storage [5], catalysis [6], field-effect transistors [7], and plasmonics [8–10]. Several strategies are currently used for preparing graphene, for instance, chemical vapor deposition (CVD) [11], epitaxial growth [12], liquid-phase exfoliation [13], shear-exfoliation [14], zeolite-shear exfoliation [15], and chemical exfoliation [16].

Among them, the liquid-phase exfoliation and chemical exfoliation strategies are the most practical methods for preparing graphene in large quantities. In particular, the chemical exfoliation uses strong oxidizing agents to produce graphite oxide [17,18], which under sonication yields graphene oxide (GO) [19]. After the oxidation process, GO is covered by different functional groups (hydroxyl, epoxide, carboxyl, and carbonyl groups [20,21]) which increase the interlayer spacing up to 0.87 nm [22]. Moreover, GO is a large-band gap material [23], limiting its use for electronic applications, but opening a multitude of other applications as the removal of heavy metals [24] or dyes [25] as well as GO-based hydrogels [26].

Depending on the oxidation process, e.g., Hummers [27], Marcano [28], and Chen [29], GO could present different structural defects, for example, edge-, vacancy-, and sp^3-type defects [30,31]. The study and control of such defects is of vital importance before choosing any of its widespread intended applications. In this respect, Raman spectroscopy is the most used characterization technique to scrutinize the quality of the as-made GO. The Raman spectra of graphite, single-layer graphene, few-layer graphene (FLG), and GO are widely reported in the literature, characterized by three prominent peaks, namely the D band, the G band, and the 2D band [32]. Less intense peaks called the D′, D*, D**, D+G, and G+D can also be observed [33].

The study of these bands is fundamental in characterizing graphene-derived materials. In this context, Kaniyoor et al. showed a study of the Raman spectrum of GO, considering seven different preparation strategies [33]. It is worth noting that the Raman spectrum of GO is significantly different from that of single-layer graphene (SLG) and must be carefully analyzed. Consequently, there is a lot of discrepancy in the literature over the Raman spectra of oxidized graphenes. Some reports show similarly intense D and G peaks with a highly broadened and low intense 2D band region [34–36]. In fact, in many published papers, the 2D band is neither shown nor discussed.

The different strategies for preparing GO present environmental issues during the preparation process. As an example, Hummers et al. [27] reported the strategy most used: The oxidation process is carried out by treating natural graphite with $KMnO_4$ and $NaNO_3$ in concentrated H_2SO_4. This procedure involves the generation of toxic gases, such as NO_2 and N_2O_4, limiting the large-scale production of GO. Recently, we have demonstrated a scalable eco-friendly protocol by excluding $NaNO_3$ from chemical reaction [29,37] and subjecting the resulting graphite oxide to simple purification steps to obtain GO. Therefore, an investigation of the Raman spectrum of the as-made GO is not reported yet.

In this paper, such a study is presented. Instead of giving a comparison between GO preparation methods (oxidation or reduction strategies) as reported in [33], we report the effect of drying time to obtain GO powder, by using Raman spectroscopy. This study is complemented by transmission electron microscopy (TEM) and UV-visible (UV-vis) absorption measurements. For comparison purpose, the Raman spectra of natural graphite and FLG prepared in ethanol, are also reported.

2. Materials and Method

All chemicals were used as received, without further purification. Graphite powder (<150 µm, 99.99%), sulfuric acid (H_2SO_4, ACS reagent, 95.0–98.0%), potassium permanganate ($KMnO_4$, ACS reagent, ≥99.0%), hydrochloric acid (HCl, ACS reagent, 37%), hydrogen peroxide (H_2O_2, 30%, Merk, Kenilworth, NJ, USA), and ethanol (purity ≥ 98.0%, CAS: 64-17-5) were obtained from Sigma-Aldrich (St. Louis, MO, USA).

2.1. Synthesis of GO

GO was prepared as reported in our earlier paper [37]. Briefly, 3.0 g of graphite powder was added to 70 mL H_2SO_4 while stirring in an ice-water bath. Then, 9.0 g $KMnO_4$ were added. The resulting mixture was transferred to an oil bath and agitated for about 0.5 h. After that, 150 mL distilled water was added, and the solution was stirred for 20 min. Additionally, 500 mL distilled water was added, followed by 15 mL H_2O_2 and stirred up to see a yellowish solution. The resulting graphite oxide suspension was washed with 1:10 HCl solution and distilled water eight times through centrifugation. The precipitated material was re-dispersed in water by sonication using an ultrasonic bath (Branson 2510 Ultrasonic Cleaner). The suspension was centrifugated at 1000 rpm for 0.5 h, and then, dried in a drying stove (2005142, 60 Hz, 1600 W, J.P. Selecta, Barcelona, Spain) at 80 °C for 0.5 h, 1 h, 3 h, 5 h, 24 h, taking 15 mL samples from the suspension. The obtained GO powder was used in subsequent characterization.

2.2. Synthesis of Few-Layer Graphene

To perform the sonication process as simple as possible, 100 mg of graphite was added into 100 mL of ethanol using closed tube containers. The resulting mixture was sonicated employing an ultrasonic bath (Branson 2510 Ultrasonic Cleaner, 40 kHz, 130 W) in continuous operation. The sonication time was set to 7 h, and the resulting dispersion was centrifuged for 10 min at 1000 rpm to remove non-exfoliated graphite particles. Solvent evaporation is avoided because the sonication is made in sealed containers, and the temperature of the bath is controlled by fluxing fresh water every 0.5 h.

2.3. Characterization

Raman spectra of graphite, FLG, and oxidized graphenes were obtained using a Jasco NRS-500 spectrometer with a 532 nm laser wavelength (0.3 mW, 100× objective). The surface morphologies of the samples were taken out on a transmission electron microscope (TEM, JEM 1400 Plus) operating at 80 kV. For TEM and Raman characterization, GO and FLG samples was prepared by drop casting onto formvar-coated copper grids, and glass substrates, respectively. Similarly, the GO samples subject to drying-time experiment were directly deposited on the corresponding substrates. The UV-visible measurements were recorded using a UV-vis spectrometer (Thermo Scientific, Evolution 220, Waltham, MA, USA), re-dispersing the treated GO by mid-sonication for 5 min.

3. Results and Discussion

Let us stress again, the main goal of the present work is the Raman study of the effect of the drying time on graphene oxide samples. As commented, [33] showed a detailed study of the Raman spectra of GO considering different preparations processes, and the effect of the temperature was widely understood [38]. The dependency of graphene oxide layers on drying methods was also briefly reported [39]. Very recently, the study of the effect of the Raman excitation laser on GO was reported [40].

3.1. Raman Spectrum of Graphite

Natural graphite has a crystal structure made up of flat graphene layers. These layers are stacked in a hexagonal honeycomb-like network, usually following the AB Bernal stacking or AA' stacking. The interatomic in-plane distance is 1.42 Å, while the out-of-plane distance between graphene layers (due to van der Waals interactions) can have values ranging from is 3.35 Å up to 3.70 Å.

The Raman spectrum of natural graphite is shown in Figure 1. The main feature is the first order spectrum displaying the E_{2g} in-plane optical mode (commonly called G peak) at 1577 cm^{-1} (Figure 1a). This narrow G peak appears due to the bond stretching of all pairs of sp^2 hybridized carbon atoms in both rings and chains [41]. The G* peak found at 2447 cm^{-1} is characteristic of graphitic materials. The 2D peak appears at 2720 cm^{-1}, characterized by two bands (Figure 1b), the intense $2D_{2A}$ band at 2720 cm^{-1}, and a less intense $2D_{1A}$ band at 2677 cm^{-1}. These bands are originated due to the splitting of π electrons as an effect of the interaction between stacked graphene layers.

The position and shape of the 2D peak depends, mainly, on the number of layers. Indeed, the 2D peak in SLG is fitted by a single Lorentzian function, say, the $2D_{1A}$ band which is located around 2680 cm^{-1} [42]. Moreover, the intensity ratio I_{2D}/I_G in SLG is >11 while our starting graphite shows a $I_{2D2A}/I_G \approx 0.45$, typical of any natural graphite.

Hence, the G peak is related to the C–C stretching mode in sp^2 carbon bonds and the 2D peak is a fingerprint to evaluate "qualitatively" the number of layers in the obtained graphene or graphene-derived material. The absent (or negligible intensity) of the D peak evidences a defect-free pristine graphite due to the D peak is ascribed to the basal/edge structural imperfections, corresponding with an increase in the amount of disorderly carbon and a decrease in the graphite crystal size [43].

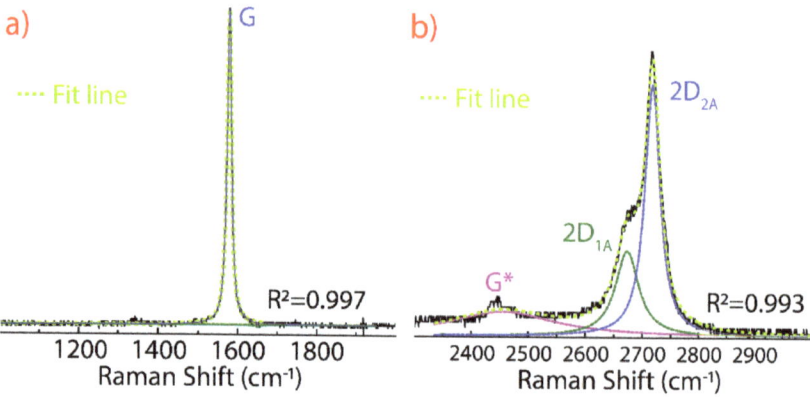

Figure 1. Raman spectrum of graphite (**a**) from 1000 to 2000 cm^{-1} and (**b**) from 2300 to 3000 cm^{-1} recorded using 532 excitation laser. The intensity was normalized by the most intense peak and the fitting of the peaks using Lorentzian functions.

3.2. Raman Spectrum of Few-Layer Graphene

The Raman spectrum of FLG prepared in ethanol by sonication was discussed in detail in Refs. [44,45]. The three significant peaks of FLG are depicted in Figure 2, the D peak at 1339 cm^{-1}, the G peak at 1578 cm^{-1}, and the 2D peak at 2719 cm^{-1}. Additionally, other less intense peaks are detected, the D** peak at 1385 cm^{-1}, the D' peak at 1615 cm^{-1}, the D+G at 2901 cm^{-1}, and the G+D' at 2961 cm^{-1}.

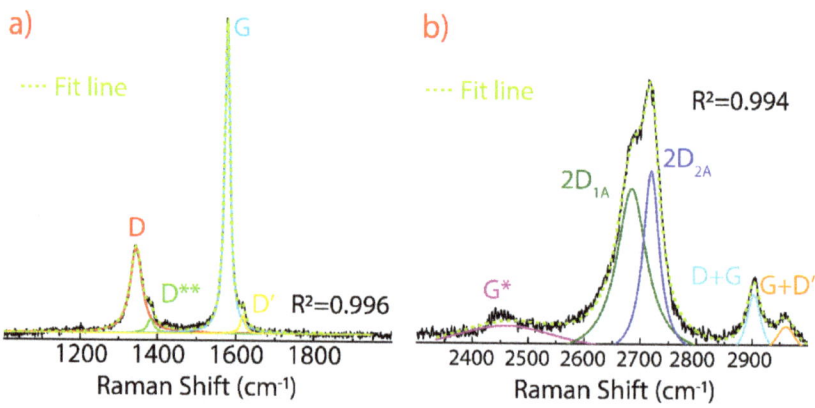

Figure 2. Raman spectrum of FLG (**a**) from 1000 to 2000 cm^{-1} and (**b**) from 2300 to 3000 cm^{-1} recorded using 532 excitation laser. The intensity was normalized by the most intense peak and the fitting of the peaks using Lorentzian functions.

Here, the D and D' peaks are activated due to (i) the induced structural damage after the sonication process and (ii) the reduction of the lateral size of the graphene sheets. However, the D peak is narrow and significantly less intense than the G peak, suggesting edge defects (folded edge samples) rather than basal defects (vacancies or impurities). The intensity ratio I_D/I_G was found to be <0.5. The type of defects can be deduced analyzing the D' peak and the intensity ratio $I_D/I_{D'}$. In particular, the value of $I_D/I_{D'}$ ~3.7 is associated to edge-type defects, but basal defects cannot be ruled out completely since the D** and D+G peaks are also present. Particularly, these peaks (with an appreciable intensity) have been found in highly disordered carbon samples.

Therefore, the origin of these low-intensity peaks (D**, and D+G) in FLG appear as an effect of the structural damage due to the sonication time, solvent type, and sonication power, crucial parameters to be controlled to attain the scalable production of graphene dispersions. On the other hand, defects do not activate the G+D' peak because of momentum conservation restrictions [46].

Compared to graphite, the 2D peak in FLG is also characterized by two bands, the $2D_{1A}$ at 2685 cm^{-1} and the $2D_{2A}$ at 2720 cm^{-1}. Three main characteristics are observed: (i) The complete shape of the 2D peak changes, (ii) the intensity of the $2D_{1A}$ band increases close to that of the $2D_{2A}$ band, and (iii) the intensity ratio of I_{2D2A}/I_{2D1A} is 1.12 (in graphite, I_{2D2A}/I_{2D1A} = 2.78). All these evidences support the transformation of graphite into FLG, likely no more than 10 layers.

3.3. Raman Spectrum of Non-Dried Graphene Oxide

We now move to the focus of this communication. The Raman spectrum of GO (obviously different from that of graphite and FLG) is shown in Figure 3, and the corresponding peak position and full-width at half maximum (FWHM) are reported in Tables 1 and 2. Notice that the G+D' peak is not detected in the wavenumber window analyzed from 1000 cm^{-1} to 3000 cm^{-1} (Figure 3b). This peak is assumed to be shifted at higher wavenumber values (>3100 cm^{-1}) because of the spectral weight of the D+G peak.

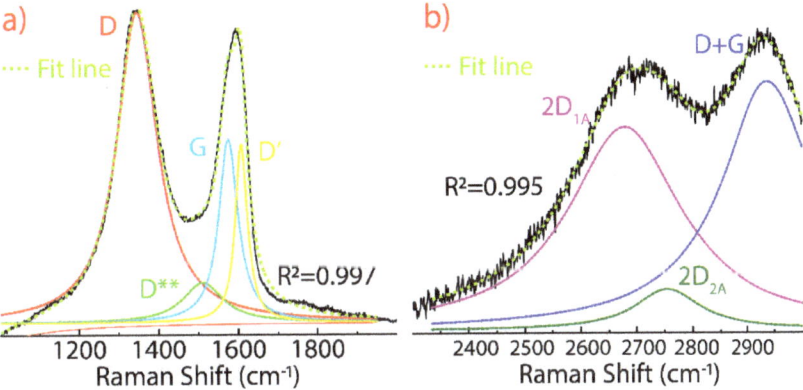

Figure 3. Raman spectra of GO (**a**) from 1000 to 2000 cm^{-1} and (**b**) from 2300 to 3000 cm^{-1} recorded using 532 excitation laser. The intensity was normalized by the most intense peak and the fitting of the peaks using Lorentzian functions.

Table 1. Peak position and full-width at half maximum (FWHM) of GO at different drying times in the region from 1000 to 2000 cm^{-1}. The FWHM was obtained using Lorentzian fitting.

	D*-FWHM (cm^{-1})	D-FWHM (cm^{-1})	D**-FWHM (cm^{-1})	G-FWHM (cm^{-1})	D'-FWHM (cm^{-1})
0.0 h	–	1348–122	1511–122	1576–58	1608–36
0.5 h	1121–61	1347–127	1498–131	1571–64	1606–39
1.0 h	1135–64	1346–126	1514–126	1576–60	1608–37
3.0 h	1135–46	1347–147	1511–129	1572–65	1605–41
5.0 h	1125–174	1346–151	1517–127	1576–63	1606–41
24 h	1124–89	1344–143	1518–130	1577–62	1607–40

Table 2. Peak position and full-width at half maximum (FWHM) of GO at different drying times in the region from 2300 to 3000 cm^{-1}. The FWHM was obtained using Lorentzian fitting.

	$2D_{1A}$-FWHM (cm^{-1})	$2D_{2A}$-FWHM (cm^{-1})	D+G-FWHM (cm^{-1})
0.0 h	2679–246	2755–140	2937–207
0.5 h	2685–259	2751–46	2930–222
1.0 h	2685–259	2751–46	2930–222
3.0 h	2669–255	2755–141	2928–251
5.0 h	2653–200	2750–200	2921–144
24 h	2653–188	2749–175	2922–131

It is well known that in GO, the D and D′ peaks are related to the presence of defects such as folded edges, vacancies, impurities (functional groups or remaining metal species), and the change from sp^2 to sp^3 hybridization [47–50]. In particular, a decrease in the D′ peak intensity can be considered as straight evidence of GO reduction. On the other hand, the intensity ratio $I_D/I_{D'}$ in our as-made GO is ~2.2, erroneously suggesting only the predominance of edge defects in its structure, even a lower value than that observed in FLG.

The D** band is due to contributions from the phonon density of states in finite size graphitic crystals, C−H vibrations in hydrogenated carbon, and hopping-like defects [51]. Depending on the preparation process, GO also displays a D* band related to the sp^3 diamond line on disordered amorphous carbons, but the broad region between ~1400 cm^{-1} and ~1650 cm^{-1} is not attributed to diamond carbon phases [31,52,53]. Although the D* band is not perceptible in non-dried GO, this band appears after the drying-time experiment (discussed below).

Interestingly enough, the 2D band region (Figure 3b) is characterized by intense $2D_{1A}$ and D+G bands and a less intense $2D_{2A}$ band in contrast to those observed in natural graphite (Figure 1b) and FLG (Figure 2b). The intensity ratio I_{2D2A}/I_{2D1A} = 0.22 in GO decreases about 12 times and 5 times compared to graphite and FLG, respectively. These outcomes are probably due to a good chemical exfoliation and reduction of the number of layers, but the resulting material has sufficient defects to activate the D, D′, D**, and D+G bands.

3.4. Raman Spectrum of Dried Graphene Oxide

Figure 4 shows the Raman spectra of GO dried at different times from 0.5 h to 24 h, keeping the temperature fixed (80 °C). The D, G, and D′ peaks are observed which are not substantially affected in position, FWHM, or intensity in the drying-time testing (Table 1, Figure 5). In fact, a thermal reduction of the material obtained is not expected, as confirmed by FT-IR results (no shown here). Interestingly, the D* band is not as intense at 0.5 h and 1 h, and can be seen as the drying time increases from 3 h to 24 h.

Two intervals are observed: (i) From 0.5 h to 3 h, the D* peak is shifted at 1135 cm^{-1} while the FWHM value remains relatively unchanged, and (ii) from 5 h to 24 h, the peak position is shifted to lower wavenumber values while the FWHM value notably increases, giving the highest value at 5 h (FWHM = 174) and 24 h (FWHM = 89). In contrast, the peak position and FWHM value of the D** peak are not clearly affected by the drying time, but its intensity increases along with the intensity of the D* band.

As seen in Figure 5, the relative intensity of the D′ (blue line) and G (orange line) bands shows a constant trend, while the D* (green line) and D** (magenta line) display a square-root-like dispersion. From the curve fitting, it can be seen that the most significant effect of drying time is ≤10 h. These outcomes show that it is not possible to transform GO into reduce graphene oxide (rGO) through drying time at 80 °C. Nevertheless, if it is possible to increase the density of sp^3-type and hopping-like defects associated with the presence of the D* and D** band, respectively.

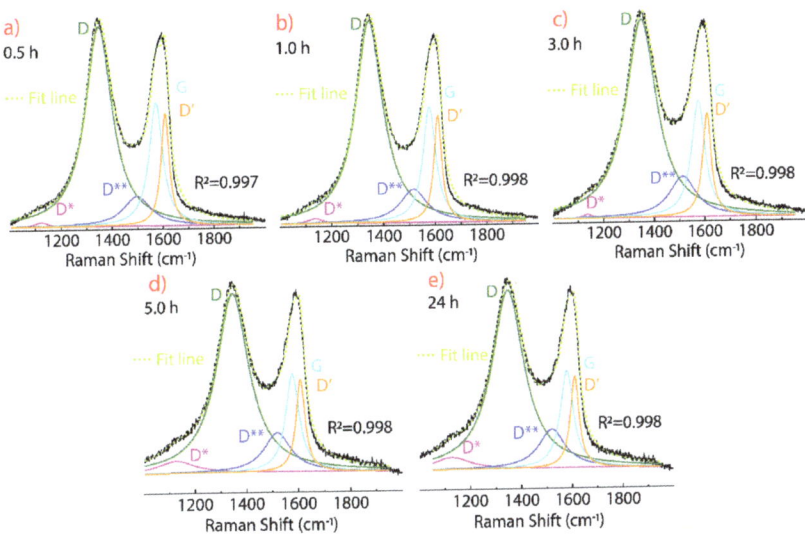

Figure 4. Raman spectra of GO from 1000 to 2000 cm^{-1} recorded using 532 excitation laser, subject to 80 °C, and considering different drying times: (**a**) 0.5 h, (**b**) 1 h, (**c**) 3 h, (**d**) 5 h, and (**e**) 24 h. The intensity was normalized by the most intense peak and the fitting of the peaks using Lorentzian functions.

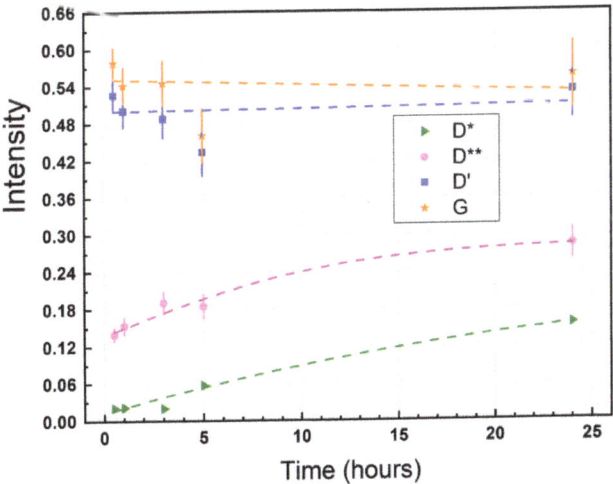

Figure 5. Intensity of the D*, D**, G, and D' peaks as function of the drying time (0.5 h, 1 h, 3 h, 5 h, 24 h). The maximum intensity was obtained from the fitting using Lorentzian functions.

Therefore, it cannot be said that there is only one type of defects in GO. Instead of doing an analysis based on the intensity ratio $I_D/I_{D'}$, we hypothesize that the study of defects in GO must be accompanied through the ratio of intensities, i.e., $I_D/I_{D'}$ for edge-, I_D/I_{D*} for sp^3-, and I_D/I_{D**} for hopping-like defects. In the present work, the corresponding intensity ratio values as function of the drying time are: $23.95 < I_D/I_{D*} < 6.05$, $6.14 < I_D/I_{D**} < 3.31$, and $1.70 < I_D/I_{D'} < 1.76$. This hypothesis and intensity ratio values should be corroborated future studies. Most importantly, it can be noted that the longer the drying time, the lower the intensity ratio of I_D/I_{D^*} and $I_D/I_{D^{**}}$ while the intensity ratio $I_D/I_{D'}$ slightly increases. The $I_D/I_{D'}$ values reasonably agree with those reported in [54].

The presence of defects in GO also affects the 2D band region which is deconvoluted in the $2D_{1A}$, $2D_{2A}$, and D+G bands (Figure 6). In the literature, these bands are not discussed in detail when talking about conventionally prepared GO because the very low intensity and only noticeable when rGO is obtained, a sign of the recovery of the graphene structure. An important feature of GO is the presence of the 2D band region, suggesting non-critical damage to graphene structure after oxidation process. Thus, non-aggressive, environmentally friendly, and highly efficient reducing agents could be used, for instance, citric acid, ascorbic acid, or citrus hystrix [55].

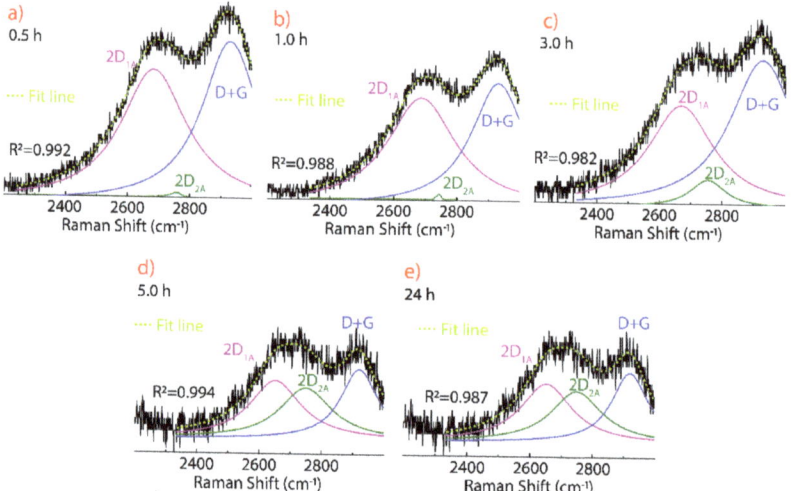

Figure 6. Raman spectra of GO from 2300 to 3000 cm^{-1} recorded using 532 excitation laser, subject to 80 °C, and considering different drying times: (**a**) 0.5 h, (**b**) 1 h, (**c**) 3 h, (**d**) 5 h, and (**e**) 24 h. The intensity was normalized by the most intense peak and the fitting of the peaks using Lorentzian functions.

The Raman spectra, peak position, FWHM, and relative intensity of the deconvoluted 2D band region as a function of drying time, are reported in Figure 6, Table 2 and Figure 7, respectively. The $2D_{1A}$ and D+G bands are clearly observed, while the $2D_{2A}$ band begins to appear as an effect of the drying time after 3 h, reaching an intensity comparable to that of the $2D_{1A}$ and D+G bands at 5 h (Figure 6d) and 24 h (Figure 6e).

The decrease of the intensity of the $2D_{2A}$ band at 0.5 h (Figure 6a) and 1 h (Figure 6b), can be interpreted as due to GO with few layers and wrinkled structure, as can be concluded from the prevalence of the $2D_{1A}$ and D+G bands, respectively. The presence of the $2D_{2A}$ band after 3 h of drying is attributed to the evaporation of water molecules between the GO layers, causing the stacking of the chemically exfoliated sheets. Trying to explain this fact, we have carried out TEM measurements (discussed below).

Although the FWHM values of the $2D_{1A}$ band does not critically change after 1 h of drying, its position is shifted at 2685 cm^{-1}. After that, the value of the FWHM decreases (e.g., FWHM = 188 at 24 h) and the peak position moves to lower wavenumber values (2653 cm^{-1} at 5 h and 24 h), even lower than that observed in graphite (2677 cm^{-1}) or FLG (2685 cm^{-1}).

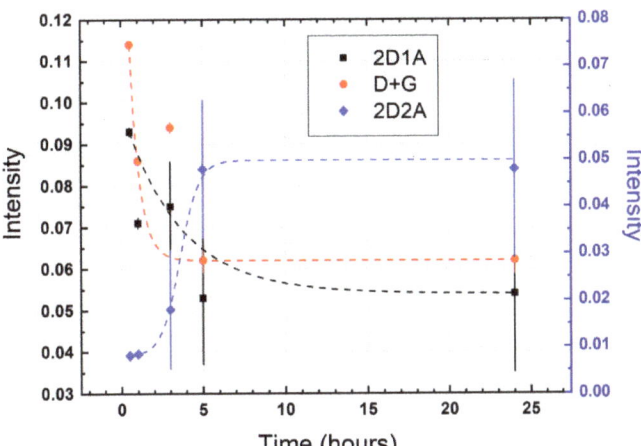

Figure 7. Intensity of the $2D_{1A}$, $2D_{2A}$, D+G peaks as function of the drying time (0.5 h, 1 h, 3 h, 5 h, 24 h). The maximum intensity was obtained from the fitting using Lorentzian functions.

Compared to the peak position of the $2D_{2A}$ band in graphite or FLG (2720 cm^{-1}), this band in GO is found at ~2750 cm^{-1} and is not affected by drying time, but the FWHM value decreases down to 46 cm^{-1}, where the intensity of the $2D_{2A}$ band is barely perceptible (0.5 h and 1 h). On the other hand, the peak position and FWHM value of the D+G band decrease (e.g., 2922 cm^{-1} and FWHM = 131 cm^{-1} at 24 h) as the drying time increases. The peak position found at 5 h and 24 h of drying are close to that of FLG (2901 cm^{-1}). The latter corroborates our statement that the D+G band moves to higher wavenumber values as an effect of the oxidation process, also causing the displacement of the G+D' band. Therefore, it is demonstrated the D+G band has a high dependence on drying time, and its effect must be carefully considered when characterizing GO samples.

The intensity ratio I_{2D2A}/I_{2D1A} increases from 0.22 to 0.89, a value close to that observed in FLG (I_{2D2A}/I_{2D1A} = 1.12) while the intensity ratio I_{D+G}/I_{2D2A} decreases from 1.21 to 1.14. The latter suggests a slight reduction of defects. Most importantly, Figure 7 shows the relative intensity of the $2D_{1A}$ (black line), $2D_{2A}$ (blue line), and D+G (red line) bands. The $2D_{1A}$ and D+G bands are characterized by an exponentially decreasing behavior while the $2D_{2A}$ band shows a sigmoid growth trend. From the curve fitting, it can be seen that the most significant effect is obtained for drying times ≤5 h. This crucial result shows that for long drying times, the water molecules and other possible oxygen-containing molecules are removed, allowing the exfoliated GO sheets to pile up, probably, to re-form graphite oxide.

With all this in mind, the Raman results showed that a GO not very disordered and with few layers can be obtained with a drying time of 1 to 3 h. After that, a highly disordered graphitic-like structure is expected, suggesting that long drying times being unnecessary in a practical large-scale GO production. Furthermore, long drying times may not facilitate the recovery of the graphene structure after even aggressive reduction processes, for example, using hydrazine. This statement motivates more extended works.

The UV-vis absorption spectra of GO subject to different drying times are depicted in Figure 8. At 0 h of drying, GO exhibits an absorption peak at ~233 nm and a shoulder at ~304 nm, which are attributed to the $\pi - \pi^*$ transition in C–C bonds and the $n - \pi^*$ transition in C=O bonds, respectively. As drying time increases, the peak and the shoulder gradually redshift at ~250 nm and at ~325 nm, respectively, suggesting that the electronic conjugation within graphene structure starts to be restored. Although it is not possible to affirm the transformation of GO into rGO (as evidenced by Raman measurements), this result offers the possibility to adapt the optical and electrical properties of GO.

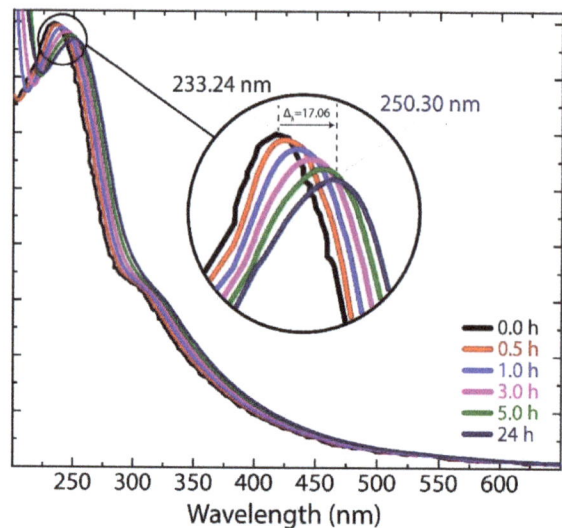

Figure 8. UV-vis spectra of GO subject to different drying times.

In addition to the results obtained by the drying time experiment, we also briefly describe the effect of rehydration in Figures S1 and S2 (Raman spectra), and Figure S3 (UV-vis measurements). The dried GO samples (at 1 h, 5 h, and 24 h) were newly redispersed in continuous stirring for 30 min. Particularly, the D* band is not perceptible in all rehydrated samples, and the intensity of the $2D_{2A}$ band substantially decreases. UV-vis results evidence a blueshift of the absorption $\pi - \pi^*$ peak. These outcomes can be attributed a well redispersion of GO due to its hydrophilic properties.

3.5. TEM Analysis of Graphene Oxide

TEM micrographs of non-dried GO and GO subjected to different drying times from 0.5 h to 24 h are shown in Figure 9. Non-dried GO sample appears as a transparent and thin nanosheet with some wrinkles and folds on the surface and edges, but not with a very disordered surface morphology as GO prepared by conventional methods. Instead of point defects, these wrinkles are associated with surface defects formed due to the folding or twisting of the exfoliated GO sheets, causing deviation from the sp^2 to sp^3 character. This wrinkled structure is a characteristic of graphene-like materials, whereas they are not present in other carbon nanostructures, e.g., amorphous carbons.

At a first approximation, the GO sample subject to 0.5 h of drying (Figure 9b) appears to be very similar to non-dried GO. However, a clear difference is observed beyond the wrinkles and folds, i.e., opaque regions are detected, which are attributed to the stacking of exfoliated GO sheets because the intercalated water molecules between the layers begin to be removed. This effect is more clearly observed after 1 h and 3 h of drying. At 5 h and 24 h, the GO samples look dark, suggesting a high removal of the intercalated molecules between layers, which causes a large stacking of the exfoliated GO sheets. This outcome supports the presence of the $2D_{2A}$ band observed in the Raman study. In particular, Figure 9f demonstrates the stacked layers after 24 h of drying, and a GO sample with a lateral size larger than the previous ones, which also allows to observe that the GO flakes are reassembling. The inset in Figure 9f shows that the flake with lateral size in the micrometers range is made of a superposition of restacked GO layers.

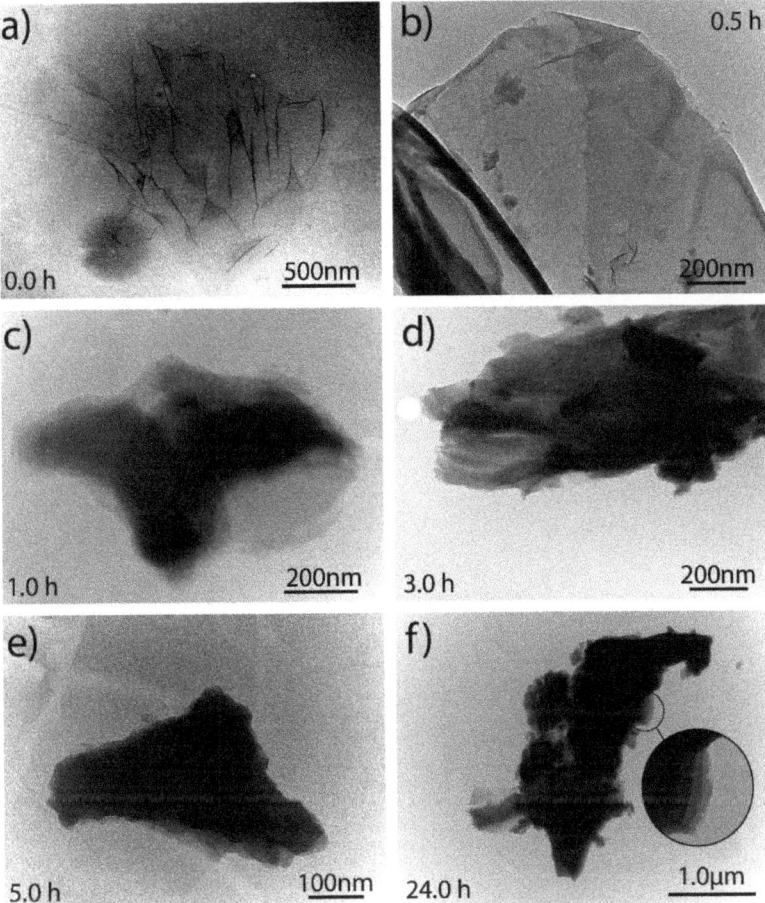

Figure 9. TEM analyses of GO subject to 80 °C and considering different drying times: (**a**) 0.0 h, (**b**) 0.5 h, (**c**) 1.0 h, (**d**) 3.0 h, (**e**) 5.0 h, and (**f**) 24 h.

4. Conclusions

In summary, we have reported the effect of drying time (from 0.0 h to 24 h) on GO by a systematic Raman study. For comparison, the Raman spectrum of graphite and FLG were also discussed. The work is complemented using UV-vis and TEM measurements. The D, G, and D' peaks were not affected by increasing the drying time, but the D*, D**, and $2D_{2A}$ peaks seemed to be very sensitive. The relative intensity of the different Raman bands and corresponding FWHM's were discussed. UV-vis results evidenced a redshift from ∼233 nm (0.0 h) to ∼250 (24 h) nm. TEM results showed the stacking and reassembly of GO sheets as a direct consequence of the drying time.

The study was carried out with a process as simple as possible, subjecting each sample to 80 °C in an oven, and characterizing them as soon as possible. For this reason, in future studies, a controlled environment should be considered as well as the time the samples remain at room temperature before being characterized. The water molecules would continue to evaporate for long periods. Additionally, we suggest to consider the type of graphite since here only graphite powder has been used. In fact, large lateral graphite or expanded graphite could present different results compared to those reported here.

Our findings are intended to contribute to the control of the technical parameters involved in the synthesis process to achieve the large-scale production of GO powder.

Supplementary Materials: The following are available online at https://www.mdpi.com/article/10.3390/nano11041035/s1, Figure S1: Raman spectra of rehydrated GO samples (1 h, 5 h, 24 h) from 1000 to 2000 cm^{-1} recorded using 532 excitation laser, Figure S2: Raman spectra of rehydrated GO samples (1 h, 5 h, 24 h) from 2300 to 3000 cm^{-1} recorded using 532 excitation laser, Figure S3: UV-vis spectra of rehydrated GO samples (1 h, 5 h, 24 h).

Author Contributions: Conceptualization, supervision, C.V.G., L.S.C. and T.T.; methodology, A.V., F.A.A. and O.S.; validation, T.T., M.G. and L.S.C.; formal analysis, C.V.G., M.A., A.S. and L.S.C.; investigation, C.V.G.; resources, T.T.; data curation, T.T. and M.G.; writing—original draft preparation, C.V.G.; writing—review and editing, C.V.G.; visualization, M.G. All authors have read and agreed to the published version of the manuscript.

Funding: This work was supported by Universidad Técnica Particular de Loja (UTPL-Ecuador), project "222-Radon adsorption on activated carbon and graphene filters" (code: PROY_INNOV_QUI_2021_3019). Part of this work has been also supported by the FONDOCyT from Ministry of Higher Education Science and Technology of the Dominican Republic (grant no. 2018-2019-3A9-139).

Institutional Review Board Statement: Not applicable.

Informed Consent Statement: Not applicable.

Data Availability Statement: The data that supports the findings of this study are available within the article. Any additional data relevant to this study are available from the author upon reasonable request.

Acknowledgments: T.T., M.G. and C.V.G. wish to thank Escuela Superior Politécnica de Chimborazo for hospitality during the completion of this work.

Conflicts of Interest: The authors declare no conflict of interest.

References

1. Soldano, C.; Mahmood, A.; Dujardin, E. Production, properties and potential of graphene. *Carbon* **2010**, *48*, 2127–2150. [CrossRef]
2. Coello-Fiallos, D.; Tene, T.; Guayllas, J.L.; Haro, D.; Haro, A.; Gomez, C.V. DFT comparison of structural and electronic properties of graphene and germanene: Monolayer and bilayer systems. *Mater. Today Proc.* **2017**, *4*, 6835–6841. [CrossRef]
3. Sindona, A.; Pisarra, M.; Gomez, C.V.; Riccardi, P.; Falcone, G.; Bellucci, S. Calibration of the fine-structure constant of graphene by time-dependent density-functional theory. *Phys. Rev. B* **2017**, *96*, 201408. [CrossRef]
4. Huang, X.; Qi, X.; Boey, F.; Zhang, H. Graphene-based composites. *Chem. Soc. Rev.* **2012**, *41*, 666–686. [CrossRef] [PubMed]
5. Raccichini, R.; Varzi, A.; Passerini, S.; Scrosati, B. The role of graphene for electrochemical energy storage. *Nat. Mater.* **2015**, *14*, 271–279. [CrossRef]
6. Huang, C.; Li, C.; Shi, G. Graphene based catalysts. *Energy Environ. Sci.* **2012**, *5*, 8848–8868. [CrossRef]
7. Britnell, L.; Gorbachev, R.V.; Jalil, R.; Belle, B.D.; Schedin, F.; Mishchenko, A.; Georgiou, T.; Katsnelson, M.I.; Eaves, L.; Morozov, S.V.; et al. Field-effect tunneling transistor based on vertical graphene heterostructures. *Science* **2012**, *335*, 947–950. [CrossRef]
8. Gomez, C.V.; Pisarra, M.; Gravina, M.; Sindona, A. Tunable plasmons in regular planar arrays of graphene nanoribbons with armchair and zigzag-shaped edges. *Beilstein J. Nanotechnol.* **2017**, *8*, 172–182. [CrossRef]
9. Sindona, A.; Pisarra, M.; Bellucci, S.; Tene, T.; Guevara, M.; Vacacela Gomez, C. Plasmon oscillations in two-dimensional arrays of ultranarrow graphene nanoribbons. *Phys. Rev. B* **2019**, *100*, 235422. [CrossRef]
10. Gomez, C.V.; Pisarra, M.; Gravina, M.; Pitarke, J.M.; Sindona, A. Plasmon modes of graphene nanoribbons with periodic planar arrangements. *Phys. Rev. Lett.* **2016**, *117*, 116801. [CrossRef]
11. Deokar, G.; Avila, J.; Razado-Colambo, I.; Codron, J.-L.; Boyaval, C.; Galopin, E.; Asensio, M.-C.; Vignaud, D. Towards high quality CVD graphene growth and transfer. *Carbon* **2015**, *89*, 82–92. [CrossRef]
12. Yang, W.; Chen, G.; Shi, Z.; Liu, C.-C.; Zhang, L.; Xie, G.; Cheng, M.; Wang, D.; Yang, R.; Shi, D.; et al. Epitaxial growth of single-domain graphene on hexagonal boron nitride. *Nat. Mater.* **2013**, *12*, 792–797. [CrossRef] [PubMed]
13. Gomez, C.V.; Tene, T.; Guevara, M.; Usca, G.T.; Colcha, D.; Brito, H.; Molina, R.; Bellucci, S.; Tavolaro, A. Preparation of few-layer graphene dispersions from hydrothermally expanded graphite. *Appl. Sci.* **2019**, *9*, 2539. [CrossRef]
14. Paton, K.R.; Varrla, E.; Backes, C.; Smith, R.J.; Khan, U.; O'Neill, A.; Boland, C.; Lotya, M.; Istrate, O.M.; King, P.; et al. Scalable production of large quantities of defect-free few-layer graphene by shear exfoliation in liquids. *Nat. Mater.* **2014**, *13*, 624. [CrossRef] [PubMed]

15. Usca, G.T.; Gomez, C.V.; Guevara, M.; Tene, T.; Hernandez, J.; Molina, R.; Tavolaro, A.; Miriello, D.; Caputi, L.S. Zeolite-assisted shear exfoliation of graphite into few-layer graphene. *Crystals* **2019**, *9*, 377. [CrossRef]
16. Zhang, L.; Liang, J.; Huang, Y.; Ma, Y.; Wang, Y.; Chen, Y. Size-controlled synthesis of graphene oxide sheets on a large scale using chemical exfoliation. *Carbon* **2009**, *47*, 3365–3368. [CrossRef]
17. Stankovich, S.; Dikin, D.A.; Piner, R.D.; Kohlhaas, K.A.; Kleinhammes, A.; Jia, Y.; Wu, Y.; Nguyen, S.T.; Ruoff, R.S. Synthesis of graphene-based nanosheets via chemical reduction of exfoliated graphite oxide. *Carbon* **2007**, *45*, 1558–1565. [CrossRef]
18. Fiallos, D.C.; Gómez, C.V.; Tubon Usca, G.; Pérez, D.C.; Tavolaro, P.; Martino, G.; Caputi, L.S.; Tavolaro, A. Removal of acridine orange from water by graphene oxide. *AIP Conf. Proc.* **2015**, *1646*, 38–45.
19. Paredes, J.I.; Villar-Rodil, S.; Martínez-Alonso, A.; Tascon, J.M.D. Graphene oxide dispersions in organic solvents. *Langmuir* **2008**, *24*, 10560–10564. [CrossRef]
20. Yan, J.-A.; Chou, M.Y. Oxidation functional groups on graphene: Structural and electronic properties. *Phys. Rev. B* **2010**, *82*, 125403. [CrossRef]
21. Gomez, C.V.; Robalino, E.; Haro, D.; Tene, T.; Escudero, P.; Haro, A.; Orbe, J. Structural and electronic properties of graphene oxide for different degree of oxidation. *Mater. Today Proc.* **2016**, *3*, 796–802. [CrossRef]
22. Sheng, Y.; Tang, X.; Peng, E.; Xue, J. Graphene oxide based fluorescent nanocomposites for cellular imaging. *J. Mater. Chem. B* **2013**, *1*, 512–521. [CrossRef]
23. Wang, Z.; Nelson, J.K.; Hillborg, H.; Zhao, S.; Schadler, L.S. Graphene oxide filled nanocomposite with novel electrical and dielectric properties. *Adv. Mater.* **2012**, *24*, 3134–3137. [CrossRef]
24. Lim, J.Y.; Mubarak, N.M.; Abdullah, E.C.; Nizamuddin, S.; Khalid, M. Recent trends in the synthesis of graphene and graphene oxide based nanomaterials for removal of heavy metals—A review. *J. Ind. Eng. Chem.* **2018**, *66*, 29–44. [CrossRef]
25. Arias Arias, F.; Guevara, M.; Tene, T.; Angamarca, P.; Molina, R.; Valarezo, A.; Salguero, O.; Vacacela Gomez, C.; Arias, M.; Caputi, L.S. The adsorption of methylene blue on eco-friendly reduced graphene oxide. *Nanomaterials* **2020**, *10*, 681. [CrossRef] [PubMed]
26. Alammar, A.; Park, S.-H.; Ibrahim, I.; Arun, D.; Holtzl, T.; Dumée, L.F.; Lim, H.N.; Szekely, G. Architecting neonicotinoid-scavenging nanocomposite hydrogels for environmental remediation. *Appl. Mater. Today* **2020**, *21*, 100878. [CrossRef]
27. William, S.; Hummers, J.R.; Offeman, R.E. Preparation of graphitic oxide. *J. Am. Chem. Soc.* **1958**, *80*, 1339.
28. Marcano, D.C.; Kosynkin, D.V.; Berlin, J.M.; Sinitskii, A.; Sun, Z.; Slesarev, A.; Alemany, L.B.; Lu, W.; Tour, J.M. Improved synthesis of graphene oxide. *ACS Nano* **2010**, *4*, 4806–4814. [CrossRef] [PubMed]
29. Chen, J.; Yao, B.; Li, C.; Shi, G. An improved Hummers method for eco-friendly synthesis of graphene oxide. *Carbon* **2013**, *64*, 225–229. [CrossRef]
30. Eckmann, A.; Felten, A.; Mishchenko, A.; Britnell, L.; Krupke, R.; Novoselov, K.S.; Casiraghi, C. Probing the nature of defects in graphene by Raman spectroscopy. *Nano Lett.* **2012**, *12*, 3925–3930. [CrossRef]
31. Kratochvílova, I.; Škoda, R.; Škarohlíd, J.; Ashcheulov, P.; Jäger, A.; Racek, J.; Taylor, A.; Shao, L. Nanosized polycrystalline diamond cladding for surface protection of zirconium nuclear fuel tubes. *J. Mater. Process. Technol.* **2014**, *214*, 2600–2605. [CrossRef]
32. Graf, D.; Molitor, F.; Ensslin, K.; Stampfer, C.; Jungen, A.; Hierold, C.; Wirtz, L. Spatially resolved Raman spectroscopy of single-and few-layer graphene. *Nano Lett.* **2007**, *7*, 238–242. [CrossRef]
33. Kaniyoor, A.; Ramaprabhu, S. A Raman spectroscopic investigation of graphite oxide derived graphene. *Aip Adv.* **2012**, *2*, 32183. [CrossRef]
34. Sun, S.; Wu, P. Competitive surface-enhanced Raman scattering effects in noble metal nanoparticle-decorated graphene sheets. *Phys. Chem. Chem. Phys.* **2011**, *13*, 21116–21120. [CrossRef] [PubMed]
35. Abdelsayed, V.; Moussa, S.; Hassan, H.M.; Aluri, H.S.; Collinson, M.M.; El-Shall, M.S. Photothermal deoxygenation of graphite oxide with laser excitation in solution and graphene-aided increase in water temperature. *J. Phys. Chem. Lett.* **2010**, *1*, 2804–2809. [CrossRef]
36. Choucair, M.; Thordarson, P.; Stride, J.A. Gram-scale production of graphene based on solvothermal synthesis and sonication. *Nat. Nanotechnol.* **2009**, *4*, 30. [CrossRef] [PubMed]
37. Tene, T.; Tubon Usca, G.; Guevara, M.; Molina, R.; Veltri, F.; Arias, M.; Caputi, L.S.; Vacacela Gomez, C. Toward Large-Scale Production of Oxidized Graphene. *Nanomaterials* **2020**, *10*, 279. [CrossRef] [PubMed]
38. Feng, H.; Cheng, R.; Zhao, X.; Duan, X.; Li, J. A low-temperature method to produce highly reduced graphene oxide. *Nat. Commun.* **2013**, *4*, 1–8. [CrossRef] [PubMed]
39. Lv, Y.N.; Wang, J.F.; Long, Y.; Tao, C.A.; Xia, L.; Zhu, H. How Graphene Layers Depend on Drying Methods of Graphene Oxide. *Adv. Mater. Res.* **2012**, *554*, 597–600. [CrossRef]
40. De Lima, B.S.; Bernardi, M.I.B.; Mastelaro, V.R. Wavelength effect of ns-pulsed radiation on the reduction of graphene oxide. *Appl. Surf. Sci.* **2020**, *506*, 144308. [CrossRef]
41. Reich, S.; Thomsen, C. Raman spectroscopy of graphite. *Philos. Trans. R. Soc. London. Ser. A Math. Phys. Eng. Sci.* **2004**, *362*, 2271–2288. [CrossRef] [PubMed]
42. Zhou, H.; Qiu, C.; Yu, F.; Yang, H.; Chen, M.; Hu, L.; Guo, Y.; Sun, L. Raman scattering of monolayer graphene: The temperature and oxygen doping effects. *J. Phys. D Appl. Phys.* **2011**, *44*, 185404. [CrossRef]
43. Tuinstra, F.; Koenig, J. Lo Raman spectrum of graphite. *J. Chem. Phys.* **1970**, *53*, 1126–1130. [CrossRef]

44. Vacacela Gomez, C.; Guevara, M.; Tene, T.; Villamagua, L. The Liquid Exfoliation of Graphene in Polar Solvents. *Appl. Surf. Sci.* **2021**, *546*, 149046. [CrossRef]
45. Cayambe, M.; Zambrano, C.; Tene, T.; Guevara, M.; Usca, G.T.; Brito, H.; Molina, R.; Coello-Fiallos, D.; Caputi, L.S.; Gomez, C.V. Dispersion of graphene in ethanol by sonication. *Mater. Today Proc.* **2020**, *37*, 4027–4030. [CrossRef]
46. Cançado, L.G.; Jorio, A.; Ferreira, E.M.; Stavale, F.; Achete, C.A.; Capaz, R.B.; Moutinho, M.V.D.O.; Lombardo, A.; Kulmala, T.S.; Ferrari, A.C. Quantifying defects in graphene via Raman spectroscopy at different excitation energies. *Nano Lett.* **2011**, *11*, 3190–3196. [CrossRef] [PubMed]
47. Claramunt, S.; Varea, A.; Lopez-Diaz, D.; Velázquez, M.M.; Cornet, A.; Cirera, A. The importance of interbands on the interpretation of the Raman spectrum of graphene oxide. *J. Phys. Chem. C* **2015**, *119*, 10123–10129. [CrossRef]
48. Villamagua, L.; Carini, M.; Stashans, A.; Gomez, C.V. Band gap engineering of graphene through quantum confinement and edge distortions. *Ric. Mat.* **2016**, *65*, 579–584. [CrossRef]
49. Eigler, S.; Dotzer, C.; Hirsch, A. Visualization of defect densities in reduced graphene oxide. *Carbon* **2012**, *50*, 3666–3673. [CrossRef]
50. Scarcello, A.; Alessandro, F.; Polanco, M.A.; Gomez, C.V.; Perez, D.C.; De Luca, G.; Curcio, E.; Caputi, L.S. Evidence of massless Dirac fermions in graphitic shells encapsulating hollow iron microparticles. *Appl. Surf. Sci.* **2021**, *546*, 149103. [CrossRef]
51. Shroder, R.E.; Nemanich, R.J.; Glass, J.T. Analysis of the composite structures in diamond thin films by Raman spectroscopy. *Phys. Rev. B* **1990**, *41*, 3738. [CrossRef]
52. Schwan, J.; Ulrich, S.; Batori, V.; Ehrhardt, H.; Silva, S.R.P. Raman spectroscopy on amorphous carbon films. *J. Appl. Phys.* **1996**, *80*, 440–447. [CrossRef]
53. Veltri, F.; Alessandro, F.; Scarcello, A.; Beneduci, A.; Arias Polanco, M.; Cid Perez, D.; Vacacela Gomez, C.; Tavolaro, A.; Giordano, G.; Caputi, L.S. Porous Carbon Materials Obtained by the Hydrothermal Carbonization of Orange Juice. *Nanomaterials* **2020**, *10*, 655. [CrossRef]
54. Venezuela, P.; Lazzeri, M.; Mauri, F. Theory of double-resonant Raman spectra in graphene: Intensity and line shape of defect-induced and two-phonon bands. *Phys. Rev. B* **2011**, *84*, 35433. [CrossRef]
55. Zhang, J.; Yang, H.; Shen, G.; Cheng, P.; Zhang, J.; Guo, S. Reduction of graphene oxide via L-ascorbic acid. *Chem. Commun.* **2010**, *46*, 1112–1114. [CrossRef] [PubMed]

Article

Waterborne Graphene- and Nanocellulose-Based Inks for Functional Conductive Films and 3D Structures

Jose M. González-Domínguez [1,*], Alejandro Baigorri [1], Miguel Á. Álvarez-Sánchez [1], Eduardo Colom [1], Belén Villacampa [2], Alejandro Ansón-Casaos [1], Enrique García-Bordejé [1], Ana M. Benito [1] and Wolfgang K. Maser [1]

[1] Instituto de Carboquímica ICB-CSIC, C/Miguel Luesma Castán 4, 50018 Zaragoza, Spain; baigorri1994@gmail.com (A.B.); maalvarez@icb.csic.es (M.Á.Á.-S.); ecolom@icb.csic.es (E.C.); alanson@icb.csic.es (A.A.-C.); jegarcia@icb.csic.es (E.G.-B.); abenito@icb.csic.es (A.M.B.); wmaser@icb.csic.es (W.K.M.)
[2] Department of Condensed Matter Physics, ICMA-CSIC, University of Zaragoza, 50009 Zaragoza, Spain; bvillaca@unizar.es
* Correspondence: jmgonzalez@icb.csic.es

Abstract: In the vast field of conductive inks, graphene-based nanomaterials, including chemical derivatives such as graphene oxide as well as carbon nanotubes, offer important advantages as per their excellent physical properties. However, inks filled with carbon nanostructures are usually based on toxic and contaminating organic solvents or surfactants, posing serious health and environmental risks. Water is the most desirable medium for any envisioned application, thus, in this context, nanocellulose, an emerging nanomaterial, enables the dispersion of carbon nanomaterials in aqueous media within a sustainable and environmentally friendly scenario. In this work, we present the development of water-based inks made of a ternary system (graphene oxide, carbon nanotubes and nanocellulose) employing an autoclave method. Upon controlling the experimental variables, low-viscosity inks, high-viscosity pastes or self-standing hydrogels can be obtained in a tailored way. The resulting inks and pastes are further processed by spray- or rod-coating technologies into conductive films, and the hydrogels can be turned into aerogels by freeze-drying. The film properties, with respect to electrical surface resistance, surface morphology and robustness, present favorable opportunities as metal-free conductive layers in liquid-phase processed electronic device structures.

Keywords: graphene; carbon nanotubes; nanocellulose; conductive inks; liquid-phase processing; film fabrication; sustainability; metal-free electrodes

Citation: González-Domínguez, J.M.; Baigorri, A.; Álvarez-Sánchez, M.Á.; Colom, E.; Villacampa, B.; Ansón-Casaos, A.; García-Bordejé, E.; Benito, A.M.; Maser, W.K. Waterborne Graphene- and Nanocellulose-Based Inks for Functional Conductive Films and 3D Structures. *Nanomaterials* **2021**, *11*, 1435. https://doi.org/ 10.3390/nano11061435

Academic Editor: Jin-Suk Chung

Received: 4 May 2021
Accepted: 27 May 2021
Published: 29 May 2021

Publisher's Note: MDPI stays neutral with regard to jurisdictional claims in published maps and institutional affiliations.

Copyright: © 2021 by the authors. Licensee MDPI, Basel, Switzerland. This article is an open access article distributed under the terms and conditions of the Creative Commons Attribution (CC BY) license (https:// creativecommons.org/licenses/by/ 4.0/).

1. Introduction

Liquid-phase processing (LPP) is currently the preferred route for building up advanced layered film devices such as environmental, physical or biological sensors, logic circuits, radiofrequency transmitters or screens [1]. The cornerstone of this field is to integrate high-performance electronic materials into functional systems in a low-cost configuration, with high performance and ease of manufacturing [2]. Therefore, the wide diversity of deposition methods in the liquid phase are very suitable for this purpose, and able to provide high-resolution patterns through liquid inks in a fully scalable manner [3,4]. Inks are complex mixtures of certain components and additives within a liquid medium (generally an organic and high boiling point solvent) whose concentrations and chemical nature determine their physical properties, namely rheology, surface tension and conductivity. The development of inks based on electrically conductive species is therefore critical to progress in this field of work [4,5].

Conductive nanomaterials have attracted considerable interest in this regard, as they offer excellent electronic properties and high compatibility with LPP [5]. Among the most studied nanostructures, those based upon carbon, such as carbon nanotubes (CNTs) and graphene derivatives, stand out. They possess extraordinary electrical, thermal and mechanical properties [5], as well as being extremely light and prone to biocompatibility.

Recent efforts of the scientific community in the area of LPP of carbon nanostructures have focused on the development of inks, showing enormous potential in technological applications and some advantages like the abundance of the source material (graphite) [6]. Beyond the graphene materials obtained by direct graphite exfoliation [7], the scientific community has also successfully employed its related chemical derivatives, such as graphene oxide (GO) or reduced GO.

Traditionally, noble metals have been incorporated into conductive inks, at concentrations of ~60% of conductive metal to reach acceptable conductivity values. In addition to the high cost involved, their high concentration generates problems of chemical stability and reaction with other neighboring species such as air and solvents. Graphene-based nanostructures can solve this drawback due to their high chemical stability [3,8]. Moreover, conductive aqueous inks with metals can cause toxicity [9], while inks based on aqueous solutions of graphene have been tested in human skin cells, resulting in neither toxicity nor morphological changes at a cellular level [10]. Therefore, these inks may be safe if the rest of the additives are harmless. In this sense, GO, as a hydrophilic derivative of graphene, has shown excellent performance when processed from water-based dispersions into graphene-based conductive films [11]. Thus, to date, graphene material-based inks have demonstrated their superiority in a wide range of functionalities, such as flexible interconnections, electrodes, transparent conductors and supercapacitors [12]. Surface electrical resistance values in the range of ~100 $\Omega/\square$ are required for acting as electrode materials in organic or perovskite solar cell devices [13,14]. Thus, in order to establish versatility in the application of a certain conductive material by use of LPP, it is necessary to control the resistivity of the deposited layer. Furthermore, its roughness and compaction mainly depend on the deposition method, influenced in turn by the viscosity of the ink. The preparation of inks made of graphene-based nanomaterials is hence a challenging task, since several rheological properties of the ink (namely density, surface tension and viscosity) have a great impact [15].

As a matter of fact, high-viscosity inks (hereafter called pastes), with viscosities higher than 500 cP, would be suitable to form thicker layers (for instance, by screen printing techniques), while those with less viscosity would be more suitable for spray or inkjet deposition, allowing for a finer patterning and, in particular, for the case of spray coating technologies, the coverage of large areas. In general, these values can be adjusted by experimental parameters such as the concentration of the conductive additive or the dispersant, and also by the solvent choice. As mentioned earlier, typical liquid media in inks are non-volatile organic solvents. The most used ones for the production graphene-based liquid suspensions are N-methyl-2-pyrrolidone (NMP), N,N-dimethylformamide (DMF) and dimethylsulfoxide (DMSO), which are the ones that behave best as liquid-phase exfoliants [16]. These are very polluting solvents, with a boiling point of >170 °C, posing a serious risk of toxicity in humans. However, the use of water does not automatically solve these problems either, since it requires the incorporation of high concentrations of surfactants and other additives [9,17,18], indefinitely remaining in the ink, possibly leading to environmental and toxicity problems. Thus, new methods compatible with lower boiling point solvents (such as water or alcohols), together with non-toxic dispersants, are in demand in order to attain a truly environmentally friendly LPP, without raising the manufacturing cost or jeopardizing the overall electric/optoelectronic device performance once the deposition method takes place. Graphene-based nanostructures have promising potential, amongst currently known nanomaterials, to fulfill this purpose [19]. In particular, the GO derivatives have advantageous features for their use in conductive inks, such as affordable commercial availability, and the facile LPP due to the huge content of oxygen groups, in turn responsible for the higher interlayer spacing between planes. As stated by many authors, the GO LPP in a myriad of solvents and media does not need the addition of stabilizers, meaning a processing benefit [19]. For this reason, we have chosen GO as a target of interest for the preparation of aqueous inks. The only drawback is that GO inks require a subsequent reduction step to turn those inks into conductive material, be it by chemical, thermal or electrochemical means, usually entailing acceptable but poorer conductive properties than pristine graphene due to a larger number of structural defects [19]. Therefore, in the present work, the joint action of GO and CNTs was pursued to optimize the eventual conductive properties.

However, if one wishes to process carbon nanostructures from the liquid phase with greener approaches, avoiding toxic solvents or surfactants, a game-changing strategy is needed. In such a scenario, nanocrystalline cellulose (NCC) in particular acquires great relevance because of its sustainability. This nanomaterial is obtained from natural cellulose sources, by selective hydrolysis of the non-crystalline domains [20], resulting in fibrillar or needle-like nanostructures with widths and thicknesses around 3–20 nm and lengths of a few hundred nanometers. Due to its intrinsic chemical nature and the sustainability of the source material, nanocellulose may be considered an environmentally friendly nanomaterial. Despite the scarcity of scientific studies, there are already some examples showing the enormous potential that nanocellulose has as an aqueous dispersing agent of reduced GO and carbon nanotubes [21,22]. The structural diversity of NCC is defined by its crystalline allomorphs, among which types I and II stand out [22]. While type I NCC (exhibiting cellulose chains parallel to each other) is dominant in nature, type II is artificially synthesized, presenting polymer chains in antiparallel arrangements and typically requiring extreme caustic conditions or recrystallization processes for its synthesis. However, we have recently implemented a method to synthesize both NCC allomorphs by one-pot acid hydrolysis with sulfuric acid without any post-treatment step [22]. We also demonstrated the feasibility of dispersing CNTs in NCC, without any previous chemical modification on the NCC, leading to very stable aqueous colloids with proven bioactivity towards colon cancer (Caco-2) cells. In fact, the combination of carbon nanostructures with nanocellulose is an emerging trend, leading to useful hybrid nanomaterials with potential applications in biomedicine [23]. Beyond their biological response, nanocellulose paves the way towards new conductive inks based on carbon nanostructures, both dispersible in water and obeying green principles.

In this work, we disclose the development of carbon nanostructure-based low-viscosity inks and high-viscosity pastes able to be processed into films, by taking advantage of the impressive properties of NCC, standing out as a sustainable and green dispersing agent in water. We herein present a parametric study of ternary inks or pastes, or even self-standing hydrogels, by combining GO, CNTs and NCC and using a hydrothermal method in an autoclave. Further, conductive films were fabricated by different deposition techniques, and also hydrogel-derived porous materials, with the potential to become reference components as LPP electrodes and interfaces in electric/optoelectronic layered film device structures, such as batteries, supercapacitors, sensors and solar cells, among others.

2. Materials and Methods
2.1. Starting Materials and Reagents

The source of cellulose used in this work was microcrystalline cellulose (MCC) powder from cotton linters with an average particle size of 20 microns (Sigma-Aldrich, San Luis, MO, USA, ref 310697). Sulfuric acid (98%) for acid hydrolysis was purchased from Labkem (Barcelona, Spain). Ultrapure water with a conductivity of 0.055 µS/cm was obtained from a Siemens Ultraclear device (München, Germany) and used in every step of the present work. Multi-walled CNTs (MWCNTs) were acquired from NANOCYL® (NC7000™ variety), produced by a catalytic chemical vapor deposition method. Before use, MWCNTs underwent mild oxidation in liquid phase with HNO_3 (1.5 M, 2 h under reflux) to render a more hydrophilic surface without compromising their structure [24]. GO came from a commercial aqueous dispersion (0.4 wt%, 4 mg/mL), purchased from Graphenea® (San Sebastián, Spain). The characterization of mildly oxidized MWCNTs is presented in the Supplementary Materials (Figures S1 and S2), while the characterization of GO is provided by the manufacturer [25].

2.2. Synthesis of NCC (Types I and II)

As discussed in the Introduction, the synthesis of NCC was carried out according to a methodology developed in our laboratory [22]. In a typical experiment, 10 g of commercial MCC were added to 45 mL of ultrapure water inside a round-bottom flask, and dispersed with the aid of an ultrasonic bath for 10 min. The flask was then placed in an ice bath at 0 °C and 45 mL of 98% sulfuric acid were added dropwise, under constant magnetic stirring. Once added, the final concentration in the flask was 64% and the mixture was quite viscous, thus requiring vigorous and constant stirring to avoid locally high concentrations which could burn the cellulose. Right after the last droplet of acid was added, the flask was removed from the ice bath and transferred to a heating plate with magnetic stirring. At this point, the procedure varied according to the type of NCC to be obtained. For NCC type I, the reaction medium was heated for 10 min at 70 °C, while for type II it was heated for 5 h at 27 °C. Once the heating step was finished, the reaction medium was poured into ultrapure cold water, least ~10 times of the initial volume (90 mL) in a 1 L beaker and left overnight at 4 °C, in order to favor sedimentation. After decanting the supernatant liquid, a dispersion was left with a very acidic pH that was neutralized by dialysis. For this, the aqueous dispersion was inserted into a dialysis membrane (SpectraPor®, Spectrum Labs, regenerated cellulose, 6–8 kDa cutoff molecular weight) immersed in 5 L of ultrapure water. The dialysis water was changed periodically until neutral pH in the washing waters was achieved. Then, the dialyzed medium was centrifuged at ~9300 rcf, and the supernatant liquid was kept and subsequently freeze-dried, in order to use NCC as a fine and light powder. Full characterization of NCC can be found in reference [22].

2.3. Hydrothermal Treatment to Obtain Inks, Pastes and Hydrogels

Obtaining aqueous formulations able to be processed into films by different deposition techniques requires a certain control over their viscosity. This has been successfully attained through hydrothermal processes carried out in an autoclave. By heating a mixture of nanomaterials in water at a constant temperature (180 °C) in pressurized containers, with controlled times and pH, the chemical crosslinking of nanomaterials is favored [26]. In our case, this led to inks and pastes of different viscosities or even self-standing hydrogels. In general, what is herein termed as ink was of the order of 60 cP, while viscous pastes exhibited values of around 1500 cP.

We have undertaken a parametric study of different hydrothermal experiments, mixing variable amounts of NCC and CNTs and a fixed amount of GO, in order to ascertain the precise conditions under which inks, pastes or hydrogels form. All samples were brought to a final volume of 10 mL, and all samples were prepared at natural pH or alkalinized by the addition of aqueous ammonia (NH_4OH), since basic pH plays a critical role by favoring the graphenic nanomaterials' aggregation during hydrothermal treatment [26]. The experimental procedure followed was: in a flat-bottom quartz vial, a specific amount of NCC (yielding final concentrations of 2.5 mg/mL or 5 mg/mL) was weighed. Unless stated otherwise, NCC refers to its type I polymorph. Then, a certain amount of CNTs were added to the vial (to attain final concentrations of 0.1 mg/mL, 0.2 mg/mL, 0.5 mg/mL, 1 mg/mL or 2 mg/mL), enabling the analysis of the influence of the amount of CNTs on the conductivity and viscosity of the resulting composites. Further, a variable volume of ultrapure water (4.8 mL or 5.0 mL) was added depending on whether or not NH_4OH (200 µL) was added, respectively. At this point, the vial was subjected to an ultrasound bath (45 kHz) for 3 min. After that, a semi-homogeneous dispersion could be observed inside the vial. Finally, a fixed volume (5.0 mL) of the aqueous dispersion of GO was added, resulting in a final GO concentration of 2 mg/mL. The resulting medium was again bath-sonicated for another 2 min, thus obtaining a homogeneous dispersion with a final volume of 10 mL.

The hydrothermal treatment consisted of placing the vials in a Teflon vessel, inserting it into a metallic autoclave, tightly closed, and then heating it in an oven. The autoclaves were placed in the oven after pre-heating it at 180 °C, with a heating ramp of 20 °C, and the treatment lasted for a specific duration. The studied times were mainly 60 min, 30 min and 15 min. After one of these times had elapsed, the autoclave was removed from the oven and allowed to cool down in ambient conditions to room temperature before taking the sample out of the autoclave. The time needed for the sample inside the autoclave to reach 180 °C was estimated to be compensated by the time needed for cooling down to a safe handling temperature, so the overall treatment times were consistent in all cases. Figure 1 shows a visual scheme of the process.

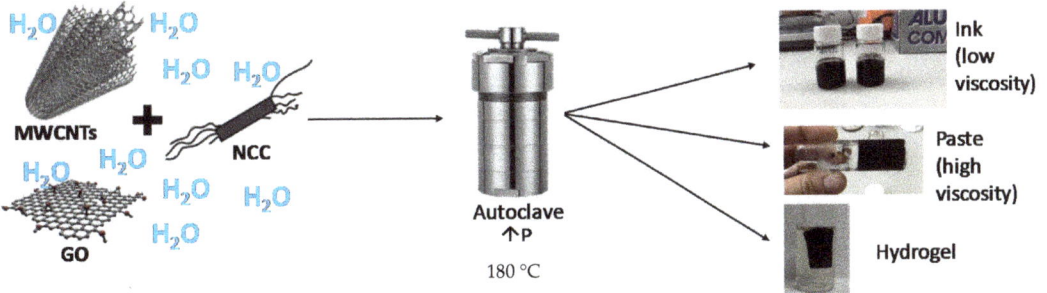

Figure 1. General scheme for the preparation of inks and pastes following the autoclave method.

2.4. Film Preparation and Characterization

Once the inks (low-viscosity formulations) and pastes (high-viscosity formulations) were obtained by the procedures described above, the different possible deposition techniques were studied, in order to obtain the optimum conductive films. Spray-coating with an airbrush onto glass substrates was the chosen approach for the deposition of inks [27]. Films obtained from the viscous pastes were deposited by means of a rod-coating method using an agate rod, also over glass substrates. In both cases, the substrate was placed on a heating plate at 60 °C. For every sample, a total of 10 mL of the ink or paste was deposited in subsequent passages on the glass substrate, leading to films prepared under comparable conditions.

In order to measure the resistivity of the films, a Keithley 4200 unit was used, working in the range from −100 to 100 mV. An in-line 4-point probe configuration with equidistant probe separations of 2.24 mm was utilized, with controlled and homogeneous pressure over the conductive film. For the resistivity measurement, the geometry of the film and the distance between the electrodes were taken into account [18].

Morphology and microstructure of the deposited films were assessed through both optical and scanning electron (SEM) microscopies. Optical images were taken with a Zeiss AXIO microscope (Jena, Germany) with 20× and 50× objective lenses (N.A. = 0.4 and 0.7, respectively). SEM images were taken using Hitachi S3400N equipment (Tokyo, Japan), working in the secondary electron mode at a voltage of 15 kV and a distance of 5 mm.

Sample thickness as well as the roughness of the surface were evaluated by using a contact DektakXT Stylus Profiler (from Bruker, Billerica, MA, USA). The radius of the stylus used in the measurements was 2.5 μm. The height of the step of each deposited layer was measured in different areas along the layer edge. The depths of grooves made in the central part of the sample (reaching the substrate) were also determined. The layer thickness was obtained as the mean value of such measurements. The profile roughness was analyzed using Dektak analytical software (from Bruker, Jena, Germany).

2.5. Aerogel Preparation and Characterization

In order to obtain aerogels from hydrogels, a unidirectional freezing was applied before lyophilization. For that purpose, hydrogels were placed in an empty quartz vessel with a bottom platform made of metal which was immersed in liquid nitrogen. Once the hydrogel was fully frozen starting from the metal base, it was placed in a Telstar Cryodos freeze-drier, working at −49 °C and 0.3 mbar [28].

In order to measure the resistivity of the aerogels, the aforementioned Keithley unit was employed in a two-probe configuration. Aerogels were held between spring-loaded copper foils. The tungsten needle probes were brought into contact with the copper sheets, avoiding damage to the aerogels. The resistivities of the aerogels were calculated, taking into account the geometry, given by the distance between the electrodes and the specimen diameter, thus assuming a perfectly cylindrical shape.

3. Results and Discussion

3.1. Hydrothermal Development of Inks, Pastes and Hydrogels: Unraveling Critical Parameters

As stated earlier, different experiments were carried out at different concentrations of NCC and CNTs, whereas the GO concentration remained fixed. Different hydrothermal treatment times were applied (always at 180 °C), and pH was also varied by adding NH_4OH or not to the medium. The systematic combination of such variables provided a roadmap for this ternary system (CNTs, GO and NCC type I) in water upon hydrothermal treatment in an autoclave. The two main possibilities pursued in the present work were liquid inks and viscous pastes, but it is worth recalling that the hydrothermal treatment of aqueous GO suspensions can also provide self-standing hydrogels, even for short times (generally beyond 30–45 min at 180 °C), based on the crosslinking of the GO sheets [28]. The viscosity of the medium increases progressively with the treatment time, also aided by the presence of basic pH. Thus, by controlling the thermal treatment time, homogeneous liquid dispersions, pastes or hydrogels can be obtained (Figures 2 and 3).

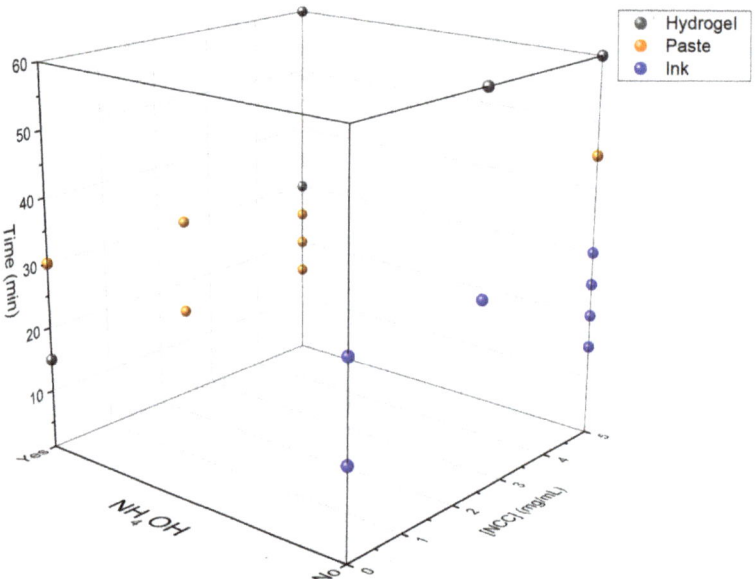

Figure 2. Three-dimensional scatter plot representing the conditions to obtain inks (blue), viscous pastes (orange) and self-standing hydrogels (gray). The effect of adding aqueous ammonia is herein represented, at different amounts of CNTs.

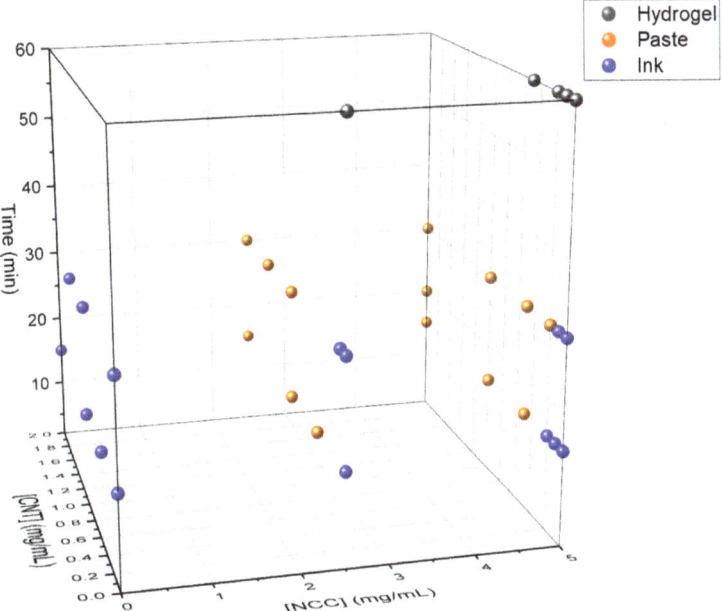

Figure 3. Three-dimensional scatter plot representing the conditions to obtain liquid inks (blue), viscous pastes (orange) and self-standing hydrogels (gray). The effect of CNTs is herein represented. All samples here had basic pH by adding aqueous ammonia.

As observed in Figures 2 and 3, hydrogels are always obtained for hydrothermal treatment times longer than 30 min, regardless of the other parameters. This suggests that treatment time is the dominant variable. This is consistent with our previous studies on the hydrothermal treatment of GO suspensions, with or without basic medium, in the absence of any other additive [28]. However, at 30 min of treatment time, the determining variable is the basicity of the medium. Samples containing NH_4OH result in pastes or hydrogels, and those without NH_4OH result in inks or pastes depending on the presence of CNTs. In fact, the joint presence of CNTs and NCC while adding NH_4OH seems to be responsible for the rise in viscosity, leading to pastes.

For the experiments carried out for 15 min, the importance of the pH is observed again. Samples containing NH_4OH give rise to pastes or hydrogels, and those without NH_4OH give rise to inks. Again, the presence of NCC is decisive, since (at a given time of hydrothermal treatment with NH_4OH) it leads to a paste, but in the absence of NCC the resulting outcome is a hydrogel. It can be reasonably postulated that the interactions between NCC and GO during hydrothermal treatment partly prevent the self-crosslinking of the GO and lead to viscous pastes instead of fully crosslinked hydrogels. Finally, it was observed that the concentration of NCC in the system was not a critical variable for the kind of aqueous formulations obtained. Therefore, in subsequent tests, this variable was set at 5 mg/mL.

In order to unravel the effect of crystalline polymorphism of NCC in the hydrothermal process, a series of experiments was also carried out with type II NCC. In this case, all experiments were performed for 15 min (since in the abovementioned results, there were no significant differences between 15 and 30 min). When NCC type II is incorporated into the system instead of type I, surprisingly, all formulations obtained in each experiment are low-viscosity inks (Table 1). This may be related to the aforesaid hindrance of GO crosslinking, as type II NCC could experience stronger interactions with the functional groups on the surface of GO, impeding its aggregation and thus leading to low-viscosity inks. This effect is stronger than that observed for type I NCC, which is consistent with the higher content of ester sulfate groups in type II NCC [22], responsible for its negative surface charge. As in the case for type I NCC, above 30 min of treatment time, in the presence of NH$_4$OH, hydrogels are always obtained with type II NCC.

Table 1. The outcome of the hydrothermal treatment and composition of the GO/CNT/NCC (type II) ternary system in water at 180 °C for 15 min.

Formulation Obtained	GO (mg/mL)	CNTs (mg/mL)	NCC (mg/mL)	V_{H2O} (mL)	V_{NH4OH} (mL)
Ink	2	0.0	5	5	0
Ink	2	0.0	5	4.8	0.2
Ink	2	0.1	5	5	0
Ink	2	0.1	5	4.8	0.2
Ink	2	0.2	5	5	0
Ink	2	0.2	5	4.8	0.2
Ink	2	0.5	5	5	0
Ink	2	0.5	5	4.8	0.2
Ink	2	1	5	5	0
Ink	2	1	5	4.8	0.2
Ink	2	2	5	5	0
Ink	2	2	5	4.8	0.2

3.2. Morphology of Films Derived from Inks and Pastes

Low-viscosity inks and high-viscosity pastes require the use of different deposition methods to attain conductive films over glass substrates. Inks were deposited by means of spray coating, while pastes were processed using the rod-coating method. The general appearance of such films is depicted in Figure 4. Films made from inks had a very compact aspect, because spray coating allowed for a tight coverage. Viscous pastes could not be deposited by spraying, with the rod coating method being the one showing the highest effectiveness, leading to films with an apparently more porous aspect. Deeper insights into the surface morphology of these films were obtained by both optical and scanning electron microscopies (Figure 5).

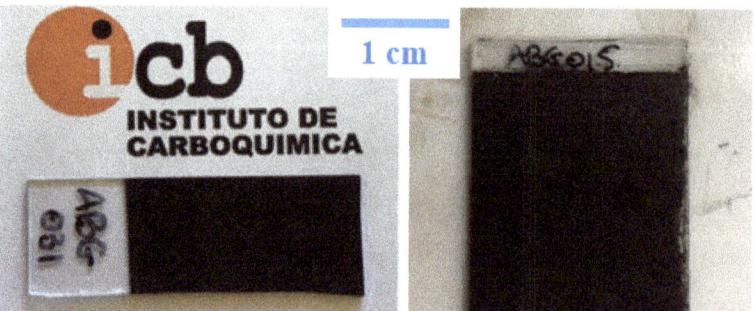

Figure 4. Photographs of films made from low-viscosity inks (**left**) and high-viscosity pastes (**right**) on glass substrates.

As confirmed by microscopy images, films derived from liquid inks and viscous pastes have a very distinct morphology both at the millimeter and the micrometer scale. According to the millimeter scale, films derived from inks exhibit no major irregularities, and a negligible presence of pores. In contrast, those coming from rod-coated pastes show the presence of large pores, together with more irregularities. These pores may be important for subsequent processing of such films, such as liquid-phase infiltration or interfacing with other species. As for the film surface microstructure, unraveled by

SEM, the observation from optical microscopy is corroborated. Films derived from inks present higher homogeneity and compaction, while those coming from pastes are more topographically irregular. In essence, both kinds of films display a very unique topography.

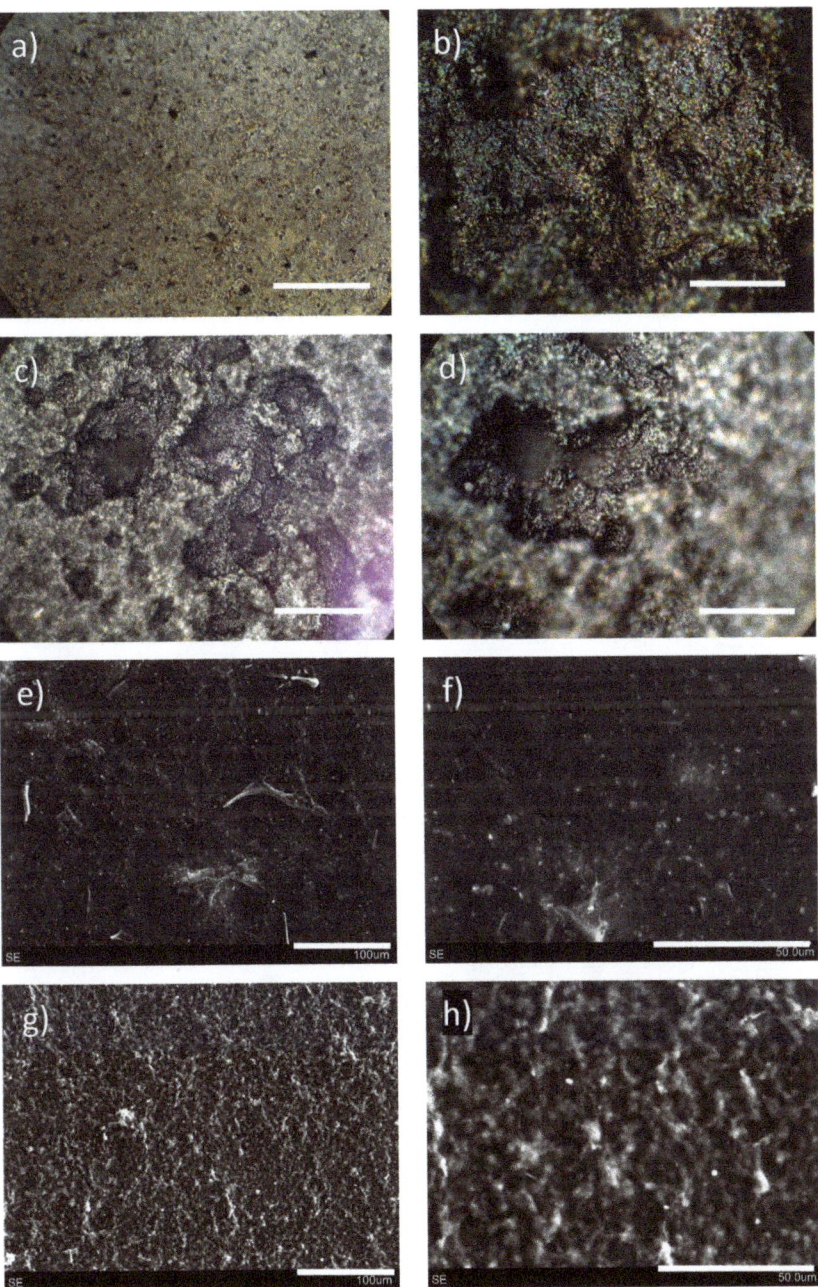

Figure 5. Optical photographs (**a**–**d**) and SEM images (**e**–**h**) of films obtained with low-viscosity inks (**a**,**b**,**e**,**f**) and high-viscosity pastes (**c**,**d**,**g**,**h**). Scale bars (in white) = 1 mm (**a**,**b**), 400 μm (**c**,**d**), 100 μm (**e**,**g**) and 50 μm (**f**,**h**). Each image (either optical or from SEM) corresponds to a random point of each sample, none is the direct magnification of another.

Additional results were obtained through the profilometer, by which thickness and roughness were quantified. The mean values for the films studied in the present work are shown in Figure 6. Inks fabricated with type I NCC are able to generate films of a low thickness (~10 µm on average) up to a certain CNT content. From 0.5 mg/mL CNTs and up, the sprayed films in identical conditions drastically increase in thickness, reaching values in the range of ~30–45 µm. Conversely, inks containing type II NCC present the opposite trend; at a null content of CNTs, the sprayed films display an average thickness of ~25 µm and progressively decrease with increasing content of CNTs. Films made from pastes showed a steady value regardless of the CNT content, with an average of ~10–15 µm thickness. Film roughness shows an identical trend as thickness for inks with type I NCC; low roughness (~1–2 µm) until reaching 0.5 mg/mL CNTs, after which this value becomes 3- or 4-fold larger. Inks from type II NCC and pastes lead to films of comparable roughness, in the range of ~2–3 µm. All of these data may be related to the different interaction that NCC has with carbon nanostructures. In a previous work [24], we reported that type I NCC hybrids with CNTs presented a discrete distribution of cellulose nanocrystals adsorbed on the sidewalls, while type II NCC led to a heavily wrapped nanohybrid. In addition, type I cellulose nanocrystals are much longer and needle-like whereas the type II ones are much shorter and slightly thicker [22]. The kind of interaction among nanostructures in the reaction mixture during the hydrothermal treatment could determine their packing upon film deposition. Spraying liquid inks with type I NCC may provide a less efficient packing of nanostructures when the CNT content is high enough, causing the thickness and roughness to increase. In contrast, type II NCC, could have led to efficient packing regardless of the CNT content, given its very distinct structure. As for pastes, the explanation could lie in the nature of the deposition technique, as rod-coating of viscous formulations may not provide a tight packing nor attain highly thick or rough surfaces. Additionally, the rod-coating method also exhibited a visible thickening of the film edges. These observations can be better understood by observing some representative profiles of each case (Figure 7). Films coming from both inks show a more regular aspect with periodic spikes in the presence of CNTs, which determine the higher or lower mean roughness, while pastes present a more irregular profile, with some visible pores in the millimeter scale, as observed by optical microscopy. The profilometer also served as a means to perform a scratching test on selected film samples (see Supplementary Materials, Video S1). The conclusions drawn from these tests is that the films have a high adhesion and mechanical resistance, since they are not damaged at all when scratched with high local pressures (3.75 MPa).

A final surface characterization of these films was conducted through Raman spectroscopy (Supplementary Materials, Figure S4). In the observed spectra, it is visible how the D-band (~1349 cm^{-1}) and the G-band (~1590 cm^{-1}) correspond to both multi-walled CNTs and reduced GO indistinguishably, but are different from the starting GO due to the better definition of the band at 2700 cm^{-1} (2D) in the films, and the rougher profile of GO. The D/G intensity ratio is lower than 1 for GO and films without CNTs, while in the presence of the latter, this D/G ratio may reach values of ≥ 1.

3.3. Characterization of Electrical Properties

The four-probe electrical measurements performed on the prepared films allowed us to obtain their surface resistivity (Figure 8). It is observed that the surface resistivity values are lower as the concentration of CNTs in the liquid formulation (ink or paste) increases. Indeed, if the presence of GO is the basis for the hydrothermal aggregation, and the incorporation of NCC exerts control over the viscosity of the medium, the control over the conductive behavior of the derived films can be ascribed to CNTs. Those films without CNTs present very similar resistivity values amongst them (~$1.5 \cdot 10^3$ Ω/□), regardless of whether they come from more or less viscous formulations before deposition, most likely owing to the lone effect of hydrothermally reduced GO. In fact, NCC does not generate any kind of char in the working conditions (Figure S3, Supplementary Materials). Regarding the inks, there are no significant differences in surface resistivity between those films coming from type I or II NCC. Both kinds of inks provide films with surface resistivities in the range of ~10^3–10^2 Ω/☐ up to 1 mg/mL CNTs, reaching further values below 100 Ω/☐ at 2 mg/mL CNTs. All these values decrease almost linearly on a logarithmic scale, typical for a behavior beyond the percolation threshold in the studied conditions. Films made from pastes exhibit a steeper decrease in the surface resistivity with an increasing amount of CNTs, attaining ~100 Ω/☐ at 1 mg/mL CNTs. This seems to be a 'plateau' value as resistivity does not change at 2 mg/mL CNTs. It is worth pointing out that these films could become perfect candidates for metal-free electrode components or current collectors in liquid-phase processed electric/optoelectronic layered film device structures. In order to account for the variation trends in film thickness with increasing CNT content (Figure 6), the bulk conductivity of the samples could also be calculated (see Figure S5, Supplementary Materials). These results also show a generally direct proportionality between both parameters.

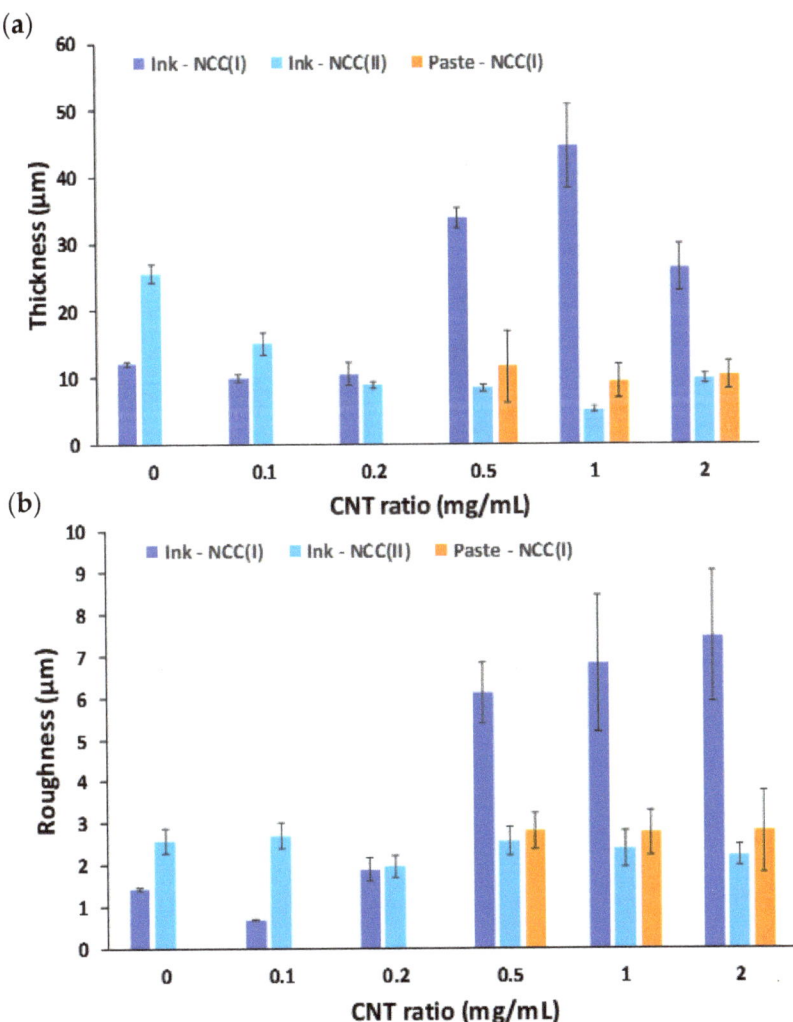

Figure 6. Film thicknesses (**a**) and roughnesses (**b**) corresponding to samples coming from sprayed inks or rod-coated pastes.

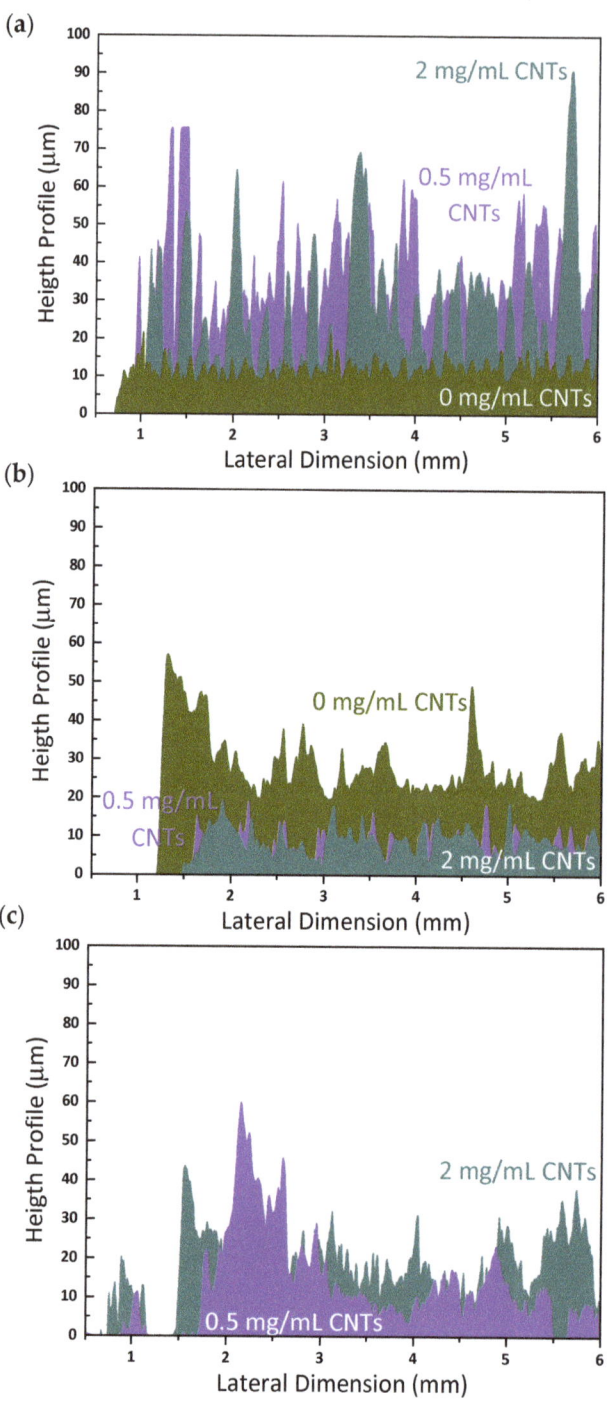

Figure 7. Topographic profiles across 6 mm distance for films made from inks with type I NCC (**a**), from inks with type II NCC (**b**) and from pastes with type I NCC (**c**), at the indicated CNT ratio in the initial formulation. All samples came from formulations with 2 mg/mL GO and 5 mg/mL NCC.

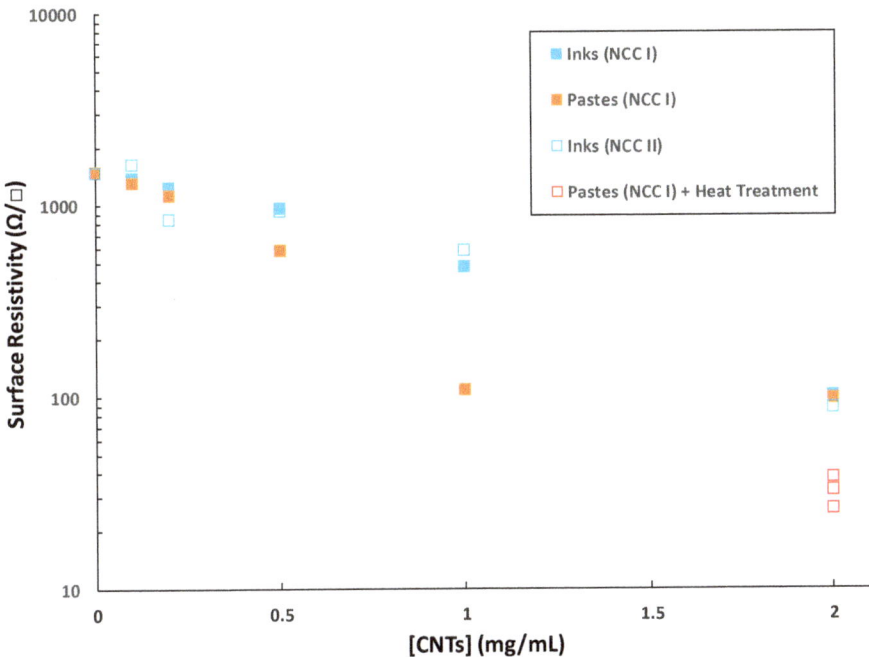

Figure 8. Surface resistivity measurements for different films prepared from inks and pastes. The concentration of CNTs refers to the one in the liquid formulation before deposition. All samples on this graph came from formulations with 2 mg/mL GO and 5 mg/mL NCC.

3.4. Post-Synthesis Processing Versatility

Up to this point, we have presented a green approach to water-based inks with carbon nanostructures. This has led to formulations of varied viscosity able to be deposited by different means, leading to conductive films with good surface resistivity, unique surface morphologies and a potential feasibility to be applied as electrodes. Nonetheless, the properties of such films may be still improved, in terms of electrical conductivity and tolerance to organics, upon heat treatment. As a proof of concept, we chose a candidate film presenting one of the lowest values of surface resistivity (97.4 Ω/□, coming from a rod-coated paste with 2 mg/mL GO, 5 mg/mL NCC and 2 mg/mL CNTs), and subjected it to heat treatment and organic solvent exposure.

In one experiment, the selected film was inserted into an oven and heated at 400 °C for 4h under air atmosphere, and the outcome was a film with intact integrity and lower surface resistivity (Figure 8). The production of three different replicas provided values in the range of ~25–35 Ω/□, being of special interest for the construction of conductive carbon layers in layered film optoelectronic devices [29]. The morphology of the films before and after heat treatment was assessed again by SEM (Figure 9). It becomes clear that the effect of heat treatment is to smoothen the surface topography, probably together with the burning of NCC. In the heat-treated films, the presence of CNTs is better discerned, very probably entailing an improved contact between CNTs and reduced GO, thus leading to lower interparticle contact resistance. These results are very encouraging for the replacement of commercial carbon pastes with many toxic additives and binders, as these pastes have been reported to peel off and deform when heated beyond 250–300 °C [30].

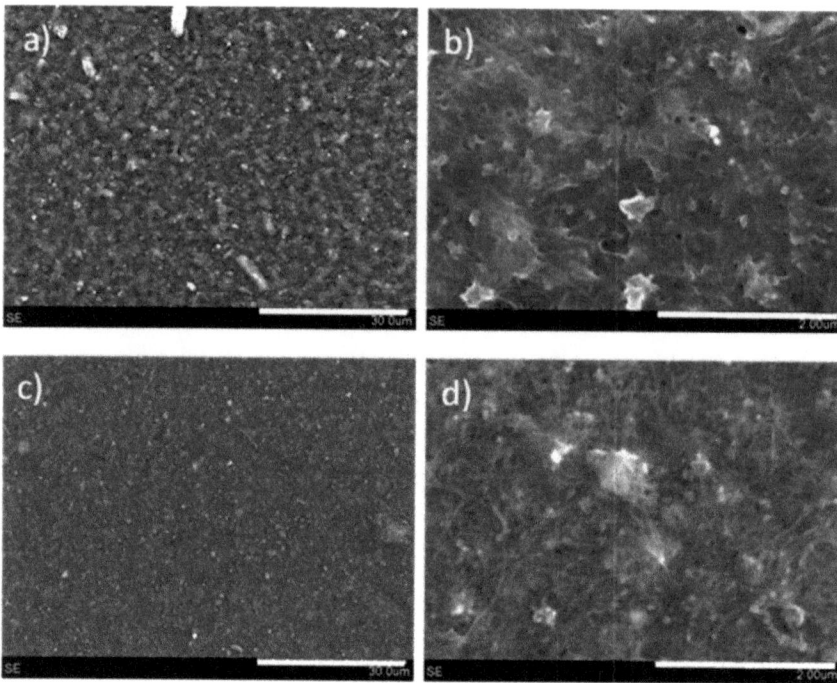

Figure 9. SEM images from films obtained with high-viscosity pastes, before (**a**,**b**) and after (**c**,**d**) heat treatment at 400 °C. Scale bars (in white) = 30 μm (**a**,**c**) and 2 μm (**b**,**d**).

An additional advantage is the stability over time of the conductive properties of the films. Along the course of this research, we corroborated the stability of the conductive properties of our films in a time frame of many hours or a few days, meaning that the measurement of the surface resistivity of a freshly prepared film is generally coincident with the measurement after a short–medium time frame. Additionally, these observations are also valid for long-term periods (months). All of this shows that the preservation of the conductive properties of these films is possible across lengthy time periods with regular shelf storage.

In another experiment, the films were subjected to a treatment with typical organic solvents in the LPP of carbon nanostructures (NMP, DMSO), in order to study the tolerance of the films to such solvents. For this, the films were heated at 60 °C for 5 min. Then, several droplets of one of these organic solvents in a volume of 45 μL were randomly scattered across different areas of the films, in order to evaluate their effect towards its integrity. The film was maintained at 60 °C for 10–15 min to ensure infiltration through the pores and placed in the oven at 60 °C for 4 h. The film perfectly resisted this treatment, as no evident damage was observed by eye nor the optical microscope, and no peeling off occurred either. In addition, its surface resistivity changed from 97.4 to 94.4 Ω/□, which could be considered negligible and within experimental error.

In essence, these tests reassert the robustness and versatility of our conductive films, as they can endure (and be improved in terms of surface resistivity by) thermal treatments at high temperatures and long durations, as well as withstand the exposure to high boiling point organic solvents in aggressive conditions without any morphological harm nor damage to the electrical properties. This latter fact is of special relevance for their future performance as electrode components, since the whole integrity and surface resistivity of the films are retained after aggressive treatments, and depending on the application in mind, the resistance of the films towards certain critical solvents could be an important advantage.

3.5. Properties of Hydrogel-Derived Aerogels

Some of the working conditions led to porous and light graphene-based aerogels [26,28]. We have taken advantage of the possibilities granted by unidirectional freezing, followed by lyophilization, which are able to create an anisotropic internal microstructure composed of parallel channels, with critical applications in energy and environmental remediation [28,31,32]. In the present case, hydrogels were prepared under specific conditions and subjected to hydrothermal treatment with the ternary GO/NCC/CNT system in water (see Section 3.1 and Figure 10), and when subjected to unidirectional freezing prior to lyophilization, they also presented an anisotropic porous microstructure.

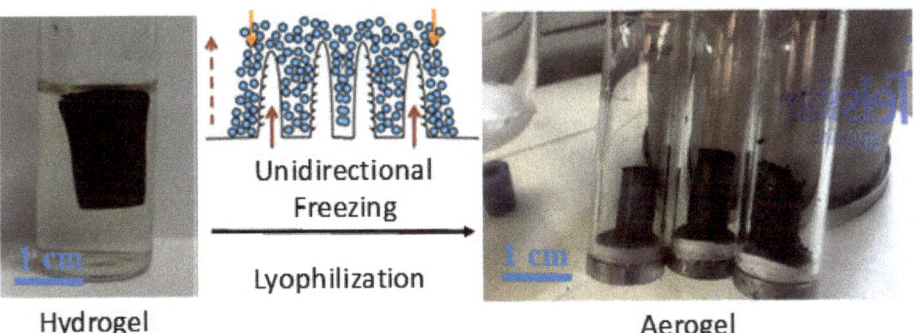

Figure 10. Real images of a hydrogel derived from hydrothermal treatment in water (**left**) and preparation scheme of aerogels by unidirectional freezing followed by lyophilization (**right**).

As inferred from the SEM images (Figure 11), aerogels resulting from the unidirectional freeze-drying of hydrogels presented an anisotropic microstructure, with continuous straight pores parallel to the aerogel's longitudinal axis. This demonstrates that not only can such a microstructure be obtained from the reduction of GO in hydrothermal conditions [28], but also in the presence of NCC and CNTs. According to measurements of weight and dimensions, we elucidated that these aerogels presented an average density of 0.028 ± 0.006 g/cm^3, and axial resistivities (measured in a two-probe configuration) in the range of 10–100 $\Omega \cdot$m, for CNT concentrations in the range of 0.1 to 0.5 mg/mL. Aerogels obtained in similar conditions but without CNTs or NCC presented lower densities (~0.005 g/cm^3), but higher electrical resistivity (~1000 $\Omega \cdot$m). Parallel to the case of inks and pastes, the presence of CNTs governs the electrical properties of freeze-dried hydrogels, but in this case, NCC has a significant influence. When only GO and NCC are present in the hydrothermal medium, the density rises by one order of magnitude (~0.036 g/cm^3) with respect to GO-only aerogels, but the axial resistivity is lowered by one order of magnitude. Therefore, all aerogels with GO, CNTs and NCC display electrical resistivities from ~10^2 down to ~10^0 in the studied range (from 0.1 to 0.5 mg/mL), revealing the importance of NCC in the process, which seems to play a role in the GO reduction during the hydrothermal treatment. NCC seems to boost the GO hydrothermal reduction and the number of interparticle contacts, hence favoring the decrease in electrical resistance at the expense of aerogel density. In summary, it is possible to produce light and porous aerogels with a GO/CNT/NCC ternary system, with an anisotropic microstructure and fairly low electrical resistivity in the absence of any post-treatment, comparable to the ones with only GO [32].

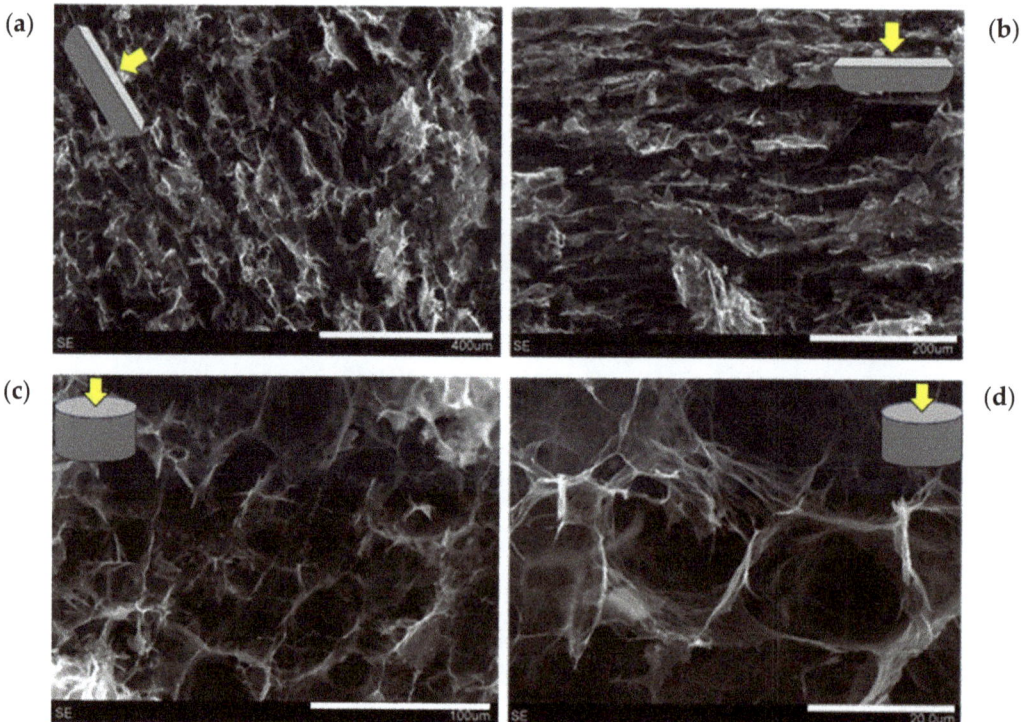

Figure 11. SEM images of aerogels derived from the unidirectional freezing followed by lyophilization of hydrothermally prepared hydrogels. Longitudinal (**a,b**) and transversal (**c,d**) cuts. Scale bars (in white): 400 µm (**a**), 200 µm (**b**), 100 µm (**c**) and 20 µm (**d**).

4. Conclusions

Nanocrystalline cellulose has been demonstrated to be a green and sustainable means to generate conductive water-based ink formulations of tailored viscosity, with carbon nanotubes and graphene oxide, as well as three-dimensional structures. Control of the processing parameters allows the preparation of inks, pastes or hydrogels using the same approach. The liquid inks and pastes have been used to fabricate films exhibiting diverse surface morphologies (depending on the deposition method and composition), with low resistivity values (<100 Ω/□). These conductive films also exhibit great robustness, being able to avoid disintegration upon aggressive treatments with organic solvents, as well as being electrically improvable by high-temperature treatments. Our studies show that formulations of this kind have great potential as metal-free electrodes in liquid-phase processed layered films, devices and structures. The green processes herein described, in addition to the capability of easily tuning the viscosity of liquid formulations, can be a great starting point for industrial development.

Supplementary Materials: The following are available online at https://www.mdpi.com/article/10.3390/nano11061435/s1, Figure S1: Thermogravimetric analysis in N_2 atmosphere, Figure S2: TEM images of mildly oxidized MWCNTs, Figure S3: Photograph of NCC aqueous colloids subjected to hydrothermal treatment. Figure S4. Raman spectra (measured with a 532nm laser, Horiba Jobin Yvon equipment) for different selected conductive films. Each spectrum is the average of at least 5 random points across the film surface. Figure S5. Electrical conductivity values for different films coming from inks or pastes with different carbon nanotube contents. Videos S1: Recording of the stylus-based scratching tests, by applying a constant pressure of 3.75 MPa with the diamond tip (in the center of the crosshead). The distance traveled was 3 mm in each case. After scratching the films, no grooves or trails were observed, and the final profile seen corresponds to the neat sample topography. This means that these films are not damaged at all when scratched at high local pressures.

Author Contributions: Conceptualization, J.M.G.-D.; experimental work, J.M.G.-D., A.B., B.V., M.Á.Á.-S.; data processing, J.M.G.-D., A.B., E.C., A.A.-C.; resources, J.M.G.-D., E.G.-B., A.M.B., W.K.M.; writing—original draft preparation, J.M.G.-D.; writing—review and editing, J.M.G.-D., E.C., B.V., E.G.-B., A.A.-C., A.M.B., W.K.M. All authors have substantially contributed to, read and agreed to the published version of the manuscript.

Funding: This research was funded by Spanish MINEICO/MICINN, under project references ENE2016-79282-C5-1-R (AEI/UE/FEDER) and PID2019-104272RB-C51/AEI/10.13039/501100011033, and the Gobierno de Aragón (Grupo Reconocido DGA-T03_20R). J.M.G.-D. greatly acknowledges Spanish MINEICO for his 'Juan de la Cierva Incorporation' contract and associated research funds (ref. IJCI-2016-27789). The APC predoctoral contracts of E.C. and M.Á.Á.-S. were funded by the Spanish MINEICO (BES2017-080020, including EU Social Funds) and the Gobierno de Aragón, respectively. The work performed at the University of Zaragoza was funded by the MICINN (PID2019-104307GB-I00/AEI/10.13039/501100011033) and Gobierno de Aragón (E47_20R).

Institutional Review Board Statement: Not applicable.

Informed Consent Statement: Not applicable.

Data Availability Statement: Data sharing is not applicable to this article.

Acknowledgments: The authors acknowledge Carlos Martínez Barón for his support.

Conflicts of Interest: The authors declare no conflict of interest.

References

1. Arias, A.C.; MacKenzie, J.D.; McCulloch, I.; Rivnay, J.; Salleo, A. Materials and applications for large area electronics: Solution-based approaches. *Chem. Rev.* **2010**, *110*, 3–24.
2. Zhong, C.; Duan, C.; Huang, F.; Wu, H.; Cao, Y. Materials and devices toward fully solution processable organic light-emitting diodes. *Chem. Mater.* **2011**, *23*, 326–340. [CrossRef]
3. Singh, M.; Haverinen, H.M.; Dhagat, P.; Jabbour, G.E. Inkjet printing-process and its applications. *Adv. Mater.* **2010**, *22*, 673–685. [CrossRef] [PubMed]
4. Aleeva, Y.; Pignataro, B. Recent advances in upscalable wet methods and ink formulations for printed electronics. *J. Mater. Chem. C* **2014**, *2*, 6436–6453. [CrossRef]
5. Kamyshny, A.; Magdassi, S. Conductive nanomaterials for printed electronics. *Small* **2014**, *10*, 3515–3535. [CrossRef]
6. Secor, E.B.; Hersam, M.C. Emerging carbon and post-carbon nanomaterial inks for printed electronics. *J. Phys. Chem. Lett.* **2015**, *6*, 620–626. [CrossRef] [PubMed]
7. González-Domínguez, J.M.; León, V.; Lucío, M.I.; Prato, M.; Vázquez, E. Production of ready-to-use few-layer graphene in aqueous suspensions. *Nat. Protoc.* **2018**, *13*, 495–506. [CrossRef] [PubMed]
8. Capasso, A.; Del Rio Castillo, A.E.; Sun, H.; Ansaldo, A.; Pellegrini, V.; Bonaccorso, F. Ink-jet printing of graphene for flexible electronics: An environmentally-friendly approach. *Solid State Commun.* **2015**, *224*, 53–63. [CrossRef]
9. Karagiannidis, P.G.; Hodge, S.A.; Lombardi, L.; Tomarchio, F.; Decorde, N.; Milana, S.; Goykhman, I.; Su, Y.; Mesite, S.V.; Johnstone, D.N.; et al. Microfluidization of graphite and formulation of graphene-based conductive inks. *ACS Nano* **2017**, *11*, 2742–2755. [CrossRef]
10. McManus, D.; Vranic, S.; Withers, F.; Sanchez-Romaguera, V.; Macucci, M.; Yang, H.; Sorrentino, R.; Parvez, K.; Son, S.K.; Iannaccone, G.; et al. Water-based and biocompatible 2D crystal inks for all-inkjet-printed heterostructures. *Nat. Nanotechnol.* **2017**, *12*, 343–350. [CrossRef]
11. Tung, T.T.; Yoo, J.; Alotaibi, F.K.; Nine, M.J.; Karunagaran, R.; Krebsz, M.; Nguyen, G.T.; Tran, D.N.H.; Feller, J.F.; Losic, D. Graphene oxide-assisted liquid phase exfoliation of graphite into graphene for highly conductive film and electromechanical sensors. *ACS Appl. Mater. Interfaces* **2016**, *8*, 16521–16532. [CrossRef]
12. Li, J.; Ye, F.; Vaziri, S.; Muhammed, M.; Lemme, M.C.; Östling, M. Efficient inkjet printing of graphene. *Adv. Mater.* **2013**, *25*, 3985–3992. [CrossRef] [PubMed]
13. Lucera, L.; Kubis, P.; Fecher, F.W.; Bronnbauer, C.; Turbiez, M.; Forberich, K.; Ameri, T.; Egelhaaf, H.-J.; Brabec, C.J. Guidelines for closing the efficiency gap between hero solar cells and roll-to-roll printed modules. *Energy Technol.* **2015**, *3*, 373–384. [CrossRef]
14. Kim, T.-H.; Yang, S.-J.; Park, C.-R. Carbon nanomaterials in organic photovoltaic cells. *Carbon Lett.* **2011**, *12*, 194–206. [CrossRef]
15. Khan, S.; Lorenzelli, L.; Dahiya, R.S. Technologies for printing sensors and electronics over large flexible substrates: A review. *IEEE Sens. J.* **2015**, *15*, 3164–3185. [CrossRef]

16. Backes, C.; Higgins, T.M.; Kelly, A.; Boland, C.; Harvey, A.; Hanlon, D.; Coleman, J.N. Guidelines for exfoliation, characterization and processing of layered materials produced by liquid exfoliation. *Chem. Mater.* **2017**, *29*, 243–255. [CrossRef]
17. Santidrián, A.; Sanahuja, O.; Villacampa, B.; Diez, J.L.; Benito, A.M.; Maser, W.K.; Muñoz, E.; Ansón-Casaos, A. Chemical postdeposition treatments to improve the adhesion of carbon nanotube films on plastic substrates. *ACS Omega* **2019**, *4*, 2804–2811. [CrossRef]
18. Ansón-Casaos, A.; Sanahuja-Parejo, O.; Hernández-Ferrer, J.; Benito, A.M.; Maser, W.K. Carbon nanotube film electrodes with acrylic additives: Blocking electrochemical charge transfer reactions. *Nanomaterials* **2020**, *10*, 1078. [CrossRef]
19. Yang, W.; Wang, C. Graphene and the related conductive inks for flexible electronics. *J. Mater. Chem. C* **2016**, *4*, 7193–7207. [CrossRef]
20. Trache, D.; Hussin, M.H.; Haafiz, M.K.M.; Thakur, V.K. Recent progress in cellulose nanocrystals: Sources and production. *Nanoscale* **2017**, *9*, 1763–1786. [CrossRef]
21. Hajian, A.; Lindström, S.B.; Pettersson, T.; Hamedi, M.M.; Wågberg, L. Understanding the dispersive action of nanocellulose for carbon nanomaterials. *Nano Lett.* **2017**, *17*, 1439–1447. [CrossRef] [PubMed]
22. González-Domínguez, J.M.; Ansón-Casaos, A.; Grasa, L.; Abenia, L.; Salvador, A.; Colom, E.; Mesonero, J.E.; García-Bordejé, J.E.; Benito, A.M.; Maser, W.K. Unique properties and behavior of nonmercerized type-II cellulose nanocrystals as carbon nanotube biocompatible dispersants. *Biomacromolecules* **2019**, *20*, 3147–3160. [CrossRef] [PubMed]
23. Bacakova, L.; Pajorova, J.; Tomkova, M.; Matejka, R.; Broz, A.; Stepanovska, J.; Prazak, S.; Skogberg, A.; Siljander, S.; Kallio, P. Applications of nanocellulose/nanocarbon composites: Focus on biotechnology and medicine. *Nanomaterials* **2020**, *10*, 1–34. [CrossRef] [PubMed]
24. Martinez, M.T.; Callejas, M.A.; Benito, A.M.; Cochet, M.; Seeger, T.; Ansón, A.; Schreiber, J.; Gordon, C.; Marhic, C.; Chauvet, O.; et al. Sensitivity of single wall carbon nanotubes to oxidative processing: Structural modification, intercalation and functionalisation. *Carbon* **2003**, *41*, 2247–2256. [CrossRef]
25. Available online: https://www.graphenea.com/products/graphene-oxide-4-mg-ml-water-dispersion-1000-ml (accessed on 28 May 2021).
26. García-Bordejé, E.; Víctor-Román, S.; Sanahuja-Parejo, O.; Benito, A.M.; Maser, W.K. Control of the microstructure and surface chemistry of graphene aerogels: Via pH and time manipulation by a hydrothermal method. *Nanoscale* **2018**, *10*, 3526–3539. [CrossRef]
27. Ansón-Casaos, A.; Mis-Fernández, R.; López-Alled, C.M.; Almendro-López, E.; Hernández-Ferrer, J.; González-Domínguez, J.M.; Martínez, M.T. Transparent conducting films made of different carbon nanotubes, processed carbon nanotubes, and graphene nanoribbons. *Chem. Eng. Sci.* **2015**, *138*, 566–574. [CrossRef]
28. Rodríguez-Mata, V.; González-Domínguez, J.M.; Benito, A.M.; Maser, W.K.; García-Bordejé, E. Reduced graphene oxide aerogels with controlled continuous microchannels for environmental remediation. *ACS Appl. Nano Mater.* **2019**, *2*, 1210–1222. [CrossRef]
29. Ryu, J.; Lee, K.; Yun, J.; Yu, H.; Lee, J.; Jang, J. Paintable carbon-based perovskite solar cells with engineered perovskite/carbon interface using carbon nanotubes dripping method. *Small* **2017**, *13*, 1–8. [CrossRef]
30. Mishra, A.; Ahmad, Z.; Zimmermann, I.; Martineau, D.; Shakoor, R.A.; Touati, F.; Riaz, K.; Al-Muhtaseb, S.A.; Nazeeruddin, M.K. Effect of annealing temperature on the performance of printable carbon electrodes for perovskite solar cells. *Org. Electron.* **2019**, *65*, 375–380. [CrossRef]
31. Rodríguez-Mata, V.; Hernández-Ferrer, J.; Carrera, C.; Benito, A.M.; Maser, W.K.; García-Bordejé, E. Towards high-efficient microsupercapacitors based on reduced graphene oxide with optimized reduction degree. *Energy Storage Mater.* **2020**, *25*, 740–749. [CrossRef]
32. Carrera, C.; González-Domínguez, J.M.; Pascual, F.J.; Ansón-Casaos, A.; Benito, A.M.; Maser, W.K.; García-Bordejé, E. Modification of physicochemical properties and boosting electrical conductivity of reduced graphene oxide aerogels by postsynthesis treatment. *J. Phys. Chem. C* **2020**, *124*, 13739–13752. [CrossRef]

Review

Developments in Synthesis and Potential Electronic and Magnetic Applications of Pristine and Doped Graphynes

Gisya Abdi [1,2], Abdolhamid Alizadeh [3], Wojciech Grochala [1] and Andrzej Szczurek [1,*]

[1] Centre of New Technologies, University of Warsaw, S. Banacha 2c, 02-097 Warsaw, Poland; abdi_gisya@yahoo.com (G.A.); w.grochala@cent.uw.edu.pl (W.G.)
[2] Academic Centre for Materials and Nanotechnology, AGH University of Science and Technology, al. A. Mickiewicza 30, 30-059 Krakow, Poland
[3] Department of Organic Chemistry, Faculty of Physics and Chemistry, Alzahra University, Tehran 1993893973, Iran; ahalizadeh2@hotmail.com
* Correspondence: a.szczurek@cent.uw.edu.pl; Tel.: +48-22-554-08-11

Abstract: Doping and its consequences on the electronic features, optoelectronic features, and magnetism of graphynes (GYs) are reviewed in this work. First, synthetic strategies that consider numerous chemically and dimensionally different structures are discussed. Simultaneous or subsequent doping with heteroatoms, controlling dimensions, applying strain, and applying external electric fields can serve as effective ways to modulate the band structure of these new sp^2/sp allotropes of carbon. The fundamental band gap is crucially dependent on morphology, with low dimensional GYs displaying a broader band gap than their bulk counterparts. Accurately chosen precursors and synthesis conditions ensure complete control of the morphological, electronic, and physicochemical properties of resulting GY sheets as well as the distribution of dopants deposited on GY surfaces. The uniform and quantitative inclusion of non-metallic (B, Cl, N, O, or P) and metallic (Fe, Co, or Ni) elements into graphyne derivatives were theoretically and experimentally studied, which improved their electronic and magnetic properties as row systems or in heterojunction. The effect of heteroatoms associated with metallic impurities on the magnetic properties of GYs was investigated. Finally, the flexibility of doped GYs' electronic and magnetic features recommends them for new electronic and optoelectronic applications.

Keywords: graphyne-like materials; synthesis and doping; electronic and magnetic properties; electronic transport; photodetectors

Citation: Abdi, G.; Alizadeh, A.; Grochala, W.; Szczurek, A. Developments in Synthesis and Potential Electronic and Magnetic Applications of Pristine and Doped Graphynes. *Nanomaterials* 2021, *11*, 2268. https://doi.org/10.3390/nano11092268

Academic Editor: José Miguel González-Domínguez

Received: 23 July 2021
Accepted: 30 August 2021
Published: 31 August 2021

Publisher's Note: MDPI stays neutral with regard to jurisdictional claims in published maps and institutional affiliations.

Copyright: © 2021 by the authors. Licensee MDPI, Basel, Switzerland. This article is an open access article distributed under the terms and conditions of the Creative Commons Attribution (CC BY) license (https://creativecommons.org/licenses/by/4.0/).

1. Introduction

Among all chemical elements, carbon exhibits the greatest flexibility of its first coordination sphere, which is usually presented in textbooks as sp, sp^2, and sp^3 hybridizations. This plasticity leads to three available types of bonds (single-, double-, and triple-bonded C atoms) that may occur in diverse, practically unlimited, connectivities. Altogether, this markedly influences the allotropy of carbon, which is the richest among all chemical elements. Familiar allotropic forms of carbon include graphite, rhombohedral graphite, diamond, lonsdaleite, amorphous carbon (soot with variable sp^2/sp^3 carbon atom contents), and a huge variety of human-made high-specific surface area carbons, carbon aerogels [1,2], carbon foams [3,4], glassy carbon [5,6], polyynes [7], diverse fullerenes from C_{12} to as large as C_{960} [8,9], a multitude of single wall nanotubes [10,11], nano-onions [12], and more. Last but not least, they include graphene, which has unique physical properties [12]. Other exotic forms such as ultra-high pressure BC8 [13,14], Po-32 [15], and ferromagnetic carbon [16] have previously been theorized [17] (C18 was even reported [18]), but some are still disputed. Nevertheless, this structural diversity and versatility of chemical bonding brings an enormous pool of physicochemical properties, reactivities, and so on; crystal structures of over 500 periodic allotropes, known and hypothesized, have been collected in a unique Sacada database (https://www.sacada.info/) [19]. Despite the long-lasting research of carbon-based materials, some fundamental issues related to the shape of the phase diagram and mutual stability of polymorphs, or even their existence, remain unresolved to this day [20–22]; e.g., it has been recently claimed that lonsdaleite is not a genuine allotropic form but a twin of cubic crystals, which raised controversy [23,24]. One illustration of the intensity of the research field of carbon materials can be provided by an inspection of the Web of Science database; this resource lists approximately 114,000 papers using the keyword 'diamond', approximately 147,000 papers discussing 'graphite', and approximately 240,000 papers featuring 'graphene'. It can be safely estimated that—even given some overlap—well over half a million of over 90 million indexed scientific works (1900–2021) have been devoted to carbon allotropes. The discovery of graphene triggered the production of diverse 2D carbonaceous materials including graphone, graphane, and graphene oxide [25–27]. Graphynes (GYs) are layered two-dimensional structures built from sp- and sp^2-hybridized carbon atoms (Figure 1A).

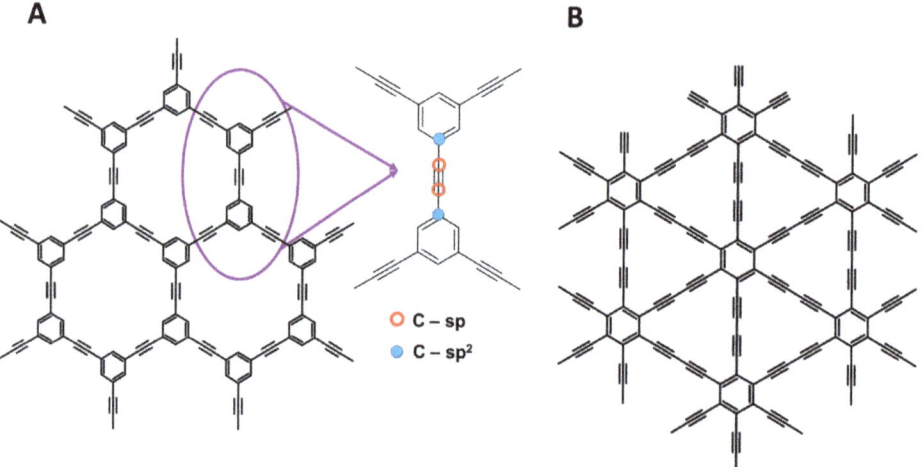

Figure 1. The structure of graphyne with indicated distribution of sp (red rings) and sp^2 (blue circles) hybridized carbons (**A**); The representation of graphdiyne structure (**B**).

Wide tunability in structural, mechanical, physical, and chemical properties make GYs fascinating candidates for use in energy storage, solar cells, electronic and spintronic devices, UV light detectors, as well as adsorbents in the separation of gases [28]. A GY can be a useful catalyst in water purification [28]. GYs have shown improved electronic properties and charge carriers in the optics and electronics industries. Graphone and graphane are hydrogenated forms of graphene—with fundamental band gaps of 2.45 and 5.4 eV, respectively, and interesting magnetic properties—that have shown potential for nanoelectronics and spintronics [29–31]. Due to structural similarities with graphene, these materials are considered to be excellent candidates for carbonaceous electronic devices, and they surely will be the subject of advanced studies for multiple and versatile applications. Scientists have been theoretically studying graphynes since the 1980s, which initially attracted attention after the discovery of fullerenes [32]. Though the structure of GY was first proposed in 1987 by Baughman et al. [33], the demanding synthesis of GYs hindered their dynamic development for more than 20 years [34]. Among all the GYs, the rising-star γ-graphdiyne (γ-GDY), seen in Figure 1B, was the first GDY member experimentally synthesized and reported by Li et al. in 2010 [35]. Because of the promising physical, optical, and mechanical features of GYs, a tremendous amount of research effort has been dedicated by theoretical, applied, and synthetic chemists. It is believed that GYs might pose competition for more common sp^2-hybridized carbon systems, particularly graphene, and meet the increasing demand for an alternative candidate to carbonaceous materials. Recent years have brought a sharp increase in research interest on the synthesis and theoretical prediction of GYs' properties in different dimensions, e.g., one-dimensional nanowires (such as nanotube arrays and ordered stripe arrays), and two-dimensional nanowalls (2D) and nanosheets (2D), and 3D frameworks [36–38]. More and more works are concerning pristine GYs' structures enriched with diverse heteroatoms (B, F, N, O, etc.) [37]. Due to the broad spectrum of intense scientific activities related to the development of new forms of GYs, this work emphasizes recent developments in this field, especially dealing with newly obtained heteroatom-doped structures and investigating the synergic effects of heteroatoms, metal oxides, and metal ions on the electronic and optoelectronic properties of doped GYs. In this work, we also focus our attention on theoretical and experimental research into the magnetism of pristine and doped GYs.

2. From Atomic Structure Suggestion to Experimental Appearance

In 1987, Baughman et al. [33] proposed a novel two-dimensional allotrope of carbon assembled by aromatic centers (sp^2) linked to each other by acetylenic bridges (sp); these were named graphynes (GYs). In the beginning, GYs (a general name given to this type of material) were designated according to the number of carbons included in the various rings forming a given network (shaped pores in the structure). For example, A, B, and C networks were called 18,18,18-graphyne (Figure 2A), 12,12,12-graphyne (Figure 2B), and 6,6,6-graphyne (Figure 2C), respectively. The latter theoretically present the lowest energy of carbon phases consisting of flat molecular sheets obtained by inserting acetylene linkers between the aromatic rings into the pristine honeycomb graphene structure. As energetically favorable, the 6,6,6-graphyne has a special place in the GY family, and the term "graphyne" without specifying the number of carbons is reserved for this phase [39].

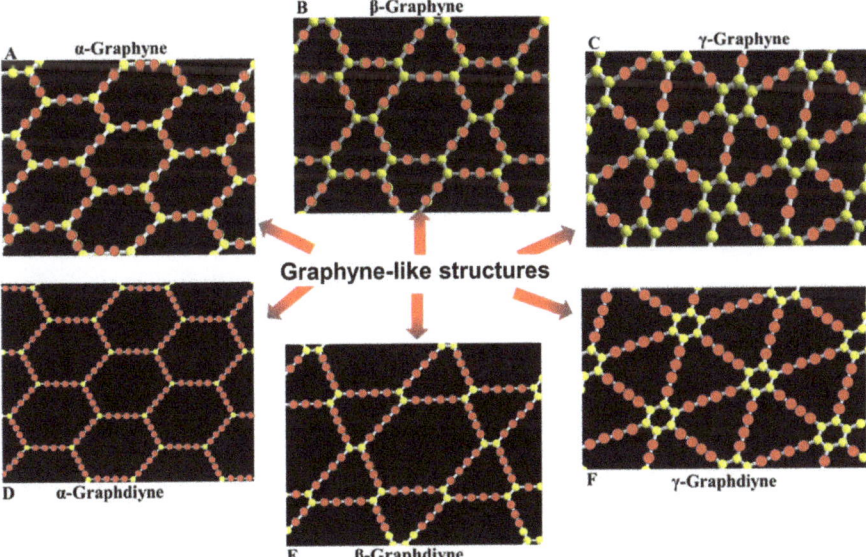

Figure 2. Representation of different graphyne and graphdiyne–like structures. α-graphyne (**A**); β-graphyne (**B**); γ-grapyne (**C**); α-graphdiyne (**D**); β-graphdiyne (**E**); γ-graphdiyne (**F**). The red dots represent acetylenic (graphynes) and diacetylenic (graphdiynes) bridges (Csp) and yellow ones—sp^2 hybridized carbons.

In 1997, Haley et al. proposed new forms of GYs called graphdiyne (GDY), a nanostructure formed from two adjacent acetylenic linkers between aromatic carbon atoms in contrast to graphyne, which is made of carbons connected by one triple bond [40]. Therefore, to investigate new structures, different GY networks designations were developed and used in the literature [39,41]. Figure 2 shows several types of GY structures named in accordance with the widespread terminology including α-graphyne (Figure 2A), β-graphyne (Figure 2B), γ-graphyne (Figure 2C), α-graphdiyne (Figure 2D), β-graphdiyne (Figure 2E), and γ-graphdiyne (Figure 2F). The last nomenclature method named γ-graphynes with various numbers of acetylenic linkers in a structure, such as graph-n-yne ($n = 1, 2, 3 \ldots$, where n is the number of triple bonds). The mentioned nomenclature methods properly operate when naming the most conventional GYs and GDYs. However, the graphyne naming system causes considerable problems and should be standardized, especially for structures containing heteroatoms. In this paper, GYs, GDYs, and their derivatives are all briefly designated as graphyne- or graphdiyne-like structures with the abbreviation "GYs" and "GDYs", respectively.

In order to prepare extended structures of GYs, chemists developed versatile synthetic methods grouped into two subsections: top-down and bottom-up [42]. The top-down methods have the potential to obtain two dimensional structures from their bulk precursors through processes such as mechanochemical synthesis and vapor-liquid-solid (VLS) growth. The bottom-up formulations result in the formation of thin films or a few layers of GYs through coupling reactions of desired substrates in carefully chosen and strictly controlled conditions. "On-surface" coupling synthesis conducted in an ultra-high vacuum (UHVS), chemical vapor deposition (CVD), thermal treatments, and wet-synthetic approaches are practical subdivisions of the bottom-up method.

By summarizing the most representative and effective synthetic methods, we hope they can create an idea and solid foundation for preparing GYs with given applications in various research fields. The wet-synthetic strategy is the most successful and high-demand method for the preparation of a wide range of applicable GYs, in both pristine and doped natures. In order to facilitate understanding and comparison, the wet-synthetic methods are classified into two categories based on the different phases where the coupling reaction of the reactants occurs amid the synthesis process: one-phase (homogenous reaction) and two-phase (interfacial reaction). Two-phase methods have three sub-divisions: liquid/liquid, solid/liquid, and gas/liquid.

We intend to summarize the developments in preparation methods resulting in different GY morphologies (nanowires, nanotubes, nanowalls, nanoribbons, nanosheets, etc.), as well as different chemical compositions. The broad-spectrum synthesis protocols employed in cross-coupling or homocoupling reactions of graphynes' subunits and derivatives were named after their inventors as Glaser [43], Eglinton [44], Hay [45], Negishi [46], Hiyama [47], and Sonogashira [48] reactions. All these methods, briefly presented in Table 1, were proven to be applicable to the preparation of valuable and scientifically important heteroatom-doped GYs.

Table 1. Synthetic methods employed in cross-coupling or homocoupling reactions for the preparation of graphyne derivatives; year when a given method was invented is indicated.

Synthetic Method	Year	Condition	Reaction	[Ref.]
Glaser coupling	1869	- Catalyst: Cu(I) - Oxidant: O_2 - Base: ammonia - Terminal alkyne	2 R–C≡C–H → R–C≡C–C≡C–R	[43]
Eglinton coupling	1959	- Catalyst: Cu(II) - Base: pyridine - Terminal alkyne	2 R–C≡C–H → R–C≡C–C≡C–R	[44]

Table 1. Cont.

Synthetic Method	Year	Condition	Reaction	[Ref.]
Hay coupling	1962	- Catalyst: Cu(I) - Base: tetramethylethylenediamine (TMEDA) - Terminal alkyne	2 R–C≡C–H → R–C≡C–C≡C–R	[45]
Negishi cross-coupling reaction	1977	- Catalyst: palladium (0) - Nickel: co-catalyst - Organic halides or triflates	RX + R'-ZnX' + ML$_n$ → R-R' X = Cl, Br, I, triflate and acetyloxy X' = Cl, Br, I R = alkenyl, aryl, allyl, alkynyl or propargyl R' = alkenyl, aryl, allyl, alkyl, benzyl, homoallyl, and homopropargyl. L = triphenylphosphine, DPPE, BINAP or chiraphos M = Ni, Pd	[46]
Hiyama coupling	1988	- Palladium (0)	R-SiR''$_3$ + R'-X → R-R' R: aryl, alkenyl or alkynyl R': aryl, alkenyl, alkynyl or alkyl R'': Cl, F or alkyl X: Cl, Br, I or OTf	[47]
Sonogashira cross-coupling reaction	2002	- Aqueous media - Mild base - Palladium (0) - Copper as co-catalyst	R$_1$-X + H–C≡C–R$_2$ → R$_1$–C≡C–R$_2$ R1: aryl R2: aryl or vinyl X: I, Br, Cl or OTf	[48]

Theoretically proposed graphynes (GYs) induced tremendous effort to find applicable routes, resulting in the fabrication of these materials [39]. The controlled oligo-trimerization of cyclo-carbons such as cyclo-C18 (cycle made of nine acetylenic groups) and cyclo-C12 (cycle made of six acetylenic groups) and the oxidative polymerization of monomeric acetylenic precursors or synthetic macrocyclic compounds under Glaser–Hay coupling conditions are potential methods for the synthesis of γ-GDY, γ-GY, β-GDY, and other derivatives [49–52]. As an alternative strategy to investigate the potential properties of GYs, Haley et al. proposed a method for the synthetic preparation of γ-GY and γ-GDY substructures [40,53]. These subunits can be next used as the first block in the construction of extended GY structures. This breakthrough in synthetic approaches to alkynyl carbon materials revealed possibilities for the synthesis of low-dimensional carbonaceous nanomaterials involving acetylenic scaffolds [40,54–57]. Since the first report of γ-GDY preparation in 2010, exciting progress has been made in the experimental preparation of GYs. In the next sections, we describe different, well-developed synthetic protocols leading to graphyne-like structures composed of C, as well as structures chemically modified by heteroatoms (e.g., N, F, Cl, H, S, and B).

2.1. Mechanochemical Synthesis

A mechanochemical synthesis route is used in one of the most effective methods. Its strength lies in its simplicity, rapidity, and repeatability. As a result, solid extended GYs or their subunits with reproducible features may be produced. Under mechanical impact and at an elevated local temperature resulting from particle collisions, selected bonds of substrates are broken and new compounds may form in the solid-state, thus overcoming the problems associated with solution-based chemistry processes.

The reaction of hexachlorobenzene (HCB; known as a persistent organic pollutant) and calcium carbide (CaC$_2$; known as an efficient and safe co-milling reagent) in a planetary ball mill at room temperature within 20 min of milling at a mass ratio of $CaC_2/HCB = 0.9$ and a rotation speed of 300 rpm, as proposed by Li et al. [58] in 2017, caught the attention of materials chemists in the preparation of GYs. After this development, a mature mechanochemical approach was applied by Li et al. in 2017 for the one-step high-yield synthesis of GY monomers [59]. One year later, Cui et al. synthesized hydrogen-substituted graphyne (H-GY) and γ-graphyne (GY) via the ball-milling-driven mechanochemical cross-coupling of 1,3,5-tribromobenzene (PhBr$_3$), hexabromobenzene (PhBr$_6$), and CaC$_2$ as precursors under a vacuum. Finally, the impurities were removed with diluted nitric acid and benzene (Figure 3) [60,61].

Figure 3. Schematic representation of mechanochemical synthesis for preparation of GY and its derivatives. Redrawn [60,61].

2.2. The Vapor–Liquid–Solid Growth

Synthesis reactions carried out with vapor-liquid-solid (VLS) growth allow one to control the production of different types of nanomaterials. In general, GY sheets are harvested on carefully prepared surfaces of monocrystalline silicon coated by metallic nanoparticles (Au, Fe, Zn, and Ni). The prepared substrate-catalyst support provides favorable physical features (surface energy, stability, and crystal structure) for the effective growth of nanostructural materials. The VLS method was applied by Li et al. for graphdiyne (GDY) film synthesis [62]. In their approach, powdered and vaporized GDY was deposited on the surface of ZnO nanorods grown on a silicon wafer (Figure 4a). Liquid nanodroplets of melted zinc (~419 °C) formed on one of the ends of the ZnO nanorods (ZnO NRs) (Figure 4b) served as energetically favorable adsorption sites of incoming vapors of GDY molecules. Due to the mixing of vapors with melted zinc, a solution of graphdiyne and Zn was formed (Figure 4c). The continuous influx of fresh portions of GDY molecules resulted in the formation of a supersaturated solution (GDY-Zn), as well as the fusion of drops and an increase in their size of droplets, thereby facilitating lateral growth due to the small edge energy of 2D materials.

2.3. Thermal Treatment

The heating of hexaethynylbenzene (HEB), N-rich precursors (2,4,6-triethynyl-1,3,5-triazine, TET), and pentaethynylpyridine (PEP) was applied by Zuo et al. to force a homocoupling reaction, resulting in GDY nanostructures with different nitrogen percentages and morphologies (nanoribbon, nanochain, and 3D-networks) [63,64]. Notably, this reaction could be carried out without using any metal catalyst. The powder of N-rich precursors were slowly delivered to the preheated conical flask (120 °C), leading to an explosive reaction whereby black GDY was obtained (Figure 5a,b). A gradual heating process (10 °C/min) to 120 °C in nitrogen converted the light-yellow HEB into black nanoribbon-like morphologies without volume change (Figure 5c,i). On the other hand, the implementation of this treatment in an air atmosphere resulted in GDY nanochains uniformly grown on the 3D network with a remarkable volume increase of 6 times (Figure 5c,ii). This means that the oxygen accelerated the dehydrogenation for the coupling reaction. However, the addition of HEB into a preheated air environment (120 °C) rapidly caused a more violent reaction, and an ultrafine nanochain with a 48-fold volume increase was obtained (Figure 5c,iii).

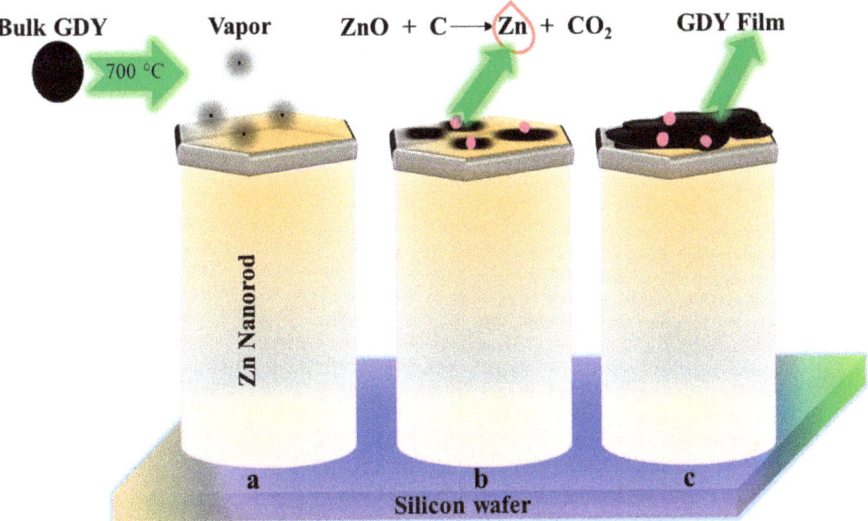

Figure 4. The illustration of the increase in a GDY layer in the VLS process on the head of a single ZnO NR (**a**) ZnO NRs in the presence of GDY vapors, (**b**) Zn droplet on the head of the ZnO NRs, and (**c**) Zn droplets scattered in GDY thin film, produced by connected several neighboring flakes of GDY. Redrawn [62].

The doping of carbonaceous materials with N atoms can be realized via the three following routes: chemical vapor deposition (CVD), the pyrolysis (annealing) of N-containing precursors, and heating with N-rich chemicals. For instance, N-doped GDY, B-doped GDY, F-doped GDY, and S-doped GDY have been prepared using an annealing strategy [65–70]. In such a case, GDY-based nanomaterials have been thermally treated with relevant chemicals, such as ammonia (N), B_2O_3 (B), NH_4F (F), and thiourea (S), at a chosen temperature [70]. The considered methods, however, suffer from drawbacks, such as the randomness of the doping sites and uncontrolled dopant's percentage. Therefore, there is still a need to develop a synthetic strategy that provides 2D carbon materials with a homogeneous distribution of atoms of the desired type and at specific desired locations.

2.4. "On-Surface" Synthesis under an Ultra-High Vacuum

The so-called "on-surface synthesis" approach is carried out under an ultra-high vacuum (UHV) and presents considerable potential in building new types of nanomaterials. In this method, the starting building blocks are deposited onto the surface of a metallic substrate (Ag and Au) [71–85]. Then, the coupling of precursors occurs, resulting in single-atom-thick 1D and 2D materials. In contrast to conventional "wet" reactions, it helps here to eliminate possible undesired influences from surroundings. The chemical character of the organic precursors and the nature of the applied substrate are crucial factors that determine the final properties of GDY nanostructures [41]. Features such as the dimensionality of the organic precursors, the reactivity of their functionalities, the geometry of the surface, and interactions occurring between the organic building blocks and substrate have significant effects on the reaction and molecular surface patterns. The reactivity and mass of the used molecules have significant impacts on the success of synthesis. High reactivity or weight may prevent them from sublimating on the surface of the substrate. On the other hand, molecules that are too small will escape from the reaction chamber. The metallic substrates play a double role as templates and catalysts of the coupling reaction. In the following sections, we discuss the most effective method for the synthesis of graphyne analog (sub-structures or infinite nanostructures) based on "on-surface" coupling reactions.

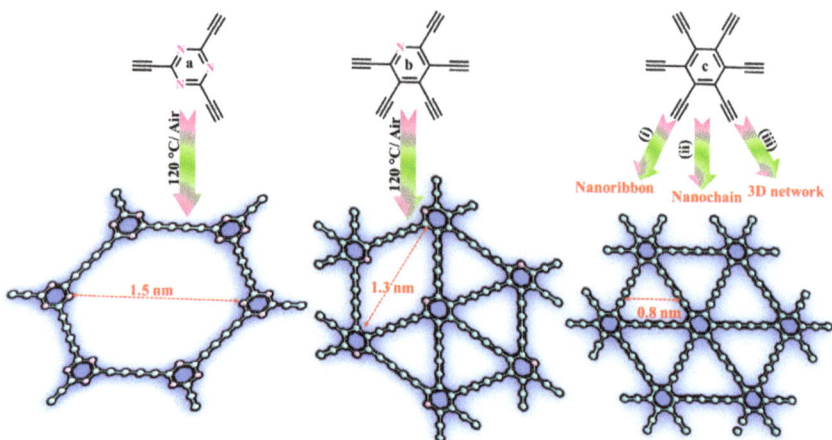

Figure 5. Thermal treatments of (**a**) TET, (**b**) PEP, and (**c**) HEB: (i) gradually heated to 120 °C/N$_2$; (ii) rapidly heated to 120 °C/Air; and (iii) gradually heated to 120 °C/Air. Redrawn [63,64].

The production of acetylenic frameworks at interfaces involving the formation of self-assembled monolayers (SAMs) followed by a cross-linking step to form linked monolayers is a direct way to create carbonaceous materials such as carbyne, graphyne, and graphdiyne [71,72]. "On-surface" homocoupling reactions requiring the detachments of halogens or hydrogen from precursors functionalized with alkynyl groups have been reported as effective fabrication methods of low-dimensional carbon-based nanostructures. An overview of the production of 1D carbonaceous nanomaterials via "on-surface" approaches was previously described in the literature [73].

Rubben et al. reported a surface-assisted dehydrogenative homocoupling reaction of terminal alkynes (Csp–H), such as triethynylbenzene (TEB) and (1,3,5-tris-(4-ethynylphenyl) benzene (Ext-TEB), that was conducted on a Ag(111) surface, wherein the hydrogen was the only by-product of the reaction. The authors stated that it was an ideal method for the synthesis of individual chemicals or polymeric structures containing a conjugated backbone (after annealing at 400 K) [74]. In 2015, Wu et al. investigated the reaction of 2,5-diethynyl-1,4-bis (phenylethynyl)-benzene (DEBPB) taking place on the surface of silver with different facets including (111), (110), and (100). The reaction was carried out with the aid of scanning tunneling microscopy (STM). The Glaser synthesis conducted on Ag(111) was dominant and yielded one-dimensional, covalently bonded wires. On the contrary, reactions conducted on Ag(110) and Ag(100) surfaces resulted in one-dimensional organometallic frameworks built on terminal alkynes and metal atoms. (Figure 6) [75].

Klappenberger et al. employed a Ag(877) support to obtain one-dimensional conjugated molecular threads as components of extended GYs with lengths reaching 30 nm [76]. Thermal dehydrogenative reactions carried out on a flat Ag(111) plate were found to be associated with several undesirable side reactions that resulted in the formation of branched, irregular nanostructures. Liu et al. induced a dehydrogenative reaction of 2,5-diethynyl-1,4-bis(4-bromophenylethynyl)benzene and noticed that bromine adatoms affected the activation of C–H groups in terminal alkynes occurring at 298 K on a Ag(111) surface [77]. The STM studies disclosed the formation of organometallic species followed by their partial conversion to covalently bonded nanostructures after annealing at 420 K.

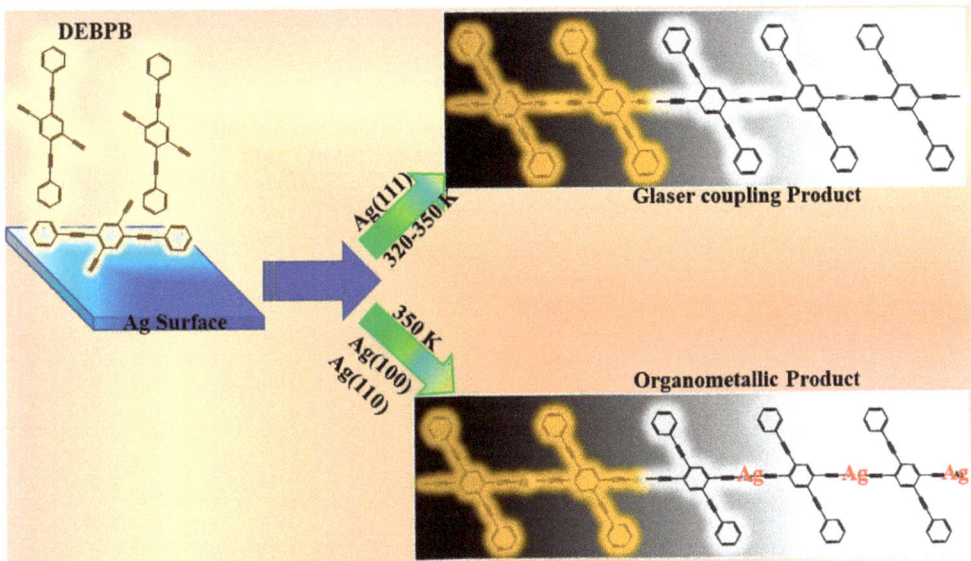

Figure 6. Schematic representation of the DEBPB coupling (Glaser reaction) carried out on silver surfaces. Redrawn [75].

In 2018, Xu et al. applied a dehalogenative homocoupling reaction to tribromoethylbenzene (TBP), 1,3-bis(tribromomethyl)benzene (bTBP), and 1,3,5-tris(tribromomethyl) benzene (tTBP), and they converted tribromomethyl functional groups (Csp^3) to form C–C triple bonds (Csp) as structural motifs of dimeric structures, such as wires or 2D networks of GYs grown on a Au(111) surface [78].

In 2020, Xu et al. created on-surface graphyne nanowires through dehalogenative homocoupling reactions via the stepwise activation of two different types of C–Br bonds (involving Csp^3–Br and Csp^2–Br) in a 1-bromo-4-(tribromomethyl)benzene (BTBMB) compound on both Au(111) and Ag(110) surfaces [79]. Sun et al. also reported the successful formation of dimer structures with acetylenic linkers (wires and networks) via the on-surface C–Br activation of alkenyl carbon atoms [80]. Two-dimensional networks with acetylenic linkages were obtained after the homocoupling reaction of 1,3,5-tris(bromoethynyl)benzene (tBEP). In the first step, the precursor was deposited on a Au(111) surface at room temperature and was slightly heated up to 320 K. As a result, organometallic networks were obtained. The further increase in temperature up to ~450 K led to the release of gold atoms and the formation of the final product. Figure 7 shows STM studies of this reaction. Considering the published results, the Ag(111) surface seems to be the most effective substrate for Glaser coupling because it causes a smaller number of side reactions [81] than Au(111) plates, for which the cyclization of the terminal alkyne to the benzene ring is common [82–84]. When used as a substrate, a Cu(111) surface presented low activity towards "on-surface" Glaser coupling. This stands in contrast to wet coupling reactions for which Cu ions show very high catalytic activity, and Cu is regarded as the most effective catalyst for such reactions [85].

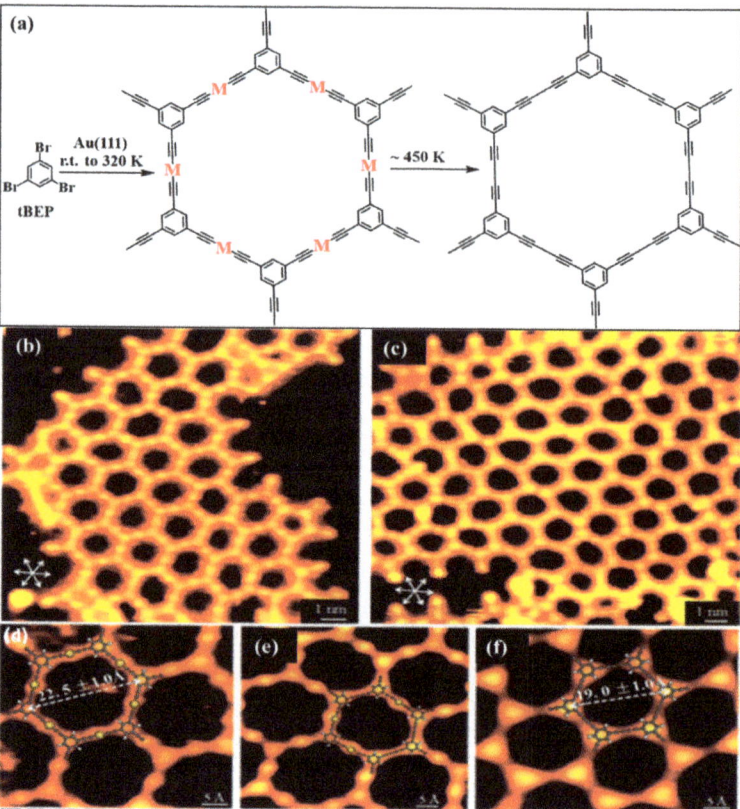

Figure 7. Schematic representation of the growth of the organometallic framework after the deposition of tribromophenyl on Au(111) surface (from RT to 320 K) and the formation of GDY after heat treatment at 450 K (**a**). Scanning tunneling microscopy micrograph revealing the creation of GDY before and after heat treatment at 450 K (**b**,**c**). Detailed STM pictures of the C–Au–C framework (**d**), the mixture of C–Au–C networks and GDY fragments (**e**), and the GDY layer (**f**). The modeled structures of all considered networks are superimposed on the STM micrographs. Reproduced with permission [80]. Copyright 2016, American Chemical Society.

2.5. "On-Surface" Synthesis by Chemical Vapor Deposition

Another synthesis strategy based on the covalent coupling of organic monomers occurring at the metal surface, so-called "on-surface" synthesis, is chemical vapor deposition (CVD).

Furthermore, the CVD method is recognized as one of the most promising routes for the creation of novel 2D materials. This approach relies on the transfer of vapors of monomers to reaction chambers and their embedding and coupling on the surface of a preheated metallic substrate. However, this method has severe limitations. As the reaction is conducted on a metal substrate without any additional catalysts, the reaction stops when the surface is fully covered with GY monolayer films (Figure 8) [86]. It has been statistically shown that silver seems to be the most efficient substrate for carrying out such reactions. In contrast to other investigated metallic substrates (such as Au and Cu), silver was found to ensure the lowest proportion of side reactions [86].

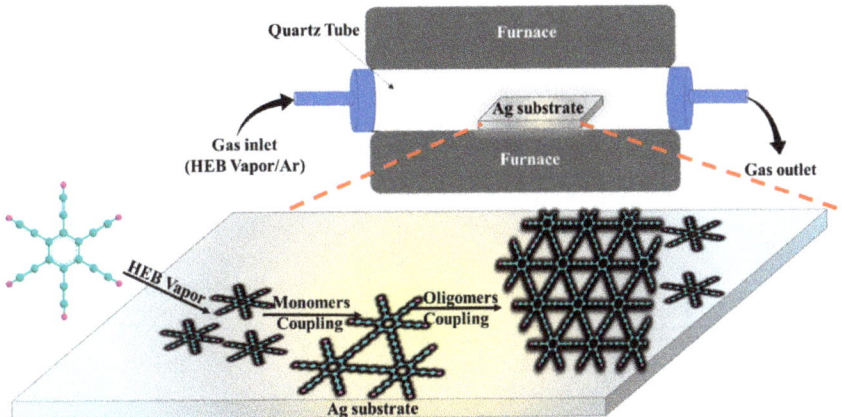

Figure 8. The formation of the GDY sheets on the silver surface with the aid of the CVD technique. Redrawn [86].

2.6. Wet Chemical Synthesis

Since the first successful fabrication of γ-GDY through a liquid/solid interfacial reaction on Cu foil as a catalytic substrate for an acceleration coupling reaction in an organic solvent, directed efforts have been carried out to prepare graphyne derivatives. Homo-coupling and cross-coupling reactions of hexaethynylbenzene (HEB) as an efficient precursor have been realized as operational pathways for the preparation of graphyne analogs (fragments, oligomers, or infinite structures) and have recently attracted immense attention. These methodologies employ metal salts (copper salts: Cu(I), Cu(II), or Pd(II)) in a homogenous conditions or on a surface template, e.g., Cu foil or other arbitrary surfaces such as graphene, Au, and 3D foam (which may have catalytic properties) to assist in a heterogeneous reaction [41].

2.6.1. Developments of Coupling Reactions (All Reactants in Solution Phase)

The coupling reaction for the synthesis of graphyne analogs proposed by Moroni et al. combined dibromoaryls (I) and 1,4-diethynylaryls (II) in the presence of $PdCl_2$, $Cu(OAc)_2$, and triphenylphosphine $(PPh)_3$ in a triethylamine/THF mixture (Figure 9A), where R1, R2, R3, and R4 could be the same or different (H, NO_2, alkyl ether, alkyl thioether, or alkyl ester). Homopolymers or copolymers with phenyl, thienyl, anthryl, or stilbene groups as aryl units were synthesized [87]. Despite this progress in synthetic methods, extended structures of GYs are still unachievable.

The combined Negishi and Sonogashira cross-coupling reactions for the formulations of various kinds of substituted hexaethynylbenzenes from chloroiodobenzenes put researchers on a fast track towards the fabrication of GYs [88,89]. In 2007, Jiang et al. obtained poly(aryleneethynylene) networks with highly developed porous structures. To do so, they applied a Pd-supported Sonogashira-Hagihara reaction, which had previously been employed to synthesize different polymeric compounds such as polymers and ligands for coordination-polymer synthesis, wires, and shape-persistent macrocycles [90]. In 2010, Dowson et al. showed that porous properties (BET surface area and pore volume) are strictly controlled by the kind of solvent used as the environment of the reaction [91]. Toluene, tetrahydrofuran (THF), N-dimethylformamide (DMF), and 1,4-dioxane were tested for these reactions. Authors showed that DMF is the most proper solvent, as its received nanostructures are characterized by the highest BET surface areas (up to 1260 m^2/g) [91,92]. In homogeneous coupling reactions, materials chemists prepared oligomers and macromolecules, but infinite structures are still elusive.

After 2010, remarkable progress in GY preparation was achieved. The synthesis of poly(aryleneethynylene)s (PAEs) using Pd and Mo/W was thoroughly investigated by

Bunz in 2010 [93]. Wu et al. utilized commercially available tris(*t*-butoxy)(2,2-dimethylpropylidyne)-tungsten(VI) as the catalyst in the synthesis of hydrogen-substituted graphyne (H-GY), as shown in Figure 9B [94]. Ding et al. prepared γ-graphyne in a homogenous ultrasound-driven reaction of hexabromobenzene ($PhBr_6$) and calcium carbide (CaC_2) in an inert atmosphere without a metal catalyst [95]. Wen et al. prepared new N-doped graphyne analogs (Figure 9C) in the reaction of nucleophilic substitution (SNAr) of cyanuric chloride and para-dilithium aromatic reagents. The process was carried out under mild conditions in a diglyme or bis(2-methoxyethyl) ether (solvents with high boiling points) solution. That designed reaction allowed them to obtain N-GYs on a gram scale [96]. The development of GDY synthesis was a breakthrough in the preparation of different morphologies such as films, nanowires, nanotube arrays, nanoribbons, nanosheets, and nanowalls of GYs with versatile properties, as well as reductions in the dimensionality [97].

Graphdiyne nanoribbons (GDYNRs) comprise a class of 1D GDY materials that stresses well-defined edges and nanometer size [97]. There have already been numerous theoretical efforts regarding GDYNRs seeking connections between their structures and properties. The results of this research are discussed in Section 3.2.3. A bottom-up chemical formulation could provide structurally uniform and well-defined nanostructures of GDYNRs. It is, however, necessary to perform the selective stepwise coupling of ethynyl groups during the synthesis procedure. Zhou et al. proposed a two-step method of intermolecular polymerization followed by the intramolecular cross-coupling of acetylenic moieties, as seen in Figure 9D (red fragment). First, the polymerization of the ethynyl units in the central part of the monomer ensures one-dimensional growth (Figure 9D, red fragment). Secondly, the intramolecular reaction of ethynyl groups on the established facing side chains and the bulky groups (such as the 3,5-di-tert-butylbenzyl group) on the outer side occurs. The latter works to sterically hinder intermolecular coupling. This strategy was applied to build GDYNRs nanostructures (Figure 9D, blue fragment) made of rhomboid units with benzene as junctions and butadiyne as linkers for the first time. The structures showed a well-defined width of ~4 nm and a length of hundreds of nanometers [97].

2.6.2. Two-Phase Methods (Interfacial Synthesis Utilizing Two Immiscible Liquids)

Atomic, ionic, or molecular compounds may be successfully applied as starting materials to the direct, bottom–up synthesis of ultrathin GDY nanostructures with in-plane periodicity. In 2017, Sakamoto et al. strived to create graphdiyne at the interface between two immiscible fluids (Figure 10A,B) [98].

The upper aqueous phase held copper (II) acetate and pyridine, which catalyzed ethynyl homocoupling (Eglinton coupling). The lower dichloromethane phase contained the HEB monomer. The continuous catalytic reaction for 24 h under an inert atmosphere at room temperature resulted in the development of a layered GDY (thickness: 24 nm; domain size: >25 μm). In 2019, Song et al. described the liquid/liquid interfacial formulation as a comprehensive way to obtain GYs via a reaction between terminal ethynyl groups and an aryl halide. The reactions catalyzed by $PdCl_2(PPh_3)_2$ and CuI resulted in various forms of GY nanostructures, including hydrogen-substituted graphyne (H-GY), methyl-substituted graphyne (Me-GY), and fluorinated graphyne (F-GY) [99]. H-GY is a framework consisting of duplicated sections of benzene rings joined by ethynyl linkers at all meta sites. Likewise, the repeating units of Me-GY (or F-GY) are 1,3,5-trimethylbenzene (or 1,3,5-trifluorobenzene) rings attached to benzene rings within acetylene bridges. Two-dimensional N-graphdiyne sheets were recently prepared via reactions conducted at the interfaces (Figure 10D,E) [100]. Nitrogen heterocycles (triazine and pyrazine) bearing terminal ethynyl groups were polymerized through Glaser coupling reactions at interfaces. This procedure was expanded to the synthesis of S-doped graphdiyne (TTF-GDY) structures comprising tetrathiafulvalene fractions (Figure 10C) and has this potential to be applied as a robust route for the synthesis of a wide range of heteroatom-rich graphyne-like structures in the future [101].

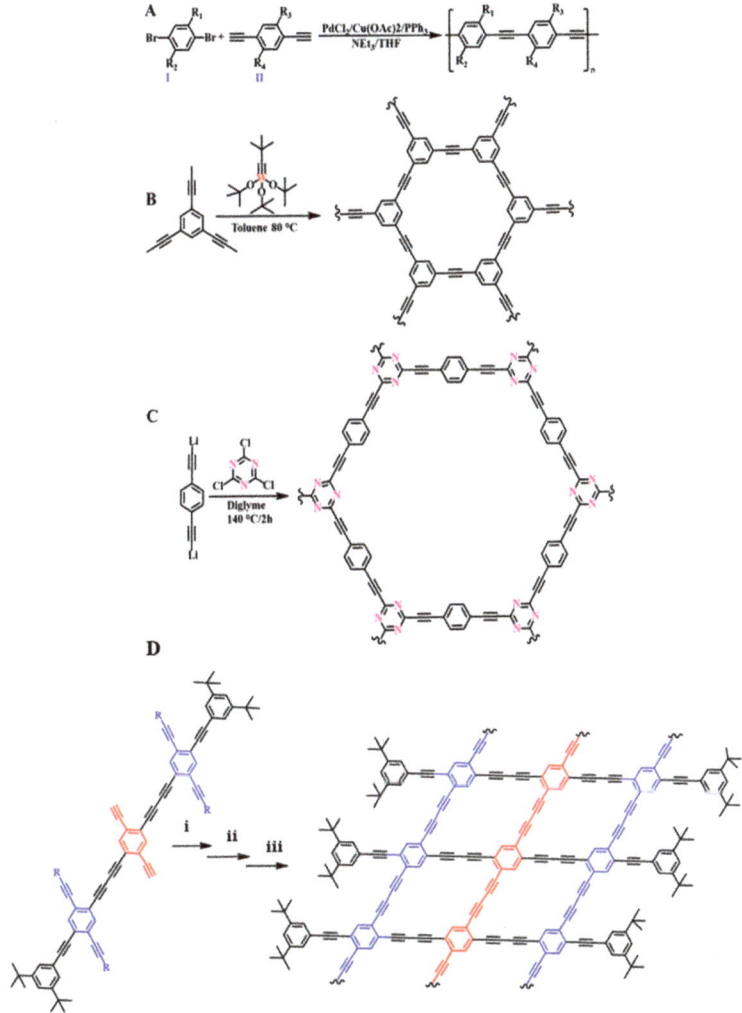

Figure 9. Homogenous coupling reactions in the preparation of graphyne derivatives (**A**). Synthetic scheme of the graphyne (**B**) and N-doped graphyne-like nanostructures (**C**). Redrawn [87,94]. (**D**) Intermolecular Glaser-Hay cross-coupling reactions of red fragments for GDY nanoribbons by copper(I) chloride, TMEDA, and acetone/tetrahydrofuran at RT (i); tetra-n-butylammonium fluoride and THF at RT (ii); intramolecular coupling reaction of blue fragments by $Cu(OAc)_2$ and pyridine, H_2O, and CH_2Cl_2 at RT (iii); R = triisopropylsilyl; TMEDA: N,N,N',N'-tetramethylethane-1,2-diamine; RT: room temperature. Redrawn [63,64].

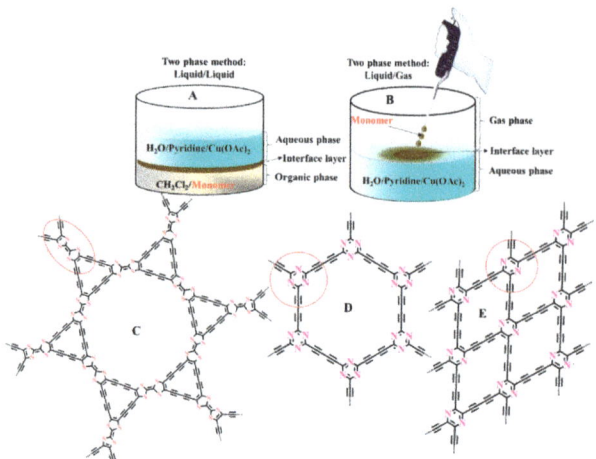

Figure 10. Homocoupling reaction in interlayer between two phases: (**A**) liquid/liquid and gas/liquid (**B**) in the preparation of sulfur-rich graphdiyne (**C**) and N-graphdiyne (**D,E**). Redrawn [100,101].

2.6.3. Developments of Heterogeneous Coupling Reactions at Liquid/Solid Interfaces on Diverse Substrates

In liquid/solid interface reactions, substrates such as copper foil, plate, foam, and walls have been applied to bring reactants together and speed up the reaction procedure [102,103]. The first γ-GDY was prepared through the heterogeneous homo-coupling reaction of hexaethynylbenzene on copper foil, playing double roles of catalyst and substrate (Figure 11A) [28]. It was reported that both the oligomer evaporation process and the kinetics of the coupling reaction were strictly controlled by temperature [104,105]. Furthermore, the factors determining the structural properties of the obtained nanostructures were found to be catalyst distribution and monomer concentration. The formation of γ-GDY nanowalls via the Glaser-Hay reaction was successfully carried out in the presence of N,N,N',N'-tetramethylethylenediamine (TMEDA) due to its ease in complexing copper ions. The copper envelope catalysis strategy was employed for the synthesis of γ-GDY nanowalls on various substrates, including one-dimensional Si nanowires; two-dimensional Au, Ni, and W foils; quartz; 3D stainless steel mesh; and 3D graphene foam (GY), as seen in Figure 11B [106]. For this purpose, these were the chosen substrates used in this method, and the target substrates were wrapped in a Cu-based envelope.

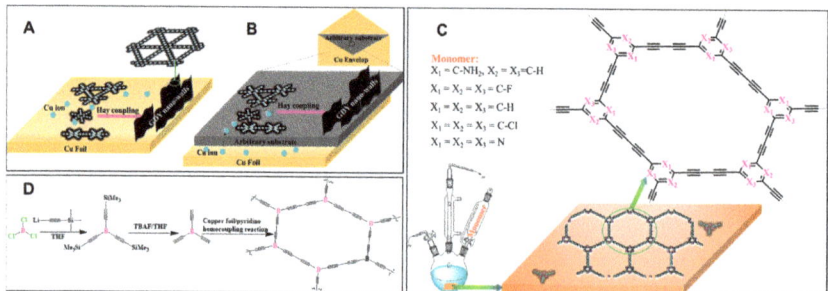

Figure 11. Proposed reaction process of GDY nanowall formation on Cu foil (**A**); envelope strategy for preparation GDY nanowalls on arbitrary substrates (**B**). Redrawn [35,106]; Glaser homocoupling reaction on Cu foil (liquid/solid method) (**C**). Redrawn [107–112]; boron-graphdiyne (B-GDY) preparation through homocoupling reaction on Cu foil (solid/liquid method) (**D**). Redrawn [113].

Well-defined films of triazine-based graphdiyne (TA-GDY), aminated-graphdiyne (NH$_2$-GDY), β-GDY, H-GY, Cl-GDY, and F-GDY were prepared through Glaser homocoupling reactions on Cu foil (Figure 11C) [107–112]. 2,4,6-triethynyl-1,3,5-triazine, 2,4,6-triethynylaniline, tetraethynylethene (TEE), 1,3,5-triethynylbenzene, 1,3,5-trichloro-2,4,6-triethynylbenzene, and 1,3,5-triethynyl-2,4,6-trifluorobenzene were the respective starting compounds (prepared from the deprotection of trimethylsilyl group by tetra-butyl ammonium fluoride (TBAF) from corresponding silylated substrates) and were applied in homocoupling reactions to prepare extended GYs on Cu foil. The resulting films were peeled off from the Cu foil by a FeCl$_3$-saturated solution and then rinsed with H$_2$O, acetone, DMF, and ethanol. Wang et al. prepared boron-doped graphdiyne (B-GDY) through the aforementioned synthetic procedure, which was recognized to be an effective technique to prepare 2D carbonaceous nanostructures with strictly controlled and well-organized chemical structures (Figure 11D) [113]. In contrast to the copper foil, the copper nanowires (CuNWs) worked as templates and delivered more reactive sites for developing γ-GDY structures [114,115]. As a result, high-quality nanostructures with well-developed surface areas were obtained. For instance, the thin films of graphdiyne (average thickness: approximately 1.9 nm) were obtained on CuNWs (100 nm in diameter) [114]. In the end, polymeric films were isolated by washing a crude product in a mixture of hydrochloric acid and FeCl$_3$.

3. Electronic Properties

3.1. Dirac Cone

A Dirac cone is a distinctive feature in an electronic arrangement in which the energy levels of the valence and conduction bands meet at one specific point in the first Brillouin zone (named Dirac points), hence setting the Fermi level; the band structure in its vicinity resembles a double cone with linear dispersion [116]. The presence of a Dirac cone renders a given material "the zero-gap semiconductor" rather than metal and results in several unusual features such as ballistic electronic transport and enormous thermal conductivity. The occurrence of Dirac cones in graphene, predicted by Wallace in 1947 and experimentally demonstrated by Novoselov et al. in 2005, has sparked unceasing research in recent years [117,118]. Some efforts have been directed towards exploring the possibility of the occurrence of the Dirac cones in GYs. The presence of Dirac cones in α-graphyne, β-graphyne, and γ-graphyne-n with hexagonal symmetry structures was demonstrated. Recently, Vines et al. discovered that 6,6,12-GY, with rectangular symmetry, has two self-doped non-equivalent and distorted Dirac cones [119,120]. These results shed new light on the electronic properties of GY-like materials and suggest that rigorous hexagonal symmetry is not a feature that determines the appearance of cones in these materials.

3.2. Electronic Band Structure

The electronic band structures of diverse GYs and GDYs have been investigated using theoretical methods, particularly density functional theory (DFT). Local density approximation (LDA) and the generalized gradient approximation (GGA) have been widely applied to study the structural, mechanical, electronic, and magnetic properties of GYs. Nevertheless, the underestimation of the band gap levels still is one of the critical problems of LDA and GGA. Hybrid Heyd-Scuseria-Ernzerhof-type functionals (HSE) have improved total energy evaluation by admixing the nonlocal Hartree-Fock exchange. They also lead to more realistic band gaps than LDA or GGA functionals. However, those conducted calculations generate significantly higher computational costs. To describe the van der Waals force in vdW-optPBE layered compounds, some extent of correction leads to improved results. Next, we review theoretical studies on GYs and compare diverse calculated parameters with the aforementioned theoretical approaches.

3.2.1. The Electronic Band Structure of GYs

The optimized geometry and electronic structures of diverse GY materials (graphyne, graphdiyne, graphyne-3, and graphyne-4) were computed by applying the full-potential linear combination of atomic orbitals (LCAO) approach by Narita et al. in 1998 [121]. The unit cell of all considered GY derivatives was similar to graphyne and is shown as a parallelogram in Figure 12. This unit cell was found to contain 12 carbons, the a and b lattice vectors were found to be equal (a = b), and the angle between them was found to be $\gamma = 120°$. The Brillouin zone of the investigated material was an equilateral hexagon. As a result of geometry optimization, all bond angles are either 120° or 180° in graphyne-n. The lattice parameters of graphyne-n structures (n = 1, 2, 3, and 4) are 6.86, 9.44, 12.02, and 14.6 Å, respectively, and binding energies are 7.95, 7.78, 7.70, and 7.66 eV, respectively. It turned out that the reported hexagon presents a bond length almost equal to those found for graphite. Moreover, it is a bit longer than the bond that extends outside a hexagon. The bridges between hexagons are not formed by cumulenic linkers =C=C=; rather, they are formed by ethynyl ones (–C≡C–). The presence of conjugated multiple bonds is a typical feature of graphyne and its derivatives.

The indirect band gap is the separation between the conduction band minimum (CB_{min}) and the valence band maximum (VB_{max}) within an electronic band structure, whereas the direct band gap denotes the smallest of gaps at one particular point in the Brillouin zone [122]. Chen et al. studied the band structures of optimized α-, β- and γ-GY structures (Figure 12A) [123]. They showed that all investigated materials were direct semiconductors. In the case of α-graphyne, the conduction band minimum meets the valence band maximum at K-point in the Brillouin zone (Figure 12Ba) and symmetric Dirac cones are formed. In the case of β- and γ-GY, the energy bands coincide at M-points in the Brillouin zone (Figure 12Bb,c) and show quasi-Dirac cone structures, albeit with a slightly open band gap. The values of band gaps were found to be equal to 0 eV (α-GY), 0.028 eV (β-GY), and 0.447 eV (γ-GY) when applying the generalized gradient approximation of the Perdew–Burke–Ernzerhof (GGA-PBE) method (Figure 12Ba–c). Moreover, γ-GDY is a direct semiconductor with a Dirac point at the zone center (Γ point of the first Brillouin zone).

The estimated band gaps of graphyne-like families (monolayer, bilayer, multilayer, nanotubes, and nanoribbons) with and without strain by different functionals are compiled in Table 2. A wide range of band gap levels were found for γ-GY [121–129] and γ-GDY [42,106,121,125–127,130–134] (0.447–2.23 eV for γ-GY and 0.44–1.21 eV for γ-GDY), which is strictly related to the applied calculation functionals, as reviewed in Table 2.

The manipulation of the band structure of GYs has been attempted through strategies including strain tuning, structural engineering (the fabrication of GYs with different dimensions), doping, and the application of external electric fields. The obtained results suggest that GYs are valuable materials for nanoelectronic and optoelectronic devices or sensors. For example, strained graphyne-n was obtained by applying different types of strain, including homogeneous biaxial (H-strain in both the x and y directions) and uniaxial strains (x-direction A-strain and y-direction Z-strain), as was reported by Li et al. in 2013 (Figure 13a) [127]. For the H-strain case, hexagonal symmetry was preserved after the strain (Figure 13b). The A-strain deformation was found to be in the direction of the propagation of the ethynyl linkers, while the Z-strain described deformation perpendicular to the acetylenic linkers with bent hexagonal symmetry (Figure 13c,d). Unlike graphene, which exhibits a band gap insensitivity to applied strain, GYs have shown band gap modulations under different straining approaches. Comprehensive studies have proven that homogeneous tensile stress expands the band gap of GYs, whereas uniaxial tensile and compressive strains lead to band gap decreases (Figure 13e). Direct band gaps at either the M or S point of the Brillouin zone have been observed for both graphyne and graphyne-3 subjected to different tensile strains. In contrast, graphyne-4 and graphdiyne have been shown to display a direct band gap established at the Γ point, whatever the nature of the implemented strain [122,127]. Subsequent studies by Qui et al. confirmed the effect of biaxial tensile stress in increasing the gap of γ-GDY within the range of 0.47–1.39 eV, while uniaxial tensile strain was found to decrease the band gap to approximately zero at the PW91 level [134].

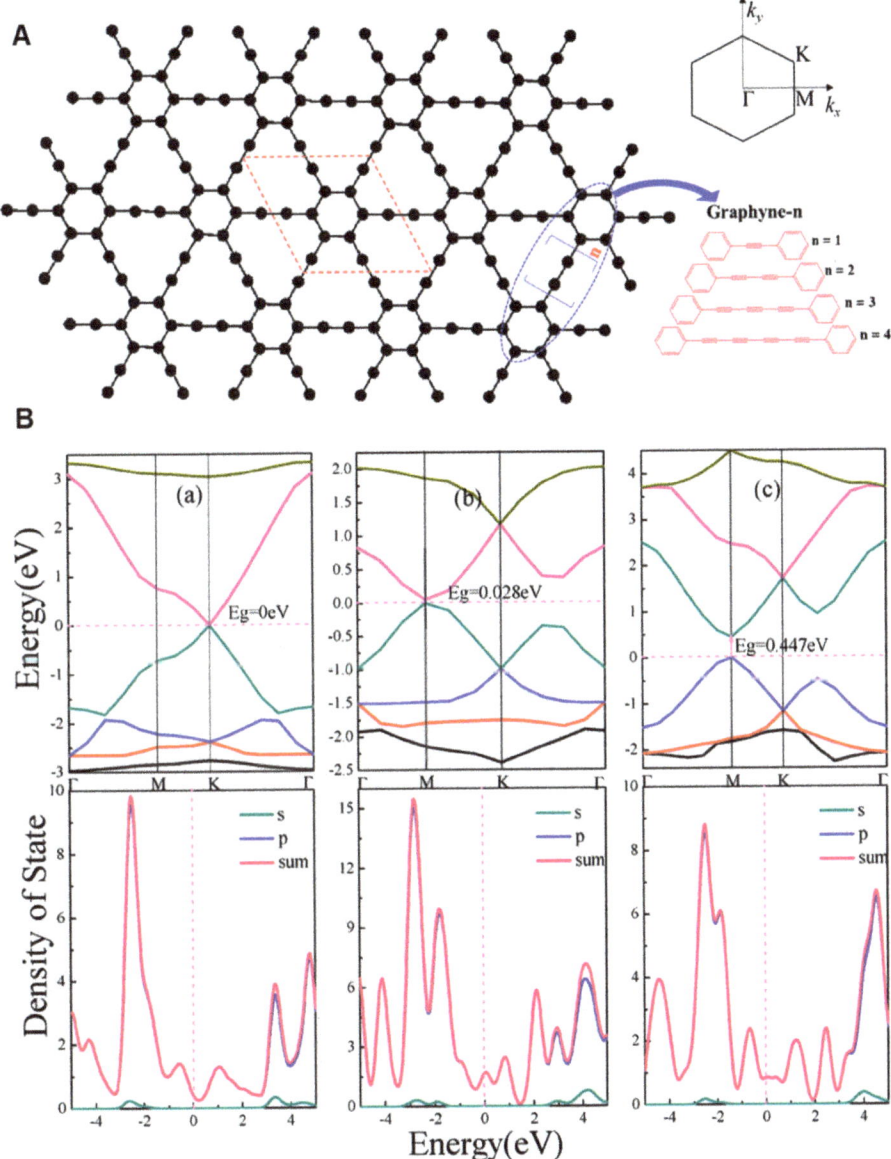

Figure 12. Illustration of graphyne-n structures. The red parallelogram shows the unit cell of GYs. The Brillouin zone is shown in the upper-right, and various acetylenic linkers are shown in the bottom-right (**A**); (**B**) band gaps (Eg) and density of states (DOS) of α-graphyne (**a**), β-graphyne (**b**), and γ-graphyne (**c**). s and p describe the partial densities of states of the s and p orbitals in carbon atoms, respectively. The sum of both elements gives a value of the total density of states. Reproduced with permission [123]. Copyright 2018, MDPI.

Table 2. Calculated band gaps of GYs based on different summarized methods.

Name/Percentage of Acetylenic Linkages %	Band Gap/Method	[Ref.]	Name	Band Gap/Method	[Ref.]
γ-GY	0.447/PBE	[122]	γ-GDY	0.5/PBE	[130]
	0.448/PBE-D	[122]		0.44/LDA	[131]
	0.45/PBE	[124]		1.10/GW	[131]
	1.2/MNDO	[33]		0.53/FP-LCAO	[121]
	0.52/FP-LCAO	[121]		1.22/HSE06	[132]
	0.47/PBE	[125]		0.46/PBE	[133]
	2.23/B3LYP	[125]		0.52/PBE	[125]
	0.46/PBE	[126]		1.18/B3LYP	[125]
	0.46/PBE	[127]		0.89/HSE06	[127]
	0.96/HSE06	[128]		0.47/PW91	[134]
	0.94 HSE06	[127]		0.9/HSE06	[134]
	0.474	[128]		1.21/HSE06	[106]
	0.454/PBE	[129]		0.485/PBE	[129]
β-GY	0.028/PBE	[122]	α-GY	0/PBE	[122]
	0.04/PBE-D	[122]		0.005/PBE-D	[122]
6,6,12-GY	0/PBE	[135] [a]	Graphyne-3	0.6 at M/FP-LCAO	[121] [b]
				0.56/PBE	[127]
				0.566/PBE	[129]
Graphyne-4	0.59/FP-LCAO	[121] [c]	Bulk-GY	0–0.5/FP-LSDA	[136] [d]
	0.54/PBE	[127]			
	0.542/PBE	[129]			
Bulk-GDY	0.05–0.74/HSE06	[137] [d]	Trilayer GDYs	0.18–0.33/PW91	[138] [d]
				0.9/HSE06	[106]
Bilayer GDYs	0.14–0.35/PW91	[138] [e]	(2,0)-AGDYNT (6.42 Å)	0.95/PBE	[139]
	0.99/HSE06	[106]			
(2,2)-ZGDYNT (10.25 Å)	0.65/PBE	[139]	(3,0)-AGDYNT (9.08 Å)	0.65/PBE	[139]
(3,3)-ZGDYNT (15.56 Å)	0.55/PBE	[139]	(4,0)-AGDYNT (12.04 Å)	0.55/PBE	[139]
(4,4)-ZGDYNT (20.81 Å)	0.5/PBE	[139]	AGYNRs (10–45 Å)	1.25–0.59/LDA	[140]
AGDYNRs (12–62 Å)	0.97–0.54/LDA	[140]	ZGYNRs (14–38 Å)	1.65–0.73/LDA	[140]
ZGYNRs (12–30 Å)	1.32–0.75/LDA	[140]	AGNRs (20 Å)	0.5/LDA	[141]
AGYNRs (20 Å)	0.8/LDA	[140]	ZGDYNRs (19.2–28.6 Å)	1.205–0.895/PBE	[133]
AGDYNRs (12.5–20.7 Å)	0.954–0.817/PBE	[133]	GY (−2 to +10% A-strain) 0.87–1.47	0.4–0.17/PBE 0.87–0.56/HSE06	[127]
GY (−2 to +10% H-strain)	0.4–0.88/PBE 0.87–1.47/HSE06	[127] [127]	GDY (−2 to +10% H-strain)	0.41–0.94/PBE 0.8–1.53/HSE06	[127] [127]
GY (−2 to +10% Z-strain)	0.37–0.3/PBE 0.83–0.71/HSE06	[127] [127]	GDY (−2 to +10% Z-strain)	0.39–0.21/PBE 0.78–0.56/HSE06	[127] [127]
GDY (−2 to +10% A-strain)	0.39–0.31/PBE 0.78–0.69/HSE06	[127] [127]	AGDYNRs (various n of nanoribbon)	0.04–0.69/VASP	[142]
α-GYs (0–10% H-strain)	0/PBE	[123]	AGDYNRs (various n of nanoribbon)	0.01–0.69/OpenMX	[142]
β-GYs (0–10% H-strain)	0.028–1.469/PBE	[123]	ZGDYNRs (28.6 Å)	0.895/PBE	[133]
γ-GYs (0–10% H-strain)	0.447–0.865/PBE	[123]	Graphene/0	0	[143]

[a] Lattice constant of a = 9.44 Å and b = 6.90 Å; [b] lattice constants of a = 12.02 Å; [c] lattice constants of a = 14.6 Å; [d] AAA configuration presents metallic band structure; [e] AA stacked structure shows also metallic behavior.

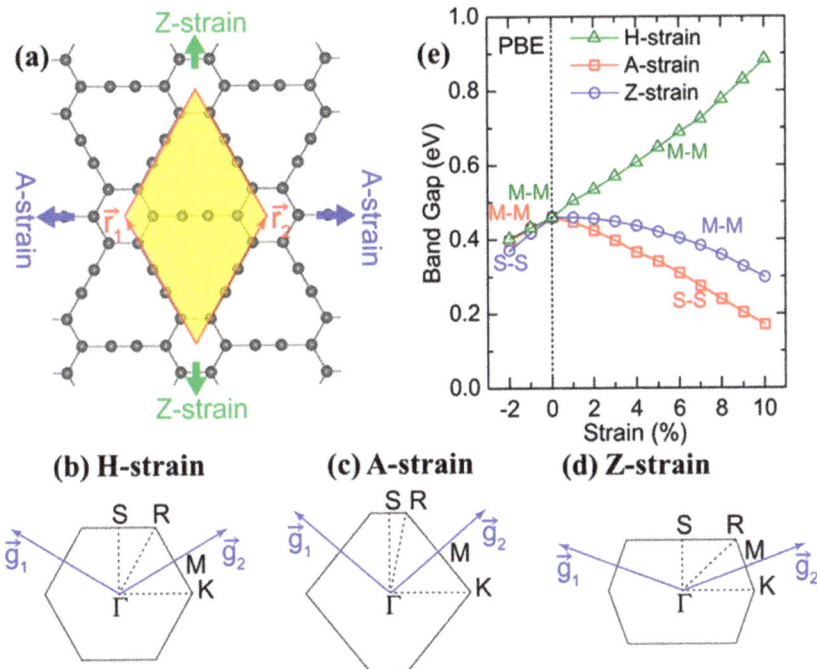

Figure 13. The structure of GY film. The yellow parallelogram indicates the unit cell (**a**). Brillouin zone with high-symmetry points marked beneath H-strain (**b**), A-strain (**c**), and Z-strain (**d**). The band gap shift under different applied strains determined from GGA-PBE (**e**). Reproduced with permission [127]. Copyright 2013, American Chemical Society.

3.2.2. The Electronic Structure of GDYs

Heteroatom doping is an effective method to alter the band structure of graphyne-like structures. The effect of the functionalization of sp-carbon atoms in γ-GDY on band structure was thoroughly investigated by Koo et al. in 2014 [130]. The hydrogenation of acetylenic linkers increased the band gap from 0.49 to a maximum of 5.11 eV, while fluorination increased the band gap up to 4.5 eV. Figure 14A–C illustrate different configurations for monolayered N-graphdiyne holding different numbers of C and N atoms varying from 24 to 42 entitled $C_{18}N_6$, $C_{24}N_4$, and $C_{36}N_6$, respectively, as reported by Singh in 2019 [144]. A hexagonal lattice was observed for $C_{18}N_6$ (a = 16.04 Å) and $C_{36}N_6$ (a = 18.66 Å), while $C_{24}N_4$ presented a rectangular unit cell (a = 15.97 Å; b = 9.67 Å). All considered crystals displayed semiconducting performance, with band gaps of 2.20 eV ($C_{18}N_6$), 0.50 eV ($C_{24}N_4$), and 1.10 eV ($C_{36}N_6$) [144].

The high-temperature treatment of carbonaceous materials is a conventional method of heteroatom doping. Until now, nitrogen-, sulfur-, and phosphorous-doped GDY derivatives have been prepared by this technique [65–69,145,146]. Chen et al. constructed X-doped graphdiynes by replacing a carbon atom (named C1, C2, and C3) (Figure 14Da) in GDY with heteroatom X, where X = B, N, P, and S. As a consequence, five models for N- (N1, N2, N3, pdN, and NH_2), three models for each B- (B1, B2, and B3), P- (as P1, P2, and P3), and S- (S1, S2, and S3) doped GDY were constructed. The pdN and NH_2 models represent the GDY structures doped with pyridinic nitrogen (Figure 14Db) and amino-derived functionalities (Figure 14Dc), respectively. The carbon atoms were divided into three classes that were labeled C1, C2, and C3 according to the arrangement of the structure. Additionally, nine adsorption sites, as indicated by red dots, were investigated in discussed studies.

The doping of pristine GDY with diverse elements (B, N, P, or S) has been found to result in a decrease in the band gap for all analyzed models when using spin-polarized DFT computations and PBE functionals [146]. When the concentration of the dopant was about 1.4 at%, N- and P-doped GDYs were found to be metallic, while doping with boron and sulfur was found to reduce the band gap of starting GDYs from 0.46 to 0.16 and 0.28 eV, respectively. The further decrease in the band gaps was induced by an increase in dopant concentration. B-doped GDYs were found to become metallic after adding 5.6 at% of boron, whereas the band gap of S-doped GDYs decreased to 0.09 eV (Model S3). The planar structure of GDY after doping with B or N atoms was found to be preserved. Due to larger atomic radii of P and S atoms, the out-of-plane distortion of planarity, except for P3 and S3 models, was observed. In terms of the cohesive energies, N, P, and S elements were found to prefer the X2 position, while boron favored the X1 position.

In 2019, Yang et al. theoretically studied the changes in electronic properties caused by the adsorption of H_2 and O_2 atoms on GDY [147]. As Figure 14E shows, there were nine possible sites for their adsorption: three top sites (T1, T2, and T3), two hollow sites (H1 and H2), and four bridge sites. It was shown that the most stable adsorption positions are those located at T2 and H2. The adsorption of atoms on the GDY surface is related to the generation of distinguished high adsorption energies (GDY/H = 3.73 eV; GDY/O = 7.53). The latter meant that strong chemical interactions occurred between the adsorbed atoms and the surface. When H atoms were adsorbed, an inadequate 0.1 eV reduction in the band gap was observed. On the contrary, the adsorption of oxygen led to a higher band gap (an increase of 0.2 eV). Furthermore, the introduction of oxygen species to the structure resulted in weak magnetism because of the broken spin degeneration. With the advancement of research with single-layer GYs, researchers have turned their attention to other geometric shapes such as one-dimensional graphyne family members (e.g., nanoribbons and nanotubes) and three-dimensional ones represented by few-layer systems with different stacking arrangements.

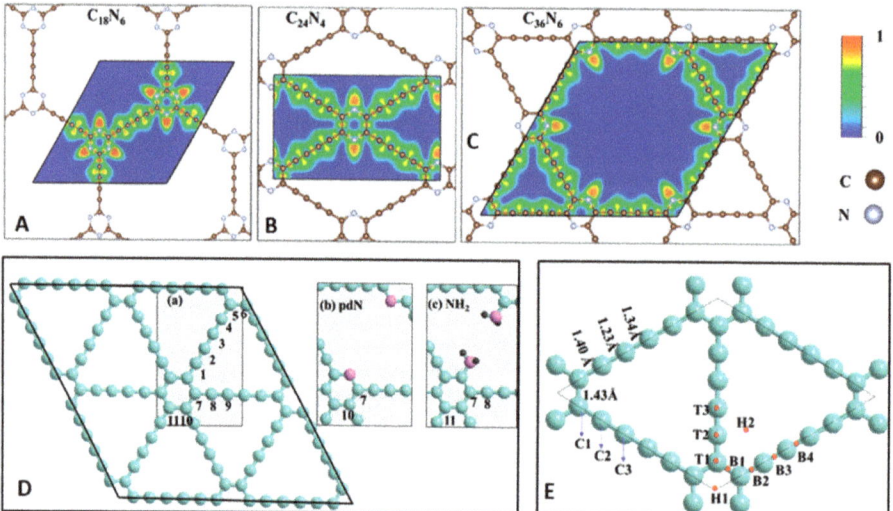

Figure 14. N-graphdiyne nanosheets unit cells of $C_{18}N_6$ (**A**), $C_{24}N_4$ (**B**), and $C_{36}N_6$ (**C**). Reproduced with permission [144]. Copyright 2018, American Chemical Society. (**D**) The available position to be occupied by dopants (Da), pyridinic N-doped GDY (pdN) (Db), and amino-group-doped GDY (NH_2) (Dc). (**E**) Optimized H_2 and O_2 adsorption sites on pristine γ-graphdiyne. Redrawn [146,147].

3.2.3. The Electronic Band Structure of GY Nanoribbons

A nanoribbon (NR), as a 1D derivative of infinite GY sheets, is an example of tuning a band gap by changing the geometry of 2D monolayer GYs. Armchair (AGYNRs) and zig-zag (ZGYNRs) graphyne nanoribbons (differing in widths) were obtained by cutting through a graphyne (or graphdiyne) film along the x and y directions terminated to a benzene ring or acetylene group [140,142]. Gao et al. investigated nanoribbons with benzene rings at their edges. Figure 15 shows AGYNR and ZGYNR arrangements, where n is the number of recurring segments. In contrast to AGYNRs, the n number in ZGYNRs can vary by a half-integer (n + 1/2).

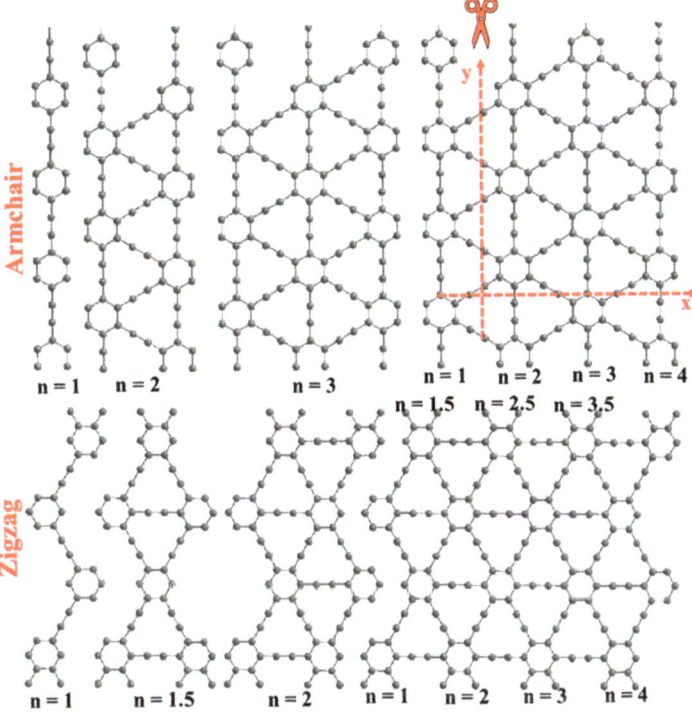

Figure 15. Armchair and zig-zag GY nanoribbons were achieved through the scissoring of the layer along the x and y directions. Redrawn [140].

The band gap of the nanoribbon decreased as the width increased because the ribbon tended to revert to a two-dimensional structure; in such a case, the expected band gap of the ribbon was 0.5 eV. The results confirmed that all considered nanoribbons showed semiconductive features, where band gaps were controlled by the widths and the nature of the edge. With the DFT-LDA method, the calculated energy gaps of AGYNR graphyne-based nanoribbons were in the range of 0.59–1.25 eV, while ZGYNRs showed band gaps ranging from 0.75 to 1.32 eV. In the case of graphdiyne nanoribbons (GDYNRs), the energy gaps were in ranges of 0.54–0.97 eV and 0.73–1.65 eV for armchair (AGDYNRs) and zig-zag (ZGYDNRs) configurations, respectively. The effects of the edge arrangement (AGDYNTs and/or ZGYDNRs) of graphdiyne nanotubes (GDYNTs) on band structure were investigated in detail by Shohany et al. [139]. All GDYNTs under investigation exhibited semiconducting behavior, with a fundamental band gap ranging from 0.65 to 0.5 for AGDYNTs and from 0.95 to 0.55 for ZGDYNTs, as summarized in Table 2.

Kang et al. reported that an intersecting electric field could provoke the giant Stark effect in one-dimensional nanostructures, leading to a diminished or even disappeared band gap [148]. In comparison with other band gap modification approaches, the experimental control of an electric field seems to be a much easier way to play with GYs' band gaps. The effect of an electric field on the band structures of GDYNRs and the band gap decreased as the electric field strength increased due to the strong localization of band-edge states. The band gap decreasing rate was found to be linearly dependent on the ribbon width.

3.2.4. The Electronic Band Structure of Bulky GYs and GDYs

It was shown that 3D graphynes, thanks to the different stack arrangements, could be considered to be both as semiconductors and metals. In 2000, Narita et al. applied the first-principle calculations using a full-potential linear combination of the atomic orbitals method and local-spin-density approximation (LSDA) to optimize the geometry and investigate the electronic properties of one AAA stacking arrangement structure represented as α and three ABA configurations denoted as β1, β2, and β3 of three-dimensional graphyne (Figure 16a) [136].

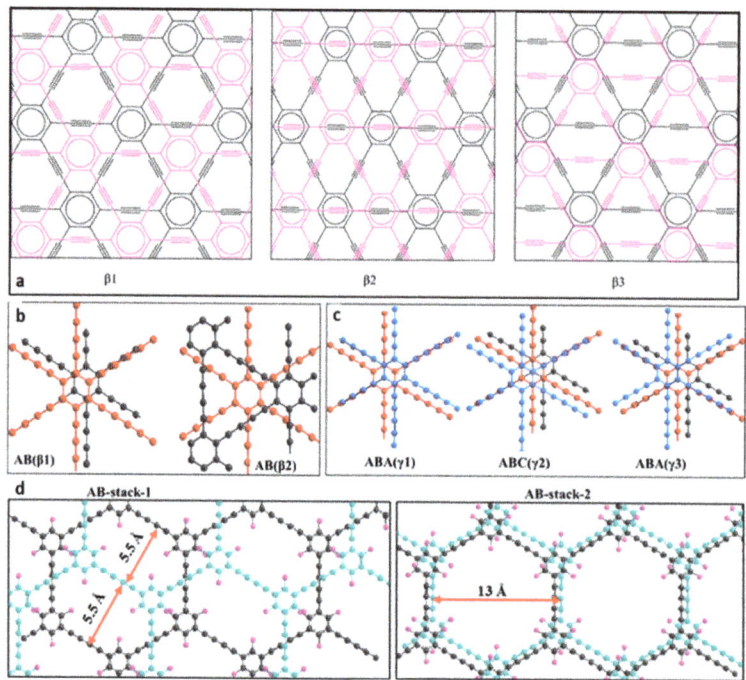

Figure 16. ABA stacking arrangements of 3D graphyne. The A and B sheets are represented by grey and pink structures (**a**). Redrawn [136]. (**b**) Optimized arrangements of double-layer GDY labeled AB(β1) and AB(β2) (red top layer and grey bottom layer, respectively); (**c**) three potential forms of the trilayer GDYs: ABA(γ1), ABC(γ2), and ABC(γ3) arrangements (blue top layer, red middle layer, and grey bottom layer). Redrawn [138]. (**d**) AB-1, AB-2, and AB-3 represent structures of bulk GDY. Red arrows show the directions of the in-plane shift of two sheets in the cell. Redrawn [108].

In Table 3, it was shown that interactions occurring between orbitals 2pπ in α and β3 were greater than those appearing in β1 and β2. As a result, a large split of the π orbitals could be observed, consequently leading to the overlapping of conduction and valence bands in α and β3. The optimized plane lattice constant (a) and bond lengths were nearly the same as in 2D graphyne. The calculated interlayer distance (d) of graphyne (~3.3 Å) was longer than that of graphite (3.17 Å). However, the greater core/core repulsion between neighboring sheets in α and β3 resulted in lower binding energies compared with those found for β1 and β2. Furthermore, the obtained values were approximately 90% of the binding energy of graphite (8.867 eV/atom). This may suggest that graphyne is metastable while it is formulated. In contrast, the more stable nanostructures of β1 and β2 exhibited semiconducting properties. The band gap of β1 was equal to 0.19 eV at the M and L points, whereas the energy gap of β2 was 0.50 eV at the L point. Thus, three-dimensional graphyne was assumed to have layered β1 and β2 organizations and semiconductive features with appropriate energy gaps.

Table 3. Optimized lattice constant, interlayer distance, binding energies, and band gap for 2D and 3D graphyne and graphite [a].

	Lattic Constant (Å)	Interlayer Distance (Å)	Binding Energy (eV/atom)	Band Gap	[Ref.]
α	6.86	3.51	7.948	0	[136]
β1	6.86	3.27	7.963	0.19	[136]
β2	6.86	3.28	7.962	0.5	[136]
β3	6.86	3.34	7.957	0	[136]
2D GY	6.86	-	7.951	0.52	[121]
graphite	2.46	3.17	8.867	0	[136]

[a] Using the projector-augmented wave method and the PBE functional.

In 2012, Zheng et al. conducted PW91 calculations and showed that the most stable bilayer and trilayer GDYs had their hexagonal rings arranged in the AB (direct Eg = 0.35 eV) and ABA (indirect Eg = 0.33 eV) configurations, respectively [138]. The decline in energy gaps, compared with monolayer species (0.46 eV), was caused by the occurrence of interlayer interactions. The application of an external electric field has been shown to be an effective technique to control the electronic and optical properties of few-layer graphyne-like materials. The two most stable arrangements for bilayer GDY (AB(β1) and AB(β2), owning band gaps of 0.35 and 0.14 eV, respectively) are depicted in Figure 16b. The stable trilayer configurations labeled ABA(γ1), ABC(γ3), and ABC(γ2) were found to present band gaps as high as 0.32, 0.33, and 0.18 eV, respectively (Figure 16c). The less stable AA and AAA configurations of GDY exhibited metallic properties.

In 2013, Nagase et al. employed a vdW-optPBE functional to investigate the relationships between the optical and electronic features of bulky GDY nanomaterials and their configurations. The obtained results were next compared with the results obtained using Heyd-Scuseria-Ernzerhof (HSE06) and LDA functionals [137]. They found that the AA configuration presented the lowest structural stability, accompanied by three AB configurations with energy gaps equal to 0.05, 0.74, and 0.35 eV, respectively. The investigations of Leenaerts et al. on the two-layer α-GY revealed that its band structure was qualitatively different from its single-layer derivative and was affected by the stacking modes of the two layers [149]. The AB staking arrangements exhibited a zero-gap feature similar to the AB configuration of bilayer graphene. It was shown again that electronic properties may be controlled by an applied electric field. The fluorinated GDY exhibited direct semiconductive behavior, with band gaps equal to 2.17 (AB stack-1) and 2.30 eV (AB stack-2), which were more than that of pristine GDY (~0.46 eV) (Figure 16d) [108]. Furthermore, the theoretically estimated band gap of the AB stack-1 configuration was consistent with the experimental value. The experimentally determined band gap of randomly fluorinated triazine-based graphyne by XeF_2, reported by Szczurek et al., ranged from 3.12 to 3.34 eV and grew with increasing fluorine concentration [150]. An increase in the band gap upon the fluorination of acetylenic bridges is consistent with the decoupling of benzenic chromophores. Multilayer boron-graphdiyne (B-GDY) was comprehensively studied by Li et al. [113]. The calculated band gap energy of the monolayer showed that B-GDY was a direct band gap semiconductor at the Γ point with a value of 1.2 eV, which corresponded well to the experimentally derived band gap (1.1 eV) of the synthetic compound.

3.3. Electronic Transport

The carrier mobility of different forms of GYs and GDys was theoretically predicted by Chen et al. In their experiments, the Boltzmann transport equation coupled with the deformation potential theory was applied to a-GY, b-GY, 6,6,12-GY, and GDY, and graphene was used as a reference. The obtained results revealed that almost all GYs and GDYs showed charge mobility values lower by one order of magnitude than graphene. 6,6,12-GY was an exception and presented a higher charge mobility (in the a direction) of around 25% for holes and 37% for electrons than the charge mobility found for graphene. Furthermore 6,6,12-GY presented a tremendous anisotropy of charge mobility along the a and b axes. The high carrier mobility of 6,6,12-graphyne might be explained by weaker electron–phonon coupling energy and longer relaxation times. Moreover, the rectangular arrangement of the 6,6,12-graphyne framework might cause the anisotropy of charge mobility in the structure [135].

In 2018, Nasri and Fotoohi theoretically investigated the electronic transport characteristics of a device built on N-doped (right electrode) and boron-doped (left electrode) α-armchair graphyne nanoribbons [151]. Four different devices differing in N and B substitution sites (sp or sp^2) were proposed (Figure 17A). The current–voltage characteristics of the considered systems revealed effective non-linear behavior that led to the generation of a p–n junctions, which, in turn, resulted in rectifying behavior. The rectification properties, however, heavily depended on the deposition site of dopant atoms. The devices with doping atoms substituted on sp^2 sites (sp^2–sp and sp^2–sp^2) showed a rectification ratio of around ten times lower than those having doping atoms attached to sp sites (sp–sp and sp–sp^2) measured in the same bias region (Figure 17B). The rectifying behaviors of the described devices may be associated with asymmetric electrode arrangements. The N and B dopants caused crucial variations in electronic structures of α-graphynes, as they generated new sub-bands in valence (VB) and conductive (CB) bands. Those sub-bands seemed to ensure effective conduction and rectification in the described devices. This finding was supported by measurements conducted on pristine α-graphyne nanoribbons. The device built on the latter showed no rectification effect due to the symmetry of both electrodes. The suitability of graphyne and its h-BN (hexagonal boron nitride) derivatives (h-BNynes) to work as a field-effect transistor (FET) was theoretically investigated by Jhon et al. in 2014 [152]. With the aid of non-equilibrium Green's function combined with the density functional theory (NEGF–DFT) method, they examined the electronic transport of graphene–graphyne–graphene devices by varying graphyne size and carbon chain length.

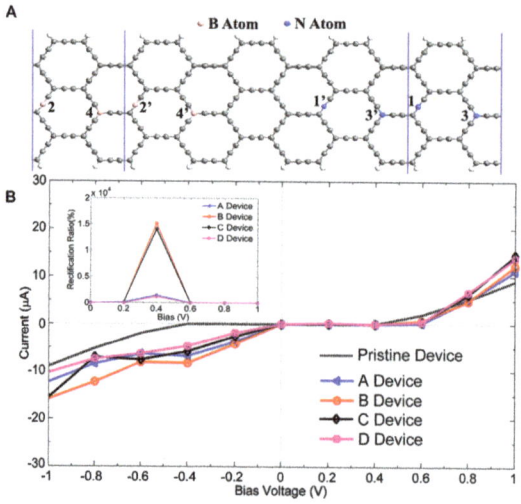

Figure 17. Cont.

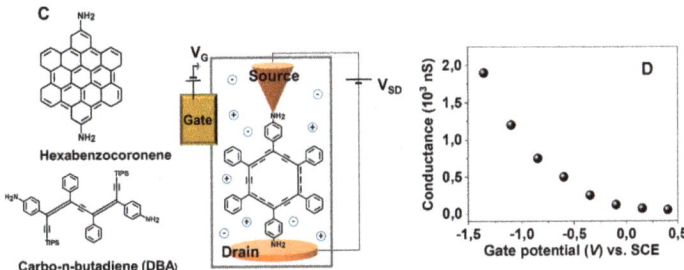

Figure 17. The scheme of devices built on an armchair graphyne nanoribbon (AGyNR) doped with boron and nitrogen atoms (**A**). The I-V curves of doped devices confronted with a pristine system; the inset figure shows the rectification ratio, RR (V), of all doped systems (**B**). Reproduced with permission [151]. Copyright 2018, Elsevier. The schemes of STM devices designed for the investigation SMC of carbobenzene, hexabenzocoronene, and carbo-n-butadiene (DBA). The molecules were attached via NH$_2$ linkers to gold STM electrodes; TIPS: triisopropylsilyl (**C**). The single molecule conductance (SMC) and gate potential (VG) relationship found for the carbo-benzene sample. In this experiment trihexyl tetradecyl-phosphonium-bis(2,4,4-trimethylphenyl)phosphinate) was applied as the electrochemical gating electrolyte and the bias voltage was constant (0.1 V); SCE: saturated calomel electrode (**D**). Redrawn [153].

The concept of such constructed transistors was based on the outstanding electronic mobility of graphene and non-zero band gap graphyne, along with structural/compositional similarities between graphene and graphyne, being robustly connected between graphene electrodes. DFT calculations revealed that both graphyne and h-BNyne-based thin-film transistors (TFTs) showed good on/off ratios on the order of 10^2–10^3. Noteworthily, the size of such transistors might be reduced to below 1 nm while maintaining good switching features. Electronic orbital analysis disclosed that, contrary to h-BNynes, electrons in the conduction and valence orbitals were considerably delocalized in graphyne TFTs. The latter finding may suggest that graphyne TFTs could offer more facilitated electron mobilities than h-BNynes.

Single molecule conductance (SMC) was experimentally and theoretically investigated by Li et al. [153]. The measurements were performed using a scanning tunneling microscope, where the α-graphyne unit was chemically attached to the Au electrode and the STM tip. Amine (NH$_2$) functionalities were used as an anchoring agent due to their ability to create molecular joints with negligible conductance aberrations (Figure 17C). The measurements conducted on α-graphyne and hexabenzocoronene representing the graphene molecular unit revealed that the α-graphyne units showed much higher conductance (106 nS, 1.9 nm) than the shorter units (1.4 nm), thus potentially being more conductive than hexabenzocoronene (14 nS). This unusual behavior originated from the electronic structure of both compounds. It turned out that the α-graphyne units had smaller HOMO-LUMO gaps than the graphene molecular units, so the transmission through the α-graphyne core was higher. The high molecular conductance of α-graphyne molecules also came from their rigid and planar structure. The measurements revealed that carbobutadiene wires showed a single molecular conductance that was 40 times smaller than that found for α-graphyne. The reason for this feature lies in the flexibility of the n-p-conjugated DBA framework, capable of adopting diverse geometries. The transport properties of different conformations of DBA molecules carried out with the NEGF-DFT model were strongly influenced by the measure of the rotation angle around the –C=C=C– sites. The increase in twist angle resulted in a higher HOMO-LUMO gap and thus a lower transmission between them. Finally, it was shown that the α-graphyne-based device showed excellent gating properties, understood as an increment of the SMC with increasing negativity of the gating potential (Figure 17D). Furthermore, the on/off ratio found for α-graphyne had an order of magnitude of ~15.

3.4. Optoelectronic Properties

The optical features of N- and B-doped graphynes were theoretically studied by Bhattacharya and Sarkar [154]. The investigated structures presented, similar to pristine graphyne, optical anisotropy independent to the direction of applied electric field (Figure 17A, first column). The authors found that below an energy of 0.4 eV, the optical response was governed by the intra-band shift coming from free charges (Figure 18A, second column). The analysis of the static dielectric tensors revealed that doped graphynes showed a better electric conductivity and higher mobility of charges than pristine GY, creating the opportunities for their application in optoelectronics. The spin-polarized optoelectronic properties of α-graphyne were theoretically investigated by Yang et al. using NEGF–DFT [155]. They showed that photocurrents were generated by irradiating investigated devices with different light wavelengths (from UV to IR), and the polarization of the formed photocurrents strictly depended on the type of contact applied. Both M1 and M2 devices (Figure 18B) produced spin-down photocurrents, whereas the M2 device could also generate spin-up ones (Figure 18C). It was revealed that photocurrents might also be guided by reversing the electrodes' magnetization. Generally speaking, two spatially separated spin photocurrents appeared on the antiparallel polarized electrodes. Functionalization is another approach to manage the optoelectronic characteristics of GY and GDY nanostructures. Theoptoelectronic application of Gdy:ZnO nanocomposites was experimentally investigated by Jin et al. [156] Their experiment involved measuring I–V characteristics in the dark and under the influence of UV radiation. When the samples were illuminated, an increase in current due to light absorption was observed. Chronoamperometry was employed to record the rise/decay time of the current measured without and with light irradiation. The obtained results revealed that the current changes strongly depended on the photochemical character of investigated devices. The junction created between the GDY and ZnO nanoparticles strongly enhanced the charge transfer between these components and, consequently, the photoresponse. The GDY:ZnO/ZnO system manifested an excellent optical response of 1260 A W^{-1}, with a rise/decay time as short as 6.1/2.1 s. The device built only on zinc oxide nanoparticles held a responsivity of 174 A W^{-1} and a much more extended rise/decay time (32.1/28.7 s). Li et al. proposed deep UV photodetectors (Figure 18D) based on TiO2:GDY nanocomposites [157]. The published results proved that the considered composites were suitable as UV photodetectors, disclosing an outstanding optical response of 76 mAW^{-1} and a rise/decay time of 3.5/2.7 s. The cited examples show the promising and efficient properties of graphyne derivatives. Their high charge mobility, on/off ratio, and photoresponse makes them prospective materials for constructing nanoscale electronics and optoelectronics.

It is noteworthy that metal oxide dopants in GDY nanocomposites, the optoelectronic amplifiers, do not seem to be mandatory additives. Zhang et al. constructed metal-free, flexible photodetectors built on GDY:PET composites. Those built devices were characterized by excellent mechanical, electronic, and optoelectronic properties. The experimentally investigated responsivity and photocurrent reached outstanding values of 1086.96 µA W^{-1} and 5.98 µA cm^{-2}, respectively. Moreover, the considered composites showed good photoresponses, even after undergoing 1000 bending and twisting cycles. The recorded loss in photocurrent was 25.6% (bending force) and 35% (twisting force) [158].

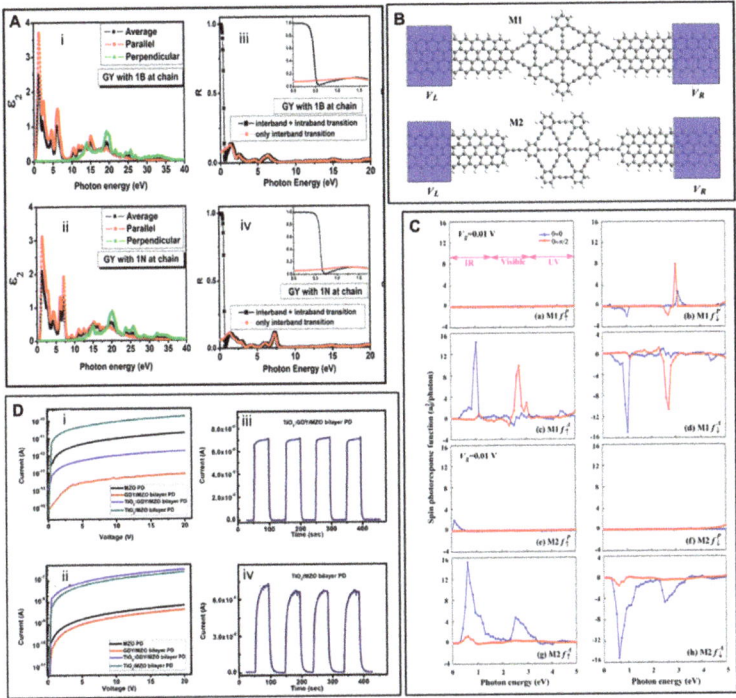

Figure 18. The 1B (**Ai**) and 1N (**Aii**) doped GYs' relationships between the imaginary part of the dielectric function and photon energies taken for different directions of electric fields. The reflectivities of inter- and intraband transitions found for 1B (**Aiii**) and 1N (**Aiv**) doped GYs. Reproduced with permission [154]. Copyright 2016, American Chemical Society. Scheme of the designed devices (**B**). The parallel and antiparallel spin-polarized photoresponse of M1 (**Ca–d**) and M2 (**Ce-h**) devices. Reproduced with permission [155]. Copyright 2017, IOP Publishing Ltd. I–V sweeps of the considered photodetectors working without (**Di**) and (**Dii**) with UV light (254 nm) irradiation. The on/off features of the TiO2:GDY/MZO bilayer PD (**Diii**) and the TiO2/MZO bilayer PD (**Div**). Reproduced with permission [157]. Copyright 2020, Elsevier.

4. Magnetism of Pure and Doped Graphyne-Like Materials
4.1. Theoretically Investigated Magnetic Properties

Local defects and substituents in carbonaceous materials might provoke superconducting or ferromagnetic features that can even appear at ambient temperature. Until now, magnetism in carbon has been experimentally revealed for (i) interacting radicals, (ii) carbons with a mixed hybridization ($sp^2 + sp^3$), (iii) amorphous carbonaceous materials doped with trivalent atoms (P, N, or B), (iv) diverse nanostructures (graphite, diamond, and foams), and (v) fullerenes [159–162]. Apart from that, vicinity to metallic ferromagnets or the contamination of transition metals (Fe, Cr, or V) has been shown to generate spin polarization in carbon structures [159,160,162]. Localized magnetism and the zig-zag rib of graphene nanoribbons or neighboring vacancies and dangling bonds in pure carbon structure are representative of this concept. The theoretical considerations and fruitful synthesis of monolayer graphyne have led to considerable interest in research on the magnetic properties of pure or/and doped graphyne structures.

4.1.1. Metal-Doped GYs

The magnetic features found in materials with electrons occupying only s and p orbitals, rather than the traditional d or f ones, might be remarkably appealing to spintronic applications. However, the source of magnetism in pure carbonaceous materials is not fully understood. In this section, we report works addressing the magnetic properties of GYs. The effect of dopant distribution and functional groups on adjusting the electronic or magnetic properties of GYs might lead to new promising electronic, optoelectronic, and spintronic devices. The calculations revealed that unmodified γ-graphyne is a nonmagnetic semiconductor [128], while the adsorption of transition metal (TM) atoms might drastically change the electronic structure of and add ferromagnetic features to GY nanostructures [163,164]. In 2012, He et al. theoretically investigated (DFT + U) the electronic structure and magnetism of GDY and GY doped with single $3d$ transition metals (V, Cr, Mn, Fe, Co, and Ni) [164] The adsorption of metal atoms on the GDY and GY surfaces generated charge transfer between metallic adatoms and polymeric sheets. Moreover, the adsorption might generate electron redistribution in the s, p, and d orbitals of transition metal atoms. Except for vanadium, the mentioned factors caused decreases in the magnetic moments of adsorbed metals (Cr, Mn, Fe, and Co). Additionally, they were ranked as follows: Cr > Mn > V > Fe > Co for TM-GDY and Cr > V > Mn > Fe > Co for TM-GY; see Figure 19A. The energy of spin polarization (ΔE_{spin}), taken as the difference between the nonmagnetic and magnetic states, was higher than 1.1 eV, suggesting the substantial stability of the spin-polarized states of transition for metal-doped GDY and GY nanostructures.

In 2017, Lee et al. applied DFT calculations to investigate the doping efficiency of the transition metals of $3d$, $4d$, and $5d$ groups deposited on a γ-GY surface [165]. For this purpose, different high-symmetry adsorption sites such as top (T), bridge (B), and hollow (H) were chosen, as depicted in Figure 19B. It was shown that adatoms typically occupy the H1 sites over ethynyl rings of γ-graphyne (taking Fe (μB: ~2.08) as an example). The lower atomic radii of Co (μB: ~1), Ni (μB: ~0), and Cu (μB: ~0), as well as Re (μB: ~1), caused those dopant atoms to be preferably placed at the H3 sites of the acetylenic rings. Magnetic moments of metals of the $4f$ group were found to be higher in comparison with $3d$ transition metals, and lanthanides with sufficiently large atomic radii are good candidates to be introduced in GY rings. Ren et al. used comprehensive first-principle calculations to study the magnetic properties of β-GY doped with different rare-earth (RE) atoms (La, Ce, Pr, Nd, Pm, Sm, and Eu) [166]. The β-GY was found to undergo a transition from semiconductor to metal. The introduction of external atoms such as neodymium, promethium, samarium, and europium (local magnetic moments in the range of 4.1–7.3) were able to translate into higher values of magnetic moments for metal-graphyne complexes (>4.1 μB). As expected, the carbon atoms neighboring dopant atoms were found to have a modest contribution to generated magnetic moments. In 2014, Alaei et al. studied two zig-zag graphyne nanotubes (ZGYNTs) and two armchair graphyne nanotubes (AGYNTs) doped with iron, cobalt, and nickel [167]. It was shown that a 12-membered ring (12-C), a hollow site surrounded by acetylenic linkers in GY, was the most preferable (the most stable) site for the deposition of those metals. The adatoms nested in the plane of the ethynyl rings and formed bonds with adjacent carbons. Complexes of Fe (μB: ~2.06) and Co (μB: ~1) with different GYNTs were magnetic and showed many features typical of metals, semimetals, half semimetals, and half-semiconductors [168]. Ni complexes (μB: ~0.01) were found to be nonmagnetic semiconductors exhibiting energy gaps narrower than those found for starting nanotubes (Figure 19A).

4.1.2. Non-Metal Doped GYs and GDYs

The electronic properties and magnetism of GYs can be also driven by applying diverse non-metal doping agents. In 2014, Drogar et al. stated that widening the band gap (~2 eV) and provoking a magnetic moment (~1 μB) of α-GY could be realized via the simple hydrogenation reaction of the latter due to the cleavage of π-bonds and the creation of unpaired electrons [169]. Subsequently, in 2018, Wang et al. studied the tuning of mono- and bilayer GY features after the hydrogenation of different carbon atoms in GY [170]. Unlike the distribution of μB in TM-GY, in which the magnetism derives from the d- or f-electrons, the unpaired 2p electrons essentially provide the magnetic moments at the non-hydrogenated carbons of GY. As indicated in Figure 19C, there are two sites (C1 (T) and C2 (B)) on which hydrogenation can occur. Twelve hydrogen atoms are located at Ti sites in the aromatic ring and the B1 sites in acetylenic linkers of GY nanosheets. It was observed that the magnetic moment of the monolayer reached a maximal value (1.59 µB) for three hydrogen atoms, whereas the magnetic moments progressively declined to 0 µB when the number of hydrogen atoms varied from 4 to 6. These results suggest that the hydrogenation of half C atoms (sp- or/and sp^2-hybridized) may result in a maximal magnetic moment, similarly to what was found for graphene [171]. Theoretical calculations of the total magnetic moments of bilayers with different stacking arrangements yielded values larger than 1.52 µB. DFT is the most frequently utilized computation method to investigate the electronic properties and magnetism of single-layer graphdiyne (GDY) doped with non-metallic elements, such as boron, nitrogen, oxygen, phosphorous, or sulfur [172]. The position of dopant atoms has a profound impact on the relevant characteristics of GDYs. Three possible places for doping are aromatic carbons (X-b) and two carbons in ethynyl linkers (X-1 and X-2), depicted in Figure 19D. Considering the cohesion energies found for the studied structures, one can conclude that boron and sulfur prefer to exchange the sp^2 aromatic carbons. In turn, the N, O, and P atoms favor substituting the carbons in the ethynyl bridges. It was shown that nitrogen atoms deposited on both GDYs and GY monolayers did not modify their magnetic properties, and all of them remained nonmagnetic [154]. In contrast, GDY structures doped with B, O, P, or S (structures X-1 and X-2) deposited at acetylenic chains offered spin polarization and were magnetic (µB varying from 0.31 to 1.26 depending on the type of dopant and applied functionals). The deposition of these elements at the aromatic site did not alter magnetic features, and doped GDYs were nonmagnetic.

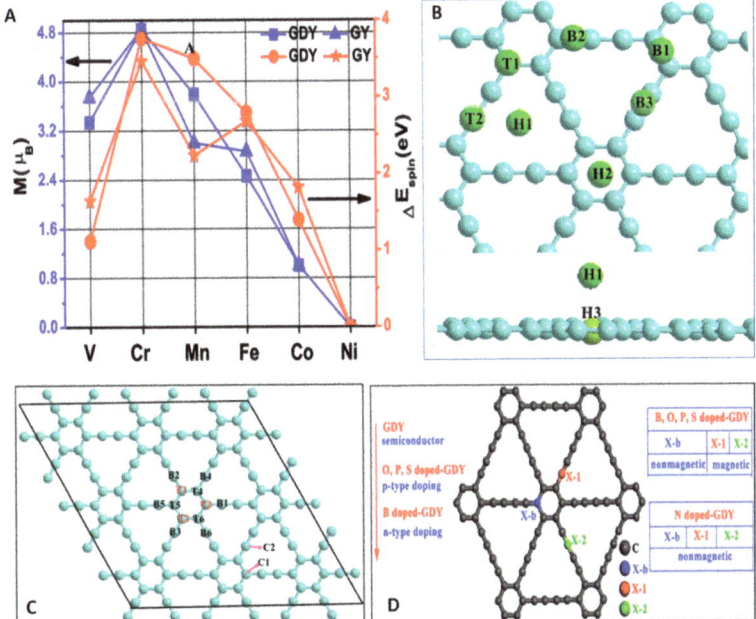

Figure 19. The magnetic moment (µB) and ΔE_{spin} (**A**). Reproduced with permission [164] Copyright 2012, American Chemical Society. The adsorption sites (green spheres) of transition metal atoms on a γ-graphyne film seen from the top (upper image); the in-plane case of H3 and its relation to the H1 site seen from a side (bottom image), blue points represent carbons (**B**). Redrawn [165]. Possible sites on GY for hydrogenation (**C**). Redrawn based on [171]. Possible site for doping by X = B, N, O, P, and S atoms (**D**). Reproduced with permission [172]. Copyright 2020, Elsevier.

4.2. Experimentally Investigated Magnetic Properties

Following comprehensive theoretical studies on the magnetic properties of modified GYs, efforts have been directed to experimentally explore and measure magnetism in GYs [173]. In 2017, Huang et al. investigated the impact of N doping on the paramagnetic properties of GDYs and demonstrated the crucial role of nitrogen atoms deposited on the benzene ring in building the local magnetic moment [174]. Based on the M-H curves (Figure 20A,B) measured for GDY and N-GDY at 2 K, the authors showed that N-GDY and GDY presented low magnetization values of 0.96 and 0.51 emu/g, respectively. The magnetization obtained for N-GDY containing 5.29% of N atoms was on the same level as the value found for fluorinated RGO [175]. These results contradicted recent theoretical findings, which claimed that N in a chain or ring is non-paramagnetic. As such, more research is required to find the origination of paramagnetism [154,172].

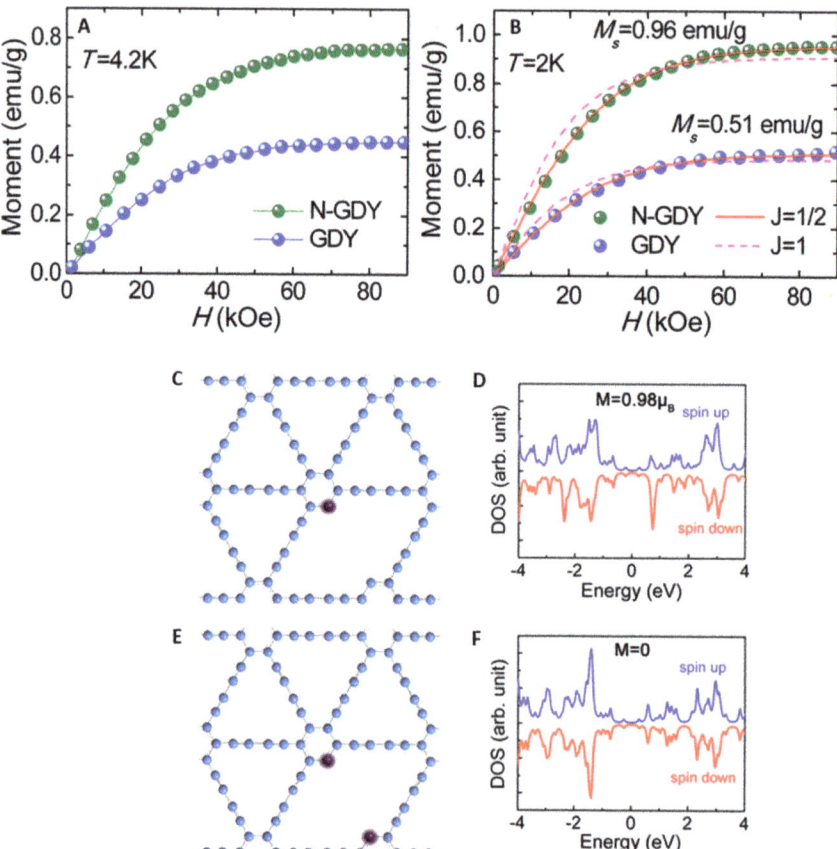

Figure 20. M-H curves obtained at 4.2 K (**A**) and 2 K (**B**) for GDY and N-GDY. Obtained results were fitted to the Brillouin function with J = 1/2 (solid line) and J = 1 (dashed line) (**B**). Spin-resolved DOS of N-GDY films with nitrogen atoms deposited on the benzene ring in GDY. The upper panel (**C,D**) shows the paramagnetic S = $\frac{1}{2}$ system, while the bottom one the antiferromagnetic one (**E,F**). Reproduced with permission [174]. Copyright 2017, Springer Nature.

The spin-polarized calculations proved that a distinct local magnetic moment (μ_B = 0.98) originates from nanostructures with deposited asymmetric pyridinic nitrogen (Py-1N), seen in Figure 20C,D. The structures bearing symmetric pyridinic nitrogen substitution (Py-2N) or N atoms attached to ethynyl linkers appeared to be nonmagnetic (Figure 20E,F).

Finally, the investigated systems did not show any ordered ferromagnetic or ferrimagnetic properties.

Further research aiming to discover the relationship between doped heteroatoms and magnetization in GY structures continued. In 2017, Zheng et al. [176] reported that, contrary to pyridinic-N, the vacancy, carbonyl, and hydroxyl functionalities of GDY contribute to the magnetic properties of thermally treated GDYs. Furthermore, DFT calculations indicated that the OH groups at the chain of the GDY layer are a considerable source of unpaired electrons and may favor antiferromagnetism in annealed GDY. Moreover, the annealing of GDY at 600 °C (GDY-600) may lead to complex magnetic properties, depending on the applied measurement temperatures. Paramagnetic characteristics could be noticed below 50 K, and a hump seen in the range of 50–200 K demonstrated that both paramagnetic and antiferromagnetic phases coexist in the considered materials.

In 2019, Huang et al. investigated the effect of sulfurization on the induction of ferromagnetic characteristics into GDYs. They found that S-doped GDY presented strong residual magnetization (>0.047 emu/g) at ambient temperatures. The investigated systems were characterized by the transition temperature being close to 460 K [177]. The local magnetic moment and electron interactions occurring between C and S atoms were found to be responsible for the appearance of the ordered internal ferromagnetism in the investigated materials. The ferromagnetic behavior was also confirmed by magnetic hysteresis (M-H) loop measurements with different temperatures. In order to investigate the influence of sulfur doping on magnetic properties of GDY, temperature-dependent magnetic susceptibility χ-T curves (the applied magnetic field H = 500 Oe) and magnetization M-H curves for GDY350 and S-GDY by VSM were measured (Figure 21a–d) [177]. These findings are very promising in terms of S-doped GDY suitability in magnetic storage devices.

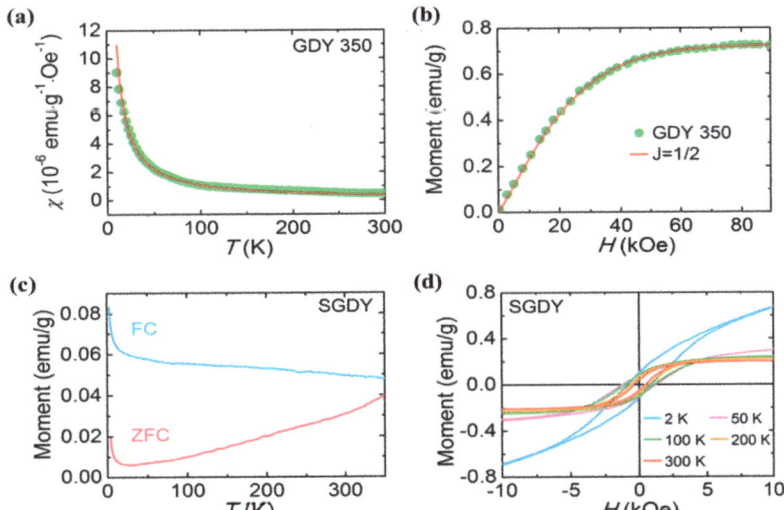

Figure 21. The χ-T relationship examined in a temperature range of 2–300 K (**a**) and an M-H curve taken at 2 K for GDY 350 (**b**). Temperature–magnetization relationship, where FC is field cooled mode and ZFC–zero field cooled mode (**c**) and hysteresis loops recorded for S-doped GDY (**d**). Reproduced with permission [177]. Copyright 2019, American Chemical Society.

5. Mechanical Properties of Graphynes

Graphene with an intrinsic tensile strength of 130 GPa, and Young's modulus equal to 1TPa is recognized as the strongest material ever tested [178]. Zhang et al. theoretically investigated mechanical properties of different forms of GYs (α-, β-, γ-, and 6,6,12-GYs), as well as graphene used as a reference [179]. The tensile stress (~125 GPa) and Young's modulus (0.99 TPa) obtained by molecular dynamic calculations confirmed the extraordinary mechanical resistance of graphene. In contrast, all considered graphynes presented 50–70% lower mechanical strength than graphene, whereas the decline of tensile stress and Young's modulus of graphynes was strictly related to an increasing amount of acetylenic linkage. A similar dependence of elastic properties on the proportion of acetylenic bridges in α-, β-, γ-graphynes was also observed by Hou et al. in 2014 [123]. Noteworthy, a deviation from the hexagonal structure of graphene and graphynes, as it happens for 6,6,12-graphyne, resulted in the appearance of anisotropy of mechanical properties measured along with x (zig-zag) and y (armchair) directions [179,180]. The anisotropy in mechanical and electronic properties makes 6,6,12-graphyne useful in diverse potential applications.5.

6. Conclusions and Outlook

Synthetic methods for the preparation of graphyne-like structures doped with heteroatoms and controllable size and dimensions were collected and summarized. Doping (i.e., the replacement of carbon atoms or covalent bond formation with foreign atoms) methods such as heating and annealing techniques have crucial drawbacks in the preparation of carbon materials; uncertainty in the values and position of heteroatoms, along with the destruction of the intrinsic properties of pristine materials, are inevitable disadvantages. To avoid the abovementioned flaws, engineering the surface of GYs to include desired heteroatom is necessary. Different methodologies have been successfully applied with high reproducibility to prepare a wide range of GYs with ordered structures in which heteroatom occupy predictable positions on the surface. The adjustment of morphology (such as 1D, 2D, and 3D) and composition (doping with N, P, F, S, Cl, H, O, B, etc.) are influential approaches to modulate the band structures and (subsequently) electric, optical, and magnetic properties of graphyne-like structures. The size- and composition-dependent band structure, electronic transport, spin-polarized optoelectronic properties, or photosensitivity, and appearing magnetic moment of these allotropes of carbon material have been investigated. There is an understanding that research into the magnetic properties of carbon materials is still in its infancy stage, and there is a substantial gap between calculations and experimental findings. Though this topic needs many more firm, particularly experimental discoveries, it is already clear that this type of material could be exploited in spintronics technology. Benefiting from the high specific capacity, remarkable cycle performance, and inflated adsorption capacity of ions and gases, GYs could also have a promising future for applications in electrochemical energy storage such as batteries, capacitors, and hydrogen fuel cells. Theoretical research has proved GY's usefulness for not only understanding but also predicting the structures and properties of new modified systems as well as for preselecting those that most merit experimental study.

Author Contributions: G.A. writing—original draft preparation; A.A. writing—original draft; W.G. writing—review and editing, and supervision; A.S. writing—review and editing, project administration, and funding acquisition. All authors have read and agreed to the published version of the manuscript.

Funding: This research was funded by the National Science Centre (NCN), Sonata Bis project (2016/22/E/ST5/00529).

Acknowledgments: *In memoriam* to Andrzej Maria Huczko (1949–2021).

Conflicts of Interest: The authors declare no conflict of interest.

References

1. Szczurek, A.; Amaral-Labat, G.; Fierro, V.; Pizzi, A.; Masson, E.; Celzard, A. The use of tannin to prepare carbon gels. Part I: Carbon aerogels. *Carbon* **2011**, *49*, 2773–2784. [CrossRef]
2. Szczurek, A.; Fierro, V.; Medjahdi, G.; Celzard, A. Carbon aerogels prepared by autocondensation of flavonoid tannin. *Carbon Resour. Convers.* **2019**, *2*, 72–84. [CrossRef]
3. Szczurek, A.; Fierro, V.; Pizzi, A.; Celzard, A. Mayonnaise, whipped cream and meringue, a new carbon cuisine. *Carbon* **2013**, *58*, 245–248. [CrossRef]

4. Celzard, A.; Fierro, V. "Green", innovative, versatile and efficient carbon materials from polyphenolic plant extracts. *Carbon* **2020**, *167*, 792–815. [CrossRef]
5. Sharma, S. Glassy Carbon: A Promising Material for Micro- and Nanomanufacturing. *Materials* **2018**, *11*, 1857–1908. [CrossRef] [PubMed]
6. Szczurek, A.; Fierro, V.; Plyushch, A.; Macutkevic, J.; Kuzhir, P.; Celzard, A. Structure and Electromagnetic Properties of Cellular Glassy Carbon Monoliths with Controlled Cell Size. *Materials* **2018**, *11*, 709. [CrossRef] [PubMed]
7. Chalifoux, W.; Tykwinski, R. Synthesis of polyynes to model the *sp*-carbon allotrope carbyne. *Nat. Chem.* **2010**, *2*, 967–971. [CrossRef]
8. Heidari, A. Computational Study on Molecular Structures of C20, C60, C240, C540, C960, C2160 and C3840 Fullerene Nano Molecules under Synchrotron Radiations Using Fuzzy Logic. *J. Mater. Sci. Eng.* **2016**, *5*, 2169-0022. [CrossRef]
9. Rietmeijer, F.J.M.; Rotundi, A.; Heymann, D. C60 and Giant Fullerenes in Soot Condensed in Vapors with Variable C/H_2 Ratio. *Fuller. Nanotub. Carbon Nanostruct.* **2004**, *12*, 659–668. [CrossRef]
10. Hersam, M.C. Progress towards monodisperse single-walled carbon nanotubes. *Nat. Nanotech.* **2008**, *3*, 387–394. [CrossRef] [PubMed]
11. Jain, A.; Homayoun, A.; Bannister, C.W.; Yum, K. Single-walled carbon nanotubes as near-infrared optical biosensors for life sciences and biomedicine. *Biotechnol. J.* **2015**, *10*, 447–459. [CrossRef] [PubMed]
12. Ahlawat, J.; Asil, S.M.; Barroso, G.G.; Nurunnabi, M.; Narayan, M. Application of carbon nano onions in the biomedical field: Recent advances and challenges. *Biomater. Sci.* **2021**, *9*, 626–644. [CrossRef]
13. Yin, M.; Cohen, M.L. Will diamond transform under megabar pressures? *Phys. Rev. Lett.* **1983**, *50*, 2006. [CrossRef]
14. Yin, M. Si-III (BC-8) crystal phase of Si and C: Structural properties, phase stabilities, and phase transitions. *Phys. Rev. B* **1984**, *30*, 1773. [CrossRef]
15. Yang, X.; Wang, Y.; Xiao, R.; Liu, H.; Bing, Z.; Zhang, Y.; Yao, X. A new two-dimensional semiconducting carbon allotrope with direct band gap: A first-principles prediction. *J. Phys. Condens. Matter* **2020**, *33*, 045502. [CrossRef]
16. Höhne, R.; Esquinazi, P. Can Carbon Be Ferromagnetic? *Adv. Mater.* **2002**, *14*, 753–756. [CrossRef]
17. Maździarz, M.; Mrozek, A.; Kuś, W.; Burczyński, T. Anisotropic-cyclicgraphene: A new two-dimensional semiconducting carbon allotrope. *Materials* **2018**, *11*, 432. [CrossRef]
18. Kaiser, K.; Scriven, L.M.; Schulz, F.; Gawel, P.; Gross, L.; Anderson, H.L. An sp-hybridized molecular carbon allotrope, cyclo [18] carbon. *Science* **2019**, *365*, 1299–1301. [CrossRef]
19. Hoffmann, R.; Kabanov, A.A.; Golov, A.A.; Proserpio, D.M. Homo Citans and Carbon Allotropes: For an Ethics of Citation. *Angew. Chem. Int. Ed.* **2016**, *55*, 10962–10976. [CrossRef]
20. Correa, A.A.; Bonev, S.A.; Galli, G. Carbon under extreme conditions: Phase boundaries and electronic properties from first-principles theory. *Proc. Natl. Acad. Sci. USA* **2006**, *103*, 1204–1208. [CrossRef] [PubMed]
21. Grochala, W. Diamond: Electronic ground state of carbon at temperatures approaching 0 K. *Angew. Chem. Int. Ed.* **2014**, *53*, 3680–3683. [CrossRef] [PubMed]
22. White, M.A.; Kahwaji, S.; Freitas, V.L.S.; Siewert, R.; Weatherby, J.A.; Ribeiro da Silva, M.D.M.C.; Verevkin, S.P.; Johnson, E.R.; Zwanziger, J.W. The Relative Thermodynamic Stability of Diamond and Graphite. *Angew. Chem. Int. Ed.* **2020**, *60*, 1546–1549. [CrossRef]
23. Németh, P.; Garvie, L.A.; Aoki, T.; Dubrovinskaia, N.; Dubrovinsky, L.; Buseck, P.R. Lonsdaleite is faulted and twinned cubic diamond and does not exist as a discrete material. *Nat. Commun.* **2014**, *5*, 1–5. [CrossRef] [PubMed]
24. McCulloch, D.G.; Wong, S.; Shiell, T.B.; Haberl, B.; Cook, B.A.; Huang, X.; Boehler, R.; McKenzie, D.R.; Bradby, J.E. Investigation of Room Temperature Formation of the Ultra-Hard Nanocarbons Diamond and Lonsdaleite. *Small* **2020**, *16*, 2004695. [CrossRef]
25. Tiwari, S.K.; Mishra, R.K.; Ha, S.K.; Huczko, A. Evolution of graphene oxide and graphene: From imagination to industrialization. *ChemNanoMat* **2018**, *4*, 598–620. [CrossRef]
26. Tiwari, S.K.; Kumar, V.; Huczko, A.; Oraon, R.; Adhikari, A.D.; Nayak, G. Magical allotropes of carbon: Prospects and applications. *Crit. Rev. Solid State Mater. Sci.* **2016**, *41*, 257–317. [CrossRef]
27. Abdi, G.; Alizadeh, A.; Khodaei, M.M. Highly carboxyl-decorated graphene oxide sheets as metal-free catalytic system for chemoselective oxidation of sulfides to sulfones. *Mater. Chem. Phys.* **2017**, *201*, 323–330. [CrossRef]
28. Huang, C.S.; Li, Y.J.; Wang, N.; Xue, Y.R.; Zuo, Z.C.; Liu, H.B.; Li, Y.L. Progress in Research into 2D Graphdiyne-Based Materials. *Chem. Rev.* **2018**, *118*, 7744–7803. [CrossRef]
29. Boukhvalov, D. Stable antiferromagnetic graphone. *Phys. E* **2010**, *43*, 199–201. [CrossRef]
30. Lebegue, S.; Klintenberg, M.; Eriksson, O.; Katsnelson, M. Accurate electronic band gap of pure and functionalized graphane from GW calculations. *Phys. Rev. B* **2009**, *79*, 245117. [CrossRef]
31. Feng, L.; Zhang, W. The structure and magnetism of graphone. *AIP Adv.* **2012**, *2*, 042138. [CrossRef]
32. Peng, Q.; Dearden, A.K.; Crean, J.; Han, L.; Liu, S.; Wen, X.; De, S. New materials graphyne, graphdiyne, graphone, and graphane: Review of properties, synthesis, and application in nanotechnology. *Nanotechnol. Sci. Appl.* **2014**, *7*, 1. [CrossRef]
33. Baughman, R.; Eckhardt, H.; Kertesz, M. Structure-property predictions for new planar forms of carbon: Layered phases containing sp 2 and sp atoms. *J. Chem. Phys.* **1987**, *87*, 6687–6699. [CrossRef]
34. Solis, D.; Woellner, C.F.; Borges, D.D.; Galvao, D.S. Mechanical and Thermal Stability of Graphyne and Graphdiyne Nanoscrolls. *MRS Adv.* **2017**, *2*, 129–134. [CrossRef]

35. Li, G.; Li, Y.; Liu, H.; Guo, Y.; Li, Y.; Zhu, D. Architecture of graphdiyne nanoscale films. *Chem. Commun.* **2010**, *46*, 3256–3258. [CrossRef]
36. Majidi, R. Electronic properties of porous graphene, α-graphyne, graphene-like, and graphyne-like BN sheets. *Can. J. Phys.* **2016**, *94*, 305–309. [CrossRef]
37. Xie, C.; Wang, N.; Li, X.; Xu, G.; Huang, C. Research on the Preparation of Graphdiyne and Its Derivatives. *Chem. Eur. J.* **2020**, *26*, 569–583. [CrossRef]
38. Li, Y.; Xu, L.; Liu, H.; Li, Y. Graphdiyne and graphyne: From theoretical predictions to practical construction. *Chem. Soc. Rev.* **2014**, *43*, 2572–2586. [CrossRef] [PubMed]
39. Puigdollers, A.R.; Alonso, G.; Gamallo, P. First-principles study of structural, elastic and electronic properties of alpha-, beta- and gamma-graphyne. *Carbon* **2016**, *96*, 879–887. [CrossRef]
40. Haley, M.M.; Brand, S.C.; Pak, J.J. Carbon networks based on dehydrobenzoannulenes: Synthesis of graphdiyne substructures. *Angew. Chem. Int. Ed.* **1997**, *36*, 836–838. [CrossRef]
41. Serafini, P.; Milani, A.A.; Proserpio, D.M.; Casari, C.S. Designing all graphdiyne materials as graphene derivatives: Topologically driven modulation of electronic properties. *arXiv* **2021**, arXiv:2103.13085.
42. Zhou, J.; Li, J.; Liu, Z.; Zhang, J. Exploring Approaches for the Synthesis of Few-Layered Graphdiyne. *Adv. Mater.* **2019**, *31*, 1803758. [CrossRef] [PubMed]
43. Glaser, C. Untersuchungen über einige Derivate der Zimmtsäure. *Justus Liebigs Ann. Chem.* **1870**, *154*, 137–171. [CrossRef]
44. Eglinton, G.; Galbraith, A. 182. Macrocyclic acetylenic compounds. Part I. Cyclo tetradeca-1: 3-diyne and related compounds. *J. Chem. Soc.* **1959**, *182*, 889–896. [CrossRef]
45. Hay, A.S. Oxidative coupling of acetylenes. III. *J. Org. Chem.* **1962**, *27*, 3320–3321. [CrossRef]
46. King, A.O.; Okukado, N.; Negishi, E.-I. Highly general stereo-, regio-, and chemo-selective synthesis of terminal and internal conjugated enynes by the Pd-catalysed reaction of alkynylzinc reagents with alkenyl halides. *J. Chem. Soc. Chem. Commun.* **1977**, *19*, 683–684. [CrossRef]
47. Hatanaka, Y.; Hiyama, T. Cross-coupling of organosilanes with organic halides mediated by a palladium catalyst and tris(diethylamino)sulfonium difluorotrimethylsilicate. *J. Org. Chem.* **1988**, *53*, 918–920. [CrossRef]
48. Sonogashira, K. Development of Pd–Cu catalyzed cross-coupling of terminal acetylenes with sp^2-carbon halides. *J. Organomet. Chem.* **2002**, *653*, 46–49. [CrossRef]
49. Vollhardt, K.P.C. Cobalt-Mediated [2 + 2 + 2]-Cycloadditions: A Maturing Synthetic Strategy [New Synthetic Methods (43)]. *Angew. Chem. Int. Ed.* **1984**, *23*, 539–556. [CrossRef]
50. Rubin, Y.; Knobler, C.B.; Diederich, F. Tetraethynylethene. *Angew. Chem. Int. Ed.* **1991**, *30*, 698–700. [CrossRef]
51. Diederich, F.; Rubin, Y. Synthetic Approaches toward Molecular and Polymeric Carbon Allotropes. *Angew. Chem. Int. Ed.* **1992**, *31*, 1101–1123. [CrossRef]
52. Diederich, F. Carbon scaffolding: Building acetylenic all-carbon and carbon-rich compounds. *Nature* **1994**, *369*, 199–207. [CrossRef]
53. Wan, W.B.; Brand, S.C.; Pak, J.J.; Haley, M.M. Synthesis of Expanded Graphdiyne Substructures. *Chem. Eur. J.* **2000**, *6*, 2044–2052. [CrossRef]
54. Tahara, K.; Yamamoto, Y.; Gross, D.E.; Kozuma, H.; Arikuma, Y.; Ohta, K.; Koizumi, Y.; Gao, Y.; Shimizu, Y.; Seki, S.; et al. Syntheses and Properties of Graphyne Fragments: Trigonally Expanded Dehydrobenzo 12 annulenes. *Chem. Eur. J.* **2013**, *19*, 11251–11260. [CrossRef]
55. Iyoda, M.; Yamakawa, J.; Rahman, M.J. Conjugated Macrocycles: Concepts and Applications. *Angew. Chem. Int. Ed.* **2011**, *50*, 10522–10553. [CrossRef]
56. Diederich, F.; Kivala, M. All-carbon scaffolds by rational design. *Adv. Mater.* **2010**, *22*, 803–812. [CrossRef]
57. Haley, M.M. Synthesis and properties of annulenic subunits of graphyne and graphdiyne nanoarchitectures. *Pure Appl. Chem.* **2008**, *80*, 519–532. [CrossRef]
58. Li, Y.; Liu, Q.; Li, W.; Lu, Y.; Meng, H.; Li, C. Efficient destruction of hexachlorobenzene by calcium carbide through mechanochemical reaction in a planetary ball mill. *Chemosphere* **2017**, *166*, 275–280. [CrossRef]
59. Li, Y.; Liu, Q.; Li, W.; Meng, H.; Lu, Y.; Li, C. Synthesis and Supercapacitor Application of Alkynyl Carbon Materials Derived from CaC_2 and Polyhalogenated Hydrocarbons by Interfacial Mechanochemical Reactions. *ACS Appl. Mater. Interfaces* **2017**, *9*, 3895–3901. [CrossRef] [PubMed]
60. Lee, J.; Li, Y.; Tang, J.; Cui, X. Synthesis of hydrogen substituted graphyne through mechanochemistry and its electrocatalytic properties. *Acta Phys. Chim. Sin.* **2018**, *34*, 1080–1087. [CrossRef]
61. Li, Q.; Li, Y.; Chen, Y.; Wu, L.; Yang, C.; Cui, X. Synthesis of γ-graphyne by mechanochemistry and its electronic structure. *Carbon* **2018**, *136*, 248–254. [CrossRef]
62. Qian, X.; Liu, H.; Huang, C.; Chen, S.; Zhang, L.; Li, Y.; Wang, J.; Li, Y. Self-catalyzed growth of large-area nanofilms of two-dimensional carbon. *Sci. Rep.* **2015**, *5*, 1–7. [CrossRef]
63. Zuo, Z.; Shang, H.; Chen, Y.; Li, J.; Liu, H.; Li, Y.; Li, Y. A facile approach for graphdiyne preparation under atmosphere for an advanced battery anode. *Chem. Commun.* **2017**, *53*, 8074–8077. [CrossRef] [PubMed]
64. Shang, H.; Zuo, Z.; Zheng, H.; Li, K.; Tu, Z.; Yi, Y.; Liu, H.; Li, Y.; Li, Y. N-doped graphdiyne for high-performance electrochemical electrodes. *Nano Energy* **2018**, *44*, 144–154. [CrossRef]

65. Zhang, S.; Du, H.; He, J.; Huang, C.; Liu, H.; Cui, G.; Li, Y. Nitrogen-doped graphdiyne applied for lithium-ion storage. *ACS Appl. Mater. Interfaces* **2016**, *8*, 8467–8473. [CrossRef] [PubMed]
66. Shen, X.; Li, X.; Zhao, F.; Wang, N.; Xie, C.; He, J.; Si, W.; Yi, Y.; Yang, Z.; Li, X. Preparation and structure study of phosphorus-doped porous graphdiyne and its efficient lithium storage application. *2D Mater.* **2019**, *6*, 035020. [CrossRef]
67. Shen, X.; Yang, Z.; Wang, K.; Wang, N.; He, J.; Du, H.; Huang, C. Nitrogen-Doped Graphdiyne as High-Capacity Electrode Materials for Both Lithium-Ion and Sodium-Ion Capacitors. *ChemElectroChem* **2018**, *5*, 1435–1443. [CrossRef]
68. Yang, Z.; Cui, W.; Wang, K.; Song, Y.; Zhao, F.; Wang, N.; Long, Y.; Wang, H.; Huang, C. Chemical Modification of the sp-Hybridized Carbon Atoms of Graphdiyne by Using Organic Sulfur. *Chem. Eur. J.* **2019**, *25*, 5643–5647. [CrossRef]
69. Du, H.; Zhang, Z.; He, J.; Cui, Z.; Chai, J.; Ma, J.; Yang, Z.; Huang, C.; Cui, G. A Delicately Designed Sulfide Graphdiyne Compatible Cathode for High-Performance Lithium/Magnesium–Sulfur Batteries. *Small* **2017**, *13*, 1702277. [CrossRef]
70. Zhang, S.; Cai, Y.; He, H.; Zhang, Y.; Liu, R.; Cao, H.; Wang, M.; Liu, J.; Zhang, G.; Li, Y. Heteroatom doped graphdiyne as efficient metal-free electrocatalyst for oxygen reduction reaction in alkaline medium. *J. Mater. Chem. A* **2016**, *4*, 4738–4744. [CrossRef]
71. Abyazisani, M.; Jayalatharachchi, V.; MacLeod, J. Directed on-surface growth of covalently-bonded molecular nanostructures. In *Comprehensive Nanoscience and Nanotechnology*, 2nd ed.; Andrews, D., Nann, T., Lipson, R., Eds.; Academic Press: Cambridge, MA, USA, 2019; Volume 2, pp. 299–326.
72. Schultz, M.J.; Zhang, X.; Unarunotai, S.; Khang, D.-Y.; Cao, Q.; Wang, C.; Lei, C.; MacLaren, S.; Soares, J.A.; Petrov, I. Synthesis of linked carbon monolayers: Films, balloons, tubes, and pleated sheets. *Proc. Natl. Acad. Sci. USA* **2008**, *105*, 7353–7358. [CrossRef]
73. Kang, F.; Xu, W. On-Surface Synthesis of One-Dimensional Carbon-Based Nanostructures via C−X and C−H Activation Reactions. *ChemPhysChem* **2019**, *20*, 2251–2261. [CrossRef] [PubMed]
74. Zhang, Y.-Q.; Kepčija, N.; Kleinschrodt, M.; Diller, K.; Fischer, S.; Papageorgiou, A.C.; Allegretti, F.; Björk, J.; Klyatskaya, S.; Klappenberger, F. Homo-coupling of terminal alkynes on a noble metal surface. *Nat. Commun.* **2012**, *3*, 1–8. [CrossRef] [PubMed]
75. Liu, J.; Chen, Q.; Xiao, L.; Shang, J.; Zhou, X.; Zhang, Y.; Wang, Y.; Shao, X.; Li, J.; Chen, W.; et al. Lattice-Directed Formation of Covalent and Organometallic Molecular Wires by Terminal Alkynes on Ag Surfaces. *ACS Nano* **2015**, *9*, 6305–6314. [CrossRef] [PubMed]
76. Klappenberger, F.; Zhang, Y.-Q.; Björk, J.; Klyatskaya, S.; Ruben, M.; Barth, J.V. On-surface synthesis of carbon-based scaffolds and nanomaterials using terminal alkynes. *Acc. Chem. Res.* **2015**, *48*, 2140–2150. [CrossRef] [PubMed]
77. Liu, J.; Chen, Q.; He, Q.; Zhang, Y.; Fu, X.; Wang, Y.; Zhao, D.; Chen, W.; Xu, G.Q.; Wu, K. Bromine adatom promoted C–H bond activation in terminal alkynes at room temperature on Ag(111). *Phys. Chem. Chem. Phys.* **2018**, *20*, 11081–11088. [CrossRef]
78. Sun, Q.; Yu, X.; Bao, M.; Liu, M.; Pan, J.; Zha, Z.; Cai, L.; Ma, H.; Yuan, C.; Qiu, X. Direct formation of C−C triple-bonded structural motifs by on-surface dehalogenative homocouplings of tribromomethyl-substituted arenes. *Angew. Chem. Int. Ed.* **2018**, *57*, 1035–4038. [CrossRef]
79. Yu, X.; Cai, L.; Bao, M.; Sun, Q.; Ma, H.; Yuan, C.; Xu, W. On-surface synthesis of graphyne nanowires through stepwise reactions. *Chem. Commun.* **2020**, *56*, 1685–1688. [CrossRef]
80. Sun, Q.; Cai, L.; Ma, H.; Yuan, C.; Xu, W. Dehalogenative homocoupling of terminal alkynyl bromides on Au (111): Incorporation of acetylenic scaffolding into surface nanostructures. *ACS Nano* **2016**, *10*, 7023–7030. [CrossRef]
81. Gao, H.Y.; Wagner, H.; Zhong, D.; Franke, J.H.; Studer, A.; Fuchs, H. Glaser coupling at metal surfaces. *Angew. Chem. Int. Ed.* **2013**, *52*, 4024–4028. [CrossRef]
82. Zhou, H.; Liu, J.; Du, S.; Zhang, L.; Li, G.; Zhang, Y.; Tang, B.Z.; Gao, H.-J. Direct visualization of surface-assisted two-dimensional diyne polycyclotrimerization. *J. Am. Chem. Soc.* **2014**, *136*, 5567–5570. [CrossRef]
83. Gao, H.-Y.; Franke, J.R.-H.; Wagner, H.; Zhong, D.; Held, P.-A.; Studer, A.; Fuchs, H. Effect of metal surfaces in on-surface glaser coupling. *J. Phys. Chem. C* **2013**, *117*, 18595–18602. [CrossRef]
84. Liu, J.; Ruffieux, P.; Feng, X.; Müllen, K.; Fasel, R. Cyclotrimerization of arylalkynes on Au (111). *Chem. Commun.* **2014**, *50*, 11200–11203. [CrossRef] [PubMed]
85. Wang, T.; Lv, H.; Feng, L.; Tao, Z.; Huang, J.; Fan, Q.; Wu, X.; Zhu, J. Unravelling the Mechanism of Glaser Coupling Reaction on Ag(111) and Cu(111) Surfaces: A Case for Halogen Substituted Terminal Alkyne. *J. Phys. Chem. C* **2018**, *122*, 14537–14545. [CrossRef]
86. Liu, R.; Gao, X.; Zhou, J.; Xu, H.; Li, Z.; Zhang, S.; Xie, Z.; Zhang, J.; Liu, Z. Chemical vapor deposition growth of linked carbon monolayers with acetylenic scaffoldings on silver foil. *Adv. Mater.* **2017**, *29*, 1604665. [CrossRef] [PubMed]
87. Moroni, M.; Le Moigne, J.; Luzzati, S. Rigid rod conjugated polymers for nonlinear optics: 1. Characterization and linear optical properties of poly(aryleneethynylene) derivatives. *Macromolecules* **1994**, *27*, 562–571. [CrossRef]
88. Petrosyan, A.; Ehlers, P.; Surkus, A.-E.; Ghochikyan, T.V.; Saghyan, A.S.; Lochbrunner, S.; Langer, P. Straightforward synthesis of tetraalkynylpyrazines and their photophysical properties. *Org. Biomol. Chem.* **2016**, *14*, 1442–1449. [CrossRef] [PubMed]
89. Sonoda, M.; Inaba, A.; Itahashi, K.; Tobe, Y. Synthesis of Differentially Substituted Hexaethynylbenzenes Based on Tandem Sonogashira and Negishi Cross-Coupling Reactions. *Org. Lett.* **2001**, *3*, 2419–2421. [CrossRef] [PubMed]
90. Jiang, J.X.; Su, F.; Trewin, A.; Wood, C.D.; Campbell, N.L.; Niu, H.; Dickinson, C.; Ganin, A.Y.; Rosseinsky, M.J.; Khimyak, Y.Z. Conjugated microporous poly (aryleneethynylene) networks. *Angew. Chem. Int. Ed.* **2007**, *46*, 8574–8578. [CrossRef]
91. Dawson, R.; Laybourn, A.; Khimyak, Y.Z.; Adams, D.J.; Cooper, A.I. High surface area conjugated microporous polymers: The importance of reaction solvent choice. *Macromolecules* **2010**, *43*, 8524–8530. [CrossRef]

92. Tan, D.; Fan, W.; Xiong, W.; Sun, H.; Cheng, Y.; Liu, X.; Meng, C.; Li, A.; Deng, W.Q. Study on the morphologies of covalent organic microporous polymers: The role of reaction solvents, Macromol. *Chem. Phys.* **2012**, *213*, 1435–1440. [CrossRef]
93. Bunz, U.H.F. Poly(aryleneethynylene)s: Syntheses, Properties, Structures, and Applications. *Chem. Rev.* **2000**, *100*, 1605–1644. [CrossRef] [PubMed]
94. Wu, B.; Li, M.; Xiao, S.; Qu, Y.; Qiu, X.; Liu, T.; Tian, F.; Li, H.; Xiao, S. A graphyne-like porous carbon-rich network synthesized via alkyne metathesis. *Nanoscale* **2017**, *9*, 11939–11943. [CrossRef] [PubMed]
95. Ding, W.; Sun, M.; Gao, B.; Liu, W.; Ding, Z.; Anandan, S. A ball-milling synthesis of N-graphyne with controllable nitrogen doping sites for efficient electrocatalytic oxygen evolution and supercapacitors. *Dalton Trans.* **2020**, *49*, 10958–10969. [CrossRef]
96. Chen, T.; Li, W.-Q.; Chen, X.-J.; Guo, Y.-Z.; Hu, W.-B.; Hu, W.-J.; Yahu, A.; Liu, Y.A.; Yang, H.; Wen, K. A Triazine-Based Analogue of Graphyne: Scalable Synthesis and Applications in Photocatalytic Dye Degradation and Bacterial Inactivation. *Chem. Eur. J.* **2020**, *26*, 2269–2275. [CrossRef]
97. Zhou, W.; Shen, H.; Zeng, Y.; Yi, Y.; Zuo, Z.; Li, Y.; Li, Y. Controllable Synthesis of Graphdiyne Nanoribbons. *Angew. Chem. Int. Ed.* **2020**, *59*, 4908–4913. [CrossRef]
98. Matsuoka, R.; Sakamoto, R.; Hoshiko, K.; Sasaki, S.; Masunaga, H.; Nagashio, K.; Nishihara, H. Crystalline graphdiyne nanosheets produced at a gas/liquid or liquid/liquid interface. *J. Am. Chem. Soc.* **2017**, *139*, 3145–3152. [CrossRef]
99. Song, Y.; Li, X.; Yang, Z.; Wang, J.; Liu, C.; Xie, C.; Wang, H.; Huang, C. A facile liquid/liquid interface method to synthesize graphyne analogs. *Chem. Commun.* **2019**, *55*, 6571–6574. [CrossRef]
100. Kan, X.; Ban, Y.; Wu, C.; Pan, Q.; Liu, H.; Song, J.; Zuo, Z.; Li, Z.; Zhao, Y. Interfacial Synthesis of Conjugated Two-Dimensional N-Graphdiyne. *ACS Appl. Mater. Interfaces* **2018**, *10*, 53–58. [CrossRef]
101. Pan, Q.; Chen, S.; Wu, C.; Zhang, Z.; Li, Z.; Zhao, Y. Sulfur-Rich Graphdiyne-Containing Electrochemical Active Tetrathiafulvalene for Highly Efficient Lithium Storage Application. *ACS Appl. Mater. Interfaces* **2019**, *11*, 46070–46076. [CrossRef] [PubMed]
102. Zhao, F.; Wang, N.; Zhang, M.; Sápi, A.; Yu, J.; Li, X.; Cui, W.; Yang, Z.; Huang, C. In situ growth of graphdiyne on arbitrary substrates with a controlled-release method. *Chem. Commun.* **2018**, *54*, 6004–6007. [CrossRef]
103. Zhou, J.; Xie, Z.; Liu, R.; Gao, X.; Li, J.; Xiong, Y.; Tong, L.; Zhang, J.; Liu, Z. Synthesis of ultrathin graphdiyne film using a surface template. *ACS Appl. Mater. Interfaces* **2018**, *11*, 2632–2637. [CrossRef]
104. Zhou, J.; Gao, X.; Liu, R.; Xie, Z.; Yang, J.; Zhang, S.; Zhang, G.; Liu, H.; Li, Y.; Zhang, J.; et al. Synthesis of Graphdiyne Nanowalls Using Acetylenic Coupling Reaction. *J. Am. Chem. Soc.* **2015**, *137*, 7596–7599. [CrossRef]
105. He, J.; Bao, K.; Cui, W.; Yu, J.; Huang, C.; Shen, X.; Cui, Z.; Wang, N. Construction of Large-Area Uniform Graphdiyne Film for High-Performance Lithium-Ion Batteries. *Chem. Eur. J.* **2018**, *24*, 1187–1192. [CrossRef] [PubMed]
106. Gao, X.; Li, J.; Du, R.; Zhou, J.; Huang, M.-Y.; Liu, R.; Li, J.; Xie, Z.; Wu, L.-Z.; Liu, Z.; et al. Direct Synthesis of Graphdiyne Nanowalls on Arbitrary Substrates and Its Application for Photoelectrochemical Water Splitting Cell. *Adv. Mater.* **2017**, *29*, 1605308. [CrossRef]
107. Yang, Z.; Zhang, C.; Hou, Z.; Wang, X.; He, J.; Li, X.; Song, Y.; Wang, N.; Wang, K.; Wang, H.; et al. Porous hydrogen substituted graphyne for high capacity and ultra-stable sodium ion storage. *J. Mater. Chem. A* **2019**, *7*, 11186–11194. [CrossRef]
108. He, J.; Wang, N.; Yang, Z.; Shen, X.; Wang, K.; Huang, C.; Yi, Y.; Tu, Z.; Li, Y. Fluoride graphdiyne as a free-standing electrode displaying ultra-stable and extraordinary high Li storage performance. *Energy Environ. Sci.* **2018**, *11*, 2893–2903. [CrossRef]
109. Wang, N.; He, J.; Tu, Z.; Yang, Z.; Zhao, F.; Li, X.; Huang, C.; Wang, K.; Jiu, T.; Yi, Y. Synthesis of Chlorine-Substituted Graphdiyne and Applications for Lithium-Ion Storage. *Angew. Chem. Int. Ed.* **2017**, *129*, 10880–10885. [CrossRef]
110. Li, J.; Xie, Z.; Xiong, Y.; Li, Z.; Huang, Q.; Zhang, S.; Zhou, J.; Liu, R.; Gao, X.; Chen, C. Architecture of β-Graphdiyne-Containing Thin Film Using Modified Glaser–Hay Coupling Reaction for Enhanced Photocatalytic Property of TiO_2. *Adv. Mater.* **2017**, *29*, 1700421. [CrossRef]
111. Yang, Z.; Liu, R.; Wang, N.; He, J.; Wang, K.; Li, X.; Shen, X.; Wang, X.; Lv, Q.; Zhang, M.; et al. Triazine-graphdiyne: A new nitrogen-carbonous material and its application as an advanced rechargeable battery anode. *Carbon* **2018**, *137*, 442–450. [CrossRef]
112. Wu, L.M.; Dong, Y.Z.; Zhao, J.L.; Ma, D.T.; Huang, W.C.; Zhang, Y.; Wang, Y.Z.; Jiang, X.T.; Xiang, Y.J.; Li, J.Q.; et al. Kerr Nonlinearity in 2D Graphdiyne for Passive Photonic Diodes. *Adv. Mater.* **2019**, *31*, 1807981. [CrossRef]
113. Wang, N.; Li, X.; Tu, Z.; Zhao, F.; He, J.; Guan, Z.; Huang, C.; Yi, Y.; Li, Y. Synthesis and Electronic Structure of Boron-Graphdiyne with an sp-Hybridized Carbon Skeleton and Its Application in Sodium Storage. *Angew. Chem. Int. Ed.* **2018**, *130*, 4032–4037. [CrossRef]
114. Shang, H.; Zuo, Z.; Li, L.; Wang, F.; Liu, H.; Li, Y.; Li, Y. Ultrathin graphdiyne nanosheets grown in situ on copper nanowires and their performance as lithium-ion battery anodes. *Angew. Chem. Int. Ed.* **2018**, *57*, 774–778. [CrossRef]
115. Li, G.; Li, Y.; Qian, X.; Liu, H.; Lin, H.; Chen, N.; Li, Y. Construction of tubular molecule aggregations of graphdiyne for highly efficient field emission. *J. Phys. Chem. C* **2011**, *115*, 2611–2615. [CrossRef]
116. Huang, H.; Duan, W.; Liu, Z. The existence/absence of Dirac cones in graphynes. *New J. Phys.* **2013**, *15*, 023004. [CrossRef]
117. Novoselov, K.S.; Geim, A.K.; Morozov, S.V.; Jiang, D.; Katsnelson, M.I.; Grigorieva, I.; Dubonos, S.; Firsov, A.A. Two-dimensional gas of massless Dirac fermions in graphene. *Nature* **2005**, *438*, 197–200. [CrossRef] [PubMed]
118. Wallace, P.R. The band theory of graphite. *Phys. Rev.* **1947**, *71*, 622. [CrossRef]
119. Malko, D.; Neiss, C.; Vines, F.; Görling, A. Competition for graphene: Graphynes with direction-dependent dirac cones. *Phys. Rev. Lett.* **2012**, *108*, 086804. [CrossRef]

120. Xi, J.; Wang, D.; Shuai, Z. Electronic properties and charge carrier mobilities of graphynes and graphdiynes from first principles. *Wiley Interdiscip. Rev. Comput. Mol. Sci.* **2015**, *5*, 215–227. [CrossRef]
121. Narita, N.; Nagai, S.; Suzuki, S.; Nakao, K. Optimized geometries and electronic structures of graphyne and its family. *Phys. Rev. B* **1998**, *58*, 11009. [CrossRef]
122. Ebadi, M.; Reisi-Vanani, A.; Houshmand, F.; Amani, P. Calcium-decorated graphdiyne as a high hydrogen storage medium: Evaluation of the structural and electronic properties. *Int. J. Hydrogen Energy* **2018**, *43*, 23346–23356. [CrossRef]
123. Hou, X.; Xie, Z.; Li, C.; Li, G.; Chen, Z. Study of electronic structure, thermal conductivity, elastic and optical properties of α, β, γ-graphyne. *Materials* **2018**, *11*, 188. [CrossRef] [PubMed]
124. Nulakani, N.V.R.; Subramanian, V. Cp-Graphyne: A Low-Energy Graphyne Polymorph with Double Distorted Dirac Points. *ACS Omega* **2017**, *2*, 6822–6830. [CrossRef]
125. Srinivasu, K.; Ghosh, S.K. Graphyne and Graphdiyne: Promising Materials for Nanoelectronics and Energy Storage Applications. *J. Phys. Chem. C* **2012**, *116*, 5951–5956. [CrossRef]
126. Kang, J.; Li, J.; Wu, F.; Li, S.-S.; Xia, J.-B. Elastic, Electronic, and Optical Properties of Two-Dimensional Graphyne Sheet. *J. Phys. Chem. C* **2011**, *115*, 20466–20470. [CrossRef]
127. Yue, Q.; Chang, S.; Kang, J.; Qin, S.; Li, J. Mechanical and Electronic Properties of Graphyne and Its Family under Elastic Strain: Theoretical Predictions. *J. Phys. Chem. C* **2013**, *117*, 14804–14811. [CrossRef]
128. Yun, J.; Zhang, Z.; Yan, J.; Zhao, W.; Xu, M. First-principles study of B or Al-doping effect on the structural, electronic structure and magnetic properties of γ-graphyne. *Comput. Mater. Sci.* **2015**, *108*, 147–152. [CrossRef]
129. Singh, N.B.; Bhattacharya, B.; Sarkar, U. A first principle study of pristine and BN-doped graphyne family. *J. Struct. Chem.* **2014**, *25*, 1695–1710. [CrossRef]
130. Koo, J.; Park, M.; Hwang, S.; Huang, B.; Jang, B.; Kwon, Y.; Lee, H. Widely tunable band gaps of graphdiyne: An ab initio study. *Phys. Chem. Chem. Phys.* **2014**, *16*, 8935–8939. [CrossRef]
131. Luo, G.; Qian, X.; Liu, H.; Qin, R.; Zhou, J.; Li, L.; Gao, Z.; Wang, E.; Mei, W.-N.; Lu, J. Quasiparticle energies and excitonic effects of the two-dimensional carbon allotrope graphdiyne: Theory and experiment. *Phys. Rev. B* **2011**, *84*, 075439. [CrossRef]
132. Jiao, Y.; Du, A.; Hankel, M.; Zhu, Z.; Rudolph, V.; Smith, S.C. Graphdiyne: A versatile nanomaterial for electronics and hydrogen purification. *Chem. Commun.* **2011**, *47*, 11843–11845. [CrossRef]
133. Long, M.; Tang, L.; Wang, D.; Li, Y.; Shuai, Z. Electronic Structure and Carrier Mobility in Graphdiyne Sheet and Nanoribbons: Theoretical Predictions. *ACS Nano* **2011**, *5*, 2593–2600. [CrossRef] [PubMed]
134. Cui, H.-J.; Sheng, X.-L.; Yan, Q.-B.; Zheng, Q.-R.; Su, G. Strain-induced Dirac cone-like electronic structures and semiconductor–semimetal transition in graphdiyne. *Phys. Chem. Chem. Phys.* **2013**, *15*, 8179–8185. [CrossRef]
135. Chen, J.; Xi, J.; Wang, D.; Shuai, Z. Carrier Mobility in Graphyne Should Be Even Larger than That in Graphene: A Theoretical Prediction. *J. Phys. Chem. Lett.* **2013**, *4*, 1443–1448. [CrossRef] [PubMed]
136. Narita, N.; Nagai, S.; Suzuki, S.; Nakao, K. Electronic structure of three-dimensional graphyne. *Phys. Rev. B* **2000**, *62*, 11146. [CrossRef]
137. Luo, G.; Zheng, Q.; Mei, W.-N.; Lu, J.; Nagase, S. Structural, Electronic, and Optical Properties of Bulk Graphdiyne. *J. Phys. Chem. C* **2013**, *117*, 13072–13079. [CrossRef]
138. Zheng, Q.; Luo, G.; Liu, Q.; Quhe, R.; Zheng, J.; Tang, K.; Gao, Z.; Nagase, S.; Lu, J. Structural and electronic properties of bilayer and trilayer graphdiyne. *Nanoscale* **2012**, *4*, 3990–3996. [CrossRef] [PubMed]
139. Shohany, B.G.; Roknabadi, M.R.; Kompany, A. Computational study of edge configuration and the diameter effects on the electrical transport of graphdiyne nanotubes. *Phys. E* **2016**, *84*, 146–151. [CrossRef]
140. Pan, L.D.; Zhang, L.Z.; Song, B.Q.; Du, S.X.; Gao, H.-J. Graphyne- and graphdiyne-based nanoribbons: Density functional theory calculations of electronic structures. *Appl. Phys. Lett.* **2011**, *98*, 173102. [CrossRef]
141. Son, Y.-W.; Cohen, Y.-W.; Louie, S.G. Energy Gaps in Graphene Nanoribbons. *Phys. Rev. Lett.* **2006**, *97*, 216803. [CrossRef]
142. Chen, C.; Li, J.; Sheng, X.-L. Graphdiyne nanoribbons with open hexagonal rings: Existence of topological unprotected edge states. *Phys. Lett. A* **2017**, *381*, 3337–3341. [CrossRef]
143. Xu, X.; Liu, C.; Sun, Z.; Cao, T.; Zhang, Z.; Wang, E.; Liu, Z.; Liu, K. Interfacial engineering in graphene bandgap. *Chem. Soc. Rev.* **2018**, *47*, 3059–3099. [CrossRef] [PubMed]
144. Makaremi, M.; Mortazavi, B.; Rabczuk, T.; Ozin, G.A.; Singh, C.V. Theoretical investigation: 2D N-graphdiyne nanosheets as promising anode materials for Li/Na rechargeable storage devices. *ACS Appl. Nano Mater.* **2018**, *2*, 127–135. [CrossRef]
145. Qiu, H.; Xue, M.M.; Shen, C.; Zhang, Z.H.; Guo, W.L. Graphynes for Water Desalination and Gas Separation. *Adv. Mater.* **2019**, *31*, 1803772. [CrossRef] [PubMed]
146. Gu, J.; Magagula, S.; Zhao, J.; Chen, Z. Boosting ORR/OER activity of graphdiyne by simple heteroatom doping. *Small Methods* **2019**, *3*, 1800550. [CrossRef]
147. Yang, Z.; Zhang, Y.; Guo, M.; Yun, J. Adsorption of hydrogen and oxygen on graphdiyne and its BN analog sheets: A density functional theory study. *Comput. Mater. Sci.* **2019**, *160*, 197–206. [CrossRef]
148. Kang, J.; Wu, F.; Li, J. Modulating the bandgaps of graphdiyne nanoribbons by transverse electric fields. *J. Phys. Condens. Matter.* **2012**, *24*, 165301. [CrossRef]
149. Leenaerts, O.; Partoens, B.; Peeters, F. Tunable double dirac cone spectrum in bilayer α-graphyne. *Appl. Phys. Lett.* **2013**, *103*, 013105. [CrossRef]

150. Abdi, G.; Filip, A.; Krajewski, M.; Kazimierczuk, K.; Strawski, M.; Szarek, P.; Hamankiewicz, B.; Mazej, Z.; Cichowicz, G.; Leszczyński, P.J.; et al. Toward the synthesis, fluorination and application of N–graphyne. *RSC Adv.* **2020**, *10*, 40019–40029. [CrossRef]
151. Haji-Nasiri, S.; Fotoohi, S. Doping induced diode behavior with large rectifying ratio in graphyne nanoribbons device. *Phys. Lett. A* **2018**, *382*, 2894–2899. [CrossRef]
152. Jhon, Y.I.; Jhon, M.S. Electron Transport Properties of Graphene-Graphyne-Graphene Transistors: First Principles Study. *arXiv* **2013**, arXiv:1307.4374.
153. Li, Z.; Smeu, M.; Rives, A.; Maraval, V.; Chauvin, R.; Ratner, M.A.; Borguet, E. Towards graphyne molecular electronics. *Nat. Commun.* **2015**, *6*, 6321. [CrossRef] [PubMed]
154. Bhattacharya, B.; Sarkar, U. The Effect of Boron and Nitrogen Doping in Electronic, Magnetic, and Optical Properties of Graphyne. *J. Phys.Chem. C* **2016**, *120*, 26793–26806. [CrossRef]
155. Yang, Z.; Ouyang, B.; Lan, G.; Xu, L.C.; Liu, R.; Liu, X. The tunneling magnetoresistance and spin-polarized optoelectronic properties of graphyne-based molecular magnetic tunnel junctions. *J. Phys. D Appl. Phys.* **2017**, *50*, 075103. [CrossRef]
156. Jin, Z.; Zhou, Q.; Chen, Y.; Mao, P.; Li, H.; Liu, H.; Wang, J.; Li, Y. Graphdiyne:ZnO Nanocomposites for High-Performance UV Photodetectors. *Adv. Mater.* **2016**, *28*, 3697–3702. [CrossRef] [PubMed]
157. Li, Y.; Kuang, D.; Gao, Y.; Cheng, J.; Li, X.; Guo, J.; Yu, Z. Titania: Graphdiyne nanocomposites for high-performance deep ultraviolet photodetectors based on mixed-phase MgZnO. *J. Alloys Compd.* **2020**, *825*, 153882. [CrossRef]
158. Zhang, Y.; Huang, P.; Guo, J.; Shi, R.; Huang, W.; Shi, Z.; Wu, L.; Zhang, F.; Gao, L.; Li, C.; et al. Graphdiyne-Based Flexible Photodetectors with High Responsivity and Detectivity. *Adv.Mater.* **2020**, *32*, 2001082. [CrossRef]
159. Makarova, T. Magnetism of carbon-based materials. *arXiv* **2002**, arXiv:cond-mat/0207368.
160. Li, Z.; Sheng, W.; Ning, Z.; Zhang, Z.; Yang, Z.; Guo, H. Magnetism and spin-polarized transport in carbon atomic wires. *Phys. Rev. B* **2020**, *80*, 115429. [CrossRef]
161. Cauchy, T.; Ruiz, E.; Jeannin, O.; Nomura, M.; Fourmigué, M. Strong magnetic interactions through weak bonding interactions in organometallic radicals: Combined experimental and theoretical study. *Chem. Eur. J.* **2007**, *13*, 8858–8866. [CrossRef]
162. Ottaviani, M.F.; Cossu, E.; Turro, N.J.; Tomalia, D.A. Characterization of starburst dendrimers by electron paramagnetic resonance. 2. Positively charged nitroxide radicals of variable chain length used as spin probes. *J. Am. Chem. Soc.* **1995**, *117*, 4387–4398. [CrossRef]
163. Chen, X.; Gao, P.; Guo, L.; Wen, Y.; Zhang, Y.; Zhang, S. Two-dimensional ferromagnetism and spin filtering in Cr and Mn-doped graphdiyne. *J. Phys. Chem. Solids* **2017**, *105*, 61–65. [CrossRef]
164. He, J.; Ma, S.Y.; Zhou, P.; Zhang, C.; He, C.; Sun, L. Magnetic properties of single transition-metal atom absorbed graphdiyne and graphyne sheet from DFT+ U calculations. *J. Phys. Chem. C* **2012**, *116*, 26313–26321. [CrossRef]
165. Kim, S.; Puigdollers, A.R.; Gamallo, P.; Vines, F.; Lee, J.Y. Functionalization of γ-graphyne by transition metal adatoms. *Carbon* **2017**, *120*, 63–70. [CrossRef]
166. Ren, J.; Zhang, S.-B.; Liu, P.-P. Magnetic and Electronic Properties of β-Graphyne Doped with Rare-Earth Atoms. *Chin. Phys. Lett.* **2019**, *36*, 076101. [CrossRef]
167. Alaei, S.; Jalili, S.; Erkoc, S. Study of the influence of transition metal atoms on electronic and magnetic properties of graphyne nanotubes using density functional theory. *Fuller. Nanotub. Carbon Nanostruct.* **2015**, *23*, 494–499. [CrossRef]
168. He, J.; Zhou, P.; Jiao, N.; Chen, X.; Lu, W.; Sun, L. Prediction of half-semiconductor antiferromagnets with vanishing net magnetization. *RSC Adv.* **2015**, *5*, 46640–46647. [CrossRef]
169. Drogar, J.; Roknabadi, M.R.; Behdani, M.; Modarresi, M.; Kari, A. Hydrogen adsorption on the α-graphyne using ab initio calculations. *Superlattices Microstruct.* **2014**, *75*, 340–346. [CrossRef]
170. Wang, Y.; Song, N.; Zhang, T.; Zheng, Y.; Gao, H.; Xu, K.; Wang, J. Tuning the electronic and magnetic properties of graphyne by hydrogenation. *Appl. Surf. Sci.* **2018**, *452*, 181–189. [CrossRef]
171. Sofo, J.O.; Chaudhari, A.S.; Barber, G.D. Graphane: A two-dimensional hydrocarbon. *Phys. Rev. B* **2007**, *75*, 153401. [CrossRef]
172. Feng, Z.; Ma, Y.; Li, Y.; Li, R.; Liu, J.; Li, H.; Tang, Y.; Dai, X. Importance of heteroatom doping site in tuning the electronic structure and magnetic properties of graphdiyne. *Phys. E* **2019**, *114*, 113590. [CrossRef]
173. Ma, Y.; Foster, A.S.; Krasheninnikov, A.V.; Nieminen, R.M. Nitrogen in graphite and carbon nanotubes: Magnetism and mobility. *Phys. Rev. B* **2005**, *72*, 205416. [CrossRef]
174. Zhang, M.; Wang, X.; Sun, H.; Wang, N.; Lv, Q.; Cui, W.; Long, Y.; Huang, C. Enhanced paramagnetism of mesoscopic graphdiyne by doping with nitrogen. *Sci. Rep.* **2017**, *7*, 1–10. [CrossRef] [PubMed]
175. Feng, Q.; Tang, N.; Liu, F.; Cao, Q.; Zheng, W.; Ren, W.; Wan, X.; Du, Y. Obtaining high localized spin magnetic moments by fluorination of reduced graphene oxide. *ACS Nano* **2013**, *7*, 6729–6734. [CrossRef]
176. Zheng, Y.; Chen, Y.; Lin, L.; Sun, Y.; Liu, H.; Li, Y.; Du, Y.; Tang, N. Intrinsic magnetism of graphdiyne. *Appl. Phys. Lett.* **2017**, *111*, 033101. [CrossRef]
177. Zhang, M.; Sun, H.; Wang, X.; Du, H.; He, J.; Long, Y.; Zhang, Y.; Huang, C. Room-Temperature Ferromagnetism in Sulfur-Doped Graphdiyne Semiconductors. *J. Phys. Chem. C* **2019**, *123*, 5010–5016. [CrossRef]
178. Lee, C.; Wei, X.; Kysar, J.W.; Hone, J. Measurement of the Elastic Properties and Intrinsic Strength of Monolayer Graphene. *Science* **2008**, *321*, 385–388. [CrossRef] [PubMed]

179. Zhang, Y.Y.; Pei, Q.X.; Wang, C.M. Mechanical properties of graphynes under tension: A molecular dynamics study. *Appl. Phys. Lett.* **2012**, *101*, 081909. [CrossRef]
180. Wang, S.; Si, Y.; Yuan, J.; Yang, B.; Chen, H. Tunable thermal transport and mechanical properties of graphyne heterojunctions. *Phys. Chem. Chem. Phys.* **2016**, *18*, 24210–24218. [CrossRef]

Article

Cationic Pollutant Removal from Aqueous Solution Using Reduced Graphene Oxide

Talia Tene [1], Stefano Bellucci [2], Marco Guevara [3,4], Edwin Viteri [5], Malvin Arias Polanco [6,7], Orlando Salguero [7], Eder Vera-Guzmán [7], Sebastián Valladares [7], Andrea Scarcello [7,8,9], Francesca Alessandro [7,8], Lorenzo S. Caputi [7,8] and Cristian Vacacela Gomez [3,7,*]

1. Grupo de Investigación Ciencia y Tecnología de Materiales, Universidad Técnica Particular de Loja, Loja 110160, Ecuador; tbtene@utpl.edu.ec
2. INFN-Laboratori Nazionali di Frascati, Via E. Fermi 54, I-00044 Frascati, Italy; Stefano.Bellucci@lnf.infn.it
3. School of Physical Sciences and Nanotechnology, Yachay Tech University, Urcuquí 100119, Ecuador; mvguevara@yachaytech.edu.ec
4. ITECA—Instituto de Tecnologías y Ciencias Avanzadas, Villarroel y Larrea, Riobamba 060104, Ecuador
5. Faculty of Mechanical Engineering, Escuela Superior Politécnica de Chimborazo, Riobamba 060155, Ecuador; eviteri@espoch.edu.ec
6. Instituto Tecnológico de Santo Domingo, Área de Ciencias Básicas y Ambientales, Av. Los Próceres, Santo Domingo 10602, Dominican Republic; melvin.arias@intec.edu.do
7. UNICARIBE Research Center, University of Calabria, I-87036 Rende (CS), Italy; orlando.salguero@yachaytech.edu.ec (O.S.); eder.vera@yachaytech.edu.ec (E.V.-G.); sebastian.valladares@yachaytech.edu.ec (S.V.); andrea.scarcello@unical.it (A.S.); francesca.alessandro@unical.it (F.A.); lorenzo.caputi@fis.unical.it (L.S.C.)
8. Surface Nanoscience Group, Department of Physics, University of Calabria, Via P. Bucci, Cubo 33C, I-87036 Rende, Italy
9. INFN, Sezione LNF, Gruppo Collegato di Cosenza, Via P. Bucci, Rende, Cosenza, Italy
* Correspondence: cvacacela@yachaytech.edu.ec

Abstract: Reduced graphene oxide (rGO) is one of the most well-known graphene derivatives, which, due to its outstanding physical and chemical properties as well as its oxygen content, has been used for wastewater treatment technologies. Particularly, extra functionalized rGO is widely preferred for treating wastewater containing dyes or heavy metals. Nevertheless, the use of non-extra functionalized (pristine) rGO for the removal of cationic pollutants is not explored in detail or is ambiguous. Herein, pristine rGO—prepared by an eco-friendly protocol—is used for the removal of cationic pollutants from water, i.e., methylene blue (MB) and mercury-(II) (Hg-(II)). This work includes the eco-friendly synthesis process and related spectroscopical and morphological characterization. Most importantly, the investigated rGO shows an adsorption capacity of 121.95 mg g^{-1} for MB and 109.49 mg g^{-1} for Hg (II) at 298 K. A record adsorption time of 30 min was found for MB and 20 min for Hg (II) with an efficiency of about 89% and 73%, respectively. The capture of tested cationic pollutants on rGO exhibits a mixed physisorption–chemisorption process. The present work, therefore, presents new findings for cationic pollutant adsorbent materials based on oxidized graphenes, providing a new perspective for removing MB molecules and Hg(II) ions.

Keywords: graphene oxide; reduce graphene oxide; dyes; heavy metals; pollutant removal

Citation: Tene, T.; Bellucci, S.; Guevara, M.; Viteri, E.; Arias Polanco, M.; Salguero, O.; Vera-Guzmán, E.; Valladares, S.; Scarcello, A.; Alessandro, F.; et al. Cationic Pollutant Removal from Aqueous Solution Using Reduced Graphene Oxide. *Nanomaterials* **2022**, *12*, 309. https://doi.org/10.3390/nano12030309

Academic Editor: José Miguel González-Domínguez

Received: 27 December 2021
Accepted: 13 January 2022
Published: 18 January 2022

Publisher's Note: MDPI stays neutral with regard to jurisdictional claims in published maps and institutional affiliations.

Copyright: © 2022 by the authors. Licensee MDPI, Basel, Switzerland. This article is an open access article distributed under the terms and conditions of the Creative Commons Attribution (CC BY) license (https://creativecommons.org/licenses/by/4.0/).

1. Introduction

Water pollution is one of the world's most serious problems due to a large amount of wastewater being produced and poured into the water bodies every year [1]. Among different types of wastewaters, water contaminated with dyes and heavy metals deserves significant attention. With the continuous development of industrialization processes, large amounts of dyes or heavy metals are released into the environment, making water unsafe for human use and disrupting aquatic ecosystems [2–4]. A large amount of dye pollutants in the wastewater brings a huge risk to both aquatic organisms and humans because they can reduce sunlight transmission and normally contain toxic substances such as heavy metals [5], causing mutagenicity, carcinogenicity as well as the dysfunction of the kidney, liver, brain, reproductive system, and central nervous system [6]. On the other hand, agricultural processes or mining activities have increased the concentration of heavy metals in water around the world, which has led to immediate legislation by various governments [7]. When these toxic metal ions enter the food chain and then the human body, they accumulate in an organ above the allowed limits, originating serious health-related diseases, for instance: skin irritation, vomiting, stomach cramps, and cancer of the lungs and kidney [8]. Additionally, infants may have delays in physical or mental development, children may have deficits in attention span and learning activities, and adults may have high blood pressure [9].

To remove dyes or heavy metals, different physicochemical and electrochemical methods have been proposed. Physicochemical processes include membrane filtration [10], ion-exchange [11], and adsorption [12]. Electrochemical processes include electrocoagulation [13], electroflotation [14], and electrodeposition [15]. Among all these possible methods, including on-site sensing ones [16,17], those cost-effective, environmentally friendly and no further pollutant features are required. Therefore, adsorption is one of the most prominent approaches to water and wastewater decontamination. Due to this fact, many adsorbents with different structural conformations and compositions have been prepared or modified, for instance, clays/zeolites [18], biosorbents [19], agricultural solid wastes [20], and industrial by-products [21]. In practical operations, activated carbon is one of the most used adsorbents, thanks to its excellent adsorption performance [22]. However, activated carbon has been limited by the high cost and the complicated regeneration processes. To tackle the aforementioned limitations, carbon-based nanomaterials, such as carbon nanotubes [23], graphene oxide (GO) [24], and reduced graphene oxide (rGO) [25], have been proposed.

Recently, rGO has attracted increased interest as an effective adsorbent for dyes or heavy metal ions [25]. While the properties of rGO are quite different from those of pristine graphene or beyond-graphene materials (i.e., unique electronic, optical, mechanical, plasmonic, and thermal properties [26–29] as well as promising applications in hybrid capacitors [30]), rGO is characterized by interesting hydrophilic and semiconducting properties [31]. Nowadays, the synthesis of rGO follows an eco-friendly and cost-effective preparation method that can be used for large-scale water treatment technologies [32,33]. Moreover, the presence of oxygen functional groups (mainly hydroxyl and epoxide groups [34]) on the rGO surface allows covalent modifications with strong chelating groups, which, in turn, present a high affinity to metal ions or more complex organic/inorganic molecules.

There is extensive literature on the use of extra functionalized rGO or GO (e.g., GONR [35], S-GO [36], GO-TSC [37], S-doped g-C_3N_4/LGO [38], GSH-$NiFe_2O_4$/GO [39], HT-rGO-N [40]) for treating water and wastewater; however, the use of pristine rGO is scarce (and sometimes unclear) when removing cationic heavy metals and cationic dyes from aqueous media. In this work, such a comparative study is presented, considering methyl blue (MB) as a cationic dye and mercury (II) (Hg(II)) as cationic heavy metal. These cationic pollutants have been selected, in particular, because MB can cause various poisoning problems and methemoglobinemia [41] and Hg (II) can cause substantial neurodevelopmental risk in fetuses, newborns, and children [42]. Various kinetics, isotherms, and thermodynamic studies are carried out to demonstrate the adsorption of MB and Hg(II) on as-made rGO. Additionally, the present work includes the synthesis of adsorbent material and the corresponding morphological and spectroscopical characterizations of raw graphite, GO, and rGO.

2. Materials and Method

Although there are several methods for synthesizing graphene and its derivatives (e.g., liquid exfoliation [43,44], zeolite-assisted exfoliation [45], hydrothermal exfoliation [46], and microwave-assisted exfoliation [47–53]), the present study focuses on the eco-friendly oxidation-reduction protocol [33] to transform as-made GO into rGO.

2.1. Materials

All chemicals were used as received, without further purification:
- Graphite powder (<150 μm, 99.99%, Sigma-Aldrich, Burlington, MA, USA);
- Sulfuric acid (H_2SO_4, 95.0–98.0%, Sigma-Aldrich, Burlington, MA, USA);
- Potassium permanganate ($KMnO_4$, ≥99.0%, Sigma-Aldrich, Burlington, MA, USA);
- Hydrochloric acid (HCl, 37%, Sigma-Aldrich, Burlington, MA, USA);
- Citric acid ($C_6H_8O_7$, ≥99.5%, Sigma-Aldrich, Burlington, MA, USA);
- Methyl blue (MB, $C_{16}H_{18}N_3ClS$, Sigma-Aldrich, Burlington, MA, USA);
- Hydrogen peroxide (H_2O_2, 30%, Merk, Darmstadt, Germany);
- Sodium hydroxide (NaOH, 1310-73-2, 40.00 g/mol, Merk, Darmstadt, Germany);
- Mercury (II) oxide (HgO, 21908-53-2, 219.59 g/mol, Merk, Darmstadt, Germany).

2.2. Preparation of Oxidized Graphenes

A round-bottom flask was charged with graphite (1.5 g), H_2SO_4 (35 mL), maintaining a uniform and moderate circular agitation. The mixture was located in an ice-water bath, and then $KMnO_4$ (4.5 g) was slowly added. The resulting mixture was agitated on a stirring plate while adding 75 mL of distilled water, being careful not to exceed 363 K.

Additionally, 250 mL of distilled water was added, followed by 7.5 mL of H_2O_2. The resulting solution was distributed to be washed by centrifugation with HCl solution and distilled water several times to adjust the pH~6 [33] and then dried (drying stove, 60 Hz, 1600 W) at 353 K for 2 h to obtain graphite oxide flakes.

After the oxidation process, 50 mg of graphite oxide flakes were dispersed in 500 mL of distilled water by sonication for 0.5 h [25]. The resulting solution was centrifuged to separate GO from non-exfoliated graphite oxide particles [32]. Under agitation, 1.0 g citric acid (CA) was added to the centrifuged suspension, setting the reduction temperature at 368 K. The precipitated rGO was collected, washed with distilled water by centrifugation, and dried at 353 K for 2 h to obtain rGO powder [33]. The resulting rGO was used for the cationic pollutant removal.

2.3. Characterization of GO and rGO

The surface morphology of raw graphite, GO, and rGO were carried out on a transmission electron microscope (TEM, JEM 1400 Plus, Akishima, Tokyo, Japan) operating at 80 kV, and a scanning electron microscope (SEM, JSM-IT100 InTouchScope, Akishima, Tokyo, Japan) equipped with a JEOL dispersive X-ray spectrometer (EDS) (Billerica, MA, USA), with the accelerating voltage of 15 kV. Raman spectra of graphite and oxidized graphene were obtained using a Jasco NRS-500 spectrometer (Oklahoma City, OK, USA), with a 532 nm laser wavelength (0.3 mW, 100X objective). Infrared spectra were collected using a Fourier transform infrared spectrometer (Jasco FT/IR 4000, Oklahoma City, OK, USA). UV–visible measurements were recorded using a UV–vis spectroscopy (Thermo Scientific, Evolution 220, Waltham, MA, USA). X-ray diffraction measurements were performed using an X-ray diffractometer (PANalytical Pro X-ray, Malvern, UK) in the diffraction angle (2θ) window of 5–90° using Cu Kα irradiation under the acceleration voltage of 60 kV and a current of 55 mA. The thermal stability of GO and rGO was examined using thermogravimetric analysis (TGA, PerkinElmer simultaneous thermal analyzer, STA 6000, Waltham, MA, USA).

SEM samples were mounted on aluminum substrates with adhesive, coated with 40–60 nm of metal such as Gold/Palladium and then observed in the microscope. TEM samples were arranged by drop-casting onto formvar-coated copper grids once the samples were cut into very thin cross-sections, allowing electrons to pass directly through the sample. Similarly, Raman samples were deposited directly by drop-casting on glass substrates and dried for a few seconds with the incident beam. To record the UV-visible spectra in the window range 190–1000 nm, the samples were redispersed in distilled water by mid-sonication for 5 min.

2.4. Preparation of MB Solutions and Experimental Set-Up of MB Adsorption on rGO

MB was dissolved in ultra-pure water to obtain a stock solution of 1000 mg L^{-1}, and the working solutions were used in the test through serial dilutions. The pH of the solutions was adjusted using HCL and NaOH and controlled by a pH meter (HI221 Hanna Instruments).

The adsorption experiments were carried out in triplicate. In total, 500 mg of rGO was added into 250 mL of the MB solution with a concentration of 100 mg L^{-1} to evaluate the adsorption kinetics and contact time effect (batch test) [25]. The resulting mixture was agitated up to 60 min at 298 K. Adsorption isotherms were obtained from batch experiments by adding 200 mg rGO in 50 mL of MB solutions, considering different concentrations in the range of 10–100 mg L^{-1} and three different temperatures (298, 313, and 333 K). pH studies were carried out by adding 200 mg rGO in 100 mL of MB solution with a concentration of 100 mg L^{-1}. Various aliquots were extracted from the solution to be evaluated by UV–vis spectroscopy. In all adsorption experiments (except in the study of the effect of pH), the pH was fixed to 6.02 ± 0.07. To investigate the pH effect, the pH of MB solutions was adjusted by HCl (0.1 M) and NaOH (0.1 M), and immediately, rGO was added.

2.5. Preparation of Hg(II) Solutions and Experimental Set-Up of Hg(II) Adsorption on rGO

rGO was placed in a dilute aqueous solution of HgO. The adsorption kinetics studies were performed by adding 300 mL aqueous HgO (pH = 6.41 ± 0.05) to a falcon tube. Then, 200 mg rGO was added to form a slurry. The mixture was stirred at 298 K for 60 min. After that, the mixture was filtered at intervals through a 0.45 mm membrane, and then the filtrated samples were analyzed by using an AAS-cold vapor to determinate the remaining Hg(II) content (standard methods 3112-B; 3111-B.4b) [13]. Adsorption isotherms were obtained by adding 2.5 mg rGO to each falcon tube containing 50 mL of HgO solution with different concentrations from 10 to 100 mg L^{-1} and considering three different temperatures (298, 313, and 333 K). The resulting mixtures were stirred at room temperature for 30 min and then filtered separately through a 0.45-mm membrane filter. The filtrates were analyzed by using AAS-cold vapor to determine the remaining Hg(II) content (standard methods 3112-B; 3111-B.4b) [13]. The pH effect was investigated by adjusting the pH of HgO solutions with HCl (0.1 M) and NaOH (0.1 M) at room temperature, and immediately, rGO was added.

3. Results and Discussion

3.1. Characterization of Graphite

Graphite has a crystal structure made up of stacked graphene layers in which the separation distance of the layers is 3.35 Å, whereas the separation of atoms within a layer is 1.42 Å. At the microscale, the starting (powder) graphite shows an irregular bulk structure with a lateral size ranging from 2 μm to 50 μm (Figure 1a). The XRD pattern of raw graphite is shown in Figure 1b. The most intense peak at $2\theta = 26.73°$ corresponds to the graphite stacking crystallinity (002) [43]. The less intense peak at $2\theta = 55.82°$ displays the long-range order of stacked graphene layers (004) [43].

Figure 1c,d show the Raman spectrum of raw graphite. The main features observed are: (i) the absence of the D peak demonstrating a defect-free starting graphite, (ii) the G peak at 1577 cm^{-1} is ascribed to the C-C strC stretching mode in sp^2 carbon bonds, and (iii) the 2D peak at 2720 cm^{-1} is characterized by two bands, the intense 2D$_{2A}$ band at 2720 cm^{-1} and a shoulder 2D$_{1A}$ band at 2677 cm^{-1}. In particular, these bands originate as the effect of the splitting of π electron bands due to the interaction between stacked graphene layers. The G* peak found at 2447 cm^{-1} is characteristic of carbon-based materials with a graphitic-like structure.

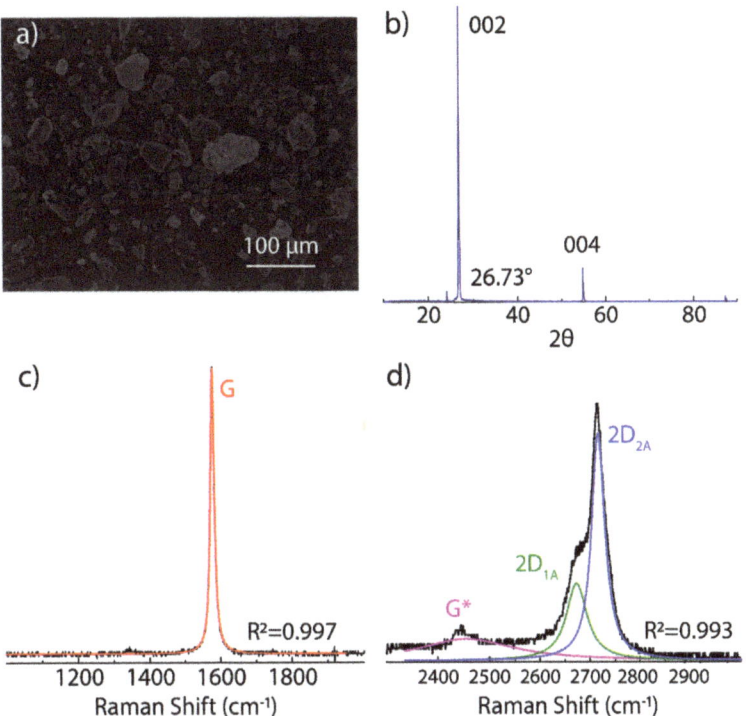

Figure 1. Characterization of starting graphite source: (**a**) SEM morphology, (**b**) XRD measurement, and (**c**,**d**) Raman spectrum from 1000 to 3000 cm^{-1} recorded using 532 excitation laser. The intensity was normalized by the most intense peak. The Raman spectrum was fitted using Lorentzian functions.

Compared to graphite, the Raman spectrum of oxidized graphenes shows a highly broadened and very-low intense 2D peak. With this in mind, the 2D band region of GO and rGO is not analyzed here, and instead, we focus on the region from 1000 to 2000 cm^{-1} to scrutinize the crystallinity and, most importantly, the basal/edge defects of the obtained materials after the oxidation-reduction process (discussed below).

3.2. Characterization of GO and rGO

While, in the present work, rGO is used for the adsorption of cationic pollutants, it is extremely important to discuss its transformation from GO. SEM micrographs of GO and rGO are shown in Figure 2a,c, respectively. The surface morphology of GO indicates a face-to-face stacking of flakes as well as randomly aggregated flakes with wrinkles and folds on the surface (Figure 2a). Instead, rGO shows a surface morphology with mesopores and micropores and randomly organized flakes (Figure 2c). The highly distorted porous surface of rGO can avoid the face-to-face stacking of flakes, as observed in GO.

EDS measurements were carried out to determine the elemental composition of GO and rGO, considering a bombarded region large enough. Then, the carbon and oxygen content were C: 49.7% and O: 50.3% for GO (after oxidation process) and C: 62.9% and O: 37.1% for rGO (after de reduction process). The oxygen content decreased by 26.2% using CA as an alternative green-reducing agent, confirming the (partial) removal of oxygen functional groups.

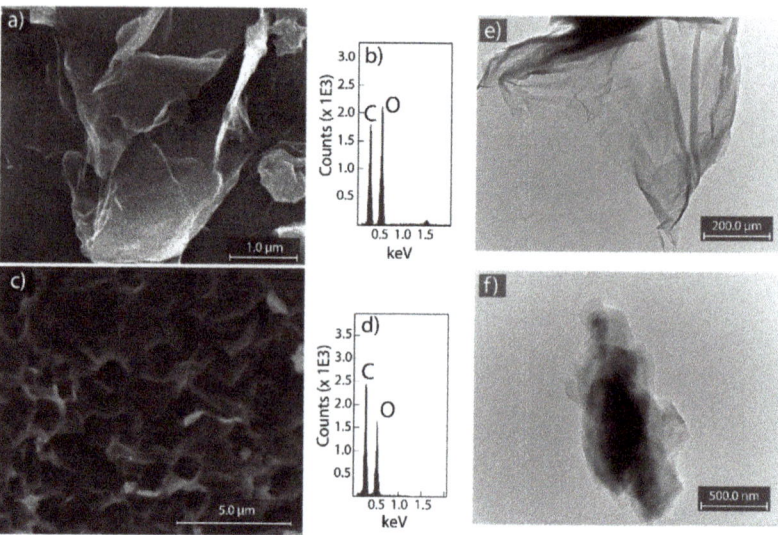

Figure 2. SEM morphology of (**a**) GO and (**c**) rGO. EDS measurements of (**b**) GO and (**d**) rGO. TEM images of (**e**) GO and (**f**) rGO.

Representative TEM graphs of GO and rGO are shown in Figure 2e,f, respectively. GO looks like a semi-transparent thin nanosheet with various wrinkles and folds on the surface and edges (Figure 2e). The observed wrinkled/folded structure is attributed to surface defects because of the deviation from sp^2 to sp^3 hybridization as the effect of a high density of oxygen-containing functional groups [54]. After the reduction process with CA, well-defined and impurity-free nanosheets with slightly wrinkled regions are observed in rGO, suggesting the recovery of sp^2 hybridization by the removal of functional groups. The observed regular surface allows concluding that rGO did not undergo severe in-plane disruption compared to GO.

Raman analyses were performed to further corroborate the transformation of GO into rGO (Figure 3a,b, respectively). As is typical for oxidized graphenes, two characteristic peaks are observed in GO and rGO: (i) the D peak at ~1349 cm^{-1} is attributed to the breathing mode of aromatic carbon rings, which is Raman active by structural defects [32], and (ii) the G peak at ~1588 cm^{-1} is due to the C-C stretching mode in the sp^2 hybridized carbon structure [46]. A detailed analysis using Lorentz functions shows the existence of three prominent bands: the D band (yellow line), the G band (green line), and the D' band (blue line). In particular, the D' band confirms the presence of basal/edge defects, and a decrease in the D' band intensity is a direct indication of GO reduction, which is observed in the Raman spectrum of rGO (Figure 3b). On the other hand, the I_D/I_G intensity ratio can be used as an indicator of the density of structural defects in the obtained oxidized graphenes [43]. It was found that the intensity ratio of GO (2.2) is larger than that of rGO (1.65), indicating that the size of the graphene-like domains increases after the reduction process.

The absorbance spectra of GO and rGO are shown in Figure 3c,d, respectively. Using the Lorentzian function, GO has the main absorbance band at 230 nm (darker green line) and a shoulder band at 329 nm (yellow line), which are related to the $\pi - \pi^*$ transitions of C-C bonds and $n - \pi^*$ transitions of C=O bonds, respectively. To confirm the transformation of GO into rGO, two characteristics are needed: (i) a redshift of the main absorbance band and (ii) the loss of the shoulder band. After the reduction process, rGO only meets the first point when the main absorbance band shifts to 261 nm, but the second one is observed at 324 nm, suggesting a close content of oxygen-containing functional groups, particularly hydroxyl and epoxide groups.

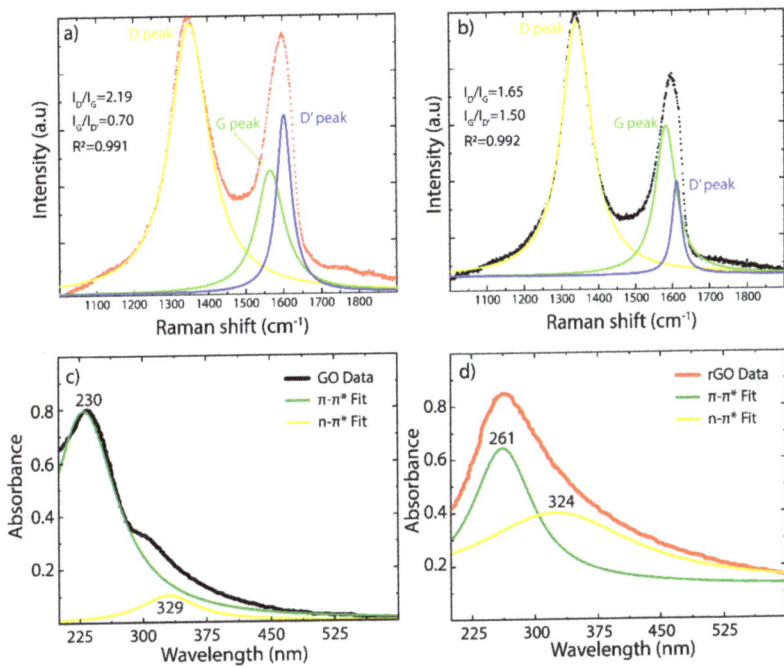

Figure 3. Raman spectra of (**a**) GO and (**b**) rGO from 1000 to 2000 cm^{-1}. The intensity was normalized by the most intense peak. The Raman spectrum was fitted using Lorentzian functions. UV–Vis spectra of (**c**) GO and (**d**) rGO and the absorbance spectra were fitted by Lorentzian functions.

The presence and type of oxygen functional groups are confirmed by the FTIR analysis (Figure 4a). It is widely accepted that the hydroxyl and epoxide groups are attached to the basal in-plane of the graphene, whereas the carboxyl and carbonyl groups are located at the edges. The FTIR spectrum of GO shows the following bands: C-O-C at 1044 cm^{-1}, C-O at 1222 cm^{-1}, C=C at 1644 cm^{-1}, and C=O at 1729 cm^{-1}. The broadband observed at ~3426 cm^{-1} is due to the presence of the hydroxyl groups (C-H) as well as adsorbed water molecules between GO flakes. The latter provides a hydrophilic characteristic in GO to be highly dispersible in water. It is worth noting that a higher hydrophilic property could interfere with the removal of pollutants from aqueous media, giving a poor adsorption process. After the reduction, these bands are significantly attenuated and weakened in the rGO spectrum, evidencing the removal of oxygen-containing functional groups [33].

To determine the thermal stability of as-made oxidized graphenes and the effect on the oxygen-containing functional groups, we carried out TGA analyses on GO and rGO (Figure 4b). In GO, the weight loss below 100 °C is ascribed to the loss of water molecules [33]. The significant weight loss in the region of 200–300 °C is attributed to the pyrolysis of unstable molecules (such as CO, CO_2, and H_2O) [33]. In the region of 300–600 °C, the weight loss is due to the removal of stable oxygen functional groups [33]. Instead, rGO shows relative thermal stability, but the observed TGA curve follows a similar trend as GO, confirming a reduced density of oxygen functional groups.

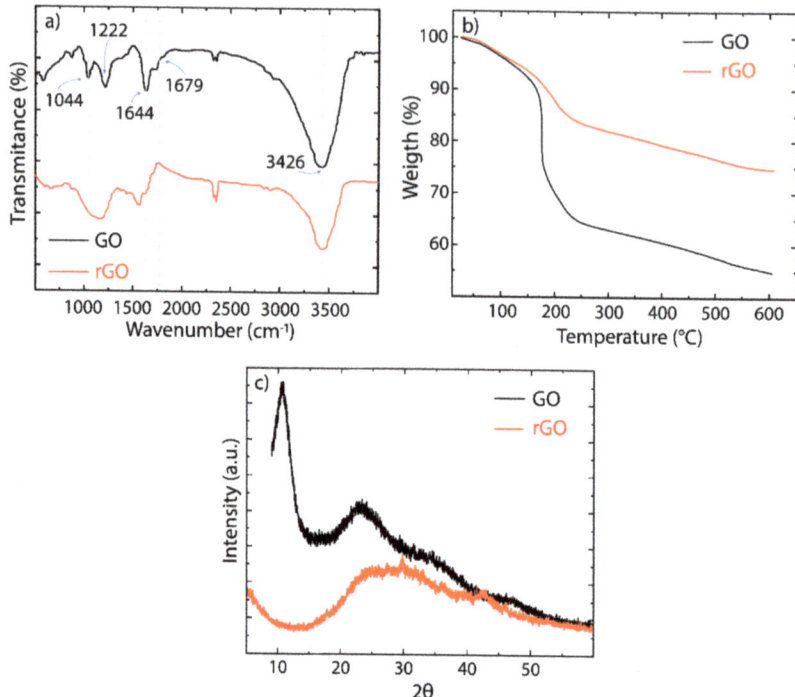

Figure 4. (**a**) Infrared spectra, (**b**) thermogravimetric study, and (**c**) XRD patterns of GO (black) and rGO (red), respectively.

Finally, the crystallinity changes from GO to rGO were revealed by XRD analysis (Figure 4c). As mentioned, graphite is characterized by an intense crystalline peak at $2\theta = 26.73°$ related to a lattice spacing of 0.334 nm, which corresponds to the (002) interplane distance [43] (Figure 1b). In GO, this peak is found at $2\theta = 10.93°$ with a lattice spacing of 0.81 nm, indicating the oxidation of graphite. The increased interlayer spacing appears as an effect of the intercalation of water molecules and oxygen functional groups. Additionally, the very low width of this peak demonstrates an ordered stacking along the out-of-plane axis. After the reduction process, the peak becomes broader due to the partial breakdown of the long-range order, and it shifts towards higher angles, $2\theta = 22°$, showing a decrease in the lattice spacing (~0.39 nm) [43].

All these facts and pieces of evidence demonstrate the transformation of GO into rGO, which will be used for the removal of cationic pollutants from aqueous media, i.e., MB and Hg(II).

3.3. Adsorption Kinetics

We begin analyzing the effectiveness of rGO for removing MB and Hg(II) from water by using the following expression:

$$q_t = \frac{(C_0 - C_t)V}{W} \qquad (1)$$

where C_0 and C_t are the initial pollutant concentration (mg L^{-1}) and the pollutant concentration at time t, respectively. W is the adsorbent mass (g), and V represents the volume of the aqueous solution (L). At the equilibrium, the equilibrium concentration is $C_e = C_t$, and the equilibrium adsorption capacity is $q_e = q_t$.

The removal efficiency ($RE\%$) of the as-made rGO material can be defined by the following simple equation:

$$RE\% = \frac{(C_0 - C_e)}{C_0} \times 100 \qquad (2)$$

Figure 5 shows the adsorption kinetics of MB and Hg (II) onto rGO at 298 K considering a contact time of up to 60 min. It can be seen that rGO rapidly captures MB molecules after 30 min (Figure 5a), while for Hg (II), the equilibrium time of adsorption is 20 min (Figure 5b). These results highlight the effectiveness of rGO for removing cationic pollutants from aqueous solutions compared with conventional benchmark sorbents [35–40]. In particular, the effectiveness of rGO can be attributed to the recovered surface area after the reduction process as well as the presence of oxygen functional groups.

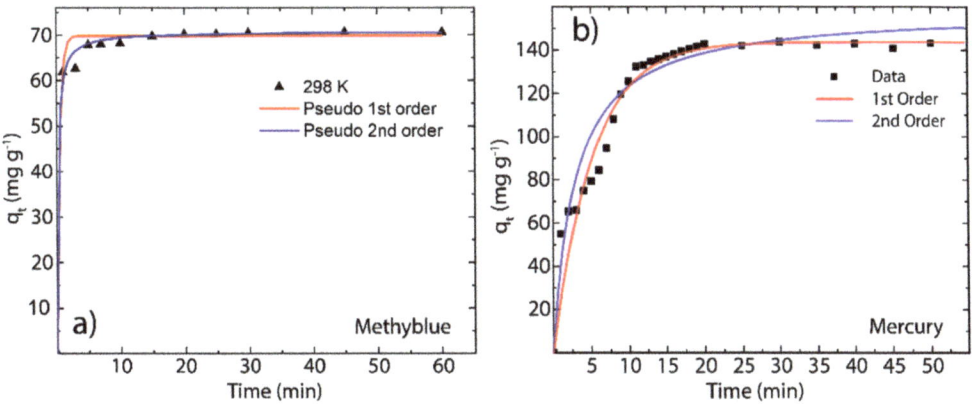

Figure 5. Adsorption kinetics of (**a**) MB on rGO and (**b**) Hg(II) on rGO as a function of contact time (60 min) at 298 K.

The parameters of the adsorption kinetic process were determined by the pseudo-first-order model and pseudo-second-order model. Specifically, Tene et al. stated that the first model assumes that the rate of change of the adsorption capacity is proportional to the concentration of available active sites per unit mass of adsorbent material [25], whereas Arias et al. stated that the second model assumes that the rate of change of the concentration of occupied active sites per unit mass of the adsorbent material is proportional to the square of the concentration of free active sites per unit mass of sorbent [13].

The pseudo-first-order model (red line) and pseudo-second-order model (blue line) can be described as follows:

$$\log(q_e - q_t) = \log q_t - \frac{k_1}{2.303} t \qquad (3)$$

and

$$\frac{t}{q_t} = \frac{1}{k_2 q_e^2} + \frac{1}{q_e} t \qquad (4)$$

where k_1 and k_2 are the pseudo-first-order and pseudo-second-order rate constants, respectively. The estimated parameters of the adsorption kinetics are summarized in Table 1.

Table 1. Estimated parameters at 298 K of the pseudo-first-order model and the pseudo-second-order model.

Parameters	MB	Hg(II)
$q_e(exp)$ (mg g^{-1})	68.21	142.26
Pseudo-first-order model		
q_e (mg g^{-1})	69.82 ± 0.05	143.70 ± 5.70
k_1 (min^{-1})	2.166 ± 0.03	0.19 ± 0.03
SSE	13.38	1826
R^2	0.997	0.949
RMSE	1.157	8.546
Pseudo-second-order model		
q_e (mg g^{-1})	70.72 ± 0.08	158.30 ± 9.45
k_2 (g mg^{-1} min^{-1})	0.075 ± 0.005	0.002 ± 0.001
SSE	5.239	2480
R^2	0.999	0.931
RMSE	0.724	9.960

A close picture of the pseudo-first-order and pseudo-second-order parameters shows that, in the case of MB, the calculated values of the equilibrium adsorption capacity ($q_{e(cal)}$ = 69.82 mg g^{-1} and $q_{e(cal)}$ = 70.72 mg g^{-1}, respectively) are very close to the experimental value ($q_{e(exp)}$ = 68.21 mg g^{-1}). In the case of Hg(II), the calculated adsorption capacity ($q_{e(cal)}$ = 143.71 mg g^{-1}) from the pseudo-first-model is close enough to the experimental value ($q_{e(exp)}$ = 142.26 mg g^{-1}). However, the pseudo-second-order model overestimates the equilibrium adsorption capacity ($q_{e(cal)}$ = 151.32 mg g^{-1}). By the comparison of SSE and R^2 metrics, the adsorption kinetics of MB onto rGO are more in line with the pseudo-second-order model (SSE = 5.24, R^2 = 0.999), whereas the adsorption kinetics of Hg(II) onto rGO are more in line with the pseudo-first-order model (SSE = 1826, R^2 = 0.949).

3.4. Intraparticle Diffusion Study

The diffusion process of any pollutant into porous solid materials, such as our rGO (Figure 2c), mostly involves several steps characterized by different rates. This fact can be calculated by the intraparticle diffusion (IPD) model [13,25], which is given by the following expression:

$$q_t = k_p t^{0.5} + C \quad (5)$$

where k_p is the intraparticle diffusion rate constant (g mg^{-1} min^{-1}), and the intercept C reflects the boundary layer or surface adsorption. The respective plot and estimated parameters of the IPD model are shown in Figure 6 and Table 2.

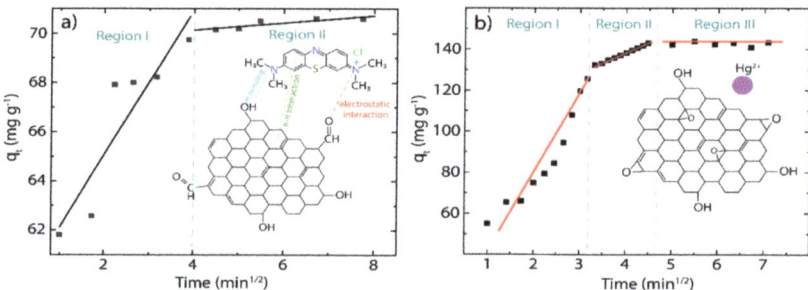

Figure 6. Intraparticle diffusion (IPD) study of (**a**) rGO+Mb and (**b**) rGO+Hg(II) at 298 K, showing different regions of linearity (MB concentration 100 mg L^{-1} and Hg(II) concentration 150 mg L^{-1}).

Table 2. Estimated parameters of the intraparticle diffusion (IPD) model at 298 K.

Parameters	MB Value	Hg(II) Value
K_p (mg g^{-1} min$^{-1/2}$)	2.95 ± 0.67	7.82 ± 1.25
C (mg g^{-1})	59.17 ± 1.75	44.28 ± 7.75
R_i	0.132	0.69
R^2	0.829	0.963

As Ofomaja et al. [55] stated, the larger the intercept value, the greater the contribution of the surface in the adsorption process. Indeed, the values observed in MB ($C = 59.17$) and Hg(II) ($C = 44.28$) indicate that a greater amount of surface adsorption occurred, leading to a decrease in the rate of diffusion of MB molecules and Hg(II) ions from the adsorbent external surface to the adsorbent internal structure. From the linearized plot of the IPD model, different regions are observed: (i) the initial region (faster stage) is related to the movement of the pollutant from the solution to the rGO surface, (ii) the second region (intermediate stage) is related to the gradual diffusion of the pollutant into the large pores of the rGO structure, and (iii) the final region (lower stage) involves a very slow diffusion of the pollutant from larger pores to smaller ones.

Interestingly, the adsorption mechanism of MB on rGO is characterized by only two regions, regions I and II (Figure 6a), while all three regions are observed when Hg(II) becomes adsorbed onto rGO. In light of understanding this fact, we hypothesize that the size of pollutants plays a significant role in the diffusion procedure, i.e., as the MB molecules exhibit larger sizes compared to Hg(II) ions, MB cannot reach region III, particularly from larger to smaller pores.

The initial adsorption factor (R_i) can be estimated to further understand the above-mentioned regions (Table 2) as follows:

$$R_i = \frac{q_{ref} - C}{q_{ref}} \quad (6)$$

where q_{ref} is the final adsorption amount at the longest time. In the MB-rGO system, the estimated R_i value is much less than 0.5, which confirms that most of the adsorption of MB occurs on the surface of rGO. In contrast, for the Hg(II)-rGO system, the value of $R_i \sim 0.49$ indicates a limit between the strong initial adsorption (related to region I) and intermediate initial adsorption (related to region II), which means that the adsorption process of Hg(II) ions could occur at almost the same time in both regions.

3.5. Adsorption Isotherms

Adsorption isotherms were carried out to analyze the interaction between MB molecules or Hg(II) ions and rGO considering a contact time of 30 min for MB and 20 min for Hg(II). The experimental data can be fitted using the Langmuir model and Freundlich model using the following equations, respectively:

$$q_e = \frac{q_m K_L C_e}{1 + K_L C_e} \quad (7)$$

and

$$q_e = K_F C_e^{1/n} \quad (8)$$

where q_m represents the maximum adsorption capacity (mg g^{-1}), K_L is the Langmuir constant (L g^{-1}), K_F is Freundlich constant (mg L^{-1}), and n is the surface heterogeneity of adsorbent material. The corresponding results and estimated parameters at different temperatures (298, 313, 333 K) are shown in Figures 7 and 8 and Tables 3 and 4.

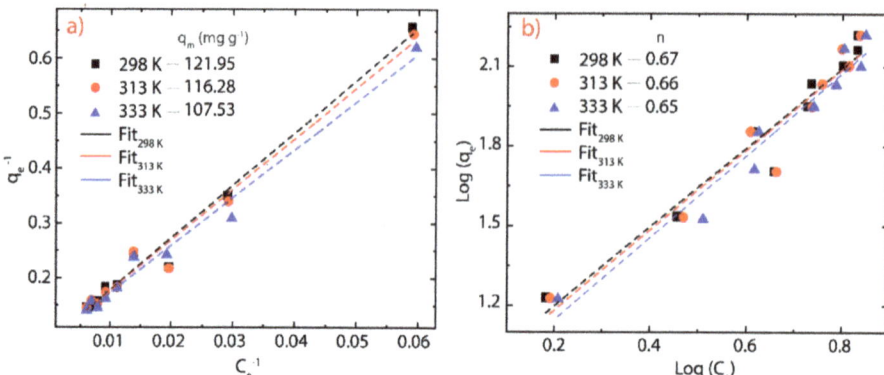

Figure 7. Adsorption isotherms of MB on rGO considering three different temperatures (289–333 K). (**a**) Langmuir model and (**b**) Freundlich model.

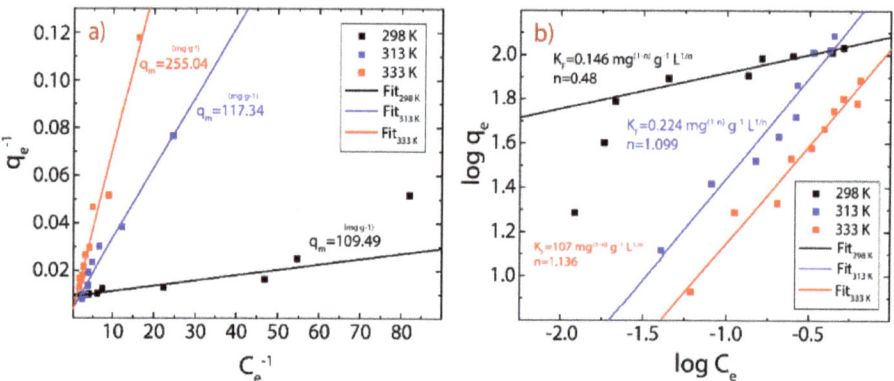

Figure 8. Adsorption isotherm of Hg(II) on rGO considering three different temperatures (289–333 K). (**a**) Langmuir model and (**b**) Freundlich model.

Table 3. Parameters of the Langmuir and Freundlich models for the adsorption isotherms of MB onto rGO considering three different temperatures.

T (K)	Langmuir Model			Freundlich Model		
	K_L (L g^{-1})	$q_{m(cal)}$ (mg g^{-1})	R^2	K_F (mg$^{(1-n)}$ g^{-1} L$^{1/n}$)	n	R^2
298	0.079	121.95	0.982	7.956	0.671	0.945
313	0.081	116.28	0.980	7.568	0.661	0.936
333	0.082	107.53	0.984	6.869	0.646	0.955

Table 4. Parameters of the Langmuir and Freundlich models for the adsorption isotherms of Hg(II) onto rGO considering three different temperatures.

T (K)	Langmuir Model			Freundlich Model		
	K_L (L g^{-1})	$q_{m(cal)}$ (mg g^{-1})	R^2	K_F (mg$^{(1-n)}$ g^{-1} L$^{1/n}$)	n	R^2
298	0.047	109.493	0.934	0.146	0.480	0.918
313	0.017	217.344	0.973	0.224	1.099	0.947
333	0.006	255.037	0.968	0.107	1.136	0.956

Taking the high correlation R^2 values (Tables 3 and 4), it can be seen that the measured points are more in line with the Langmuir model. Although the temperature does not dramatically modify the chemical composition of rGO at temperatures below 100 °C (Figure 4b), it seems to be an important parameter in the adsorption process because when the temperature increases, in the case of MB on rGO, a slight decrease in the maximum adsorption capacity is observed from 121.95 to 107.53 mg g^{-1}. In contrast, in the case of Hg (II) on rGO, a significant increase in the maximum adsorption capacity is detected from 109.49 to 255.04 mg g^{-1}. The temperature is a key point to be considered if rGO is used to treat water or wastewater at an industrial scale.

From the Freundlich model, the estimated values of n for MB-rGO (Table 3) or Hg(II)-rGO (Table 4) systems indicate that the adsorbent heterogeneity tends to be homogeneous as the temperature rises. Indeed, values of n close to zero (<0.1) indicate strong surface heterogeneity. The affinity of the tested cationic pollutants for rGO can also be determined by the K_L parameter, where the estimated values were found to be much less than 0.1, suggesting a good affinity of rGO to capture cationic pollutants, i.e., MB molecules and Hg(II) ions. However, this statement motivates more extended work for testing more cationic and non-cationic pollutants.

3.6. Effect of pH and Initial Concentration

To scrutinize the effect of the pH on the process of cationic pollutant removal, the experiments were carried out at different pH values ranging from 2 to 12 and setting the temperature at 298 K.

For MB on rGO (Figure 9a), the adsorption increases, starting from a removal percentage of about 76% at pH = 3 up to 92% at pH = 6. The removal percentage remains relatively constant from pH = 6 to pH = 8. The removal percentage decreases down to 83% for pH ≥ 10. To understand this fact, the effect of pH can be divided into three different regions: (i) from pH = 2 to pH = 4, the acid region is rich in cations which are captured together with the cationic MB molecules; (ii) the (relatively) neutral region from pH = 6 to pH = 8 is free from cations in the medium, and therefore, only the cationic dye molecules are captured by rGO; and (iii) the basis region (pH ≥ 10) is characterized by an excess of OH$^-$ ions that interact with the cationic dye molecules, remaining suspended in the aqueous media.

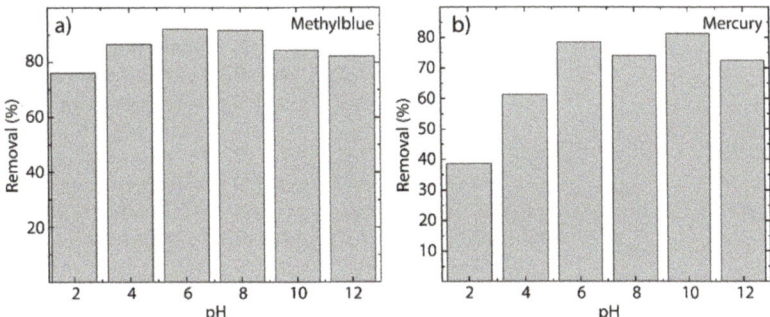

Figure 9. Removal percentage as a function of the pH (from 2 to 12) at 298 K of (**a**) MB on rGO and (**b**) Hg(II) on rGO (MB concentration 100 mg L^{-1} and Hg(II) concentration 100 mg L^{-1}).

For Hg(II) on rGO (Figure 9b), a removal percentage of about 39% at pH = 2 is observed, and the maximum removal percentage is found at pH = 10 (~82%). The removal percentage remains relatively constant for pH ≥ 6, with an average value of 76.58%. To understand these results, a similar description can be given: (i) for pH ≤ 4, the cations in the acidic medium fight with the mercury cations for the active sites of rGO; (ii) in the neutral region, mercury cations easily reach the active sites of rGO; and, interestingly, (iii) for pH ≥ 8, mercury cations sometimes prefer to interact with the active sites of rGO rather than OH$^-$ ions due to the variation of the removal percentage when the pH increases.

The adsorption capacity (q_e) of rGO increases quite linearly with the initial concentration of MB in solution (C_0), almost in the range from 10 to 80 mg L^{-1}; however, at high concentrations ($\geq$90 mg L^{-1}), a deviation from linearity does occur (Figure 10a). Similarly, the q_e values of rGO increase linearly with the initial concentration of Hg(II) in the solution, from 10 to 50 mg L^{-1}, and at concentrations $\geq$50 mg L^{-1}, a deviation from linearity is also observed (Figure 10b). These results suggest that rGO has a finite amount of active adsorbent sites, which is fixed by its quality and the experimental conditions, such as temperature, pH, and solution volume/adsorbent mass ratio. To further emphasize, at the beginning of the adsorption process, rGO has a vast number of active sites, increasing the q_e value as long as free active sites are available on rGO. Then, if all the active sites are involved, the saturation, and therefore the maximum adsorbent capacity (q_m), is attained [13,25].

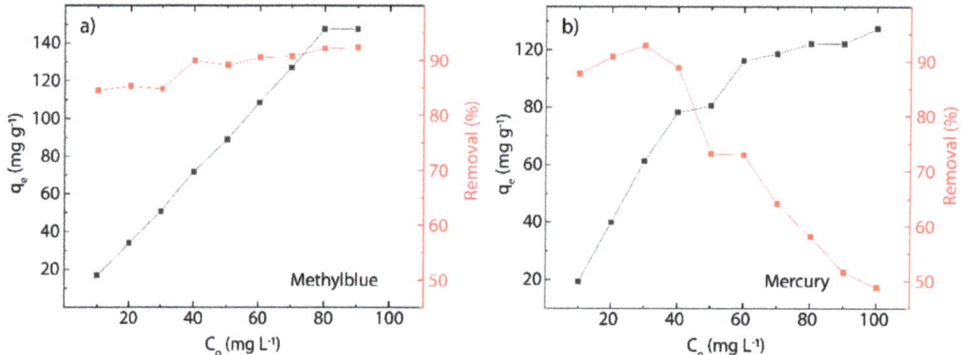

Figure 10. Effect of the initial concentration on the adsorption process at 298 K of (**a**) MB or rGO and (**b**) Hg(II) on rGO. Adsorption capacity (black markers) and removal percentage (red markers).

The adsorption effectiveness of rGO—defined as the percentage of cationic pollutant removal from water—is almost independent of C_0 in the adsorption of MB onto rGO, assuming an average value of 89.21%. Interestingly, a clear dependence on C_0 is observed for the adsorption of Hg(II) onto rGO, i.e., an abrupt drop from 92.89% (C_0 = 30 mg L^{-1}) to 48.85% (C_0 = 100 mg L^{-1}), giving an average value of mercury removal of 72.93%.

3.7. Adsorption Thermodynamics

To acquire information about the energy changes due to the involved adsorption process [13], the Gibbs free energy (ΔG^0), enthalpy change (ΔH^0), and entropy change (ΔS^0) were calculated by the following expressions:

$$K_d = \frac{q_e}{C_0} \qquad (9)$$

$$\ln K_d = \frac{\Delta S^0}{R} - \frac{\Delta H^0}{R T} \qquad (10)$$

$$\Delta G^0 = -R T \ln K_d \qquad (11)$$

where K_d represent the distribution coefficient [13]. ΔH^0 and ΔS^0 were calculated from the slope and intercept of Van't Hoff plot of $\ln K_d$ as a function of T^{-1} [25]. The Van't Hoff plot and estimated parameters are shown in Figure 11 and Table 5, respectively.

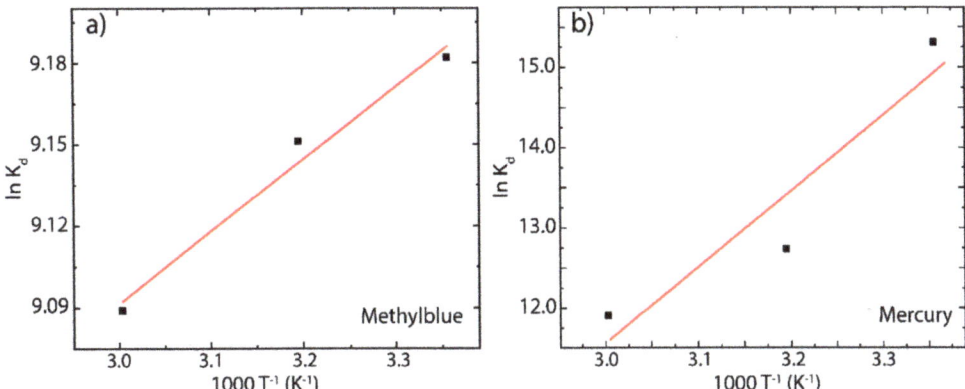

Figure 11. Van't Hoff study for the adsorption of (**a**) MB on rGO and (**b**) Hg(II) on rGO.

Table 5. Thermodynamics parameters for Mb and Hg(II) adsorption on rGO at three different temperatures.

T (K)	ΔG^0(kJ mol^{-1})	ΔH^0(kJ mol^{-1})	ΔS^0(kJ mol^{-1}K^{-1})
		MB	
298	−22.75		
313	−23.81	−2.20	0.069
333	−25.16		
		Hg(II)	
298	−39.84		
313	−31.55	−0.14	0.079
333	−32.97		

The negative ΔG^0 values observed at different temperatures indicate spontaneous adsorption of MB molecules and Hg(II) ions onto the rGO surface. It is worth noting that, for ΔG^0 values in the range from 0 to −20 kJ mol^{-1}, the adsorption process is assigned to physisorption or multilayer adsorption [25], while in the range from −80 to −400 kJ mol^{-1}, the adsorption is assigned to chemisorption or monolayer adsorption [25]. The region from −20 to −80 kJ mol^{-1} remains unclear, and a combined adsorption process can be assumed. With this in mind, the estimated ΔG^0 values for MB onto rGO (−22.75, −23.81, and −25.16 kJ mol^{-1}) and Hg(II) onto rGO (−39.84, 31.55, 32.97 kJ mol^{-1}) suggest that the adsorption process of tested cationic pollutants on rGO is governed by a mixed physisorption–chemisorption process. Interestingly, for MB on rGO, the ΔG^0 value increases by 5% at 313 K and by 10% at 333 K. In contrast, an inversely proportional relationship is observed for Hg(II) on rGO; say, the ΔG^0 value decreases by 21% at 313 k and by 17% at 333 K.

The negative ΔH^0 values indicate the exothermic nature of the adsorption process, i.e., a negative enthalpy implies that the temperature increase had a negative impact, particularly on the adsorption of MB ($\Delta H^0 = -2.20$ kJ mol^{-1}). However, in the adsorption of Hg(II) on rGO, the value observed ($\Delta H^0 = -0.14$ kJ mol^{-1}) is very small and could be considered negligible since increasing the temperature significantly increases the maximum adsorption capacity of rGO, as evidenced by the Langmuir model (Figure 8a and Table 4). The positive values of $\Delta S^0 = 0.069$ kJ mol^{-1} · K^{-1} and $\Delta S^0 = 0.079$ kJ mol^{-1} · K^{-1} corroborate the affinity of MB molecules and Hg(II) ions toward the rGO surface.

3.8. Final Remarks

Table 6 shows the estimated q_m values for MB ($q_m = 121.95$ mg g^{-1}) and Hg(II) ($q_m = 109.49$ mg g^{-1}) at 298 K, which are compared to those of recent studies.

Table 6. Comparative adsorption capacity of several adsorbents for the removal of MB and Hg(II).

	MB		Hg(II)	
Adsorbents	Adsorption Capacity (mg g^{-1})	Adsorbents		Adsorption Capacity (mg g^{-1})
Graphene/SrAl2O3:Bi^{3+} [56]	42.93	GONR [35]		33.02
ß-cyclodextrin/MGO [57]	93.97	S-GO [36]		3490
g-C$_3$N$_4$ (Urea) [58]	2.51	GO-TSC [37]		231
Magnetic carboxyl functional nanoporous polymer [59]	57.74	S-doped g-C$_3$N$_4$/LGO [38]		46
Ag-Fe$_3$O$_4$-polydopamine [60]	45.00	GSH-NiFe$_2$O$_4$/GO [39]		272.94
Citrus hystrix-rGO [61]	276.06	HT-rGO-N [40]		75.8
This work	121.95	This work		109.49

The estimated q_m value of the dye adsorption is higher than those previously reported and only surpassed by GO reduced by Citrus hystrix (q_m = 276.06 mg g^{-1}), suggesting that as-made (non-extra functionalized) rGO are excellent platforms to replace conventional adsorbent materials. In the case of heavy metal adsorption, the estimated q_m value is higher than some functionalized/decorated GO and rGO. However, S-GO seems to be more profitable to be used for the removal of mercury (q_m = 3490 mg g^{-1}), but this is due to the fact that, obviously, the presence of sulfur improves the affinity and specificity for Hg (II) ions in any adsorbent material.

Although, in the present work, the regeneration of rGO was not studied, which motivates more extended work, we propose the following well-known techniques or processes: (i) the adsorbed rGO-pollutant system can be separated from aqueous media by filtration using filters with a pore size less than 1 µm since rGO is within the order of few micrometers, (ii) pollutant can be released from rGO by applying the concept of ionic force, i.e., by applying buffer solutions, and (iii) the isolated MB molecules or Hg(II) ions can be extracted by sulfide precipitation.

4. Conclusions

In summary, we have demonstrated the effective and efficient removal of cationic pollutants (i.e., MB molecules and Hg(II) ions) from aqueous solutions using an eco-friendly and as-made rGO. The adsorbent material shows fast adsorption with a saturation capacity of 121.95 mg g^{-1} and 109.49 mg g^{-1} at 298 K, suggesting a good affinity for MB molecules and Hg(II) ions. These results are superior to those recently reported for other graphene-based benchmark materials [35–40]. By means of several chemical physics analyses, we have also shown that rGO keeps a good efficiency over a wide range of initial cationic pollutant concentrations and a broad range of pH values. Specifically, the maximum removal percentage as a function of pH was found in the range of 6 to 8 for MB and 6 to 10 for Hg(II). Our results allowed us to conclude that the MB-rGO and Hg(II)-rGO adsorption interaction follows a combined physisorption–chemisorption process due to the fact that the Gibbs free energy was found from −22.75 to −25.16 kJ mol^{-1} for MB and from −39.84 to −32.97 kJ mol^{-1} for Hg(II). The present study proposes non-extra functionalized rGO as a potential green adsorbent for wastewater decontamination.

Author Contributions: Conceptualization, S.B.; Data curation, T.T., M.G. and C.V.G.; Formal analysis, L.S.C. and C.V.G.; Funding acquisition, T.T.; Investigation, T.T., L.S.C. and C.V.G.; Methodology, E.V., M.A.P., O.S., E.V.-G., S.V., S.B., F.A., A.S. and L.S.C.; Validation, M.G.; Visualization, M.G.; Writing—original draft, S.B. and C.V.G. All authors have read and agreed to the published version of the manuscript.

Funding: This work was supported by Universidad Técnica Particular de Loja (UTPL-Ecuador), project "222-Radon adsorption on activated carbon and graphene filters" (code: PROY_INNOV_QUI-_2021_3019), and by the FONDOCyT from the Ministry of Higher Education Science and Technology of the Dominican Republic (grant no. 2018-2019-3A9-139).

Data Availability Statement: Not applicable.

Acknowledgments: T.T., M.G. and C.V.G. wish to thank Escuela Superior Politécnica de Chimborazo for its hospitality during the completion of this work.

Conflicts of Interest: The authors declare no conflict of interest.

References

1. Hairom, N.H.H.; Soon, C.F.; Mohamed, R.M.S.R.; Morsin, M.; Zainal, N.; Nayan, N.; Zulkifli, C.Z.; Harun, N.H. A review of nanotechnological applications to detect and control surface water pollution. *Environ. Technol. Innov.* **2021**, *24*, 102032. [CrossRef]
2. Zamora-Ledezma, C.; Negrete-Bolagay, D.; Figueroa, F.; Zamora-Ledezma, E.; Ni, M.; Alexis, F.; Guerrero, V.H. Heavy metal water pollution: A fresh look about hazards, novel and conventional remediation methods. *Environ. Technol. Innov.* **2021**, *22*, 101504. [CrossRef]
3. Benkhaya, S.; M'rabet, S.; El Harfi, A. Classifications, properties, recent synthesis and applications of azo dyes. *Heliyon* **2020**, *6*, e03271. [CrossRef]
4. Zhou, Y.; Lu, J.; Zhou, Y.; Liu, Y. Recent advances for dyes removal using novel adsorbents: A review. *Environ. Pollut.* **2019**, *252*, 352–365. [CrossRef] [PubMed]
5. Ayodhya, D.; Venkatesham, M.; Santoshi Kumari, A.; Reddy, G.B.; Ramakrishna, D.; Veerabhadram, G. Photocatalytic degradation of dye pollutants under solar, visible and UV lights using green synthesised CuS nanoparticles. *J. Exp. Nanosci.* **2016**, *11*, 418–432. [CrossRef]
6. Tang, A.Y.L.; Lo, C.K.Y.; Kan, C. Textile dyes and human health: A systematic and citation network analysis review. *Color. Technol.* **2018**, *134*, 245–257. [CrossRef]
7. Zhang, D.; Tong, Z.; Zheng, W. Does designed financial regulation policy work efficiently in pollution control? Evidence from manufacturing sector in China. *J. Clean. Prod.* **2021**, *289*, 125611. [CrossRef]
8. Fu, Z.; Xi, S. The effects of heavy metals on human metabolism. *Toxicol. Mech. Methods* **2020**, *30*, 167–176. [CrossRef] [PubMed]
9. Yang, F.; Massey, I.Y.; Al Osman, M. Exposure routes and health effects of heavy metals on children. *Biometals* **2019**, *32*, 563–573.
10. Efome, J.E.; Rana, D.; Matsuura, T.; Lan, C.Q. Experiment and modeling for flux and permeate concentration of heavy metal ion in adsorptive membrane filtration using a metal-organic framework incorporated nanofibrous membrane. *Chem. Eng. J.* **2018**, *352*, 737–744. [CrossRef]
11. Dąbrowski, A.; Hubicki, Z.; Podkościelny, P.; Robens, E. Selective removal of the heavy metal ions from waters and industrial wastewaters by ion-exchange method. *Chemosphere* **2004**, *56*, 91–106. [CrossRef] [PubMed]
12. Arias, F.E.A.; Beneduci, A.; Chidichimo, F.; Furia, E.; Straface, S. Study of the adsorption of mercury (II) on lignocellulosic materials under static and dynamic conditions. *Chemosphere* **2017**, *180*, 11–23. [CrossRef]
13. Al-Shannag, M.; Al-Qodah, Z.; Bani-Melhem, K.; Qtaishat, M.R.; Alkasrawi, M. Heavy metal ions removal from metal plating wastewater using electrocoagulation: Kinetic study and process performance. *Chem. Eng. J.* **2015**, *260*, 749–756. [CrossRef]
14. Belkacem, M.; Khodir, M.; Abdelkrim, S. Treatment characteristics of textile wastewater and removal of heavy metals using the electroflotation technique. *Desalination* **2008**, *228*, 245–254. [CrossRef]
15. Rajamani, A.R.; Ragula, U.B.R.; Kothurkar, N.; Rangarajan, M. Nano- and micro-hexagons of bismuth on polycrystalline copper: Electrodeposition and heavy metal sensing. *Cryst. Eng. Comm.* **2014**, *16*, 2032–2038. [CrossRef]
16. Umapathi, R.; Ghoreishian, S.M.; Sonwal, S.; Rani, G.M.; Huh, Y.S. Portable electrochemical sensing methodologies for on-site detection of pesticide residues in fruits and vegetables. *Coord. Chem. Rev.* **2022**, *453*, 214305. [CrossRef]
17. Umapathi, R.; Sonwal, S.; Lee, M.J.; Rani, G.M.; Lee, E.S.; Jeon, T.J.; Kang, S.M.; Oh, M.-H.; Huh, Y.S. Colorimetric based on-site sensing strategies for the rapid detection of pesticides in agricultural foods: New horizons, perspectives, and challenges. *Coord. Chem. Rev.* **2021**, *446*, 214061. [CrossRef]
18. Cheng, T.-H.; Sankaran, R.; Show, P.L.; Ooi, C.W.; Liu, B.-L.; Chai, W.S.; Chang, Y.-K. Removal of protein wastes by cylinder-shaped NaY zeolite adsorbents decorated with heavy metal wastes. *Int. J. Biol. Macromol.* **2021**, *185*, 761–772. [CrossRef]
19. Joo, G.; Lee, W.; Choi, Y. Heavy metal adsorption capacity of powdered Chlorella vulgaris biosorbent: Effect of chemical modification and growth media. *Environ. Sci. Pollut. Res.* **2021**, *28*, 25390–25399. [CrossRef]
20. Sabir, A.; Altaf, F.; Batool, R.; Shafiq, M.; Khan, R.U.; Jacob, K.I. Agricultural Waste Absorbents for Heavy Metal Removal. In *Green Absorbents to Remove Metals, Dyes and Boron from Polluted Water*; Springer: Berlin/Heidelberg, Germany, 2021; pp. 195–228.
21. Penido, E.S.; de Oliveira, M.A.; Sales, A.L.R.; Ferrazani, J.C.; Magalhães, F.; Bianchi, M.L.; Melo, L.C.A. Biochars produced from various agro-industrial by-products applied in Cr (VI) adsorption-reduction processes. *J. Environ. Sci. Health Part A* **2021**, 1–10. [CrossRef]
22. Mariana, M.; HPS, A.K.; Mistar, E.M.; Yahya, E.B.; Alfatah, T.; Danish, M.; Amayreh, M. Recent advances in activated carbon modification techniques for enhanced heavy metal adsorption. *J. Water Process Eng.* **2021**, *43*, 102221. [CrossRef]

23. Hoang, A.T.; Nižetić, S.; Cheng, C.K.; Luque, R.; Thomas, S.; Banh, T.L.; Pham, V.V.; Nguyen, X.P. Heavy metal removal by biomass-derived carbon nanotubes as a greener environmental remediation: A comprehensive review. *Chemosphere* **2021**, *287*, 131959. [CrossRef] [PubMed]
24. Velusamy, S.; Roy, A.; Sundaram, S.; Kumar Mallick, T. A Review on Heavy Metal Ions and Containing Dyes Removal Through Graphene Oxide-Based Adsorption Strategies for Textile Wastewater Treatment. *Chem. Rec.* **2021**, *21*, 1570. [CrossRef] [PubMed]
25. Arias Arias, F.; Guevara, M.; Tene, T.; Angamarca, P.; Molina, R.; Valarezo, A.; Salguero, O.; Vacacela Gomez, C.; Arias, M.; Caputi, L.S. The adsorption of methylene blue on eco-friendly reduced graphene oxide. *Nanomaterials* **2020**, *10*, 681. [CrossRef]
26. Sindona, A.; Pisarra, M.; Gomez, C.V.; Riccardi, P.; Falcone, G.; Bellucci, S. Calibration of the fine-structure constant of graphene by time-dependent density-functional theory. *Phys. Rev. B* **2017**, *96*, 201408. [CrossRef]
27. Villamagua, L.; Carini, M.; Stashans, A.; Gomez, C.V. Band gap engineering of graphene through quantum confinement and edge distortions. *Ric. Mat.* **2016**, *65*, 579–584. [CrossRef]
28. Sindona, A.; Pisarra, M.; Bellucci, S.; Tene, T.; Guevara, M.; Gomez, C.V. Plasmon oscillations in two-dimensional arrays of ultranarrow graphene nanoribbons. *Phys. Rev. B* **2019**, *100*, 235422. [CrossRef]
29. Gomez, C.V.; Pisarra, M.; Gravina, M.; Riccardi, P.; Sindona, A. Plasmon properties and hybridization effects in silicene. *Phys. Rev. B* **2017**, *95*, 85419. [CrossRef]
30. Han, C.; Wang, X.; Peng, J.; Xia, Q.; Chou, S.; Cheng, G.; Huang, Z.; Li, W. Recent Progress on Two-Dimensional Carbon Materials for Emerging Post-Lithium (Na^+, K^+, Zn^{2+}) Hybrid Supercapacitors. *Polymers* **2021**, *13*, 2137. [CrossRef]
31. Fossum, J.; Lindholm, F.; Shibib, M. The importance of surface recombination and energy-bandgap arrowing in p-n-junction silicon solar cells. *IEEE Trans. Electron Devices* **1979**, *26*, 1294–1298. [CrossRef]
32. Tene, T.; Guevara, M.; Valarezo, A.; Salguero, O.; Arias Arias, F.; Arias, M.; Scarcello, A.; Caputi, L.S.; Vacacela Gomez, C. Drying-Time Study in Graphene Oxide. *Nanomaterials* **2021**, *11*, 1035. [CrossRef] [PubMed]
33. Tene, T.; Tubon Usca, G.; Guevara, M.; Molina, R.; Veltri, F.; Arias, M.; Caputi, L.S.; Vacacela Gomez, C. Toward Large-Scale Production of Oxidized Graphene. *Nanomaterials* **2020**, *10*, 279. [CrossRef] [PubMed]
34. Fiallos, D.C.; Gómez, C.V.; Tubon Usca, G.; Pérez, D.C.; Tavolaro, P.; Martino, G.; Caputi, L.S.; Tavolaro, A. Removal of acridine orange from water by graphene oxide. *AIP Conf. Proc.* **2015**, *1646*, 38–45.
35. Sadeghi, M.H.; Tofighy, M.A.; Mohammadi, T. One-dimensional graphene for efficient aqueous heavy metal adsorption: Rapid removal of arsenic and mercury ions by graphene oxide nanoribbons (GONRs). *Chemosphere* **2020**, *253*, 126647. [CrossRef]
36. Bao, S.; Wang, Y.; Yu, Y.; Yang, W.; Sun, Y. Cross-linked sulfydryl-functionalized graphene oxide as ultra-high capacity adsorbent for high selectivity and ppb level removal of mercury from water under wide pH range. *Environ. Pollut.* **2021**, *271*, 116378. [CrossRef]
37. Sitko, R.; Musielak, M.; Serda, M.; Talik, E.; Zawisza, B.; Gagor, A.; Malecka, M. Thiosemicarbazide-grafted graphene oxide as superior adsorbent for highly efficient and selective removal of mercury ions from water. *Sep. Purif. Technol.* **2021**, *254*, 117606. [CrossRef]
38. Li, M.; Wang, B.; Yang, M.; Li, Q.; Calatayud, D.G.; Zhang, S.; Wang, H.; Wang, L.; Mao, B. Promoting mercury removal from desulfurization slurry via S-doped carbon nitride/graphene oxide 3D hierarchical framework. *Sep. Purif. Technol.* **2020**, *239*, 116515. [CrossRef]
39. Khorshidi, P.; Shirazi, R.H.S.M.; Miralinaghi, M.; Moniri, E.; Saadi, S. Adsorptive removal of mercury (II), copper (II), and lead (II) ions from aqueous solutions using glutathione-functionalized $NiFe_2O_4$/graphene oxide composite. *Res. Chem. Intermed.* **2020**, *46*, 3607–3627. [CrossRef]
40. Yap, P.L.; Tung, T.T.; Kabiri, S.; Matulick, N.; Tran, D.N.H.; Losic, D. Polyamine-modified reduced graphene oxide: A new and cost-effective adsorbent for efficient removal of mercury in waters. *Sep. Purif. Technol.* **2020**, *238*, 116441. [CrossRef]
41. Golden, P.J.; Weinstein, R. Treatment of high-risk, refractory acquired methemoglobinemia with automated red blood cell exchange. *J. Clin. Apher. Off. J. Am. Soc. Apher.* **1998**, *13*, 28–31. [CrossRef]
42. Al-Saleh, I.; Moncari, L.; Jomaa, A.; Elkhatib, R.; Al-Rouqi, R.; Eltabache, C.; Al-Rajudi, T.; Alnuwaysir, H.; Nester, M.; Aldhalaan, H. Effects of early and recent mercury and lead exposure on the neurodevelopment of children with elevated mercury and/or developmental delays during lactation: A follow-up study. *Int. J. Hyg. Environ. Health* **2020**, *230*, 113629. [CrossRef]
43. Gomez, C.V.; Guevara, M.; Tene, T.; Villamagua, L.; Usca, G.T.; Maldonado, F.; Tapia, C.; Cataldo, A.; Bellucci, S.; Caputi, L.S. The liquid exfoliation of graphene in polar solvents. *Appl. Surf. Sci.* **2021**, *546*, 149046. [CrossRef]
44. Cayambe, M.; Zambrano, C.; Tene, T.; Guevara, M.; Usca, G.T.; Brito, H.; Molina, R.; Coello-Fiallos, D.; Caputi, L.S.; Gomez, C.V. Dispersion of graphene in ethanol by sonication. *Mater. Today Proc.* **2020**, *37*, 4027–4030. [CrossRef]
45. Tubon Usca, G.; Vacacela Gomez, C.; Guevara, M.; Tene, T.; Hernandez, J.; Molina, R.; Tavolaro, A.; Miriello, D.; Caputi, L.S. Zeolite-Assisted Shear Exfoliation of Graphite into Few-Layer Graphene. *Crystals* **2019**, *9*, 377. [CrossRef]
46. Vacacela Gomez, C.; Tene, T.; Guevara, M.; Tubon Usca, G.; Colcha, D.; Brito, H.; Molina, R.; Bellucci, S.; Tavolaro, A. Preparation of Few-Layer Graphene Dispersions from Hydrothermally Expanded Graphite. *Appl. Sci.* **2019**, *9*, 2539. [CrossRef]
47. Pierantoni, L.; Mencarelli, D.; Bozzi, M.; Moro, R.; Moscato, S.; Perregrini, L.; Micciulla, F.; Cataldo, A.; Bellucci, S. Broadband microwave attenuator based on few layer graphene flakes. *IEEE Trans. Microw. Theory Tech.* **2015**, *63*, 2491–2497. [CrossRef]
48. Maffucci, A.; Micciulla, F.; Cataldo, A.; Miano, G.; Bellucci, S. Bottom-up realization and electrical characterization of a graphene-based device. *Nanotechnology* **2016**, *27*, 095204. [CrossRef]

49. Ivanov, E.; Velichkova, H.; Kotsilkova, R.; Bistarelli, S.; Cataldo, A.; Micciulla, F.; Bellucci, S. Rheological behavior of graphene/epoxy nanodispersions. *Appl. Rheol.* **2017**, *27*, 1–9.
50. Bellucci, S.; Maffucci, A.; Maksimenko, S.; Micciulla, F.; Migliore, M.D.; Paddubskaya, A.; Pinchera, D.; Schettino, F. Electrical permittivity and conductivity of a graphene nanoplatelet contact in the microwave range. *Materials* **2018**, *11*, 2519. [CrossRef]
51. Bellucci, S.; Bovesecchi, G.; Cataldo, A.; Coppa, P.; Corasaniti, S.; Potenza, M. Transmittance and reflectance effects during thermal diffusivity measurements of GNP samples with the flash method. *Materials* **2019**, *12*, 696. [CrossRef]
52. Cataldo, A.; Biagetti, G.; Mencarelli, D.; Micciulla, F.; Crippa, P.; Turchetti, C.; Pierantoni, L.; Bellucci, S. Modeling and Electrochemical Characterization of Electrodes Based on Epoxy Composite with Functionalized Nanocarbon Fillers at High Concentration. *Nanomaterials* **2020**, *10*, 850. [CrossRef]
53. Cabrera, H.; Korte, D.; Budasheva, H.; Abbasgholi, N.A.B.; Bellucci, S. Through-Plane and In-Plane Thermal Diffusivity Determination of Graphene Nanoplatelets by Photothermal Beam Deflection Spectrometry. *Materials* **2021**, *14*, 7273. [CrossRef]
54. Gomez, C.V.; Robalino, E.; Haro, D.; Tene, T.; Escudero, P.; Haro, A.; Orbe, J. Structural and electronic properties of graphene oxide for different degree of oxidation. *Mater. Today Proc.* **2016**, *3*, 796–802. [CrossRef]
55. Ofomaja, A.E.; Naidoo, E.B.; Pholosi, A. Intraparticle diffusion of Cr (VI) through biomass and magnetite coated biomass: A comparative kinetic and diffusion study. *S. Afr. J. Chem. Eng.* **2020**, *32*, 39–55.
56. Oliva, J.; Martinez, A.I.; Oliva, A.I.; Garcia, C.R.; Martinez-Luevanos, A.; Garcia-Lobato, M.; Ochoa-Valiente, R.; Berlanga, A. Flexible graphene composites for removal of methylene blue dye-contaminant from water. *Appl. Surf. Sci.* **2018**, *436*, 739–746. [CrossRef]
57. Ma, Y.-X.; Shao, W.-J.; Sun, W.; Kou, Y.-L.; Li, X.; Yang, H.-P. One-step fabrication of β-cyclodextrin modified magnetic graphene oxide nanohybrids for adsorption of Pb (II), Cu (II) and methylene blue in aqueous solutions. *Appl. Surf. Sci.* **2018**, *459*, 544–553. [CrossRef]
58. Zhu, B.; Xia, P.; Ho, W.; Yu, J. Isoelectric point and adsorption activity of porous g-C3N4. *Appl. Surf. Sci.* **2015**, *344*, 188–195. [CrossRef]
59. Su, H.; Li, W.; Han, Y.; Liu, N. Magnetic carboxyl functional nanoporous polymer: Synthesis, characterization and its application for methylene blue adsorption. *Sci. Rep.* **2018**, *8*, 6506. [CrossRef]
60. Wu, M.; Li, Y.; Yue, R.; Zhang, X.; Huang, Y. Removal of silver nanoparticles by mussel-inspired Fe_3O_4@polydopamine core-shell microspheres and its use as efficient catalyst for methylene blue reduction. *Sci. Rep.* **2017**, *7*, 42773. [CrossRef] [PubMed]
61. Wijaya, R.; Andersan, G.; Santoso, S.P.; Irawaty, W. Green reduction of graphene oxide using kaffir lime peel extract (Citrus hystrix) and its application as adsorbent for methylene blue. *Sci. Rep.* **2020**, *10*, 667. [CrossRef]

MDPI
St. Alban-Anlage 66
4052 Basel
Switzerland
Tel. +41 61 683 77 34
Fax +41 61 302 89 18
www.mdpi.com

Nanomaterials Editorial Office
E-mail: nanomaterials@mdpi.com
www.mdpi.com/journal/nanomaterials